D1545024

STEREOCHEMISTRY

New Comprehensive Biochemistry

Volume 3

General Editors

A. NEUBERGER
London

L.L.M. van DEENEN
Utrecht

ELSEVIER BIOMEDICAL PRESS
AMSTERDAM · NEW YORK · OXFORD

Stereochemistry

Editor

Ch. TAMM

Basel

1982

ELSEVIER BIOMEDICAL PRESS
AMSTERDAM · NEW YORK · OXFORD

ISBN for the series: 0444 80303 3
ISBN for the volume: 0444 80389 0

Published by:
Elsevier Biomedical Press
Molenwerf 1, P.O. Box 1527
1000 BM Amsterdam, The Netherlands

Sole distributors for the U.S.A. and Canada:
Elsevier Science Publishing Company Inc.
52 Vanderbilt Avenue
New York, NY 10017, U.S.A.

Library of Congress Cataloging in Publication Data
Main entry under title:

Stereochemistry.

(New comprehensive biochemistry ; v. 3)
Includes bibliographies and index.
1. Stereochemistry. I. Tamm, Christoph, 1923-
II. Series.
QD415.N48 vol. 3 [QD481] 540s [541.2'23] 82-7369
ISBN 0-444-80389-0 (U.S.) AACR2

Printed in The Netherlands

Preface

The past years have witnessed a rapid development of biochemistry and molecular biology. The chemical structures of many complex biopolymers such as proteins and nucleic acids have been elucidated. They are strongly interrelated with the enzymatic reactions that regulate all processes in the living cell. The understanding of the stereochemical details of many important transformations catalyzed by enzymes has greatly increased. An essential prerequisite is a clear conception of the geometry of the molecules serving as substrates and hence of definitions and nomenclature. Very often biochemists and biologists are not familiar enough with the symmetry of molecules, isomeric structures, problems of chirality and conformations. It is the purpose of Chapter 1 to stress these very basic points which reflect the structural complexity of biomolecules. For the investigation of the stereochemistry of enzymic reactions, well-established chemical methods have been refined and new procedures developed. These are treated in Chapter 2, which also settles notions like classification of reaction types and selectivities, thus providing the basis for the determination of configurations of both chiral and prochiral elements. Selected examples of widely occurring types of enzymic reactions are discussed in subsequent chapters. Chapter 3 deals with the various dehydrogenases with special emphasis on the problems of how stereospecificity arises. Chapter 4 is devoted to the stereochemistry of pyridoxal phosphate-catalyzed reactions such as transamination, racemization, decarboxylation and reactions occurring at the β- and γ-carbon atoms. The recent advances in the fascinating field of the stereochemistry of enzymatic substitution at phosphorus, including chiral phosphothioates, phosphates and metal nucleotides, are reviewed in Chapter 5. Coenzyme B_{12} catalyzes many types of rearrangement whose stereochemistry has been elucidated recently; these are described in Chapter 6. In this connection the stereochemistry of enzymes that are involved in the biosynthesis of corrins are mentioned. Chapter 7 summarizes the new insights that have been gained very recently into the process of vision. These involve very complicated spectroscopic and stereochemical problems.

This book attempts to give a comprehensive account of all aspects of molecule structure and the stereochemical implications of the dynamics of the most important enzymic reactions. The editor hopes that the volume will not only be of interest to specialists, but will also provide general information useful to organic chemists, biochemists and molecular biologists. Future problems can only be resolved by close interdisciplinary collaboration of scientists in these various fields.

Ch. Tamm

Basel, March 1982

Contents

Preface v

Chapter 1. The geometry of molecules: Basic principles and nomenclatures, by B. Testa *1*

1. Introduction: The concept of chemical structure 1
2. Symmetry 3
3. The classification of isomeric structures 7
   (1) Geometry-based classifications of isomeric molecules 7
   (b) Energy-based classification of stereoisomers 9
   (c) Steric relationships between molecular fragments 10
4. Tri-, tetra-, penta- and hexacoordinate centers of stereoisomerism 10
   (a) Chiral tricoordinate centers 11
   (b) Chiral tetracoordinate centers 12
   (c) Pentacoordinate centers 14
   (d) Hexacoordinate centers 15
5. Axes and planes of chirality; helicity 16
   (a) The chiral axis 17
   (b) The chiral plane 18
   (c) Helicity, propellers, chiral cages 19
6. Diastereoisomerism 20
   (a) π-Diastereoisomerism 20
   (b) Stereoisomerism resulting from several centers of chirality in acyclic molecules 22
   (c) Diastereoisomerism in cyclic molecules 23
7. Prostereoisomerism 24
   (a) Homotopic groups and faces 25
   (b) Enantiotopic groups and faces 26
   (c) Diastereotopic groups and faces 27
8. The conformation of linear systems 29
   (a) Rotation about sp^3–sp^3 carbon–carbon bonds 29
   (b) Rotation about sp^3–sp^2 and sp^2–sp^2 carbon–carbon single bonds 33
   (c) Rotation about carbon–heteroatom and heteroatom–heteroatom single bonds 34
9. The conformation of cyclic systems 36
   (a) Non-substituted carbocycles 37
   (b) Substituted carbocycles 38
   (c) Heterocycles 40
10. Conclusion: The structural complexity of biomolecules 42

References 45

Chapter 2. Chemical methods for the investigation of stereochemical problems in biology, by R. Bentley *49*

1. Introduction 49
2. The basis for the biological recognition of chirality and prochirality 51

(a) Differentiation at chiral positions 51
(b) Enzymes reacting with both enantiomeric forms of a substrate 59
(c) Differentiation at prochiral positions 61
3. Classification of reaction types and selectivities 64
(a) Constitutional isomers 65
(b) Stereoisomers 66
(i) Enantioface differentiation 71
(ii) Enantiotopos differentiation 72
(iii) Enantiomer differentiation 72
(iv) Diastereoface differentiation 73
(v) Diastereotopos differentiation 73
(vi) Diastereoisomer differentiation 74
4. The determination of configuration 77
(a) For chiral elements 77
(b) For prochiral elements 78
(i) Compounds containing a hydrogen isotope 78
(ii) The configuration of NADH and NADPH 83
(iii) The configuration of citric acid 87
5. The study of chiral methyl groups 98
6. Epilogue 109
References 109

Chapter 3. Stereochemistry of dehydrogenases, by J. Jeffery *113*

1. The enzymes and what they do 113
(a) Introduction 113
(b) General characteristics 114
(i) Flavin involvement 114
(ii) Solely nicotinamide coenzymes 116
(c) Chemical comparisons 117
(d) Definitive descriptions of stereospecificity 117
(e) Dehydrogenase reaction mechanisms 118
2. How the stereospecificity arises 119
(a) Reactions involving flavin coenzymes 119
(i) Glutathione reductase (EC 1.6.4.2) 119
(ii) *p*-Hydroxybenzoate hydroxylase (EC 1.14.13.2) 120
(b) Reactions with direct transfer of hydrogen between nicotinamide coenzyme and substrate 121
(i) Dihydrofolate reductase (EC 1.5.1.3) 121
(ii) 6-Phosphogluconate dehydrogenase (EC 1.1.1.44) 126
(iii) Lactate dehydrogenase (EC 1.1.1.27) 127
(iv) Malate dehydrogenase (EC 1.1.1.37) 128
(v) Glyceraldehyde-3-phosphate dehydrogenase (EC 1.2.1.12) 128
(vi) Glycerol-3-phosphate dehydrogenase (EC 1.1.1.8) 130
(vii) Glutamate dehydrogenase (EC 1.4.1.2–4) 134
(viii) Alanine dehydrogenase (EC 1.4.1.1) 134
(ix) Saccharopine dehydrogenase (EC 1.5.1.7) 136
(x) Octopine dehydrogenase (EC 1.5.1.11) 137
(xi) Alcohol dehydrogenase (EC 1.1.1.1) 137
(xii) Aldehyde reductase (EC 1.1.1.2) and similar enzymes 142
3. Do particular structural features fulfil similar functions in different dehydrogenases? 148
4. Why are the structures related? 154
5. Conclusions 155
References 156

Chapter 4. Stereochemistry of pyridoxal phosphate-catalyzed reactions, by H.G. Floss and J.C. Vederas *161*

1. Introduction 161
2. Stereochemical concepts of pyridoxal phosphate catalysis 163
3. Results on the stereochemistry of pyridoxal phosphate enzymes 165
(a) Reactions at the α-carbon 165
(i) Transaminases 165
(ii) Racemases 170
(iii) Decarboxylases 172
(iv) Enzymes catalyzing α,β-bond cleavage or formation 175
(b) Reactions at the β-carbon 178
(i) Stereochemistry at C-β in nucleophilic β-replacements and α,β-eliminations 178
(ii) Tryptophan synthase 182
(iii) Tryptophanase and tyrosine phenol-lyase 185
(iv) Electrophilic displacement at C-β 186
(c) Reactions at the γ-carbon 188
(d) Other pyridoxal phosphate-catalyzed reactions 193
4. Common stereochemical features of pyridoxal phosphate enzymes 194
References 195

Chapter 5. Stereochemistry of enzymatic substitution at phosphorus, by P.A. Frey *201*

1. Introduction 201
(a) Enzymatic substitution at phosphorus 201
(b) Stereochemistry and mechanisms of substitution in phosphates 202
(c) Stereochemistry and metal–nucleotide complexes 204
2. Methodologies of stereochemical investigations 205
(a) Chiral phosphorothioates 206
(i) Synthesis 206
(ii) Configuration assignments 214
(iii) Phosphorothioates as substrates 219
(b) Chiral phosphates 221
(i) Synthesis 222
(ii) Configuration assignments 224
(c) Chiral metal–nucleotides 227
(i) Synthesis and separation 228
(ii) Configurations of metal–nucleotides 228
3. Selected stereochemical investigations 229
(a) Phosphohydrolases 230
(b) Phosphotransferases 234
(c) Nucleotidyltransferases 237
(d) ATP-dependent synthetases 240
(e) Structure of enzyme-bound nucleotides 241
4. Conclusions 243
References 246

Chapter 6. Vitamin B_{12}: Stereochemical aspects of its biological functions and of its biosynthesis, by J. Rétey *249*

1. The stereochemical course of the coenzyme B_{12}-catalysed rearrangement 249
(a) Dioldehydratase 251

(b) Methylmalonyl-CoA mutase 261
(c) β-Lysine mutase 265
(d) Ethanolamine ammonia lyase 268
(e) Conclusions 271
2. Stereospecificity of some enzymes in the biosynthesis of the corrin nucleus 271
(a) General outline of corrin biosynthesis 271
(b) The use of stereospecifically labelled precursors 275
(i) Labelled glycine 275
(ii) Doubly labelled succinate 277
(ii) Chiral [*methyl*-$^{2}H_{1}$,^{3}H]methionine 278
(c) Conclusions 279
References 280

Chapter 7. The stereochemistry of vision, by V. Balogh-Nair and K. Nakanishi *283*

1. Introduction 283
(a) The properties of visual pigments 285
(b) Bleaching and bleaching intermediates 288
(c) The binding of retinal to opsin 292
2. In vitro regeneration of visual pigments 292
3. The primary event 296
(a) Low temperature studies of the primary event 296
(b) Ultrafast kinetic spectroscopy of bleaching intermediates at room temperature 299
(c) Resonance Raman studies of the primary event 300
(d) Visual pigment analogs and the involvement of *cis–trans* isomerization in the primary event 302
(i) Deuterated retinals 302
(ii) Visual pigment analogs versus proton translocation in primary event 302
(iii) Non-bleachable rhodopsins retaining the full natural chromophore 303
4. Conformation of the chromophore 304
5. Visual pigment analogs 307
(a) Visual pigment analogs from retinal isomers other than 11-*cis*-retinal 307
(b) Isotopically labeled retinal derivatives 308
(c) Alkylated and dealkylated retinals 309
(d) Halogenated retinals 310
(e) Allenic rhodopsins and the chiroptical requirements of the binding site 310
(f) Retinals with modified ring structures 311
(g) Modified retinals for photoaffinity labeling of rhodopsin 313
(h) Modified retinals not forming visual pigment analogs 315
6. Models proposed to account for molecular changes in the primary event 315
(a) Proton translocation models directly involving the Schiff base nitrogen 315
(b) Proton translocation models involving charge stabilization 316
(c) Electron transfer model 317
(d) Models involving *cis–trans* isomerization in the primary event 317
(e) Summary 322
7. Models to account for the color and wavelength regulation in visual pigments 322
(a) The retinylic cation 323
(b) Anionic groups close to the ionone ring and a twist of the chromophore 323
(c) Inductive or field-effect perturbation of the positive charge of the nitrogen in the iminium bond by substituents attached to it 323
(d) Microenvironmental polarizability models 323
(e) Distance of the counterion from the protonated Schiff base nitrogen 324

(f) The charge-transfer model 324
(g) Point-charge perturbation models 324
References 329

Subject Index 335

The geometry of molecules: Basic principles and nomenclatures

BERNARD TESTA

Department of Medicinal Chemistry, School of Pharmacy, University of Lausanne, Lausanne, Switzerland

1. Introduction: The concept of chemical structure

"Information is made up of a support and semantic.... In biology there are two main languages, molecular and electrical.... In the case of molecular language, the support is the molecule, and the semantic.... is the effect on the receptor.... The macromolecular language is that of polynucleotides, polypeptides and polysaccharides. The language of micromolecules is that of coactones, pheromones, hormones and different substrates, intermediates and terminal products of metabolic sequences."

These extracts from the courageous book of Schoffeniels [1] convey to us the critical role of molecules as support of biological information. More specifically, it is the chemical structure of a molecule which determines its effects on 'receptors', hence the semantic.

The concept of chemical structure, although frequently used, is not always defined or comprehended with sufficient breadth. More than often, the term is taken as designating the geometry of chemical entities, be it simply the manner in which the constituting atoms are connected (atom connectivity, two-dimensional structure); or the geometry viewed as a frozen object in space (configuration). At these levels of modellization, molecules are considered as rigid geometrical objects. However, the concept of chemical structure extends far beyond this limited description, since to begin with molecules are more or less flexible entities. Their three-dimensional geometry will thus vary as a function of time (intramolecular motions, conformation) [2].

The time dependency of molecular geometry is under the influence of electronic properties. These are of paramount importance for a more realistic view of chemical structure since it can be stated that the geometric skeleton of a molecule is given flesh and shape in its electronic dimensions. The problem of the 'true' shape of a molecule, and of the fundamental differences existing between a geometric and an electronic modellization of molecules, has fascinated a number of scientists. Thus, Jean and Salem [3] have compared electronic and geometric asymmetry. An enlight-

Tamm (ed.) Stereochemistry

TABLE 1
The description of chemical structure

Dimensionality	Conceptual level	Properties considered	Examples of representations
Low	Geometric	2-Dimensional structure (atom connectivity)	Simple diagrams
		3-Dimensional (spatial) structure (configuration, 'steric' properties)	Perspective diagrams, molecular models
Higher	+ Electronic	Spatio-temporal structure (flexibility, conformation)	Conformational energy diagrams, computer display
		Electronic properties (electron distribution, polarizability, ionisation)	Molecular orbitals, electrostatic potential maps
	+ Interaction with the environment	Solvation, hydration, partitioning, intermolecular interactions	Computer display

ening discussion has been published by Mislow and Bickart [4] on the differences between molecules treated as real objects and as high-level abstractions. In a previous edition of this work, Bernal [5] has presented systematic considerations on molecular structure and shape. The reader may find much interest in a recent controversy on the problem of molecular structure and shape and its morphogenesis [6,7]; particularly fruitful in this respect appears the theory of quantum topology [7].

Geometric and electronic properties are obviously mutually interdependent. These also influence, and are influenced by, the interaction of chemical entities with their environment (e.g., solvent). A number of molecular properties which are accessible by experiment result from, or are markedly influenced by, interactions with the environment (e.g., solvation, ionisation, partitioning, reactivity). For these reasons, the concept of chemical structure must be extended to include interaction with the environment. Table 1 summarizes the above discussion and may help broaden the intuitive grasp of the concept of chemical structure. Table 1 is also useful in that it allows a delineation of the matters to be discussed in this chapter. As indicated by the title, we will consider molecules at the geometric levels of modellization, either as rigid (configurational aspects) or as flexible geometric objects (conformational aspects). Broader conceptual levels (electronic features, interaction with the environment) lie outside the scope of this chapter and will be considered only occasionally.

2. *Symmetry*

Terms such as 'symmetrical', 'dissymmetric', 'asymmetric', are frequently encountered in descriptions of molecular structures. At the intuitive level of comprehension, there appears to exist some form of relationship between the degree of 'order' and of 'symmetry' displayed by a molecule, namely that the more ordered molecular structures are the more symmetrical. At the mathematical level, symmetry elements and symmetry operations have been devised which allow to describe rigorously a number of geometrical properties displayed by molecular entities, or for that matter by any object. A short description of symmetry as a mathematical tool will be given in this section, and the interested reader is referred to a number of valuable monographs [8–14] for more extensive treatments.

TABLE 2
Elements and operations of symmetry

Symmetry elements	Symbol	Symmetry operations
Proper (simple) axes of rotation	C_n	Rotations
Planes of symmetry	σ	Reflections
Center of symmetry (of inversion)	i	Inversion
Rotation–reflection axes (mirror axes, improper axes, alternating axes)	S_n	Rotation–reflections

Symmetry elements provide the basis of *symmetry operations*. Thus, a molecule 'A' is said to contain a given element of symmetry when the derived symmetry operation transforms 'A' into a molecule to which it is superimposable. Elements and operations of symmetry are presented in Table 2, with the exception of the pseudo-operation of identity which will not be considered. Table 2 shows that corresponding elements and operations of symmetry share the same symbol, and indeed these two terms lack independent meaning.

A molecule is said to have a *symmetry axis* C_n of order n (n-fold axis of symmetry) if a *rotation* of $360°/n$ around this axis yields an arrangement which cannot be distinguished from the original. Benzene (I) has a C_6 axis perpendicular to

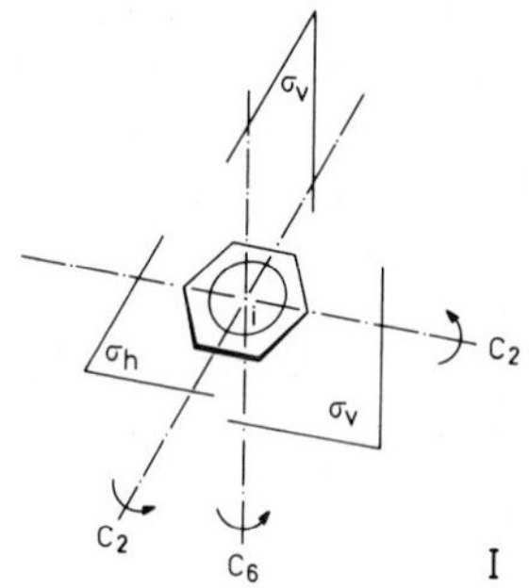

I

the plane of the molecule and passing through the geometric center, and 6 additional C_2 axes lying in the molecular plane. In this example, C_6 is the principal axis, it having the higher order. An extreme case is represented by linear molecules such as acetylene for which n can take an infinite number of values (C_∞) since any angular rotation about this C_∞ axis will yield an orientation indiscernible from the original.

When a plane divides a molecule into two symmetrical halves, it is called a *plane of symmetry* σ. By definition, σ is a mirror plane passing through the molecule in such a way that the *reflection* of all atoms through the plane yields a three-dimensional arrangement which is indistinguishable from the original one. In a molecule having a plane of symmetry, the atoms can either be in the plane or out of it; in the latter case, they exist in pairs. Planes of symmetry can be perpendicular to the principal axis, being labelled σ_h (h = horizontal), or they may contain the principal axis, in which case they are labelled σ_v (v = vertical). For example, benzene (I) has a σ_h axis which contains all the atoms of the molecule and which is the molecular plane. Benzene in addition also displays six σ_v planes, each of which contains the C_6 axis and one C_2 axis.

A *center of symmetry* i exists in a molecule in which every atom has a symmetrical counterpart with respect to this center. In such a case, *inversion* of all atoms relatively to the center of symmetry results in a three-dimensional structure indistinguishable from the original. For benzene (I), the center of symmetry is at the intercept of C_6 and of the 6 C_2. It must be noted that no more than one center of symmetry can exist per molecule.

Molecules possessing an *axis of rotation–reflection* (S_n) are said to display reflection symmetry, meaning that they are superimposable on their reflection or mirror image. This property is tested by means of the symmetry operation known as rotation–reflection; the latter operation involves two manipulations, namely rotation of $360°/n$ about an axis designating S_n, followed by reflection through a mirror plane perpendicular to S_n. Thus, *trans*-dichloroethylene (II) possesses an S_2 axis since

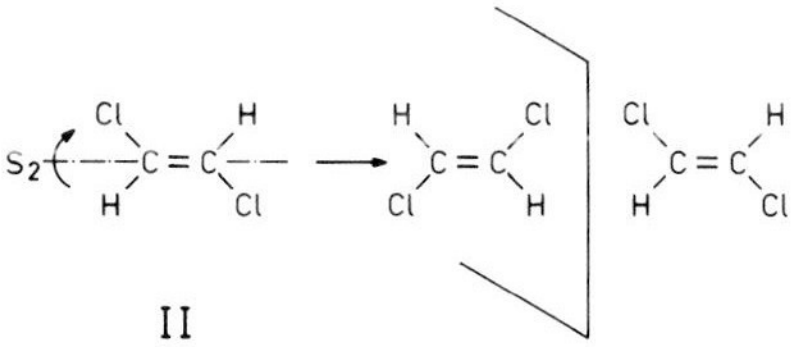

II

a rotation of 180° around S_2 followed (or preceded) by reflection in a mirror plane restores the original orientation. It must be noted that *trans*-dichloroethylene possesses neither a C_2 nor σ_h.

Molecules may possess no, one, or a number of elements of symmetry. Although the number of molecules is immense, the possible combinations of symmetry operations are relatively few. These combinations are called *point groups* (they must leave a specific point of the molecule unchanged). The point group of a molecule is thus the ensemble of all symmetry operations which transform that molecule into an indistinguishable orientation. Point groups are classified into two main categories depending whether they exclude or include reflection symmetry.

TABLE 3
Principal point groups

Chiral groups		Achiral groups	
Point group	Elements	Point group	Elements
C_1	No symmetry element (asymmetric)	C_s	σ
		S_n	S_n (n even)
C_n	C_n ($n>1$) (dissymmetric) (axial symmetry)	C_{nv}	C_n, $n\sigma_v$
		C_{nh}	C_n, σ_h
D_n	C_n, nC_2 (dissymmetric) (dihedral symmetry)	D_{nd}	C_n, nC_2, $n\sigma_v$
		D_{nh}	C_n, nC_2, $n\sigma_v$, σ_h
		T_d	$4\ C_3$, $3\ C_2$, 6σ (tetrahedral symmetry)
		O_h	$3\ C_4$, $4\ C_3$, $6\ C_2$, 9σ (octahedral symmetry)
		K_h	all symmetry elements (spherical symmetry)

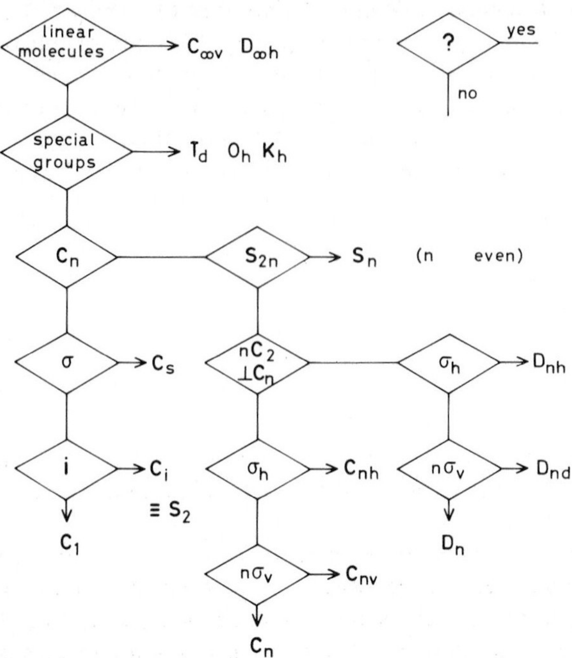

Fig. 1. A scheme for the selection of point groups (reproduced from [15], with permission from Marcel Dekker Inc., New York).

Molecules without reflection symmetry (no σ plane) are called *dissymmetric* or *chiral. Chirality* (from the Greek 'cheir', hand) is the property displayed by any object (e.g., a hand) which is nonsuperimposable on its mirror image. If a C_n ($n > 1$) is also absent the structure lacks all elements of symmetry and is called *asymmetric* (point group C_1). A carbon atom bearing four different substituents (asymmetric carbon atom) is a classical example of this point group.

Molecules possessing one or more C_n can be dissymmetric but not asymmetric. They build point groups C_n and D_n (Table 3).

Molecules displaying reflection symmetry are *nondissymmetric* or *achiral*, rather than the ambiguous term 'symmetric'. These molecules can belong to a number of point groups, the principal of which are presented in Table 3. A scheme for the selection of point groups [15] is presented in Fig. 1. Recently, a powerful procedure has been presented by Pople [16] to classify molecular symmetry. Based on the novel concept of framework group, it specifies not only the geometrical symmetry operations of the point group but also the location of the nuclei with respect to symmetry subspaces such as central points, rotation axes, and reflection planes. An extensive list of point groups, together with all possible framework groups for small molecules, is given in this publication [16].

3. *The classification of isomeric structures*

(a) Geometry-based classifications of isomeric molecules

Isomers can be defined as molecules which closely resemble each other, but fail to be identical due to one difference in their chemical structure. Thus, *structural isomers* are chemical entities which share the same molecular formula (i.e., the same atomic composition), but which differ in one aspect. When they differ in their constitution (i.e., in the connectivity of their atoms), they are called *constitutional isomers*, for example 1-propanol and 2-propanol. When structural isomers have identical constitution but differ in the spatial arrangement of their atoms, they are designated as *stereoisomers*.

To the concepts of constitutional isomerism and stereoisomerism correspond those of *regiochemistry* and *stereochemistry*, respectively. Epiotis [17] has put forward the proposal to collectively describe regiochemistry and stereochemistry by the term 'chorochemistry' (Greek 'choros' = space).

A fundamental subclassification is that of stereoisomers, which can be divided into *enantiomers* and *diastereoisomers*. Either two stereoisomers are related to each other as object and nonsuperimposable mirror image, or they are not. In the former case, they share an enantiomeric relationship. This implies that the molecules are dissymmetric (chiral), and chirality is the necessary and sufficient condition for the existence of enantiomers. An example of an enantiomeric relationship is illustrated in diagram III which shows the (R)- and (S)-enantiomers (see Section 4.b) of

HO C H CH_3 (R)-(+) | H C OH CH_3 (S)-(−)

III

1-phenylethanol. Enantiomers are also referred to as optical isomers since they show optical rotations of opposite signs and ideally of identical amplitude. Note however that this optical activity may be too small to be detected, as is known in a few cases.

Stereoisomers which are not enantiomers are diastereoisomers. While a given molecule may have one and only one enantiomer, it can have several diastereoisomers. However, two stereoisomers cannot at the same time be enantiomers and diastereoisomers of each other. Enantiomeric and diastereoisomeric relationships are thus mutually exclusive. Diagram IV shows the (E)- and (Z)-diastereoisomers (also

Cl Cl C=C H H (Z) | Cl H C=C H Cl (E)

IV

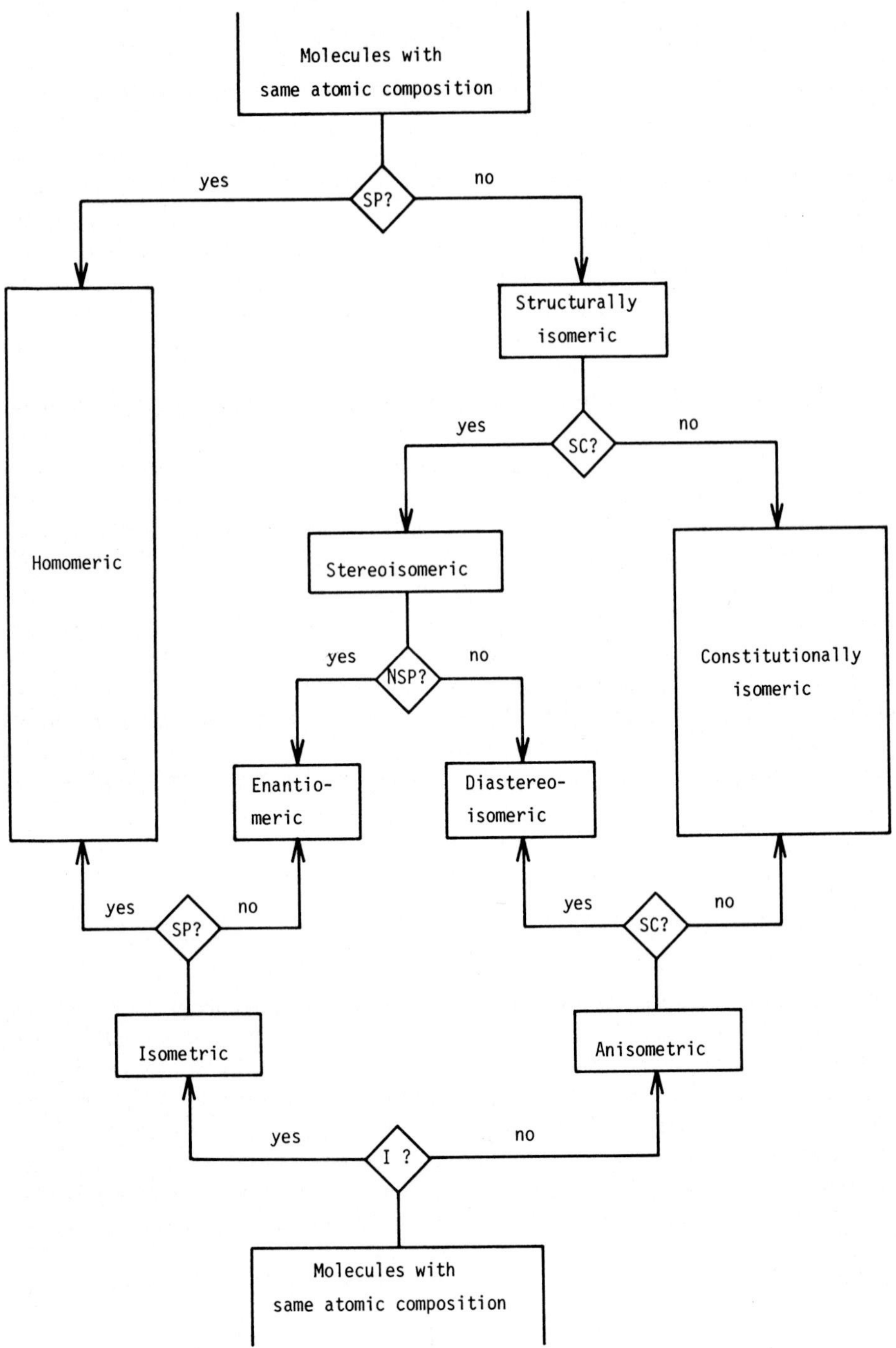

Fig. 2. Geometry-based classification of isomeric molecules. Upper half: the conventional classification. Lower half: the isometry-based classification. SP, superimposable; SC, same constitution; NSP, nonsuperimposable mirror images; I, isometric. Adapted from [18] and [19].

called *cis* and *trans*, see Section 6.a) of 1,2-dichloroethylene.

The above-discussed classification of isomers is depicted schematically in the upper half of Fig. 2. Such a classification, which is considered classical and widely accepted, nevertheless fails to be fully satisfactory, as aptly demonstrated by Mislow [18]. Thus, this classification considers diastereoisomers to be more closely related to enantiomers than to constitutional isomers. In fact, diastereoisomers resemble constitutional isomers in that their energy content is different, and therefore they differ in their chemical and physical properties. In this perspective, diastereoisomers differ from enantiomers which have identical energy contents and thus display identical physical and chemical properties.

Mislow [18] has proposed a classification of isomers based not on the bonding connectivity of atoms as above, but on the pairwise interactions of *all* atoms (bonded and nonbonded) in a molecule. The operation of comparison of all pairwise interactions is called *isometry* (for detailed explanations, see [19]). Isomers in which all corresponding pairwise interactions are identical are said to be *isometric*, and they are *anisometric* if this condition is not fulfilled. Isometric molecules may be superimposable, in which case they are identical (homomeric), or they may be nonsuperimposable, in which case they share an enantiomeric relationship. As regards anisometric molecules, they are categorized as diastereoisomers or constitutional isomers, depending on whether their constitution is identical or not. This discussion is schematically summarized in the lower half of Fig. 2.

Fig. 2 is offered as a scheme allowing immediate comparison of the conventional and isometry-based classifications. These lead by distinct dichotomic pathways to the same four classes, namely homomers, enantiomers, diastereoisomers and constitutional isomers. Note however that the isometry-based classification has the disadvantage of not explicating stereoisomers as a class of isomers.

(b) Energy-based classification of stereoisomers

The geometry-based classification of stereoisomers, as discussed above, discriminates two mutually exclusive categories, namely enantiomers amd diastereoisomers.

Independently from this classification, stereoisomers can be discriminated according to the energy necessary to convert one stereoisomer into its isomeric form; here, the energy barrier separating two stereoisomers becomes the criterion of classification. In qualitative terms, a 'high'-energy barrier separates *configurational isomers*, while a 'low'-energy barrier separates *conformational isomers* (*conformers*).

The configuration–conformation classification of stereoisomers lacks a well defined borderline. In the continuum of energy values, intermediate cases exist which are difficult to classify. In the author's opinion (see also [19]), the boundary between configuration and conformation should be viewed as a broad energy range encompassing the value of 80 kJ/mol (ca. 20 kcal/mol), which is the limit of fair stability under ambient conditions.

The classification of stereoisomers according to the two independent criteria of symmetry and energy is presented graphically in Fig. 3. Representing all cases of

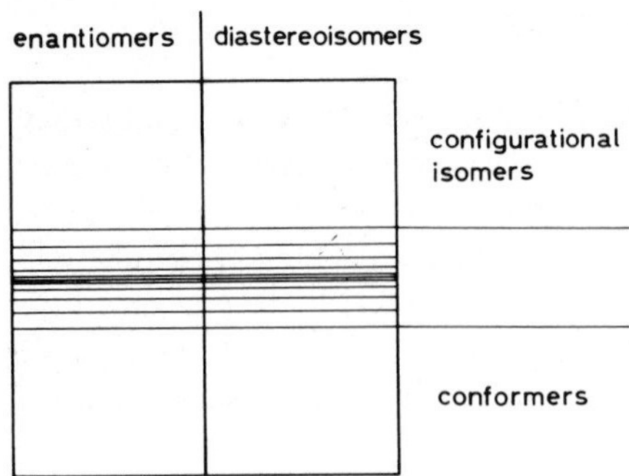

Fig. 3. Summary of the classification of stereoisomers (reproduced from [19], with permission from Marcel Dekker Inc., New York).

stereoisomerism by a square box, a sharp division discriminates between enantiomers and diastereoisomers, while a broad division separates conformers and configurational isomers, with allowance for some overlap between the two fields.

(c) Steric relationships between molecular fragments

Molecular fragments, like whole molecules, may display steric relationships, as pioneered by Hanson [20] and Mislow [18]. When such fragments are considered in isolation, namely separated from the remainder of the molecule, *morphic relationships* arise. When the partial structures are considered in an intact molecule or in different intact molecules, one speaks of *topic relationships*.

A scheme analogous to the upper part of Fig. 2 has been presented for topic and morphic relationships [18,20]. Thus, fragments of the same atomic composition may be homotopic or heterotopic, depending on whether they are superimposable or not. If the latter have the same constitution, they are stereoheterotopic, in the other case they are constitutionally heterotopic. Stereoheterotopic fragments are enantiotopic or diastereotopic. Morphic analysis yields the corresponding classification (see [19]).

Topic relationships are of fundamental importance when considering prostereoisomerism, and they will be discussed again and illustrated in this context (see Section 7).

4. *Tri-, tetra-, penta- and hexacoordinate centers of stereoisomerism*

Pyramidal tricoordinate and tetrahedral tetracoordinate centers are centers of chirality when all substituents of the central atom are different. In contrast, penta- and hexacoordinate centers generate far more complex situations and may be elements of diastereoisomerism as well as enantiomerism. Selected cases will be considered.

(a) Chiral tricoordinate centers

A tricoordinate center where the central atom is coplanar with the three substituents (Va) is obviously not chiral, since a plane of symmetry exists in the molecule.

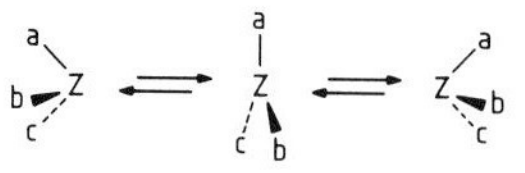

Vb Va Vc

However, deviation from full planarity results in a pyramidal geometry which is dissymmetric when the three substituents are different. This is represented in diagrams Vb and Vc, the interconversion of the two forms occurring via the planar transition state Va.

Generally, chiral tricoordinate centers are configurationally stable when they are derived from *second-row elements*. This is exemplified by sulfonium salts, sulfoxides and phosphines. In higher rows, stability is documented for arsines and stibines. In contrast, tricoordinate derivatives of carbon, oxygen, and nitrogen (*first-row atoms*) experience fast inversion and are configurationally unstable; they must therefore be viewed as conformationally chiral (see Fig. 3, Section 3.b). Oxonium salts show very fast inversion, as do carbanions. Exceptions such as the cyclopropyl anion are known. Carbon radicals and carbenium ions are usually close to planarity and tend to be achiral independently of their substituents [21–23].

The *tricoordinate nitrogen atom* has retained much interest. Fast inversion is the rule for amines, but the barrier of inversion is very sensitive to the nature of the substituents. When two of these substituents are part of a cyclic system, the barrier may in some cases be markedly increased. Thus, the enantiomers of 2-disubstituted aziridines (VI) can be discriminated at low temperature. Even more noteworthy is

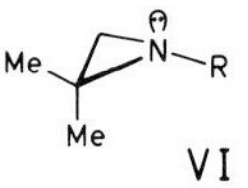

VI 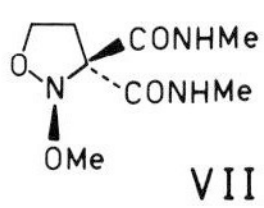 VII

the configurational stability of (2*S*)-(+)- and (2*R*)-(−)-methoxyisoxazolidine-3,3-dicarboxylic acid bis-methylamide [24]. The former enantiomer is depicted in diagram VII. The electroattractive nature of two of the nitrogen substituents certainly account for the configurational stability of VII, in agreement with the high inversion barrier of NCl_3.

When the nitrogen atom is at a ring junction in bridged systems, pyramidal inversion is impossible without bond cleavage. An asymmetrically substituted nitrogen atom then becomes a stable center of chirality, a common situation in alkaloid chemistry.

Most of the tricoordinate atoms discussed above bear an unshared electron pair which formally occupies the position of a fourth substituent. These systems therefore show clear geometric similarities with the tetracoordinate centers to be considered next, and the same stereochemical descriptors (e.g., the R and S nomenclature) can be used.

(b) Chiral tetracoordinate centers

An atom bearing four different substituents lies at the center of a chiral tetrahedral structure. Such an assembly is asymmetric (group C_1) and has one, and only one, stereoisomer which is its enantiomeric form (VIII). The interconversion of the two

VIII

enantiomers involves coplanarity of the central atom and of three ligands, while the fourth substituent swings around after its bond with the central atom has been either cleaved or elongated. This process is typically a high-energy one, meaning that isolated enantiomers containing a chiral tetracoordinate center are configurationally stable at room temperature.

The absolute configuration at a chiral tetracoordinate center can be described using the D and L nomenclatures (but there are drawbacks, see [25]), or with the R and S nomenclature to be summarized below.

The *R and S nomenclature* was first presented in 1951 by Cahn and Ingold [26], and then consolidated and extended by Cahn, Ingold and Prelog [27,28]. The essential part of this nomenclature (also called the CIP nomenclature) of chiral centers is the *sequence rule*, i.e., a set of arbitrary but consistent rules which allow a hierarchical assignment of the substituents (a > b > c > d).

By convention, the chiral center is viewed with a, b and c pointing toward the observer, and d pointing away. The path a to b to c to a can be either clockwise (IX) in which case the configuration is designated (R) (rectus), or counterclockwise (X) which means an (S) configuration (sinister).

IX

(S)

X

The sequence rule contains five subrules which are applied in succession until a decision is reached. First, the four atoms adjacent to the central atom are given a rank according to atomic number, e.g., I > Br > Cl > S > P > F > O > N > C > H > free electron pair. More than often, however, two of these adjacent atoms are identical as exemplified by structure XI. In such a case, one proceeds outwards from the two identical atoms to consider the once-removed atoms, finding C(C,C,H) and C(C,C,H). The two sets of once-removed atoms are arranged in order of preference

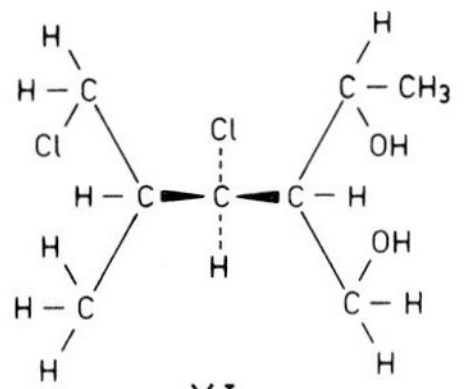

XI

and compared pairwise. The decision is reached at the first difference. In case XI, however, the once-removed atoms show no difference, and the exploration is continued further. The two sub-branches on the left-hand side are arranged in the order C(Cl,H,H) > C(H,H,H), and on the right-hand side we find C(O,C,H) > C(O,H,H). The senior sub-branches are compared, and a decision is reached at this stage: C(Cl,H,H) > C(O,C,H). Therefore, there is no need to compare the junior sub-branches or to continue the exploration; the left-hand side ligand has preference over the right-hand side ligand, and structure XI has (*S*) configuration.

Double and triple bonds are split by the sequence rule into two and three single bonds, respectively. The duplicated or triplicated atoms are considered carrying no substituents and are drawn in brackets. Diagrams XII give a few examples. Aromatic

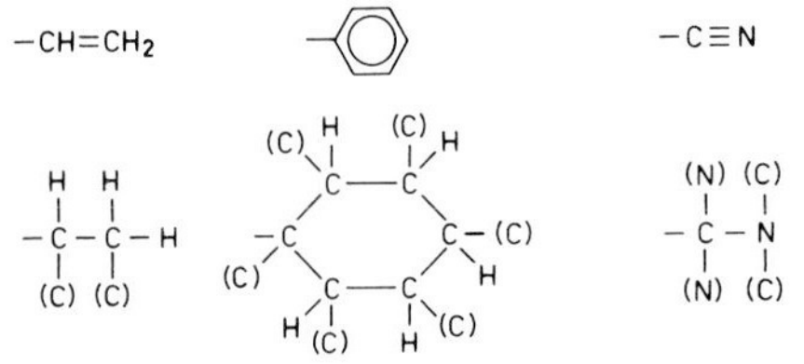

$-CH{=}CH_2$ $\qquad -C{\equiv}N$

XII

heterocycles require additional conventions. A useful list of 76 groupings properly classified can be found in the IUPAC Recommendations [29].

The second subrule states that isotopic substituents are classified according to mass number (e.g., $^3H > ^2H > ^1H$), as exemplified by (*S*)-(+)-α-d-ethylbenzene (XIII) ([30], see also [31]). Other known cases of chirality due to isotopic substitution include $^{12}CH_3$ versus $^{13}CH_3$ and CH_3 versus C^3H_3 [32].

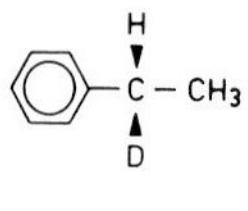

XIII

The most frequently encountered chiral tetracoordinate center is the *carbon atom* bearing four different substituents, as exemplified above. Another element which has significance as a chiral center is the *nitrogen atom*. The quaternary nitrogen is chiral and configurationally stable when as depicted in diagram XIV (Z = N) $a \neq b \neq c \neq d \neq H$. Amine oxides (*N*-oxides, XV) offer another case of configurational stability. Other tetrahedral centers include silicon (silanes), and germanium (germanes) derivatives, as well as phosphonium (XIV, Z = P) and arsonium (XIV, Z = As) salts.

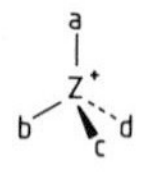

XIV 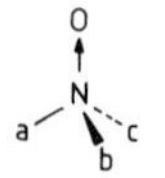 XV

Other possible cases of chiral tetracoordinate centers exist beside the asymmetric centers (point group C_1) discussed above. Indeed, *chiral molecules of higher symmetry* exist; these may contain a chiral center of the types $Z(a_2b_2)$ (point group C_2), $Z(a_3b)$ (point group C_3), or $Z(a_4)$ (point group D_2). An example of C_2 chirality is provided by (*S*)-(−)-spiro[4.4]nonane-1,6-dione (XVI) [33]. An in-depth discussion

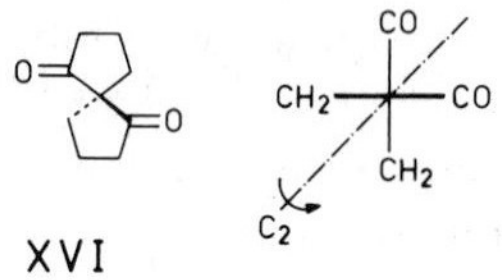

XVI

of C_2 chiral centers can be found in [34]. Some years ago, optically active compounds designated vespirenes have been synthetized as first examples of molecules containing a single center of chirality of the type $Z(a_4)$ [35]. The structure of (*R*)-(−)-[6.6]vespirene is shown in diagram XVII.

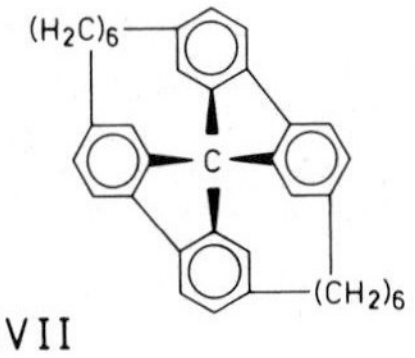

XVII

(c) Pentacoordinate centers

Pentacoordinate centers are stereochemically far more complex than tetracoordinate centers. Idealized geometries for such centers are the *trigonal-bipyramidal* (XVIII) and *tetragonal-pyramidal* (XIX) arrangements.

XVIII 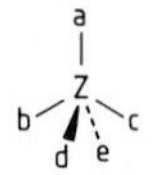 XIX

Pentacoordinate centers are mainly exemplified by phosphorus, whereas some pentacoordinate sulfur derivatives exist, and a few other elements can be envisaged [36]. In the case of phosphorus derivatives (phosphoranes), the trigonal-bipyramidal arrangement is the low-energy geometry, while interconversion of isomers occurs by pseudorotation through the tetragonal-pyramidal transition state [23,36].

The number of stereoisomers generated by a pentacoordinate center varies with the number of chemically different substituents carried by the central atom (e.g., Z(abcde), Z($a_2$bcd), Z(a_2b_2c), etc.). Nourse has given a general scheme allowing to count the number of isomers possible and to determine the number of potentially differentiable modes for the degenerate rearrangement of a molecular skeleton with a set of identical ligands [37]. In the present discussion, we will limit ourselves to the case where all five substituents are different, and follow the treatment given by Mislow [36]. If we consider a chiral tetrahedral structure (diagram XX), a fifth ligand e can attack at any one of six different edges, or at four different faces. In the former case (XXI), the new ligand is in an *equatorial* position, while in the latter case (XXII) it occupies an *apical* position. Ten stereoisomers are thus generated from the

XXI XX XXII

tetrahedral structure XX, and ten more from the enantiomer of XX. Thus, twenty stereoisomers (ten diastereoisomeric pairs of enantiomers) exist for the pentacoordinate center with five different substituents in a trigonal-bipyramidal geometry. A proposal for the extension of the *R* and *S* nomenclature to such systems has been presented [38]. It has been applied to designate as (*S*) the absolute configuration of the dextrorotatory sulfurane of structure XXIII. This compound indeed has trigonal-bipyramidal geometry, the sulfur atom however being tetracoordinate [38].

XXIII

(d) Hexacoordinate centers

Hexacoordinate centers are exemplified by a considerable variety of coordination compounds in which a central metallic cation is bound to six ligands. Three basic arrangements of the ligands are conceivable, namely *hexagonal planar* (the central atom and all ligands lying in the same plane), *trigonal prismatic* (the ligands occupy the vertices of a prism of triangular basis) and *octahedral* (XXIV). Depending on the

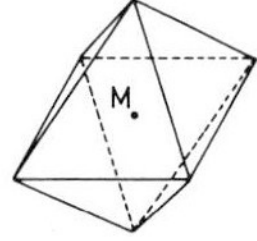

XXIV

number of identical and different ligands, various isomers are possible. For example, in the case of $[Ma_4b_2]$ complexes, only two isomers exist (XXV), one which is designated as *trans* since the two ligands b are opposed to one another about the central atom, and the other which is designated as *cis* since the ligands b are neighbours. These two isomers share a diastereoisomeric relationship.

In the case of $[Ma_3b_3]$ complexes, there are again two diastereoisomeric forms (diagrams XXVI). The former has each triplet of identical ligands occupying the

XXV: trans, cis — XXVI: fac, mer

vertices of one triangular face of the octahedron, and it is designated as *facial* (*fac*). The other isomer is the *meridional* form (*mer*); two ligands a are opposed to one another, as are two ligands b [39].

The stereochemical aspects of hexacoordinate centers become much more complex when bi- or polydentate ligands replace monodentate ligands. Many cases can be discriminated depending on the type of ligands, e.g., tris(bidentate) complexes with five-, six- or seven-membered chelate rings, terdentate complexes, quadridentate complexes, sexidentate complexes. An extensive review on the stereochemistry of chelate complexes has been published by Saito [40]. As an example, let us consider tris(bidentate) complexes formed from a ligand having two identical binding groups (e.g., ethylenediamine). In such a case, two enantiomers can be formed (diagrams XXVII), the absolute configuration of which is designated Δ and Λ. For example,

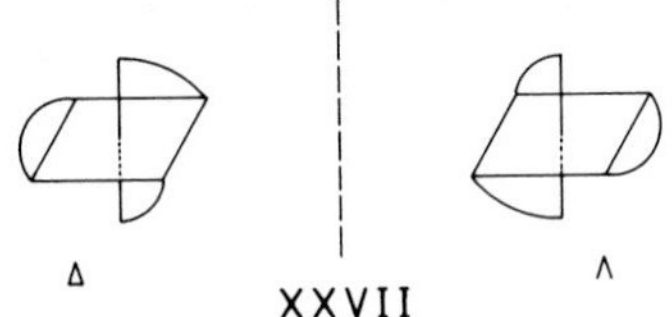

$(+)_{589}$-$[Co(ethylenediamine)_3]^{3+}$ has the configuration Λ; this enantiomer is depicted in diagram XXVIII omitting the hydrogen atoms [39,40].

XXVIII

5. Axes and planes of chirality; helicity

Molecules containing a single tetrahedral center of chirality exist only in two enantiomeric forms (Section 4). Molecules containing several centers of chirality and existing in a number of stereoisomeric forms will be discussed in the following section. In contrast, the present section considers chiral molecules which have either no or several centers of chirality, but which can exist only in two enantiomeric forms. Such molecules can display axes or planes of chirality, they can exhibit helicity, and they can be chiral cages.

(a) The chiral axis

Molecules display an axis of chirality when two structural conditions are met, namely (a) that they have four groupings occupying the vertices of an elongated tetrahedron, and (b) that these groupings meet the condition $a \neq b$ (diagrams XXIX). If these two conditions are fulfilled, the XY axis becomes an axis of

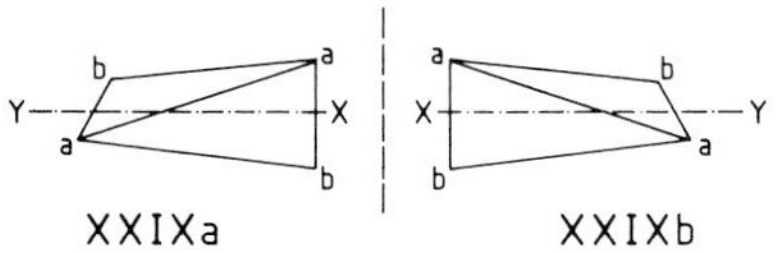

XXIXa XXIXb

chirality, and the molecule is dissymmetric without possessing a center of chirality (diagrams XXIX). The sequence rule has been extended in order to describe the absolute configuration of axially chiral molecules [41]. An additional rule however is necessary in such a case, stating that the two near groups precede the two far groups. As a result, viewing XXIXa from either the X or the Y end is equivalent and yields (assuming $a > b$) the sequence shown in XXXa which has the (*R*) configuration. Similarly, XXIXb is equivalent to XXXb and has the (*S*) configuration.

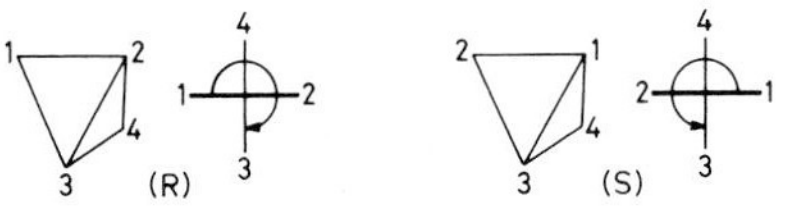

XXXa XXXb

Among the several molecular assemblies able to display axial chirality, well known examples include allenes (XXXI), spiranes (XXXII), and biphenyls (XXXIII).

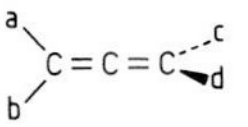

XXXI 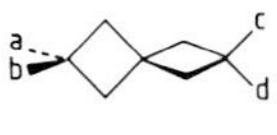 XXXII 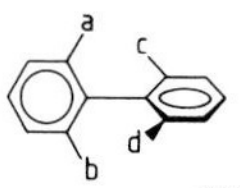 XXXIII

These molecules have a C_2 axis when $a = c \neq b = d$, as exemplified by the chiral allene derivative (*R*)-(−)-glutinic acid (diagram XXXIV) [33].

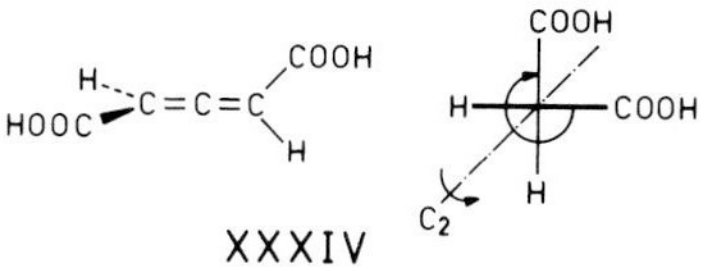

XXXIV

TABLE 4
Barrier of rotation (racemization) for biphenyls (diagram XXXIII) (from [42,43])

Substituent		Barrier (kcal/mol)
a=c	b=d	
OCH_3	H	13.7
CH_3	H	17.4
$OCOCH_3$	H	18.5
CH_2OCOCH_3	H	20.2
OCH_3	$COOCH_3$	25.0
OCH_3	$CONH_2$	27.8

Biphenyls (diagram XXXIII) are torsional isomers about a single bond. In the absence of sufficiently bulky *ortho*-substituents, the rotation about the single bond is a low-energy one and thus resorts to conformational isomerism (Section 8.b). Only with adequately sized *ortho*-substituents is the rotation sufficiently hindered to allow for manageable stability, under ordinary conditions, of the isolated enantiomers. A small series of biphenyls ranging from unresolvable (barrier of rotation clearly below 20 kcal/mol) to resolvable (barrier of rotation clearly above 20 kcal/mol) under ordinary conditions is presented in Table 4. Such a series aptly illustrates the progressive transition between conformational and configurational isomerism. Enantiomers resulting from restricted rotation about a single bond are labeled *atropisomers*.

(b) The chiral plane

A plane of chirality is encountered in molecules in which a molecular plane is 'desymmetrized' by a bridge (ansa compounds and analogs). Examples include the *paraphane derivatives* XXXV and XXXVI, and *trans*-cycloalkenes (XXXVII).

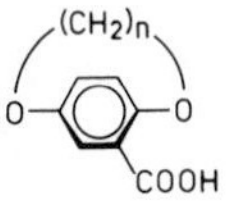

XXXV 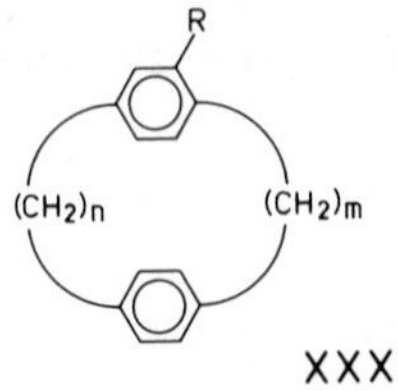 XXXVI 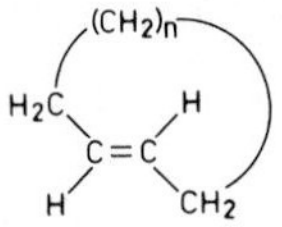 XXXVII

The R and S nomenclature has also been extended to the chiral plane (see for example [44]). Thus, structure XXXV has the (S) absolute configuration. Some confusion between chiral axes and chiral planes has led to the proposal of a new definition of the chiral plane and of a new procedure for specifying planar chirality [45].

When in XXXV $n = 8$, the compound is configurationally stable, whereas for $n = 9$ it can be racemized above 70°C, and for $n = 10$ it is no longer resolvable [46].

(c) Helicity, propellers, chiral cages

Helices are chiral objects often encountered in nature, for example helical shells. The absolute configuration of these objects is designated *P* (plus) and *M* (minus) for right- and left-handed helices, respectively. A number of chemical structures resort to helicity, the most famous example being the class of molecules known as *helicenes*. Thus, the compound shown in diagram XXXVIII is (*P*)-(+)-hexahelicene [33].

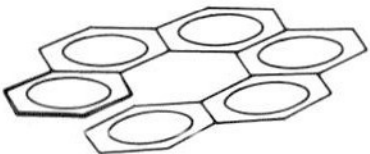

XXXVIII

A particular case of helicity is that displayed by *molecular propellers* [47], aptly described by this picturesque name. An example of a 3-bladed propeller structure is provided by tri-*o*-thymotide, the (−)-enantiomer of which has now been shown to be the left-handed form (*M* configuration) (diagram XXXIX) [48].

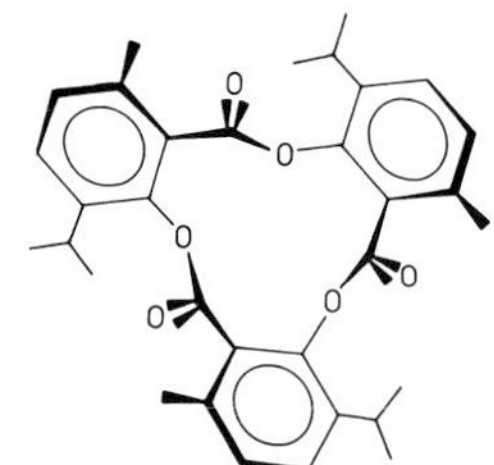

XXXIX

The molecular helices and propellers discussed above contain no center of chirality, and the *P* and *M* nomenclature is thus the only way of describing their absolute configuration. This nomenclature, however, is also applicable to some series of chiral compounds which display several centers of chirality. As will be discussed in Section 6, the presence in a molecule of two or more centers of chirality usually implies the existence of several stereoisomers, but steric reasons may reduce down to two the possible number of stereoisomeric forms. Thus, 2,3-epoxycyclohexanone contains two asymmetric carbon atoms, but for steric reasons only two stereoisomers, namely the (2*S*;3*S*)-(−)- and the (2*R*;3*R*)-(+)-enantiomer, exist; the former is depicted in diagram XL [49].

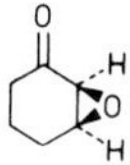

XL

Chiral cages are such examples of molecules containing several centers of chirality but existing only in two stereoisomeric (enantiomeric) forms the absolute configuration of which can be described according to helicity rules. Thus, the two enantiomers of 4,9-twistadiene are the (1*S*;3*S*;6*S*;8*S*)-(+)- and (1*R*;3*R*;6*R*;8*R*)-(−)-isomer,

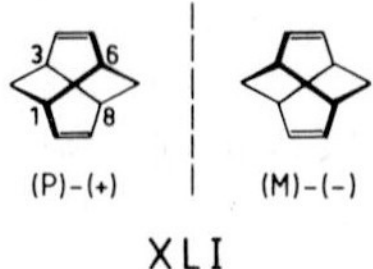

XLI

which display (P)- and (M)-helicity, respectively (XLI) [50]. The absolute configuration of these enantiomers can also be designated as (all-S) and (all-R), respectively. The enantiomers of the saturated analog twistane are also the (P)-(+)- and the (M)-(−)-form [51], as are the 2,7-dioxa analogs [50].

The cage-shaped compounds discussed above belong to point group D_2. The term 'gyrochiral' has been proposed in order to describe all chiral but not asymmetric structures [52].

6. Diastereoisomerism

Diastereoisomerism is encountered in a number of cases such as achiral molecules without asymmetric atoms, chiral molecules with several centers of chirality, and achiral molecules with several centers of chirality (*meso* forms). Such cases can be encountered in acyclic and cyclic molecules alike, but for the sake of clarity these two classes of compounds will be considered separately.

(a) π-Diastereoisomerism

A molecular structure such as the one shown in diagram XLII is achiral (presence of a plane of symmetry), but when $a \neq b$ and $c \neq d$ it can exist in two diastereoisomeric forms. When the two largest or remarkable substituents are on the same side of the double bond, the isomer is *cis*, and it is designated *trans* in the other case. Ambiguities have been encountered, and it is recommended to designate as (Z) the isomer with the two sequence rule-preferred substituents on the same side of the double bond (usually the *cis*-form), and as (E) the other isomer (usually the *trans*-form) (XLIII; $a > H, c > H$) ([53], and refs. therein).

$$\underset{\text{XLII}}{(a)(b)X{=}Z(c)(d)} \qquad \underset{\text{cis, (Z)}}{(a)(H)X{=}Z(c)(H)} \quad \underset{\text{trans, (E)}}{(a)(H)X{=}Z(H)(c)}$$

XLIII

π-Diastereoisomerism is most frequently encountered with carbon–carbon double bonds (XLII, $X = Z = C$), but also with carbon–nitrogen and nitrogen–nitrogen double bonds. The term 'π-diastereoisomerism' is more useful than the usual designations of *cis–trans*-isomerism or geometrical isomerism since it conveys the chemical origin and the correct description of the stereoisomerism. It also avoids any confusion with *cis–trans*-isomerism in cyclic systems where no double bond is involved [54].

Diastereoisomers have different relationships between nonbonded atoms, and as a consequence their energy content is different. It is generally found that due to steric effects the more extended (*trans*) isomer is more stable than the *cis*-isomer by 1–10 kcal/mol. For example, (*E*)-2-butene (XLIV) is more stable than its (*Z*)-isomer by 1 kcal/mol [55]. However, through-bond and through-space attractive orbital interactions have been calculated in several cases to favor the *cis*-isomer. Thus, (*Z*)-1-methoxypropene (XLV) is more stable than its (E)-diastereoisomer by about 0.5 kcal/mol [55].

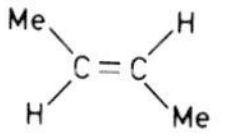

XLIV 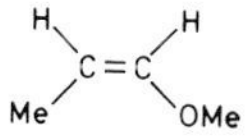 XLV

In the case of the carbon–carbon double bond, the conversion of one diastereoisomer into the other form occurs by rotation about the double bond, whereby the π-component is formally cleaved. This is typically a high-energy process. Thus, for ethylene itself, a reliable rotation barrier of 82 kcal/mol has been determined by a theoretical approach [56]. Steric, and mainly electronic, contributions of substituents greatly influence the barrier height, a lowering in bond energy resulting in a lowered barrier [56]. For example, the barrier for (*Z*)-1,2-diphenylethylene (XLVI) is 43 kcal/mol due to the partial delocalization of the double bond [56].

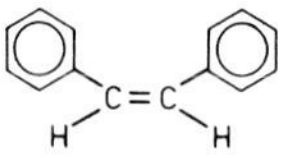

XLVI

In the case of the carbon–nitrogen double bond, the reaction of isomerization can occur via rotation about the double bond, and by nitrogen inversion (XLVII). The latter process as a rule is strongly favored over rotation, resulting in a lowered barrier of overall isomerization as compared to ethylenes. For many imines (XLVII, c = H), this barrier is in the range 20–30 kcal/mol. Electronegative substituents on the nitrogen atom increase stability toward inversion, as evidenced by the relative stability of oximes (XLVII, c = OH) and hydrazones (XLVII, c = NRR′) [56].

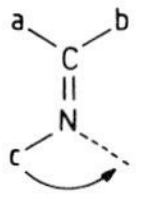

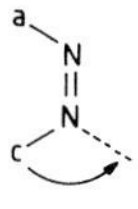

XLVII XLVIII

Isomerization at a nitrogen–nitrogen double bond (azo derivatives; XLVIII) occurs by inversion at one of the nitrogen atoms. The observed isomerization barrier of azobenzene (XLVIII, a = c = phenyl) is about 23 kcal/mol, as compared with 18 kcal/mol for the analogous imine (XLVII, a = c = phenyl, b = H) and with 43 kcal/mol for diphenylethylene (see above) [56].

(b) Stereoisomerism resulting from several centers of chirality in acyclic molecules

In a molecule containing n centers of chirality, the number of possible stereoisomers varies depending whether the molecule is constitutionally unsymmetrical (nonidentical centers of chirality) or constitutionally symmetrical.

An acyclic, *constitutionally unsymmetrical* molecule can exist as 2^n stereoisomers which are enantiomeric in pairs. In other words, such a molecule can exist as $2^{(n-1)}$ diastereoisomeric pairs of enantiomers. Any stereoisomer will thus have one enantiomer (that stereoisomer of opposed configuration on every chiral center) and $2^n - 2$ diastereoisomers. The latter may have as little as 1 and as much as $n-1$ centers of opposed configuration. Those diastereoisomers which differ in the configuration of a single chiral center (i.e., which have identical configuration on $n-1$ centers) are called *epimers*. Any stereoisomer in such a series has n epimers.

As a simple example with $n=2$, let us consider norephedrine (XLIX). Two

CH_3 / H_2N►$_2$C◄H / HO►$_1$C◄H / Ph (1S;2R)-(+)	CH_3 / H►C◄NH_2 / H►C◄OH / Ph (1R;2S)-(−)	erythro
CH_3 / H_2N►C◄H / H►C◄OH / Ph (1R;2R)-(−)	CH_3 / H►C◄NH_2 / HO►C◄H / Ph (1S;2S)-(+)	threo

XLIX

diastereoisomeric pairs of enantiomers exist, namely the *erythro* pair and the *threo* pair. Between any *erythro*-isomer and any *threo*-isomer, the relationship is that of diastereoisomerism. Indeed, such two stereoisomers have one chiral center with opposed configurations, and one with an identical configuration. Therefore, they cannot be mirror images.

Molecules with n centers of chirality are called *constitutionally symmetrical* when those centers equidistant from the geometrical center of the molecule are identically substituted. For such molecules, $2^{(n-1)}$ stereoisomers exist when n is odd, and $2^{(n-1)} + 2^{(n-2)/2}$ when n is even.

Tartaric acid (L) is a classical example for *n even*. One pair of enantiomers is

COOH / H►C◄OH / HO►C◄H / COOH	COOH / HO►C◄H / H►C◄OH / COOH	COOH / H►C◄OH / H►C◄OH / COOH
(R;R)-(+)	(S;S)-(−)	(R;S)-meso

L

($R;R$)-(+) and ($S;S$)-(−). However, the expected second pair of enantiomers ($R;S$) and ($S;R$) does not exist. Indeed, ($R;S$) and ($S;R$) are superimposable and therefore achiral and identical, as indicated also by the plane of symmetry of the molecule (L). The achiral stereoisomer is called the *meso*-form, and it shares a diastereoisomeric relationship with the two other, optically active stereoisomers. In accordance with the above rule, tartaric acid thus exists as 2 + 1 stereoisomers.

The case when *n is odd* is illustrated by trihydroxyglutaric acid (LI), for which

COOH
$_2$CHOH
$_3$CHOH
$_4$CHOH
COOH

LI

four stereoisomers are predicted. The two stereoisomers ($S;S$) and ($R;R$) differ in the configuration of C-2 and C-4; as regards their carbon-3, it carries two identical substituents which are the (S)-glycolyl moiety in the case of the ($S;S$)-stereoisomer, and the (R)-glycolyl moiety for the ($R;R$)-stereoisomer. Carbon-3 in these two stereoisomers is thus a prochiral center (Caabc, see Section 7). The ($S;S$)- and ($R;R$)-stereoisomers are enantiomeric forms having the opposite configuration on all (in this case two) their centers of chirality.

When in trihydroxyglutaric acid C-2 and C-4 have opposed configurations, a plane of symmetry renders the molecule achiral. In this case however, C-3 has four different substituents, namely H, OH, (R)-glycolyl, and (S)-glycolyl. Because of this situation, C-3 may have two opposed configurations, and the achiral molecule may exist in two distinct stereoisomeric forms both called *meso*, namely the (R;r;S)-*meso* and the (R;s;S)-*meso*. An atom like C-3 lying in a plane of symmetry and having four different substituents (two of which are thus enantiomorphic, see Section 3) is called a *pseudoasymmetric* atom. Its general expression is Ca^+a^-bc [57,58].

(c) Diastereoisomerism in cyclic molecules

Configurational isomerism is encountered in bi- and polysubstituted cyclic molecules, as well as in fused ring systems. In the simple case of *bisubstituted monocyclic systems*, *cis–trans*-isomerism exists provided that the two substituents are not geminal. Thus, 1,2-, 1,3- and 1,4-disubstituted cyclohexane derivatives (LII) show this

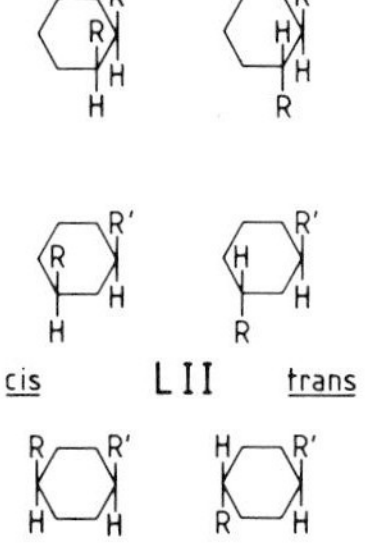

LII

cis–trans-diastereoisomerism. The planar structures drawn in LII ignore the conformational aspects of ring systems (Section 9), but they are nevertheless sufficient to unambiguously symbolize and count the various possible configurational isomers [59]. It must be added that some of the structures drawn in LII are chiral (*trans*-1,2, *trans*-1,3), some are achiral (*cis*- and *trans*-1,4), while others (*cis*-1,2, *cis*-1,3) are achiral or chiral depending whether the two substituents are identical or different, respectively.

The cases of *polysubstituted cyclic systems* are obviously more complex due to the existence of a number of possible configurational isomers. A simple *cis–trans* nomenclature is obviously not sufficient here, so the IUPAC recommends to designate the various diastereoisomers by choosing a reference substituent (the lowest numbered substituent, designated *r*) and defining its *cis* or *trans* (*c* or *t*) relationship to all other substituents (see [60]). Thus, the three diastereoisomers of LIII are *c*-2,*c*-5-dimethyl-, *t*-2,*t*-5-dimethyl and *c*-2,*t*-5-dimethyl-*r*-1-cyclopentanol.

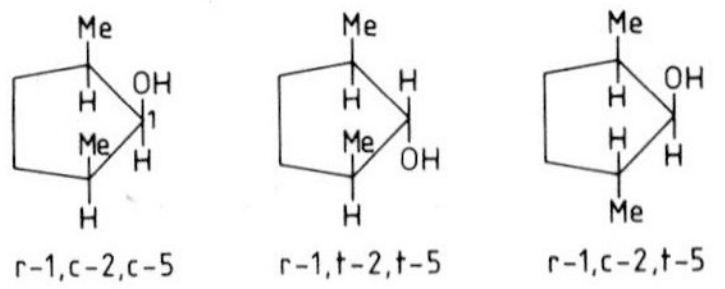

r-1,c-2,c-5 r-1,t-2,t-5 r-1,c-2,t-5

LIII

Only the latter is chiral, and this molecule therefore may exist in four possible stereoisomeric forms. In fact, 2,5-dimethyl-1-cyclopentanol is a constitutionally symmetrical molecule with an odd number of centers of chirality, and it obeys the rules discussed in Section 6.b.

Fused bicyclic systems show a fundamental stereochemical similarity with bisubstituted monocyclic systems. Here again, a *cis–trans*-diastereoisomerism may exist, depending whether the two hydrogen atoms adjacent to the two vallee carbons are on the same or on opposite sides. This is illustrated in LIV by *cis*- and *trans*-de-

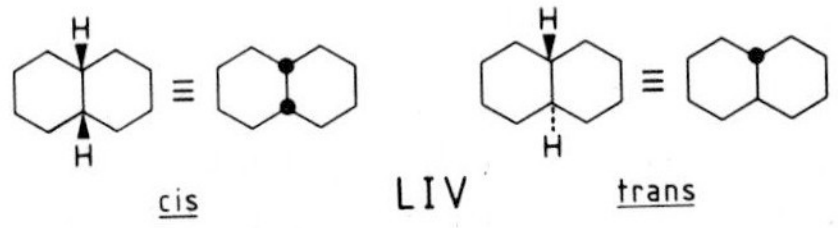

cis LIV trans

caline. Note however that *cis–trans*-isomerism resulting from ring fusion is impossible on steric grounds for the smallest rings (cyclopropane and cyclobutane). Thus, only the *cis*-forms of bicyclo[1.1.0]butane and bicyclo[2.2.0]hexane are known [61].

7. *Prostereoisomerism*

This section discusses the relationships between groups or atoms of same constitution within intact molecules (topic relationships). Such intramolecular relationships are of fundamental importance in understanding stereochemical aspects of enzymatic reactions, hence their interest in the present volume.

Three criteria are useful when assessing topic relationships, namely (a) the molecular environment, (b) symmetry considerations, and (c) the substitution criterion. When two topic groups in a molecule have stereoisomeric environments, the molecule is said to possess elements of *prostereoisomerism*. Mislow and Raban have given a definitive classification of topic relationships [62], and the following discussion is based on this classification.

(a) Homotopic groups and faces

Let us consider dichloromethane (LV), which contains one pair of chlorine atoms

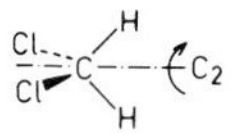

LV

and one pair of hydrogen atoms. The two identical atoms in each pair are completely equivalent, i.e., they are nondistinguishable. Indeed, the molecular environment of each pair is identical, meaning that each Cl 'feels' the other substituents in the same Cl-H-H sequence, and that each H 'feels' them in the same H-Cl-Cl sequence. Furthermore, the symmetry criterion shows that the identical groups are interchangeable by C_n ($\infty < n < 1$; in this case C_2) operation. The two hydrogen atoms in structure LV are said to be *homotopic*, as are the two chlorine atoms.

Applying the substitution criterion leads of course to the same conclusion of homotopism. For example, substituting each H in turn by another group such as ^{2}H yields two molecules which are superimposable, i.e., nondistinguishable.

1,2-Dichloroethylene (LVI) is an example similar to dichloromethane. Here, the

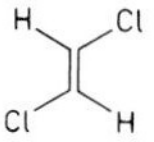

LVI

two chlorine atoms 'feel' the substituents in the same H-Cl-H sequence and the two hydrogen atoms 'feel' them in the same Cl-H-Cl sequence. Here also, a C_2 axis (perpendicular to the molecular plane) interchanges the identical atoms, and substituting the homotopic atoms in turn generates nondistinguishable molecules.

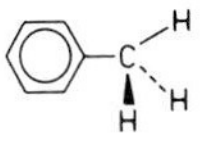

LVII

The molecule of toluene (LVII) provides a slightly more complex case since here the time factor must be taken into account. Indeed, the three hydrogen atoms of the methyl group can be considered homotopic if 'free' rotation is assumed. If the methyl rotation is fast relative to the time scale of the means of observation, then the three H atoms are indeed completely equivalent and nondistinguishable.

The molecular environment within a molecule can be defined relative to *faces* of the molecule instead of groups. For example, dichloroethylene (LVII) has two faces. Since there is no way for an observer or attacking reagent to distinguish between these two faces, they are said to be equivalent.

(b) Enantiotopic groups and faces

The three criteria of the molecular environment, of substitution, and of symmetry, show that the two hydrogen atoms in bromochloromethane (LVIIIa) are not

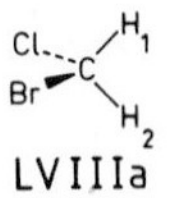

LVIIIa 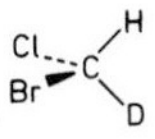 LVIIIb 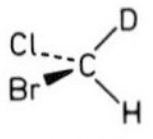 LVIIIc 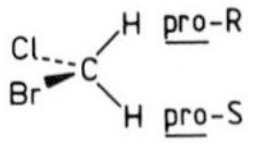 LVIIId

equivalent. Indeed, the sequence of the three atoms Br-Cl-H is clockwise when viewed from H_1, and counterclockwise from H_2. The molecular environments of H_1 and H_2 are thus enantiomeric. Furthermore, no C_n axis exists. And finally, substituting in turn H_2 and H_1 yields LVIIIb and LVIIIc, respectively, which are enantiomeric molecules. Because of these properties, the two H atoms in LVIII are designated as *enantiotopic*. If H_1 is arbitrarily preferred over H_2, an (R) configuration is obtained (e.g., LVIIIc); H_1 is therefore designated *pro-R*, while H_2 is *pro-S* [57,63] (LVIIId). As regards the molecule LVIII itself, it is not chiral since it contains a plane of symmetry. But because it bears two enantiotopic groups it is said to be prochiral. The concept of *prochirality* [63] has its origin in biological studies showing that in molecules like citric acid (LIX) two chemically identical groups such as CH_2COOH are biochemically quite distinct [64].

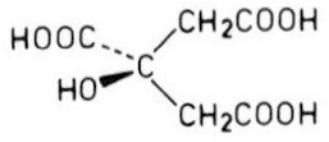

LIX

Molecules such as LVIII and LIX contain a center of prochirality, or prochiral center. It is a serious error, too often encountered in biochemical papers, to confuse enantiotopic groups or atoms and prochiral centers. Enantiotopic groups belong to prochiral centers; they are not themselves prochiral.

The presence of a center of prochirality is not an obligatory condition for a molecule to be prochiral. Indeed, other elements of prochirality exist, namely axes and planes of prochirality. Thus, the prochiral allene derivative LX displays two geminal hydrogen atoms which are enantiotopic.

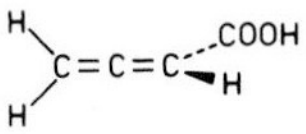

LX

The concept of prochirality can also be applied to trigonal centers, i.e., to faces of suitable molecules. In acetaldehyde (LXI), the two faces of the molecule are not

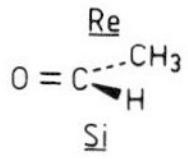

Re
$O=C(CH_3)H$
Si

LXI

equivalent, but enantiotopic. Indeed, the groups O-Me-H define a clockwise path when the molecule is viewed from above, while a view from below affords a counterclockwise path. The two face are designated *Re*-face (from *Rectus*) and *Si*-face (from *Sinister*), respectively [20,65].

(c) Diastereotopic groups and faces

Enantiotopic groups together with diastereotopic groups form the class of stereoheterotopic groups (Section 3.c). Diastereotopic groups reside in diastereoisomeric environments, cannot be interchanged by symmetry operations, and upon substitution by chiral or achiral groups generate diastereoisomeric structures.

While the presence of enantiotopic groups in a molecule necessarily implies the presence of an element of prochirality, diastereotopic groups imply prostereoisomerism as an element of prochirality, or of *proachirality*. For example, chloroethylene (LXII) contains two geminal hydrogen atoms which are diastereotopic. But no

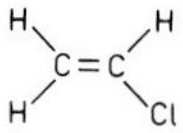

$H_2C=CHCl$

LXII

element of prochirality exists in LXII, and the carbon atom carrying the two hydrogens is a proachiral center.

1,2-Propanediol (LXIII) contains a center of chirality (C-2) and is therefore chiral. Carbon-1, on the other hand, is prochiral; replacement of one of the two adjacent hydrogen atoms would generate diastereoisomeric products. As a consequence, the two hydrogen atoms at C-1 are diastereotopic groups adjacent to a prochiral center.

Diastereotopic faces also exist, as seen in the achiral 4-methylcyclohexanone (LXIV) and in the chiral 2-methylcyclohexanone (LXV). In contrast, cyclohexanone itself has two equivalent faces.

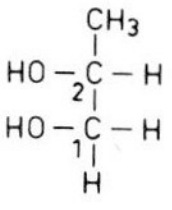

LXIII 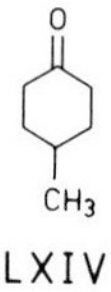 LXIV 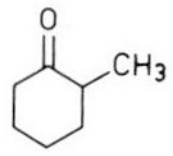 LXV

The classification and criteria discussed in this section are summarized in Table 5.

TABLE 5
Relationships between constitutionally similar groups in molecules (modified from [62] and [66])

Type of groups	Molecular environment	Symmetry criterion	Substitution with achiral or chiral test group yields	Elements of prostereoisomerism
Homotopic (equivalent)	Equivalent	Interchangeable by C_n ($\infty > n > 1$)	no isomer	None
Enantiotopic	Enantiomeric	Interchangeable by S_n only	enantiomers or diastereoisomers, respectively	Element of prochirality
Diastereotopic	Diastereoisomeric	Not interchangeable by any symmetry operation	diastereoisomers	Element of prochirality or proachirality

It must be emphasized however that only the most common and simplest aspects of prostereoisomerism have been considered here. In view of the considerable importance of this concept in biochemistry, the consultation of additional key references [34,58,67,68] is recommended.

8. The conformation of linear systems

Conformational isomerism, as already defined (Section 3.b), is a property of stereoisomers separated by a 'low' barrier of energy. The separation of isomers at room temperature requires half-lives of several hours, which correspond approximately to a free energy of activation of $\Delta G^{\neq} > 20$ kcal/mol [56]. An operational and convenient definition of conformational isomerism is thus to consider as conformers those stereoisomers which are not physically separable under ordinary conditions, in other words, which are separated by an energy barrier lower than 20 kcal/mol. Such a definition is further useful in that it sets no conditions as to the chemical process by which conformer interconversion occurs; while *bond rotation* is the most frequently encountered interconversion process, *inversion* processes are also important.

Stereoisomers in general, and conformational isomers in particular, are characterized not only by the energy barrier separating them, but also by their free energy difference ΔG°, which is related to the conformational equilibrium constant K (or conformational ratio) by the equation:

$$-\Delta G^{\circ} = RT \cdot \ln K$$

For example, energy differences of 1 and 3 kcal/mol correspond at 290°K to isomeric compositions of 85/15 and 99.5/0.5, respectively. Comprehensive tabulations covering a wide range of temperatures have been published [69]. Two classical books stand as important milestones in the development and evolution of the conformation concept [70,71].

(a) Rotation about sp^3–sp^3 carbon–carbon bonds

A suitable model molecule for discussing rotation about sp^3–sp^3 carbon–carbon bonds is *n*-butane (LXVI). Three degrees of conformational freedom are apparent,

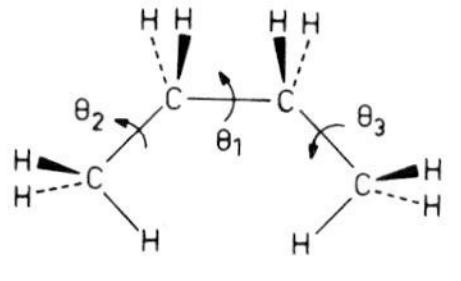

LXVI

namely rotation about the three C–C bonds; the corresponding torsion angles are labeled θ_1, θ_2 and θ_3 (LXVI).

Let us first neglect θ_2 and θ_3, and consider only rotation about C(2)–C(3). A number of remarkable conformers and their torsion angle are displayed in diagrams LXVII using Newman projection. The diagrams show three eclipsed conformations

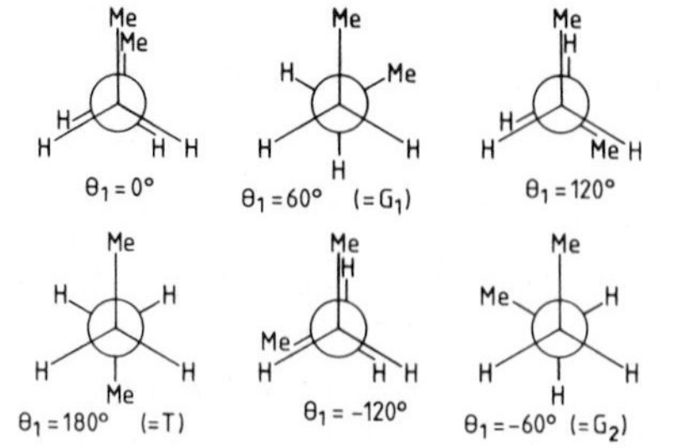

$\theta_1 = 0°$ $\theta_1 = 60°$ (= G_1) $\theta_1 = 120°$

$\theta_1 = 180°$ (= T) $\theta_1 = -120°$ $\theta_1 = -60°$ (= G_2)

LXVII

($\theta_1 = 0°$, $\pm 120°$) and three staggered conformations ($\theta_1 = \pm 60°$, 180°). The conformers $\theta_1 = \pm 60°$ have the two methyl groups *gauche* relative to each other, and these conformers are designated *gauche* (G_1 and G_2). G_1 and G_2 are enantiomeric and therefore of identical energy. The conformer $\theta_1 = 180°$ is *trans* (T).

Klyne and Prelog [72] have proposed a very useful nomenclature for the description of steric relationships across single bonds. In the system A–X–Y–B, the torsion angle about X–Y is defined as exemplified in LXVII, A and B being selected according to criteria derived from the sequence rule (for details, see [72]). The nomenclature of Klyne and Prelog is given in Table 6.

Of considerable interest is the variation of the potential energy of *n*-butane with rotation about C(2)–C(3). The energy diagram is shown in Fig. 4 for $\theta_2 = \theta_3 = 60°$, as calculated by Peterson and Csizmadia using quantum mechanics [73]. The global energy minimum is found for the *trans*-conformer, while local minima, located very close to the fully staggered *gauche*-conformers, are 0.9 kcal/mol above the global minimum. The eclipsed conformers are barriers to the interconversion of the low-energy forms. The two enantiomeric eclipsed conformers $\theta_1 = \pm 120°$ are 3.5 kcal/mol above the global minimum, while the high-energy barrier ($\theta_1 = 0°$) is 9.3 kcal/mol above this minimum [73].

TABLE 6
Description of steric relationships across single bonds [72]

Torsion angle	Description	Shorthand
$0° \pm 30°$	±synperiplanar	±sp
$+30°$ to $+90°$	+synclinal	+sc
$+90°$ to $+150°$	+anticlinal	+ac
$+150°$ to $+180°$	+antiperiplanar	+ap
$-30°$ to $-90°$	−synclinal	−sc
$-90°$ to $-150°$	−anticlinal	−ac
$-150°$ to $-180°$	−antiperiplanar	−ap

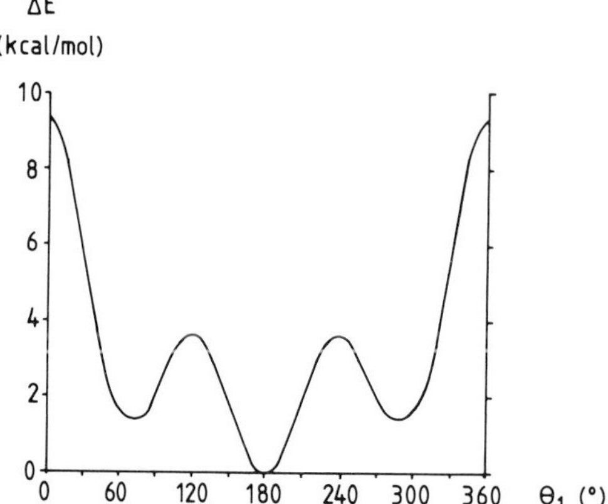

Fig. 4. Potential energy of *n*-butane central C–C rotation, with the two methyl groups held staggered ($\theta_2 = \theta_3 = 60°$) [73].

The potential energy shown in Fig. 4 is a line in a 2-dimensional space defined by the orthogonal axes ΔE and θ_1. If the three C–C bonds in *n*-butane are allowed to rotate, the potential energy of the molecule will be represented by a hypersurface in a 4-dimensional space defined by the four orthogonal axes ΔE, θ_1, θ_2 and θ_3. The topological features of this conformational hypersurface have been analyzed [73]. The global minimum is that shown in Fig. 4 (180/60/60), as are local minima (e.g., 60/60/60). The highest energy peak is (0/0/0; 44.8 kcal/mol), while a number of other maxima exist (e.g., 120/0/0). The hypersurface is also characterized by a number of saddle points (e.g., 0/60/60 and 120/60/60) which are the maxima in Fig. 4 and of super saddle points (e.g., 60/0/0; 180/0/0; 0/0/60; 120/0/60). This example illustrates the difficulty of presenting and visualizing the behavior of a molecule with more than 2 degrees of conformational freedom.

Considerable efforts have been made to understand the origin of the forces responsible for rotation barriers. In particular, the primary contributions to the barrier are found to arise from vicinal interactions between orbitals of bond and antibond type [74,75]. As a consequence, the nature of the substituents does much to influence the conformational behavior. In *n*-butane, steric repulsions between the two methyl groups is the simplest argument explaining the stability of the *trans*-conformer, and the relatively high energy of eclipsed conformers. However, replacing the methyl groups by other substituents may give rise to a variety of attractive or repulsive intramolecular interactions. Among the latter, hydrogen bonds and electrostatic interaction play a capital role. Such interactions have been presented in a stereochemical context [76] and cannot be discussed here. A simple example may suffice to illustrate some principles. Thus, 2-mercaptoethanol has been shown to exist in the gas phase in the conformation displayed (LXVIII) [77]. This conformer is designated as all-*gauche*, meaning that the molecule has a *gauche* conformation

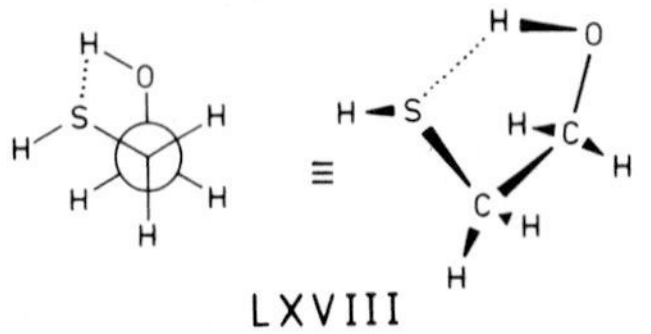

LXVIII

about the three torsion angles H-S-C-C, S-C-C-O and C-C-O-H. The predominance of this conformer is due to the existence of an intramolecular hydrogen of the type S···H-O. In the liquid state, the intramolecular H-bond is in competition with intermolecular H-bonds, and as a consequence the *trans*-conformer becomes favored [77]. A number of similar examples can be found in [78].

In biomolecules, and even in relatively simple ones, the conformational problem is often quite complex. As a consequence, many studies examine preferred conformations and their relevance to biochemistry, while the conformational dynamics of biomolecules remains largely unexplored. Citrate (LXIX) is a particularly interesting

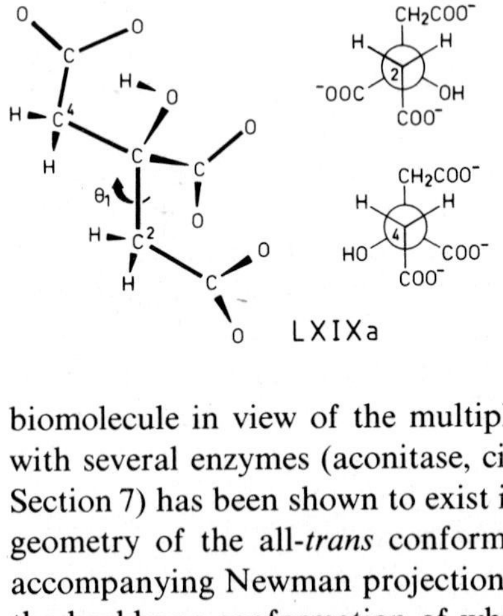

LXIXa 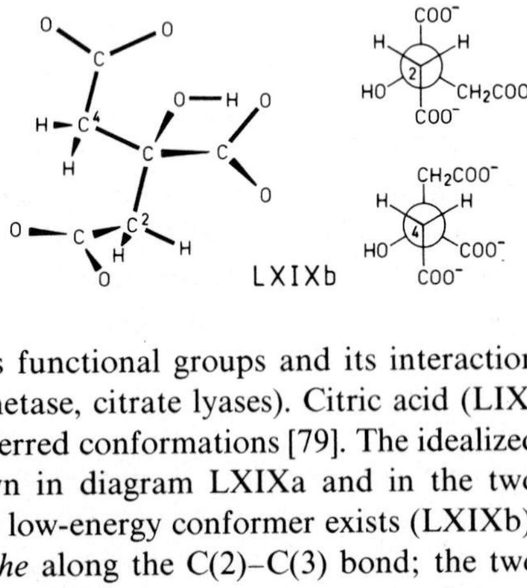 LXIXb

biomolecule in view of the multiplicity of its functional groups and its interaction with several enzymes (aconitase, citrate synthetase, citrate lyases). Citric acid (LIX, Section 7) has been shown to exist in two preferred conformations [79]. The idealized geometry of the all-*trans* conformer is shown in diagram LXIXa and in the two accompanying Newman projections. Another low-energy conformer exists (LXIXb), the backbone conformation of which is *gauche* along the C(2)–C(3) bond; the two conformers can interconvert by θ_1 variation, neglecting rearrangement rotations of the hydroxyl group and of the carboxylate functions. These conformations of citrate have relevance to its chelating capacity, and chelation of a ferrous cation is postulated to be involved in the interaction of aconitase and citrate. Diagram LXIXc shows how citrate in its extended conformation can chelate iron [79].

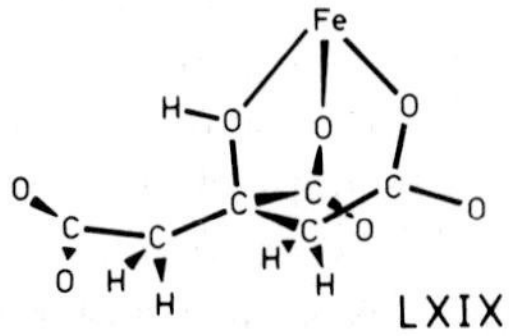

LXIXc

(b) Rotation about sp^3–sp^2 and sp^2–sp^2 carbon–carbon single bonds

The conformational behavior about sp^3–sp^2 and sp^2–sp^2 bonds is dominated by electronic forces partly different from those influencing the conformational behavior about sp^3–sp^3 bonds. This is particularly true for the through-space electrostatic and steric effects. Thus, quantum calculations [80,81] have shown that the conformation of *sp^3–sp^2 systems* is mainly controlled by one-electron attractive interactions, whereas in sp^3–sp^3 systems two-electron repulsive interactions are operative.

LXX

In toluene (LXX), like in a number of analogs, the preferred conformation shows eclipsing between a hydrogen atom and the π-system, as represented in diagram LXX. This conformation minimizes σ- and π-electron loss to the π-system (hyperconjugation) and therefore allows maximal one-electron interaction [80]. Toluene is characterized by an extremely low rotation barrier (a few cal/mol) due to the fact that this barrier is sixfold, and the latter is always found to be extremely small. Indeed, an equivalent conformation is found after every 60° rotation, and no real relief of conformational strain is obtained after a rotation of only 30°.

The rotation in toluene derivatives will be highly dependent on the size and electronic properties of the substituents. An interesting example is provided by 3,5-dichlorobenzyl alcohol, thiol and selenol (LXXI, X = O, S, and Se, respectively),

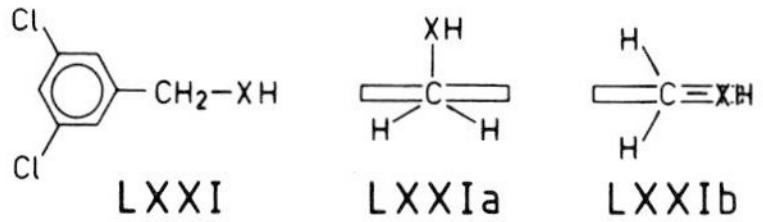

LXXI LXXIa LXXIb

the conformation of which has been investigated by a NMR method [82]. Such compounds display a twofold barrier (a 180° rotation is necessary to restore an equivalent conformation), the height of which (0.3, 1.8, and 3.8 kcal/mol, respectively) is a function of the size of X (O < S < Se). The preferred conformation of the alcohol is LXXIb, while that of the thiol and selenol is LXXIa, again indicating steric (bulk) factors to play a role beside electronic factors.

Torsional isomerism about *sp^2–sp^2 carbon–carbon single bonds* is well illustrated by molecules such as butadiene and acrolein (LXXII, R = CH_2 and O, respectively).

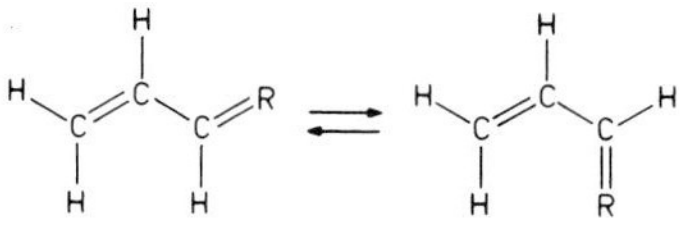

s-trans LXXII s-cis

The conjugation of the two double bonds across the single bond is most pronounced when the system is planar, hence planarity is markedly favored. In molecules such as LXXII however, two planar conformers exist, designated *s-trans* (*s* for single bond) and *s-cis*. For butadiene and acrolein, the *s-trans* form is preferred by 2.5 and 2.1 kcal/mol, respectively, while the barrier of rotation is 7.2 and 5.0 kcal/mol, respectively (for a compilation of values, see [83]).

Aromatic carbonyl derivatives (LXXIII) show degeneracy of the *s-cis* and *s-trans*

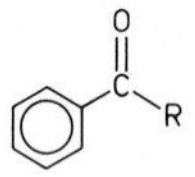

LXXIII

forms, except of course in the case of *ortho-* or *meta*-substituted derivatives. The twofold torsional barrier of benzaldehyde (LXXIII, R = H) and acetophenone (LXIII, R = CH_3) was found by NMR methods to be ca. 7.7 and 5.4 kcal/mol, respectively [84]. These barriers have also been shown to be markedly influenced by the field (inductive) and resonance effects of *para*-substituents [84].

Conformational effects as already mentioned may be influenced by the environment of the molecule, e.g., solvent. The number of examples to be found in the literature is immense, and we restrict ourselves here to mention briefly results obtained with methyl 2-fluorobenzoate (LXXIV). This molecule exists as two

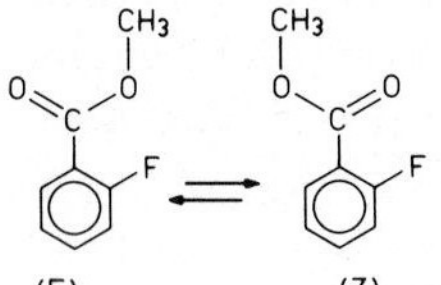

(E) LXXIV (Z)

preferred planar conformers designated (*E*) and (*Z*) (LXXIV), and the equilibrium is dominated by electronic interactions between the fluorine atom and the two oxygen atoms. This equilibrium is quite sensitive to solvent conditions. Thus in polar solvents the population of the (*E*)-conformer was found to drop under 50%, while in apolar solvents this conformer is preferred (population above 60% in hydrocarbons) [85]. The biochemical relevance of solvent effects must not escape the reader. Indeed, biological systems are characterized by regions of high and low polarity, allowing flexible molecules to change conformation in biological hydrophilic and hydrophobic environments, hence to adapt their three-dimensional structure, physicochemical properties, and reactivity.

(c) Rotation about carbon–heteroatom and heteroatom–heteroatom single bonds

The rotation barriers about sp^3 *carbon–heteroatom bonds* are usually low in the absence of strong nonbonded interactions between substituents. Factors governing the values reported in Table 7 include the carbon–heteroatom bond length, the

TABLE 7
Barrier of rotation about carbon–heteroatom bonds (compilation published in [83])

	Barrier (kcal/mol)
CH_3–CH_3	2.88
CH_3–SiH_3	1.70
CH_3–NH_2	1.94
CH_3–PH_2	1.96
CH_3–OH	1.07
CH_3–SH	1.27
CH_3–NO_2	0.006

number of H–H interactions, and partly understood electronic factors. Nitromethane provides a special case due to its sixfold barrier of rotation.

Rotation about *sp^2 carbon–heteroatom single bonds* is usually a higher energy process compared to rotation about sp^3 carbon–heteroatom single bonds due to conjugation and to resulting partial double bond character of the pivot bond. Compare the barrier of methanol (1.07 kcal/mol, Table 7) and ethanol (1.34 kcal/mol) with that of phenol (3.1 kcal/mol) [83].

One of the difficulties encountered when investigating or discussing rotation barriers about C–O and especially C–N bonds is that this process may be overshadowed by a competitive process, namely *inversion*. Inversion barriers for simple amines are in the region of 5 kcal/mol [86], thus somewhat above the rotation barrier reported in Table 7 for methylamine. However, bulky substituents may markedly increase the rotation barrier and cause inversion to become the preferred process of conformational interconversion. This is precisely what happens with *tert*-butylamines, where inversion (LXXVa–LXXVb interconversion) but not C–N bond rotation (LXXVa–LXXVc interconversion) is observed [87].

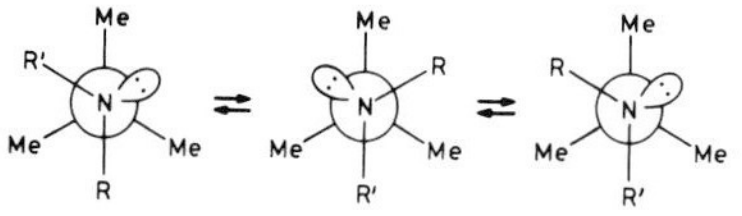

LXXVc LXXVa LXXVb

In aromatic amines, the partial double bond character of the C–N bond will tend to enhance the rotation barrier, while the inversion is found to be reduced, as compared to aliphatic amines. For example, the rotational barrier in *N*-methylaniline is just above 7 kcal/mol, while the nitrogen inversion barrier is 1.6 kcal/mol [88].

The partial double bond character of the C–N bond is even more pronounced in amides than in aromatic amines. As a result, the rotation barrier of the C–N bond in amides often lies in the range 15–20 kcal/mol. In N-monosubstituted amides (LXXVI), two low-energy conformers exist, namely the (*Z*)- and (*E*)-forms. In a number of derivatives, the (*Z*)-isomer is found to represent 90–100% of the total,

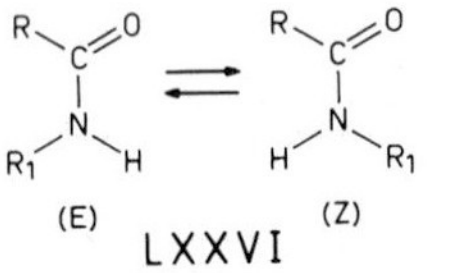

LXXVI

implying that it is preferred over (E)-form by more than 1 kcal/mol [83]. These conformational features are particularly important in understanding the stereochemistry of peptides and proteins, the backbone of which is predominantly made up of peptide bonds (also described by the general diagram LXXVI).

Rotation about *heteroatom–heteroatom single bonds* has been investigated for such cases as O–O (peroxides) and N–N (hydrazines) systems. Of greater relevance to biochemistry is the disulfide bond since its geometry is one of the determining factors in protein structure. The conformational behavior of a number of dialkyl disulfides (LXXVII) has been studied ([89], and refs. therein) showing the preferred

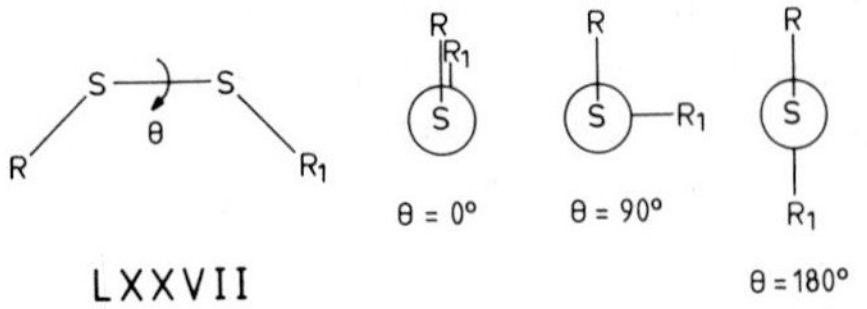

LXXVII

rotamer to have dihedral angles ranging from approximately 85° to 115° when going from dimethyl to di-*tert*-butyl disulfide. Two rotation barriers exist in such molecules, namely an eclipsed barrier ($\theta = 0°$) and a *trans* barrier ($\theta = 180°$). The latter is always of lower energy than the eclipsed barrier (2.2 and 7.0 kcal/mol, respectively, for dimethyl disulfide [90]). Increasing the size of the alkyl substituents produces a small drop in the height of the *trans* barrier, but a large increase in the height of the eclipsed barrier.

9. *The conformation of cyclic systems*

The conformational behavior of cyclic molecules is conveniently discussed separately from that of acyclic molecules. Ring closure indeed imposes conformational constraints and limitations not existing in linear systems.

Conformational isomerism resulting from bond rotation is displayed by four- and higher-membered rings. Lack of space imposes restrictions as to the types of rings to be considered here; as a consequence, only five- and six-membered rings will be examined in the following pages. The conformational aspects of several four-membered ring systems have been reviewed by Malloy et al. [91] and by Legon [92], while those of seven- and higher-membered ring systems have been summarized by a number of authors (e.g., [93–96]). The dynamic stereochemistry of many five-, six- and seven-membered rings has been given impressive coverage in a recent review [97].

(a) Nonsubstituted carbocycles

In cyclopentane, as opposed to cyclobutane and cyclopropane, the bond angles have values close to the optimum. Therefore, the strain in the molecule arises essentially from bond opposition and is partly relieved by puckered conformations. Two flexible forms of cyclopentane exist, namely the so-called *envelope* (LXXVIIIa) and *half-chair* (LXXVIIIb) forms. The former has four carbons in the same plane, and

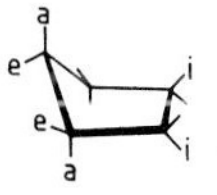

LXXVIIIa 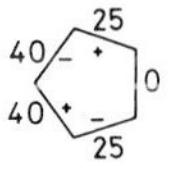 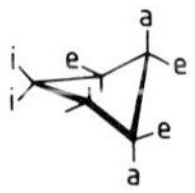 LXXVIIIb

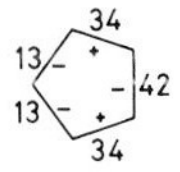

displays C_5 symmetry. In contrast, the half-chair has three coplanar carbon atoms, one above, and one below the plane; this form has C_2 symmetry. These forms are represented in diagrams LXXVIII in perspective drawing, as well as in the conventional formula (as proposed by Bucourt [96]) indicating the approximate value of the torsion angles and their sign. In these conformers, the following types of exocyclic bonds have been recognized: *equatorial* (*e*), *axial* (*a*), and the so-called *isoclinal* (*i*) or *bisectional* bonds (see [98], and refs. therein). For a general definition of ring substituent positions, the reader is referred to the excellent work of Cremer [99].

The two conformers LXXVIIIa and LXXVIIIb are extremes of symmetry in what is known as the *pseudorotational circuit* of cyclopentane. If in structure LXXVIIIa, the out-of-plane carbon (arbitrarily designated C-1 here) is pushed down together with C-2, a half-chair is obtained (C-1 above, C-2 below the plane). If the motion is continued, another envelope is reached (C-2 below the plane). The process then repeats with C-2 and C-3, and so on; 10 envelope forms and 10 half-chair forms interconvert by this process, which is not unlike a wave on a water surface ([100], and refs. therein).

The pseudorotation circuit of cyclopentane is essentially of constant strain, and therefore without maxima and minima; in contrast, the fully planar conformer is less stable by ca. 5 kcal/mol [101]. An extensive treatment of the conformational aspects of cyclopentane and of many five-membered-ring compounds has been published by Legon [92].

The preferred conformer of *cyclohexane* is the *chair* form (LXXIXa), in which all

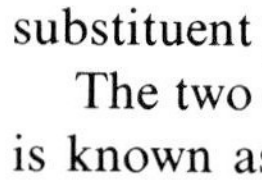

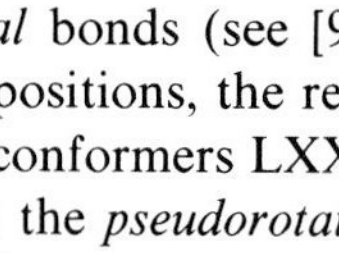

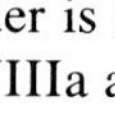

LXXIXa

dihedral angles have an absolute value close to 56°. A Newman projection shows that in this conformation all bonds (endocyclic as well as exocyclic) are staggered, thus explaining the minimum in conformational strain (see Section 8.a). The energy barrier to convert a chair form into another one is close to 10 kcal/mol; this chair–chair conversion is a complex process of cycle inversion aptly described by Bucourt [96].

The chair form of cyclohexane is considered 'rigid' since it must undergo cycle inversion of relatively high energy to yield other conformers. However, besides the rigid chair conformers, cyclohexane also exists as flexible forms which include the *boat* conformer LXXIXb and the *twist* form LXXIXc. In fact, the flexible form gives

LXXIXb

LXXIXc

rise to an infinite number of conformations by continuous variation of the torsion angles. The boat and twist forms represent the local energy maxima and minima, respectively, of the flexible form, with the former about 6 kcal/mol, and the latter about 5 kcal/mol, above the global minimum of the chair forms [93].

The boat and the chair forms of cyclohexane show only *equatorial* and *axial* exocyclic positions. In contrast, the twist conformers display *pseudoequatorial* (e'), *pseudoaxial* (a') and *isoclinal* (i) positions [102].

(b) Substituted carbocycles

In *monosubstituted cyclohexanes*, two conformers can be discriminated by the *equatorial* (LXXXa) and *axial* position (LXXXb) of the substituent. While the energy

LXXXa

LXXXb

barrier separating the two chairs is usually in the range 10–12 kcal/mol, the difference in the free energy of the two conformers varies with the substituent.

In most cases, steric effects control the *axial–equatorial* equilibrium and favor the *equatorial* position of the substituents. Indeed, an *axial* substituent experiences *gauche* interactions with C-3 and C-5, while the *equatorial* substituent is *trans* to C-3 and C-5 and thus relieves the strain of these *gauche* interactions. Furthermore, the two *axial* hydrogens on C-3 and C-5 experience steric interactions with the hydrogen atom (LXXXa) or the substituent (LXXXb) occupying the C-1 *axial* position. These *diaxial* 1/3 interactions are larger with a substituent than with H.

A compilation of classical substituents and the free energy difference associated with their *axial–equatorial* equilibrium in monosubstituted cyclohexanes is given in Table 8 [103]. All these substituents prefer the *equatorial* position, but a 20-fold variation in $-\Delta G^{\circ}$ values is apparent. However, a very few substituents are known to prefer the *axial* position. This is the case for the deuterium atom [104], and for the -HgCl substituent (0.3 kcal/mol [105]).

Considerable conformational complexity is found in disubstituted cycloalkanes,

TABLE 8
Standard free energy change for the *axial–equatorial* equilibrium in monosubstituted cyclohexane derivatives (from [103])

Substituent	$-\Delta G°$ (kcal/mol)
F	0.15
Cl	0.43
Br	0.38
I	0.43
CN	0.17
CH_3	1.70
CH_2CH_3	1.75
$CH(CH_3)_2$	2.15
Phenyl	3.0
COOH	1.35
COO^-	1.92
$COOCH_3$	1.27
OH (aprotic solvents)	0.52
(H-bond donor solvents)	0.87
OCH_3	0.60
NH_2 (aprotic solvents)	1.20
(H-bond donor solvents)	1.60
NH_3^+	1.90
NO_2	1.10

and results are currently accumulating at a rapid pace, particularly concerning *disubstituted cyclohexanes*. In the derivatives which experience an *axial/equatorial–equatorial/axial* equilibrium, namely *cis*-1,2- (LXXXI), *trans*-1,3- and *cis*-1,4-

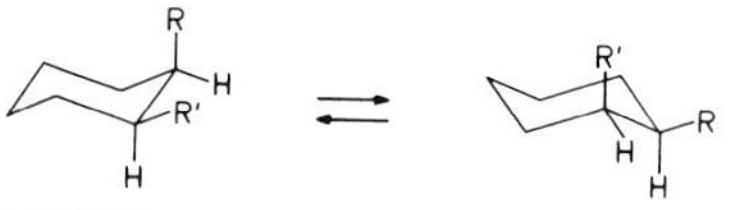

LXXXIa LXXXIb

disubstituted cyclohexanes, the exocyclic interactions are comparable in the two chair conformers, the more so the more the two substituents R and R′ resemble each other. The chair–chair energy difference is the balanced result of the influence of both substituents, and when R = R′ the two chairs become identical.

An altogether different problem arises in the derivatives displaying a *diequatorial-diaxial* equilibrium, namely *trans*-1,2- (LXXXII), *cis*-1,3- and *trans*-1,4-disubstituted

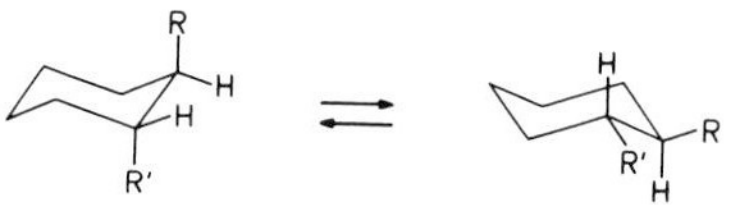

LXXXIIa LXXXIIb

TABLE 9
Free energy difference for the *diequatorial–diaxial* equilibrium in *trans*-1,2-disubstituted cyclohexanes [106] [a]

Substituents		$-\Delta G^{\circ}$ (kcal/mol)	% of *aa*-conformer
F	OH	1.65±0.23	6.1
F	OAc	1.28±0.17	10.6
F	OCH_3	0.81±0.09	20.6
F	Cl	0.59±0.08	27.3
F	Br	0.42±0.07	33.3
F	I	0.54±0.07	29.1

[a] As determined by ^{1}H-NMR in CCl_4 at 30°C.

cyclohexanes. The equilibrium is governed by a number of factors including the conformational free energy contributions of the individual substituents (see Table 8), as well as complex interactions between the two substituents. These interactions are particularly marked in the conformation which brings the two substituents close to each other, namely in the *diequatorial trans*-1,2-isomers and in the *diaxial cis*-1,3-isomers. Experimental values for the chair–chair equilibrium in *trans*-1,2-disubstituted cyclohexanes (LXXXII) is given in Table 9. A quantitative interpretation of these data shows [106] that besides a minor steric repulsion and a major electrostatic repulsion term, an additional *gauche*-attraction exists in LXXXIIb for such pairs of substituents as O/O, F/O, F/Cl, F/Br and F/I. This '*gauche*-effect' appears as a rather general, quite complex, and unsufficiently understood conformational effect [107]. Much interest has been devoted since a few years to the conformational behavior of disubstituted cycloalkanes such as cyclohexane (see above) and cyclopentane (e.g., [108]), and to polysubstituted derivatives (e.g., disubstituted cyclohexanones [109]).

(c) Heterocycles

The conformational behavior of saturated heterocycles resembles that of saturated carbocycles in many but not all aspects. Indeed, the heteroatom(s) influence(s) the relative energy of the various conformers (low-energy forms as well as transition states), and may contribute specific interactions with the substituent(s).

Six-membered rings (LXXXIII) undergo a chair–chair equilibrium; a comparison

LXXXIIIa ⇌ LXXXIIIb

of the free energies of activation for ring inversion (Table 10) shows the values to decrease with increasing size of the substituent. In fact, the barrier depends directly on the magnitude of the C–X torsional energy.

TABLE 10
Free energy of activation for ring inversion in six-membered cycles [110]

Cycle (LXXXIII)	$\Delta G^{\neq}$ (kcal/mol)
Cyclohexane (X=CH_2)	10.3
Piperidine (X=NH)	10.4
Oxane (X=O)	9.5
Thiane (X=S)	9.0
Selenane (X=Se)	8.2

When in LXXXIII the X-member in the ring is CH_2, O, S or Se, the two chairs LXXXIIIa and LXXXIIIb are identical. In the case of piperidine (LXXXIV, R = H), however, ring inversion yields the two chairs LXXXIVa and LXXXIVb

LXXXIVc ⇌ LXXXIVa ⇌ LXXXIVb

which have the 1-H atom in an *equatorial* and an *axial* position, respectively; these two conformers thus share a diastereoisomeric relationship. In addition to ring inversion, the two chairs can also interconvert by nitrogen inversion (LXXXIVa–LXXXIVc), LXXXIVb and LXXXIVc being identical. It appears that the N–H *equatorial* conformer (LXXXIVa, R = H) is favored over the N–H *axial* conformer by 0.4 ± 0.2 kcal/mol in the gas phase and nonpolar solvents, while in polar solvent the latter conformer may well be favored [111]. The conformational aspects of six-membered rings containing one heteroatom have been reviewed in detail by Lambert and Featherman [110].

Cycles containing *two or more heteroatoms* obviously give rise to many conformational possibilities some of which have been investigated (e.g., [91,92,112–114]).

The conformational aspects of *substituted heterocycles* have been a subject of intense interest for many years, and the topic is obviously a complex one due to the great number of possibilities. In nitrogen-containing heterocycles, the preferred position of a N-substituent is usually found to be the *equatorial* one. In *N*-methylpiperidine (LXXXIV, R = Me) for example the *N*-methyl group prefers the *equatorial* position by 2.7 kcal/mol [115], a value considerably larger than the energy difference for methylcyclohexane (Table 8). The *equatorial–axial* conversion appears governed by a process of *N*-inversion (LXXXIVa–LXXXIVc), with an energy barrier (*equatorial* to transition state—*eq*-to-*ts*) of 8.7 kcal/mol [115]. An endocyclic β-heteroatom decreases the *eq*-to-*ts* half barrier, but increases the *ax*-to-*ts* half barrier. In contrast, endocyclic α-heteroatoms show less regular effects, while γ-heteroatoms have a small effect [115].

In the case of oxygen- and sulfur-containing heterocycles, some electron-rich substituents adjacent to C-2 show an unusual conformational behavior. Indeed, an electronegative substituent will tend to prefer an *axial* position when vicinal to the endocyclic oxygen (or sulfur) atom (LXXXV). This *axial* preference is known as the *anomeric effect* (see [107,116], and refs. therein). The magnitude of the anomeric effect has been defined as the free energy difference between the *axial* and *equatorial* conformer (LXXXVb and LXXXVa, respectively), plus the ordinary conformational

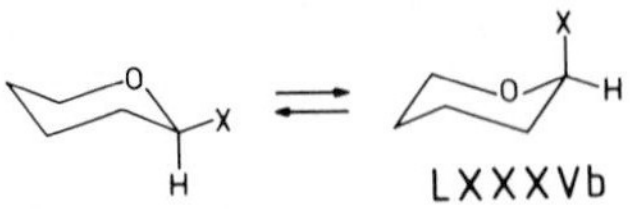

LXXXVa LXXXVb

preference of the 2-substituent (Table 8) [113]. The anomeric effect is in the range 2.5–3 kcal/mol for halogen atoms, and smaller (0.9–1.4 kcal/mol) and more solvent-dependent for the hydroxy, alkoxy, and acyloxy groups [113].

The anomeric effect is the major factor influencing the anomeric equilibrium in hexopyranoses. Thus, the anomeric equilibrium in D-glucopyranose (LXXXVI) shows the β- and α-anomer (LXXXVIa and LXXXVIb, respectively) in a 66/34

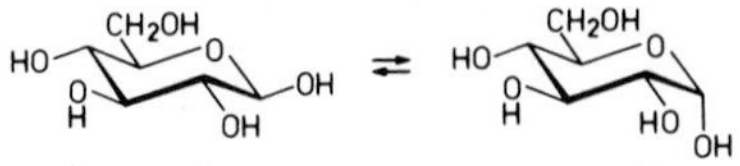

LXXXVIa LXXXVIb

ratio [116]; the high proportion of the α-anomer reflects the stabilizing influence of the anomeric effect.

The conformational equilibrium in hexopyranoses is considerably influenced by the configuration of the various centers of chirality. In β-D-glucopyranose (LXXXVIa) all substituents can simultaneously adopt an *equatorial* position, hence the considerable stability of this conformer. In contrast, α-D-isopyranose (LXXXVII)

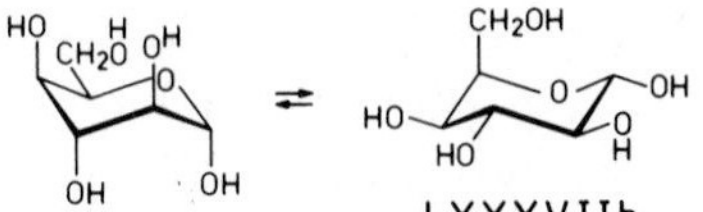

LXXXVIIa LXXXVIIb

exists as two chair conformers, the stabilities of which are comparable [116].

10. Conclusion: The structural complexity of biomolecules

The present review has been written with the aim of presenting the various aspects of molecular geometry relevant to stereoisomerism, namely symmetry, configurational and conformational isomerism, enantiomerism and diastereoisomerism. It certainly was apparent to the reader that the examples were never examined for the full spectrum of their stereochemical features, but only for that particular aspect under consideration in the relevant pages. By so doing, clarity is gained, but an integrated view is lost, a loss tolerable, however, since stereochemists often investigate model molecules displaying few elements of stereoisomerism and a limited number of degrees of conformational freedom.

As opposed to fundamental stereochemists, biochemists cannot choose or design the molecules they plan to investigate from a stereochemical viewpoint. These molecules tend to be characterized by their structural complexity, among other aspects a large number of degrees of conformational freedom, and several elements of stereoisomerism.

Biomolecules span an extremely broad range of physicochemical properties, in particular size (molecular weight) and structural organization. Relatively small biomolecules include among many others monosaccharides, amino acids, some vitamins, carboxylic acids; somewhat larger molecules include steroids, some fatty acids and derivatives (e.g., prostaglandins), porphyrins, some vitamins. Absolutely fundamental to life is the biosynthesis of polymers made of smaller molecular units. Schematically, several levels of complexity can be discriminated (Table 11). Oligomers and medium-sized molecules include oligosaccharides, nucleotides and coenzymes, polypeptides and lipids. The next level is that of the genuine macromolecules such as nucleic acids and the many proteins and enzymes. Highly organized supramolecular structures represent the next level, as exemplified by biological membranes and polyenzymatic systems (e.g., monooxygenases). This latter level is the highest one

TABLE 11
Biological levels of complexity (size/organization)

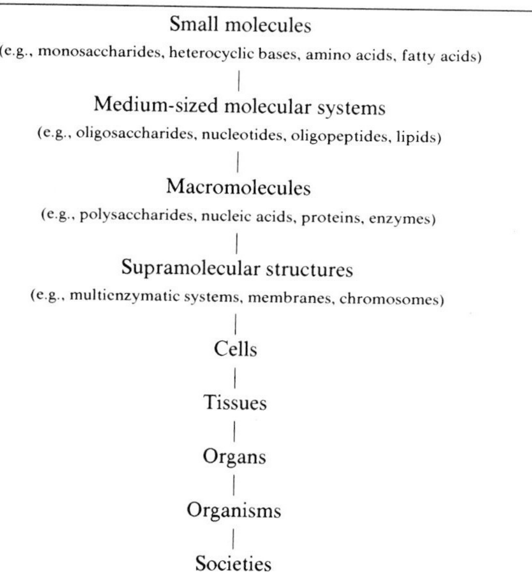

Small molecules
(e.g., monosaccharides, heterocyclic bases, amino acids, fatty acids)
|
Medium-sized molecular systems
(e.g., oligosaccharides, nucleotides, oligopeptides, lipids)
|
Macromolecules
(e.g., polysaccharides, nucleic acids, proteins, enzymes)
|
Supramolecular structures
(e.g., multienzymatic systems, membranes, chromosomes)
|
Cells
|
Tissues
|
Organs
|
Organisms
|
Societies

compatible with current stereochemical approaches and hence of interest here. Still higher levels of biological complexity obviously exist (Table 11).

Focal to the present discussion is the hypothesis that to each of the above levels of complexity corresponds a set of properties which are non-existent in the lower levels, and which allow new biological functions to be fulfilled. In other words, each level is separated from the next by a threshold where new properties emerge. Such a threshold cannot correspond to an increase in size alone but in both size and degree of organization, as combined in the term of 'complexity'. As an example, consider the porphyrin derivatives, where degree of oxidation and cyclic arrangement of the four pyrrol rings result in extensive electronic conjugation and very high energy of resonance (6 to 8 times that of an individual pyrrol ring) [117]. An additional property of the porphyrin ring system is its strong ability to chelate various metallic cations, the resulting metalloporphyrins being the active constituents of such enzymes as cytochromes, chlorophylls and hemoglobins. Clearly such a chelating ability is non-existent for pyrrol and biologically non-significant for linear tetrapyrrollic systems (e.g., bilirubin and biliverdin). It is indeed size, and organization of subunits, which account for the key biological properties of porphyrins.

One of the aspects of structural complexity in biomolecules is that of stereochemistry. To the increase in stereochemical complexity when going to larger biomolecules correspond new stereochemical features and properties. Thus, the concepts of cycloenantiomerism, cyclodiastereoisomerism and cyclostereoisomerism have been introduced by Prelog and collaborators [118–120] to describe stereochemical features of cyclic molecules displaying $2n$ centers of chirality, e.g., cyclic polypeptides.

Current stereochemical interest among biochemists is markedly concentrated on polypeptides and proteins, and on nucleotides and their polymers. In such biomolecules, the above considerations gain particular significance. The numbers of elements of stereoisomerism and of degrees of conformational freedom do not merely add arithmetically, but result in such novel features and properties as a marked stabilization of highly regular shapes (entropically highly unfavorable), the existence of large hydrophobic regions in an aqueous environment, catalytic activity and catalytic cooperativity (enzymes), highly specific chemical recognition (receptors, immunoproteins), duplication (nucleic acids), and many others non-existent in the lower levels of organization. All these properties are linked to the high-order stereochemical features of biopolymers and biological structures. Here, and as opposed to the model molecules mentioned at the beginning of this conclusion, partial stereochemical approaches will fail to lead to understanding. As a result, each biomolecule, each polymer or biological structure must be studied globally and for itself by considering and integrating all its stereochemical features. Enlightening examples are found in a review on structure and function in neuropeptides [121], and in two outstanding books [122,123]. To such a global stereochemical comprehension of biomolecular systems, the present review merely serves as a spelling book.

References

1 Schoffeniels, E. (1976) Anti-Chance (translated from the French by B.L. Reid), Pergamon, Oxford, pp. 80–85.
2 Testa, B. (1979) Principles of Organic Stereochemistry, Dekker, New York.
3 Jean, Y. and Salem, L. (1978) J. Am. Chem. Soc. 100, 5568–5569.
4 Mislow, K. and Bickart, P. (1976–1977) Isr. J. Chem. 15, 1–6.
5 Bernal, J.D. (1962) in M. Florkin and E.H. Stotz (Eds.), Comprehensive Biochemistry, Vol. 1, Elsevier, Amsterdam, pp. 113–191.
6 Woolley, R.G. (1978) J. Am. Chem. Soc. 100, 1073–1078.
7 Bader, R.F.W., Tal, Y., Anderson, S.G. and Nguyen-Dang, T.T. (1980) Isr. J. Chem. 19, 8–29.
8 Jaffé, H.H. and Orchin, M. (1965) Symmetry in Chemistry, Wiley, New York.
9 Hochstrasser, R.M. (1966) Molecular Aspects of Symmetry, Benjamin, New York.
10 Cotton, F.A. (1971) Chemical Applications of Group Theory, 2nd Edn., Wiley, New York.
11 Hollas, J.M. (1972) Symmetry in Molecules, Chapman and Hall, London.
12 Donaldson, J.D. and Ross, S.D. (1972) Symmetry and Stereochemistry, Wiley, New York.
13 Mead, C.A. (1974) Symmetry and Chirality, Topics in Current Chemistry, Vol. 49, Springer, Berlin.
14 Vincent, A. (1977) Molecular Symmetry and Group Theory, Wiley, London.
15 Ref. 2, p. 8.
16 Pople, J.A. (1980) J. Am. Chem. Soc. 102, 4615–4622.
17 Epiotis, N.D. (1978) Theory of Organic Reactions, Springer, Berlin, p. 34.
18 Mislow, K. (1977) Bull. Soc. Chim. Belg. 86, 595–601.
19 Ref. 2, pp. 33–43.
20 Hanson, K.R. (1976) Annu. Rev. Biochem. 45, 307–330.
21 Henderson, J.W. (1973) Chem. Soc. Rev. 2, 397–413.
22 Lambert, J.B. (1971) in N.L. Allinger and E.L. Eliel (Eds.), Topics in Stereochemistry, Vol. 6, Wiley, New York, pp. 19–105.
23 Mislow, K. (1971) Pure Appl. Chem. 25, 549–562.
24 Kostyanovsky, R.G., Rudchenko, V.F., D'yachenko, O.A., Chervin, I.I., Zolotoi, A.B. and Atovmyan, L.O. (1979) Tetrahedron 35, 213–224.
25 Ref. 2, pp. 49–54.
26 Cahn, R.S. and Ingold, C.K. (1951) J. Chem. Soc. 612–622.
27 Cahn, R.S., Ingold, C.K. and Prelog, V. (1956) Experientia 12, 81–94.
28 Cahn, R.S., Ingold, C.K. and Prelog, V. (1966) Angew. Chem. Int. Ed. 5, 385–415, 511.
29 IUPAC Commission on Nomenclature of Organic Chemistry (1976) 1974 Recommendations for Section E, fundamental stereochemistry, Pure Appl. Chem. 45, 11–30.
30 McMahon, R.E., Sullivan, H.R., Craig, J.C. and Pereira, W.E., Jr. (1969) Arch. Biochem. Biophys. 132, 575–577.
31 Arigoni, D. and Eliel, E.L. (1969) in E.L. Eliel and N.L. Allinger (Eds.), Topics in Stereochemistry, Vol. 4, Wiley, New York, pp. 127–243.
32 Pak, C.S. and Djerassi, C. (1978) Tetrahedron Lett. 4377–4378.
33 Krow, G. (1970) in E.L. Eliel and N.L. Allinger (Eds.), Topics in Stereochemistry, Vol. 5, Wiley, New York, pp. 31–68.
34 Hirschmann, H. and Hanson, K.R. (1971) J. Org. Chem. 36, 3293–3306.
35 Haas, G. and Prelog, V. (1969) Helv. Chim. Acta 52, 1202–1218.
36 Mislow, K. (1970) Acc. Chem. Res. 3, 321–331.
37 Nourse, J.G. (1977) J. Am. Chem. Soc. 99, 2063–2069.
38 Martin, J.C. and Balthazor, T.M. (1977) J. Am. Chem. Soc. 99, 152–162.
39 Saito, Y. (1979) Inorganic Molecular Dissymmetry, Springer-Verlag, Berlin, pp. 1–11.
40 Saito, Y. (1978) in E.L. Eliel and N.L. Allinger (Eds.), Topics in Stereochemistry, Vol. 10, Wiley, New York, pp. 95–174.
41 Ref. 2, pp. 62–63.

42 Kessler, H. (1970) Angew. Chem. 82, 237–253 (Angew. Chem. Int. Ed. 9, 219).
43 Hall, D.M. and Harris, M.M. (1960) J. Chem. Soc. 490–494.
44 Ref. 2, pp. 67–68.
45 Lemière, G.L. and Alderweireldt, F.C. (1980) J. Org. Chem. 45, 4175–4179.
46 Förster, H. and Vögtle, F. (1977) Angew. Chem. Int. Ed. 16, 429–441.
47 Mislow, K. (1976) Acc. Chem. Res. 9, 26–33.
48 Gerdil, R. and Allemand, J. (1979) Tetrahedron Lett. 3499–3502.
49 Wynberg, H. and Marsman, B. (1980) J. Org. Chem. 45, 158–161.
50 Capraro, H.G. and Ganter, C. (1980) Helv. Chim. Acta 63, 1347–1351.
51 Tichý, M. (1972) Tetrahedron Lett. 2001–2004.
52 Nakazaki, M., Naemura, K. and Yoshihara, H. (1975) Bull. Chem. Soc. Japan 48, 3278–3284.
53 Ref. 2, p. 74.
54 Pierre, J.L. (1971) Principes de Stéréochimie Organique Statique, Armand Colin, Paris, pp. 103–108.
55 Epiotis, N.D., Bjorkquist, D., Bjorkquist, L. and Sarkanen, S. (1973) J. Am. Chem. Soc. 95, 7558–7562.
56 Kalinowski, H.O. and Kessler, H. (1973) in N.L. Allinger and E.L. Eliel (Eds.), Topics in Stereochemistry, Vol. 7, Wiley, New York, pp. 295–383.
57 Hirschmann, H. and Hanson, K.R. (1974) Tetrahedron 30, 3649–3656.
58 Prelog, V. and Helmchen, G. (1972) Helv. Chim. Acta 55, 2581–2598.
59 Leonard, J.E., Hammond, G.S. and Simmons, H.E. (1975) J. Am. Chem. Soc. 97, 5052–5054.
60 Ref. 2, pp. 124–125.
61 Moriarty, R.M. (1974) in E.L. Eliel and N.L. Allinger (Eds.), Topics in Stereochemistry, Vol. 8, Wiley, New York, pp. 271–421.
62 Mislow, K. and Raban, M. (1967) in N.L. Allinger and E.L. Eliel (Eds.), Topics in Stereochemistry, Vol. 1, Wiley, New York, pp. 1–38.
63 Hanson, K.R. (1966) J. Am. Chem. Soc. 88, 2731–2742.
64 Bentley, R. (1978) Nature (London) 276, 673–676.
65 Jones, J.B. (1976) in J.B. Jones, C.J. Sih and D. Perlman (Eds.), Applications of Biochemical Systems in Organic Chemistry, Part 1, Wiley, New York, pp. 479–490.
66 Ref. 2, p. 154.
67 Hanson, K.R. (1975) J. Biol. Chem. 250, 8309–8314.
68 Hanson, K.R. and Rose, I.A. (1975) Acc. Chem. Res. 8, 1–10.
69 Gordon, A.J. and Ford, R.A. (1972) The Chemist's Companion, Wiley, New York, pp. 115–124, 156–167.
70 Eliel, E.L., Allinger, N.L., Angyal, S.J. and Morrison, G.A. (1965) Conformational Analysis, Wiley, New York.
71 Hanack, M. (1965) Conformation Theory, Academic Press, New York.
72 Klyne, W. and Prelog, V. (1960) Experientia 16, 521–523.
73 Peterson, M.R. and Csizmadia, I.G. (1978) J. Am. Chem. Soc. 100, 6911–6916.
74 Brunck, T.K. and Weinhold, F. (1979) J. Am. Chem. Soc. 101, 1700–1709.
75 Gavezzotti, A. and Bartell, L.S. (1979) J. Am. Chem. Soc. 101, 5142–5146.
76 Ref. 2, Chap. 3.
77 Sung, E.-M. and Harmony, M.D. (1977) J. Am. Chem. Soc., 99, 5603–5608.
78 Ref. 2, pp. 83–107.
79 Glusker, J.P. (1980) Acc. Chem. Res. 13, 345–352.
80 Liberles, A., O'Leary, B., Eilers, J.E. and Whitman, D.R. (1972) J. Am. Chem. Soc. 94, 6894–6898.
81 Eilers, J.E. and Liberles, A. (1975) J. Am. Chem. Soc. 97, 4183–4188.
82 Schaefer, T., Danchura, W., Niemczura, W. and Parr, W.J.E. (1978) Can. J. Chem. 56, 1721–1723.
83 Ref. 2, pp. 95–107.
84 Drakenberg, T., Sommer, J. and Jost, R. (1980) J. Chem. Soc. Perkin Trans. II, 363–369.
85 Friedl, Z., Fiedler, P. and Exner, O. (1980) Coll. Czech. Chem. Commun. 45, 1351–1360.
86 Jennings, W.B. and Worley, S.D. (1980) J. Chem. Soc. Perkin Trans. II, 1512–1515.

87 Jackson, W.R. and Jennings, W.B. (1974) Tetrahedron Lett., 1837–1838.
88 Lunazzi, L., Magagnoli, C., Guerra, M. and Macciantelli, D. (1979) Tetrahedron Lett., 3031–3032.
89 Jørgensen, F.S. and Snyder, J.P. (1979) Tetrahedron 35, 1399–1407.
90 Boyd, D.B. (1972) J. Am. Chem. Soc. 94, 8799–8804.
91 Malloy, T.B., Bauman, L.E. and Carreira, L.A. (1979) in N.L. Allinger and E.L. Eliel (Eds.), Topics in Stereochemistry, Vol. 11, Wiley, New York, pp. 97–185.
92 Legon, A.C. (1980) Chem. Rev. 80, 231–262.
93 Hanack, M. (1965) Conformation Theory, Academic Press, New York, pp. 72–171.
94 Anet, F.A.L. (1971) in G. Chiurdoglu (Ed.), Conformational Analysis, Academic Press, New York, pp. 15–29.
95 Sicher, J. (1962) in P.B.D. de la Mare and W. Klyne (Eds.), Progress in Stereochemistry, Vol. 3, Butterworths, London, pp. 202–263.
96 Bucourt, R. (1974) in E.L. Eliel and N.L. Allinger (Eds.), Topics in Stereochemistry, Vol. 8, Wiley, New York, pp. 159–224.
97 Toromanoff, E. (1980) Tetrahedron 36, 2809–2931.
98 Ref. 2, pp. 112–113.
99 Cremer, D. (1980) Isr. J. Chem. 20, 12–19.
100 Fuchs, B. (1978) in E.L. Eliel and N.L. Allinger (Eds.), Topics in Stereochemistry, Vol. 10, Wiley, New York, pp. 1–94.
101 Cremer, D. and Pople, J.A. (1975) J. Am. Chem. Soc. 99, 1358–1367.
102 Kellie, G.M. and Riddell, F.G. (1974) in E.L. Eliel and N.L. Allinger (Eds.), Topics in Stereochemistry, Vol. 8, Wiley, New York, pp. 225–269.
103 Hirsch, J.A. (1967) in N.L. Allinger and E.L. Eliel (Eds.), Topics in Stereochemistry, Vol. 1, Wiley, New York, pp. 199–222.
104 Lee, S.-F., Barth, G., Kieslich, K. and Djerassi, C. (1978) J. Am. Chem. Soc. 100, 3965–3966.
105 Anet, F.A.L., Krane, J., Kitching, W., Dodderel, D. and Praeger, D. (1974) Tetrahedron Lett., 3255–3258.
106 Zefirov, N.S., Samoshin, V.V., Subbotin, O.A., Baranenkov, V.I. and Wolfe, S. (1978) Tetrahedron 34, 2953–2959.
107 Zefirov, N.S. (1977) Tetrahedron 33, 3193–3202.
108 Fuchs, B. and Wechsler, P.S. (1977) Tetrahedron 33, 57–64.
109 Pons, A. and Chapat, J.P. (1980) Tetrahedron 36, 2219–2224.
110 Lambert, J.B. and Featherman, S.I. (1975) Chem. Rev. 75, 611–626.
111 Blackburne, I.D., Katritzky, A.R. and Takeuchi, Y. (1975) Acc. Chem. Res. 8, 300–306.
112 Moriarty, R.M. (1974) in E.L. Eliel and N.L. Allinger (Eds.), Topics in Stereochemistry, Vol. 8, Wiley, New York, pp. 271–421.
113 Romers, C., Altona, C., Buys, H.R. and Havinga, E. (1969) in E.L. Eliel and N.L. Allinger (Eds.), Topics in Stereochemistry, Vol. 4, Wiley, New York, pp. 39–97.
114 Ref. 2, pp. 128–129.
115 Katritzky, A.R., Patel, R.C. and Riddell, F.G. (1979) J. Chem. Soc. Chem. Commun., 674–675.
116 Paulsen, H. (1975) Die Stärke-Starch 27, 397–405.
117 George, P. (1975) Chem. Rev. 75, 85–111.
118 Prelog, V. and Gerlach, H. (1964) Helv. Chim. Acta 47, 2288–2294.
119 Gerlach, H., Owtschinnikow, J.A. and Prelog, V. (1964) Helv. Chim. Acta 47, 2294–2302.
120 Gerlach, H., Haas, G. and Prelog, V. (1966) Helv. Chim. Acta 49, 603–607.
121 Schwyzer, R. (1980) Proc. R. Soc. London Ser. B 210, 5–20.
122 Schulz, G.E. and Schirmer, R.H. (1979) Principles of Protein Structure, Springer-Verlag, New York.
123 Chapeville, F. and Haenni, A.-L. (Eds.) (1980) Chemical Recognition in Biology, Springer-Verlag, Berlin.

Chemical methods for the investigation of stereochemical problems in biology

RONALD BENTLEY

Department of Biological Sciences, University of Pittsburgh, Pittsburgh, PA 15260, U.S.A.

1. Introduction

One of the great scientific puzzles "is the understanding of enzymic reactions in terms of enzymic structure" [1]. To solve the puzzle, and describe in detail the interactions between enzymes, substrates, cofactors and products, requires a description of the three-dimensional structures of the various components, separately and in combination, during all phases of the reaction. Within the past few years, precise three-dimensional stereochemical information has become available for several enzymes, primarily as a result of sequence determination and X-ray crystallography. In some cases, this information now extends to combinations between an enzyme and its substrate(s) and/or cofactor(s).

Clearly, great strides have been made since Mautner's sardonic remark in 1967 [2]: "Enzyme chemists, once known as 'people, who if they spill sulfuric acid on a table will call it tablease and feel that the matter has been explained' can now consider the interaction of some enzymes with their substrates in terms of clearly defined molecules interacting in a fashion to be understood in terms of organic reaction mechanisms".

Chemical and structural investigations on the small molecules which interact with enzymes date back for more than one hundred years. Ever since Biot's discovery (1815) of rotatory polarization in naturally occurring materials such as oils of turpentine and lemon, and in solutions of camphor, sugar and tartaric acid, there has been a close relationship between the biological sciences and stereochemistry. In 1894 Emil Fischer clearly recognized that enzyme reactions show a high degree of selectivity [3]; at that time, yeast preparations (containing maltase) were found to hydrolyze methyl α-D-glucoside but not methyl β-D-glucoside, while the preparation emulsin (from almonds and containing a β-glucosidase) showed an opposite pattern.

Beginning with Pasteur's work in 1860 [4] the fields of stereochemistry and biology were dominated for almost nine decades by the phenomenon now called chirality. Chiral molecules are those for which a three-dimensional model of the molecule is not superimposable on the mirror image of the model. Since the operation determining the existence of chirality is reflection in a plane mirror, this

Tamm (ed.) Stereochemistry

area of stereochemistry is concerned with whether or not molecules possess 'reflection asymmetry'. An enormous volume of work, both chemical and biological, was carried out with chiral molecules; a high point was reached in 1933 with the publication of Freudenberg's magnum opus [5].

Despite the volume of chemical and biological work carried out for nearly a century, another aspect of stereochemistry remained almost completely hidden until 1948. In that year, Ogston realized that identical groups in some symmetrical molecules could be distinguished one from another in enzymatic reactions [6] (the differentiation is, however, not limited to enzymatic reactions). The possibility for differentiation resides in a structural property of the substrate; compounds in which differentiation is possible have 'rotational asymmetry'. Ogston's vision (for a retrospective review, see [7]) opened up the role of rotational asymmetry in stereochemistry and biology and has led to many important developments in practice, theory and nomenclature. The term, prochiral [8] is now applied to the molecular feature that makes possible the differentiation of attached identical groups (or 'faces'). Thus, there are now two major areas of interest, chirality and prochirality. The symmetry elements which prevent the differentiation of identical groups (finite simple axis of rotation greater than one) are different in kind from those preventing resolution into two object–mirror-image related structures (alternating axis of symmetry).

The contributions of chemical methods and principles to the study of stereochemical problems in biology have, of course, been prodigious. They have ranged from determinations of configurations and conformations of many compounds, through syntheses of substrates and products, to studies of model compounds and reactions. An interesting development is that increasingly, biochemical methods are being applied in organic chemistry (see, for example [9]). It would be impossible to give a complete account of all of the chemical investigations in this chapter. An attempt will be made, however, to cover some of the important work. In so doing it should be possible to give the reader a feeling for the scope of chemical methodology. In many investigations, chemical and biochemical methods have been used together. No attempt will be made to be exclusively chemical since the combination of methods is usually extremely productive.

The terminology and nomenclature used in this chapter are those generally accepted. Much of this material has already been described by Testa in the first chapter of this book. The term 'configuration' will be applied to both chiral and prochiral situations. Its meaning with respect to chirality is that of long usage—the fixed three-dimensional ordering of atoms or groups. In most cases, this will refer to the four atoms, or groups of atoms, disposed tetrahedrally around a chiral center; it could also refer to the ordering of groups about a chiral axis, as for example, in an allene. A 'prochiral configuration' will be defined by an isotopic substitution at a prochiral position. Such substitutions may involve the complete replacement of one isotope by another, or the use of only a tracer level of isotope. If the replacement is 100% it may be possible to observe an optical activity. The replacement of H by ca. 100% ^{2}H is easily achieved and such replacements at prochiral centers have given rise to many compounds in which optical activity has been measured [10,11].

Similarly, optical activity due to isotopic replacement has been observed for oxygen and carbon [12].

More commonly in biochemistry the configuration at a prochiral position is defined operationally by the enzymatic reaction or biological process, and only a tracer level of the isotope is used. The configuration of citric acid (an achiral compound) will then specify the location of an isotopic atom in one of the two CH_2COOH groups, or in one of the hydrogens of the CH_2 groups, and its selective utilization (or not) in some process. The configuration of citric acid thus provides an answer to questions such as the following—does the CH_2COOH group of citric acid which is derived from acetyl CoA occupy the *pro-R* or *pro-S* position? Projection formulae will be drawn using the usual 'Fischer conventions'. Unless otherwise indicated, evidence for the configurations of chiral compounds discussed in this chapter can be located in standard reference sources [13–15].

2. *The basis for the biological recognition of chirality and prochirality*

(a) Differentiation at chiral positions

It has been known for more than a century that living systems can distinguish between isomeric forms of many substances. In 1860, Pasteur [16] showed by polarimetry that when the ammonium salt of racemic tartaric acid ('paratartrate') was subjected to fermentation by a yeast only one of the two enantiomeric forms was consumed—"The yeast which causes the right salt to ferment leaves the left salt untouched, in spite of the absolute identity in physical and chemical properties of the right and left tartrates of ammonium....". (It is believed that Pasteur's 'yeast' was actually a species of *Penicillium*—see [17], in particular p. 699.) Although Le Bel [18] later claimed to have resolved a chiral nitrogen compound (methylethylpropylisobutylammonium iodide) by a similar process with *Penicillium glaucum*, others could not repeat this work. It was concluded that Le Bel did not have a satisfactory preparation of the ammonium salt [19].

Another kind of biological differentiation was carried out in 1886 when Piutti [20] observed that one enantiomer of asparagine had a sweet taste, while the other was 'insipid'. Commenting on this fact, Pasteur [21] suggested that the difference was due to a different action of the two enantiomers on the asymmetric constituents of the gustatory nerve. "Le corps actif dissymétrique qui interviendrait dans l'impression nerveuse, traduite par une saveur sucrée dans un cas et presque insipide dans l'autre, ne serait autre chose, suivant moi, que la matière nerveuse elle-même, matière disymmétrique comme toutes les substances primordiales de la vie; albumine, fibrine, gélatine, etc."

With the development of theories of chemical structure and the science of stereochemistry, greater sophistication became possible. As already noted, Fischer used enzymes in 1894 which differentiated between carbohydrate diastereoisomers. By about the turn of the nineteenth century, the role of three-dimensional structures began to emerge. For example, in a chapter entitled 'The Relations of Stereochemistry to Physiology', Stewart [22] discussed experiments carried out by Crum Brown and Fraser in 1868–1869. If certain alkaloids were treated with methyl iodide they gained a new curare-like activity without losing their 'previous efficiency'. Other experiments showed that when the base present in curare ('curine') was treated with methyl iodide, the methylated derivative was "two hundred and twenty-six times as poisonous as the mother substance." To account for these changes, it was proposed that the change from trivalent to pentavalent (i.e., $R_4N^+X^-$) nitrogen was critical, with the associated change from a planar structure to the tetrahedral ("the curare action appears to be due to the change from a plane to a three-dimensional arrangement of the atoms"). Stewart also cited experiments showing that other tetrahedral compounds (e.g., tetraethylphosphonium iodide) had a strong curare-like action, unlike that of the trivalent phosphorus compounds (also for As, and Sb). Other evidence in favor of the stereochemical explanation was provided by the behavior of sulfur compounds. Here again the change from a planar arrangement (sulfides) to the 'solid configuration' was accompanied by a change in physiological properties. Thus, the base $(CH_3)_3SOH$ was cited as having a pronounced curare character.

While these experiments stimulated other research, it is sad to relate that the proposed explanation is not correct. It required almost seventy years, however, before it was realized that the curare-like action of the onium salts was a result of their ionic character. Other examples of the role of spatial arrangement were discovered. For example, quoting Stewart: "Ishizuka (1897) found that maleic acid was a much stronger poison than its stereoisomer, fumaric acid; 1.94 grammes for every kilogramme in a dog's weight was a fatal dose of the former acid, while the same dose of fumaric acid was harmless." Similarly these isomers were found to have differential effects on microorganisms. Stewart also rationalized the mydriatic action of tropine and the inactivity of pseudotropine in terms of three-dimensional formulae as shown below (1 and 2). In many ways, these structural representations are close to the present-day conformational structures (pseudotropine = 3, tropine = 4):

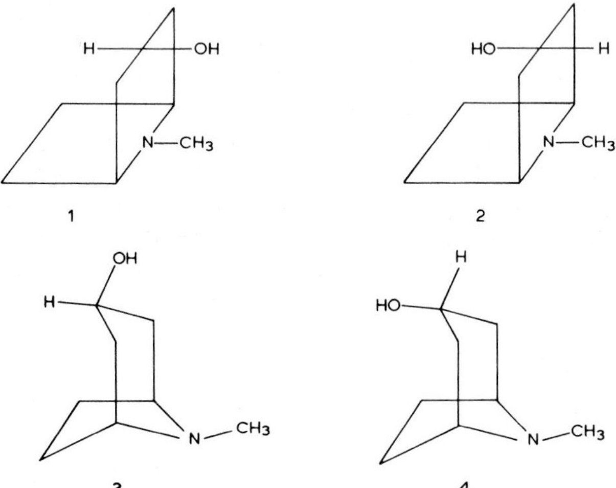

However, despite these and other early observations, it was not until considerably later that Cushny demonstrated clearly that one enantiomer of a molecule could be much more active pharmacologically than the other enantiomer [23]. Thus (−)-hyoscyamine was 12–20 times as active as the (+)-enantiomer with respect to their action on motor nerve endings. In his Charles E. Dohme Memorial Lecture (1925), Cushny [23] stated with reference to optical activity: "The chemists have long been mildly interested in this form of vital activity, the pharmacologists have shown its importance in their sphere, but biologists in general do not seem to have appreciated it as a very definite and measurable feature of living matter".

Cushny appreciated that an optically active cell receptor (represented here as +) would combine with the (+)- and (−)-enantiomers of a drug to form diastereoisomeric complexes:

(+)-drug + (+)-receptor → (+)-drug–(+)-receptor

(−)-drug + (+)-receptor → (−)-drug–(+)-receptor

Since the two products or complexes are diastereoisomers they have different physical properties and Cushny attributed the different pharmacological properties to the different physical properties. However, he seemed to ignore the overall three-dimensional structure, and Parascandola suggests that Cushny had difficulty in imagining that a receptor could actually differentiate between enantiomeric structures [24].

A clear three-dimensional visualization of the means by which a receptor could differentiate between enantiomers was provided by Easson and Stedman in 1933 [25]. They proposed that three (b, c, d) of the four groups (a, b, c, d) linked to a chiral carbon atom were concerned in the process (either by normal valence forces, or by adsorptive or other forces). The receptor possessed three groups b′, c′ and d′; for maximum physiological effect, the drug molecule must "become attached to the receptor in such a manner that the groups b, c and d in the drug coincide respectively with b′, c′ and d′ in the receptor. Such coincidence can only occur with one of the enantiomorphs and this consequently represents the more active form of the drug". The interaction (5) and non-interaction (6) were illustrated as follows:

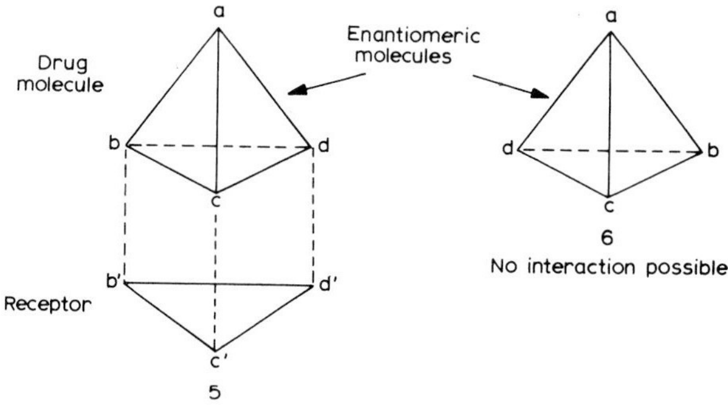

This paper thus clearly defined the possibility for a 'three-point attachment' to provide for the differentiation of enantiomeric molecules. In support of the theory, they cited results on the action of epinephrine and related compounds. Cushny had reported that (−)-epinephrine is 12–15 times as active as (+)-epinephrine. Easson and Stedman deduced that of the four groups around the chiral center, the hydrogen played no part in the drug–receptor attachment. Thus for (−)-epinephrine the complex could be represented as 7. With the (+)-enantiomer the hydroxyl group cannot be positioned on the hydroxyl receptor. It was suggested that a "less perfect combination" was possible if the hydrogen were located on the normal hydroxyl

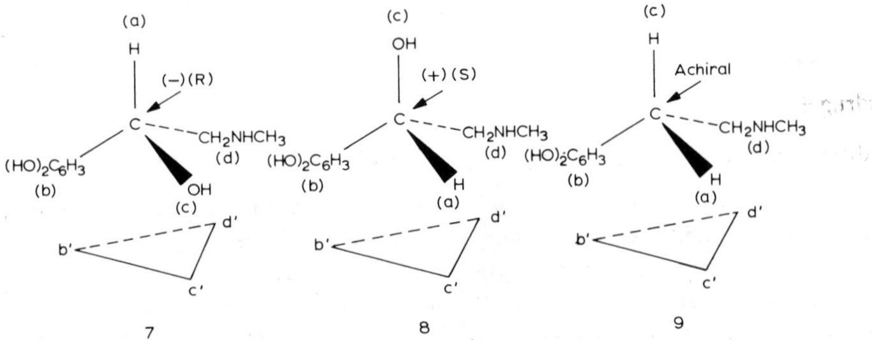

receptor, 8. Furthermore, they argued that since the achiral 3,4-dihydroxy-β-phenylethylmethylamine could adopt a similar orientation, 9, the activity of the latter achiral component should approach that of the (+)-enantiomer of epinephrine itself. Their calculations from data then available indicated that (+)-epinephrine was 1.07–0.87 times as active as 3,4-dihydroxy-β-phenylethylmethylamine "a result which corresponds with that deduced from our hypothesis much more closely than might have been anticipated".

This interesting paper of Easson and Stedman seems to have made almost no impression at the time of its publication and lay unrecognized until it was cited by Parascandola [24]. As will be seen it also foreshadows the topic of prochirality but was not known to Ogston (I am indebted to Dr. Ogston for bringing it to my attention [26]; Dr. J. Glusker had brought it to his attention).

A little later, and without reference to Easson and Stedman, Bergmann and his colleagues [27–29] discussed the "antipodal specificity of papain peptidase I". This enzyme hydrolyzed only (−)-L-benzoylleucineamide and not the D-enantiomer. They stated that the enzyme "must contain at least three different atoms or atomic groups which are fixed in space with respect to one another, these groups entering during the catalysis into relation with a similar number of different atoms or atomic groups of the substrate". This was referred to as a "polyaffinity relationship". They imagined in the enzyme substrate complex a plane formed by the active groups of the enzyme and a plane formed by the active groups of the substrate. They postulated that one or more large atomic groups could "jut out" and prevent the approach of the binding plane of the enzyme toward the binding plane of the substrate—thus catalysis failed to occur as a result of steric hindrance (the original refers to 'stearic' hindrance). For aminopeptidase acting on leucylglycylglycine the 'active groups' in the substrate were the amino group (N-terminus leucine) and various groups of the adjacent peptide linkage, 10. The enzyme was required to find the groups NH, CO, α carbon and NH_2 arranged in a clockwise order and to approach that defined side of the molecule.

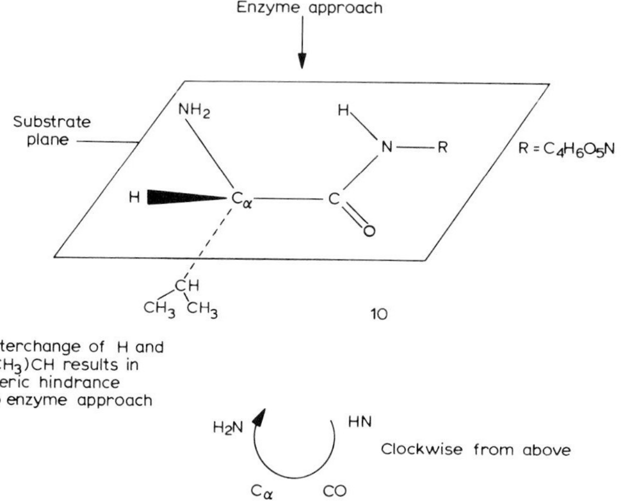

Action of aminopeptidase on L-leucylglycylglycine (redrawn from Bergman and Fruton [27]).

In the late 1940s and early 1950s the concept of three-point attachment was being generally used to rationalize a number of biological interactions. Since this theory has become something of a dogma, it is worthwhile to explore the basis for chiral differentiation in chemistry and biology a little further. In principle, a chiral reagent (+)A reacting with a racemic mixture of enantiomers (±)B can form two products (or complexes, etc.), (+)A(+)B and (+)A(−)B. Since these two materials are diastereoisomers and since diastereoisomers differ in internal energy, they will, in general, not be formed in equal amounts.

$$(+)\mathrm{A} + (\pm)\mathrm{B} \rightarrow \underset{x\%}{(+)\mathrm{A}(+)\mathrm{B}} + \underset{y\%}{(+)\mathrm{A}(-)\mathrm{B}} \qquad x \neq y \tag{1}$$

Unlike the racemic pair (+)A(+)B and (−)A(−)B, the diastereoisomeric pair (+)A(+)B and (+)A(−)B can also be separated by simple processes (which do not involve chiral agents or influences)—for example, crystallization or chromatography. This is the basis for Pasteur's 'resolution' of (±)-tartaric acid and for countless other similar resolutions. (For a recent account of strategies in optical resolutions, see [30].)

Similarly, if (+)A reacts separately with (+)B and (−)B the results will be different in the two cases since the two possible products are the same diastereoisomers just considered. In such processes the different properties or energy levels associated with diastereoisomers will lead to the production of different amounts of AB complex for the two reactions (2 and 3).

$$(+)\mathrm{A} + (+)\mathrm{B} \rightarrow (+)\mathrm{A}(+)\mathrm{B} \quad (2)$$

$$(+)\mathrm{A} + (-)\mathrm{B} \rightarrow (+)\mathrm{A}(-)\mathrm{B} \quad (3)$$

The result observed will be determined either by the probability of successful collision between (+)A and (+)B on the one hand and between (+)A and (−)B on the other (kinetic control), or by the stability of (+)A(+)B relative to (+)A(−)B if the process is reversible (thermodynamic control).

Clearly, the outcome of such processes is determined by a diastereoisomerism existing somewhere along the reaction pathway. This could, for example, be in the approach of two materials to each other, in a transition state or reaction intermediate, or in the properties of the final product. The nature of the approach, interaction or bonding, is immaterial. For example, it has been shown recently that there are real and significant differences in the energetics of aggregation of chiral ions in solution to make diastereoisomeric ion pairs [31]. As shown in Table 1, even very simple chiral ion pairs can differ by 200–500 cal/mol in their heats of formation from the free ions. Such a difference can account for the difference between a reaction yield of 50:50 and 60:40.

TABLE 1
Heats of formation of ion pairs

Base	Acid	$-\Delta H$ (kcal/mol)
(A) α-Phenylethylamine and mandelic acid		
R	*R*	7.630±0.013
S	*S*	7.612±0.035
R	*S*	7.431±0.014
S	*R*	7.374±0.024
(B) Ephedrine and mandelic acid		
1*R*,2*S*	*R*	7.294±0.017
1*S*,2*R*	*S*	7.325±0.030
1*R*,2*S*	*S*	7.550±0.029
1*S*,2*R*	*R*	7.607±0.028
(C) Pseudoephedrine and mandelic acid		
1*R*,2*R*	*R*	6.672±0.025
1*S*,2*S*	*S*	6.657±0.018
1*R*,2*R*	*S*	6.401±0.019
1*S*,2*S*	*R*	6.380±0.017

While many factors influence the binding of a substrate to an enzyme, a drug to a receptor, and similar situations, it appears that in all cases where the biological differentiation of the enantiomeric forms of a chiral ligand occurs the fundamental stereochemical basis is that the interaction is diastereoisomeric in nature. (It is assumed here that the same site is utilized by the two enantiomers.) The enzyme, receptor, membrane, etc., will be a highly chiral molecule—most usually a highly chiral macromolecule. If the chiral enzyme or receptor is designated as (+)A and the two enantiomers of the ligand as (+)B and (−)B, the possible interactions are those previously noted (see Equations 2 and 3).

It can be stated categorically that for the two reactions 2 and 3 the possibility of different results depending on whether (+)B or (−)B is used will always exist in the theoretical sense (for the possibility not to exist identical properties would be required for the diastereoisomers). Whether the potential will or will not be expressed will be dependent on many other factors.

In other words, different reactivities with the enantiomeric forms of chiral substrates or drugs, binding at the same site, would not be observed in the absence of diastereoisomeric possibilities. This is really another way of expressing three-point attachment. If the enzyme or receptor is chiral, then the classical three- and two-point arrangements 11 and 12, are, of necessity, diastereoisomeric.

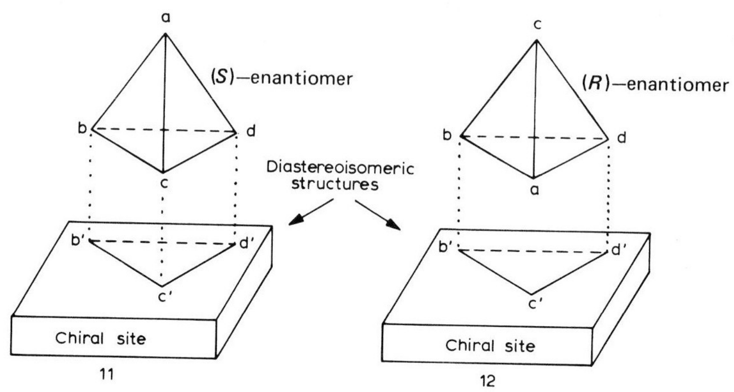

Assumed priority sequence: a>b>c>d

(Another possibility that would lead to discrimination between the enantiomers of a substrate would be utilization of different binding sites on the protein. In this event, constitutional isomerism would be involved and would provide a mechanism for the differentiation.)

In principle, three-point attachment represents overkill. The two structures shown below, 13 and 14, with a one-point attachment are still diastereoisomeric. Differences could, therefore, be observed in the two reaction possibilities.

Another difficulty with three-point attachment is the emphasis on 'attachment'. At times this word has almost seemed to imply covalent bond formation. Yet in principle, actual bond formation or strong physical attachment is not a requirement;

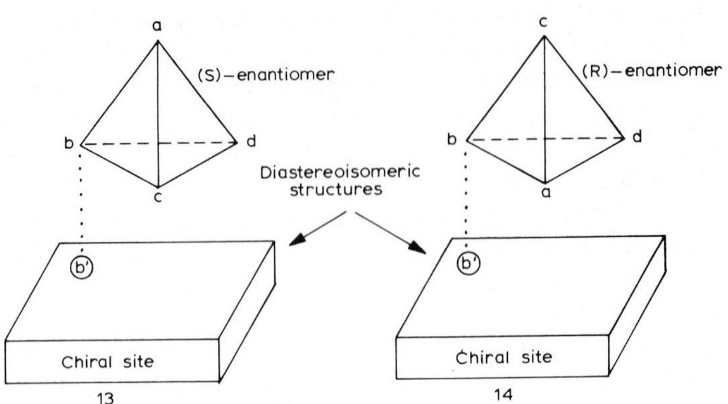

an approach of, say, an enzyme to the (R) form of a substrate will have a diastereoisomeric counterpart in the approach to the (S)-enantiomer.

Thus, in terms of stereochemical theory, the minimal requirement for the differentiation of enantiomers by a chiral reagent is a 'one-point approach' rather than a 'three-point attachment'. This idea is illuminated by the data cited earlier for the heats of neutralization between the base, α-phenylethylamine, and mandelic acid. Here the reaction does not involve a covalent bond formation; rather it is an ionization leading to an electrostatic situation. For the neutralization of (R)-base with (R)-acid, $-\Delta H = 7.630 \pm 0.013$ kcal/mol; for (R)-base with (S)-acid, $-\Delta H = 7.431 \pm 0.014$ kcal/mol—a significant difference of approx. 0.2 kcal/mol.

In an actual biological interaction a much more complex situation arises. Many enzymes have specific clefts or pockets into which the substrate fits precisely. It is likely that the substrate will be held in place by a number of interactions (including the magic number of 3 or more or less than this number). Some of these may even be covalent; about 100 enzymes are actually known to form covalent intermediates during catalysis [32].

Increasingly, very detailed information is becoming available about the interactions at the active sites of enzymes. For example, in the binding of glycyl-L-tyrosine (a substrate with one chiral carbon) to carboxypeptidase A (an enzyme with 307 amino acid residues in a single polypeptide chain), X-ray crystallographic studies [33] have revealed five interactions, 15. The C-terminal side chain of the dipeptide (that is, the tyrosyl residue) is located in a 'pocket'; the pocket contains no specific binding groups but is of a sufficient size to accommodate a tryptophan side chain. The carboxylate group of the tyrosine interacts electrostatically with a positively charged guanidinium group of the enzyme. This group is provided by the arginine residue at position 145.

The free α-amino group of the dipeptide interacts with glutamate at position 270; the γ-COOH group is involved and a water molecule is located between the amino and carboxyl groups. Interactions which directly involve the catalytic action of the enzyme are as follows: (a) the carbonyl of the peptide bond ligands to a zinc atom, which itself is further bound to two histidines (positions 69, 196) and a glutamate residue (position 72); (b) the NH component of the peptide interacts with the hydroxyl group of tyrosine at position 248.

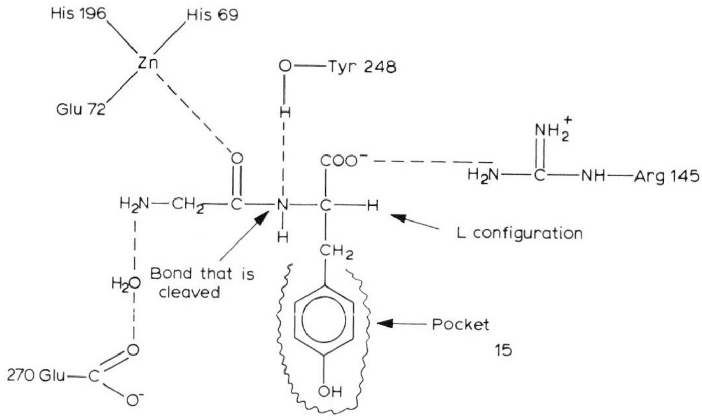

Of these interactions, that with glutamate at 270 is restricted to the dipeptide substrate. With a larger substrate, this residue interacts with the carbonyl of the peptide bond being cleaved. With larger peptides, further binding sites for the characteristic side chains are postulated. It should also be noted that conformational changes occur when the substrates bind to carboxypeptidase A. These have the result of changing the cavity from a water-filled to a hydrophobic region.

In retrospect, Cushny was one of the earliest investigators to point out the diastereoisomeric situations which occur when a chiral biopolymer interacts with the enantiomeric forms of another molecule. The three-point attachment and poly-affinity concepts provided an easily visualized picture of how the differentiation might occur. For a time, unfortunately, the possibility of differentiation by a one-point approach was not clearly recognized and three-point attachment became somewhat dogmatic. With more detailed structural investigations, it is becoming clear that interactions between enzymes and substrates, and receptors and drugs, often involve a multiplicity of interactions.

(b) Enzymes reacting with both enantiomeric forms of a substrate

Many enzymes are known to be active with both enantiomeric forms of a substrate [34] and frequently, as one would expect from a diastereoisomeric interaction, significant kinetic differences may be observed between the two enantiomers. However, in some cases, kinetic constants are similar for two enantiomers. With highly purified rat liver 4-hydroxy-2-ketoglutarate aldolase, K_M values for D- and L-substrate were, respectively, 2.2×10^{-4} and 1.1×10^{-4} M [35]. The V_{max} values for the two enantiomers were in the ratio of 1.0:1.6. Unlike the rat liver 4-hydroxy-2-ketoglutarate aldolase, the enzyme from *Escherichia coli* is strongly selective toward the L-enantiomer of the substrate [36]. These two enantiomers of 4-hydroxy-2-ketoglutarate are also substrates for the crystalline L-glutamate dehydrogenase of

TABLE 2
Kinetic constants for crystalline L-glutamate dehydrogenase

	2-Ketoglutarate	D-4-Hydroxy-2-ketoglutarate	L-4-Hydroxy-2-ketoglutarate
K_M	2×10^{-4} M	1.3×10^{-4} M	1.4×10^{-4} M
V_{max} [a]	1.0	0.07	0.12

[a] Relative to 2-ketoglutarate.

liver. When their kinetic constants were compared under the same conditions with those for 2-ketoglutarate, the K_M values were much the same in all three cases. The V_{max} values for the chiral hydroxy compounds were much lower than that of 2-ketoglutarate, and differed between themselves by a factor of 2 (see Table 2).

The racemase enzymes are also of interest in this connection. Both enantiomers function as substrates and presumably bind to the same active site.

It is worthwhile to appreciate that there is a definite free energy change associated with racemization (defined as conversion of either 100% R or 100% S to 50% R, 50% S). Since at equilibrium, equal amounts must be present the equilibrium constant is 1. Hence the standard free energy change (ΔG°) is 0. This, however, refers to the conversion of 1 mole R to 1 mole S; in other words, to inversion not racemization.

$$\Delta G^{\circ} = -RT \ln K = 0 \qquad \text{if } K = 1$$

For the actual free energy change of racemization (ΔG) it can be shown that

$$\Delta G = -RT \ln 2$$

Hence, with the reasonable assumption that $\Delta H = 0$:

$$\Delta S = R \ln 2$$

Thus, there is a driving force for racemization of about -400 cal/mol at 25°C. The entropy term for racemization is positive—the system becomes more random.

The racemase enzymes present an interesting situation in terms of chiral recognition. In terms of the classification discussed later, they lack both substrate and product selectivity. The simplest kinetic equations that could be written are as follows (D and L represent the two enantiomers of a substrate, E the enzyme):

$$\mathrm{E + D} \underset{k_2}{\overset{k_1}{\rightleftharpoons}} \text{E–D complex} \underset{k_4}{\overset{k_3}{\rightleftharpoons}} \text{E–L complex} \underset{k_6}{\overset{k_5}{\rightleftharpoons}} \mathrm{E + L} \qquad (4)$$

With such a kinetic scheme, two diastereoisomeric intermediates are involved (E–D versus E–L). Hence in the above scheme, it is probable that there will be rate differences; for instance, $k_3 \neq k_4$, or $k_1/k_2 \neq k_5/k_6$.

In a study of the highly purified alanine racemase of *E. coli*, Lambert and Neuhaus determined significant differences in the maximal velocities and the Michaelis–Menten constants of the substrates in the 'forward' (L → DL) and 'reverse' directions (D → DL) [37]. From these data the value calculated for K_{eq} is 1.11 ± 0.15. The time course of the reaction showed that in 10 min with L-alanine as substrate ca. 0.09 μmol of D-alanine were formed. With the same amount of enzyme (750 ng) and in the same time period, ca. 0.05 μmol of L-alanine were formed from D-alanine. Similar results have been reported for the same enzyme from *S. faecalis* and for proline racemases [37]. Thus, in these cases, there are definite kinetic differences, as expected for the existence of two diastereoisomers formed between enzyme and two substrate enantiomers.

Somewhat different conclusions were reached for mandelate racemase [38]. Dissociation constants for the enzyme substrate complexes in the absence (k_5) and presence (k_3) of Mn^{2+} were obtained (from proton relaxation rate titration of E–Mn^{2+} complex and other means). Evidence was obtained that in this case k_3 was approximately equal to k_4. Thus the authors concluded that the "enzyme-catalyzed racemization steps should be approximately equal for both the D- and L-enantiomers", and termed this "a rather startling prediction". Clearly, more detailed studies of the behavior of this, and other racemase enzymes would be desirable.

(c) Differentiation at prochiral positions

In 1948, in a "short but historic note" [39] to Nature, Ogston [6] pointed out that enzymes could differentiate between 'identical groups'—for example, the two a groups in molecules such as Caabc. As a result, he was able to rationalize isotopic tracer experiments on the role of citrate in the tricarboxylic acid cycle. At this date, even organic chemists were not accustomed to thinking in stereochemical terms [40] and Cornforth [39] has said that in 1948 he had "as an organic chemist interested in the synthesis of natural products, the same kind of feeling for stereochemistry that a motorist might have for a system of one-way streets—a set of rules forming one more obstacle on the way to a destination". For some time many individuals were somewhat confused [41]. Ogston's explanation was a three-point attachment between the enzyme and the prochiral substrate and, again, that phrase worked its usual magic. It is of interest that Ogston's historic act of conception "may have taken five

TABLE 3
Kinetic constants for alanine racemase from *E. coli*

	Reaction proceeding L → DL	Reaction proceeding D → DL
K_m	$(9.7 \pm 0.4) \times 10^{-4}$ M	$(4.6 \pm 0.1) \times 10^{-4}$ M
V_{max}	2.22 ± 0.08 μmol/l	0.95 ± 0.04 μmol/l

seconds, perhaps less" and that the famous note was written and sent a day or two later [42].

As noted earlier, Ogston was, in 1948, not aware of the 1933 paper by Easson and Stedman [25]. However, in that paper, those authors had taken an important conceptual step. From the three-point attachment postulated for the enantiomer of Cabcd with a receptor they "supposed that the dissymmetry of 5 is abolished by replacing the group a by a second group b, the resultant molecule, represented by 16, retains unchanged that part of the structure of 5, i.e., the base bcd of the tetrahedron, which is concerned with its attachment to the specific receptor and must therefore be considered capable, despite the absence of molecular dissymmetry, of exerting its physiological action with an intensity numerically equal to that of 5 .... It will be noted that 16 possesses a second face, shaded in the diagram, containing the groups, bcd. This face, however, corresponds with the base of 6 (the enantiomer of 5) and, like this, cannot be brought into coincidence with the receptor".

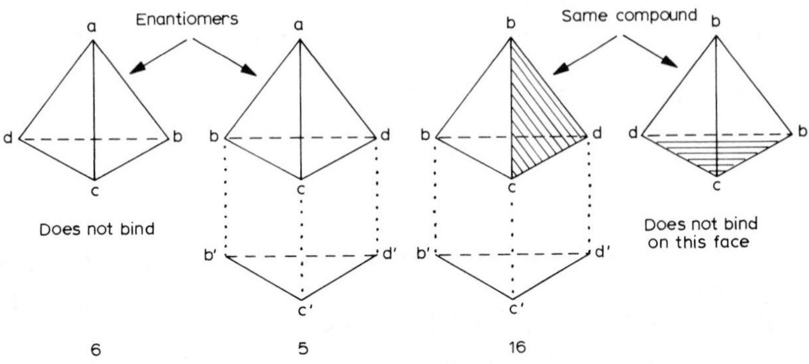

Long before the word chiral was in common use, they had taken the giant step from chirality to prochirality. As Ogston [26] has noted, "it is interesting that they have there all the essentials of prochirality and I am sure would have got on to the later aspects of 'three-point attachment': only the *question* had not then been asked, isotopic marking not then being available."

It was left to Hirschmann to clarify the subject of prochiral differentiation in a brilliant essay published in 1956 [43]. He removed the mystique of 'three-point attachment' between enzyme and prochiral substrate, and clearly noted that differentiation of the two a groups in Caabc could result even from a one-point attachment (either by covalence, electrovalence, hydrogen bonding, or some other force) or from a one-point approach of the enzyme. His paper opened up the role of rotational asymmetry in biology in the way that Pasteur's work opened up the role of reflective asymmetry. However, with a few exceptions (see, for example, Mislow's [44] clear sentence—"Structural interpretations, such as the *three-point mechanism*, are not essential"), the supposed necessity for a three-point attachment was widely quoted. Another possibility was that suggested by Prelog [45] to account for observations on the reduction of bicyclic ketones with NADH; this was a "two-plane theory" and Prelog noted that it was a variation on the Ogston three-point theory. In a subsequent comment on this two-plane theory, Westheimer stated "We have all taken over the Ogston three-point attachment and Professor Prelog now wishes to modify this to a two-plane attachment. I think we should bear in mind that, so far as the formal theory is concerned, a one-point attachment is sufficient" [46]. In fact, it is again the diastereoisomeric nature of the two possible transition states or enzyme substrate complexes which is the critical factor. Ogston himself later stated that his hypothesis of three-point attachment was not to be taken as a literal description of the forces acting between the molecules [47].

As is the case for interactions between enzymes and chiral substrates, the structures of enzyme complexes with prochiral substrates are increasingly being explored in detail. A case in point is liver alcohol dehydrogenase. The active-site topography of this enzyme was initially specified by "diamond lattice sections" in the classical work of Prelog, Ringold and their colleagues (for a review, see [48] and refs. therein) and by Jones and Beck [49]. The horse liver alcohol dehydrogenase contains separate binding sites for the pyridine nucleotide coenzyme and for substrate. The active site pocket is hydrophobic in nature, Zn ions are essential for activity, and are bound by cysteine thiol and histidine imidazole groups [50].

More recently, the active site pocket of horse liver alcohol dehydrogenase has been defined by model-building studies based on crystal-structure analysis and kinetic investigations [51]. The top of the pocket is open to the environment and the bottom is closed by groups of the active site and by the nicotinamide groups of the coenzyme. The oxygen of the substrate is directly bound to the active-site zinc. The nicotinamide ring is bound so that the A side is facing this zinc atom and the B side faces the protein surface. This binding is consistent with the known H_A specificity of this enzyme.

The bottom part of the pocket is hydrophilic and there is a compact hydrophobic belt around the pocket. Ethanol used as a substrate is oriented by interactions near the interface between the hydrophilic bottom and the hydrophobic belt.

It is difficult to summarize the observed interactions in horse liver alcohol dehydrogenase in a short space. The reader is advised to consult the original paper [51] which contains a vivid stereo diagram (one stereo diagram is equal to at least 10^3 words). However, for the present discussion it suffices to say that the situation in the enzyme substrate complex is not that of a simple three-point attachment. The observed stereochemical course is as follows:

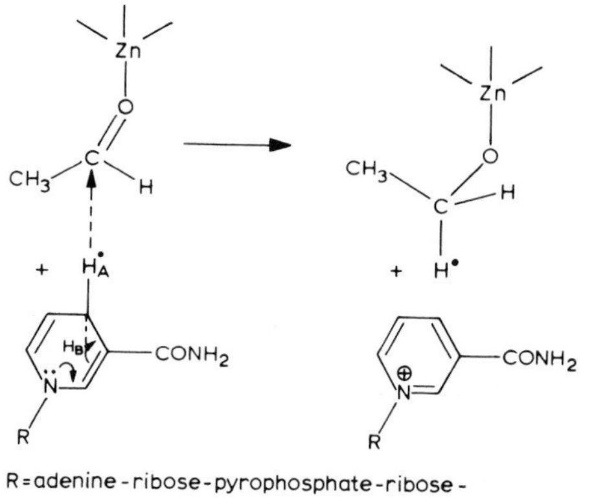

R = adenine - ribose - pyrophosphate - ribose -

3. Classification of reaction types and selectivities

Enzymes clearly have the ability to distinguish between various structures and are commonly described as highly efficient catalysts showing to a high degree the ability to be selective and specific. If an anthropomorphic approach is used, we can ascribe a deductive ability to an enzyme. In the first place, the 'thinking enzyme' shows a general chemical or reaction selectivity. That is, there is no universal enzyme, but many enzymes with the capacity to catalyze, within limitations, different chemical processes. A dehydrogenase (usually) does not have the ability to catalyze a dehydration; a ligase is different from a lyase.

In the second place, the enzyme can make distinctions between isomers. In fact, nowhere is the selection ability of enzymes so manifest as in the ability to distinguish between isomeric structures. The various possibilities for a given number of atoms may be examined systematically. Thus, we can imagine the enzyme confronted with two materially identical molecules, making yes/no decisions in accordance with a defined classification chart. The enzyme asks "Can the two molecules be superposed?" Arrows from the left sides of the diamonds indicate a yes answer; from the right sides the arrows represent no (see Scheme 1). The possibility of superposition is

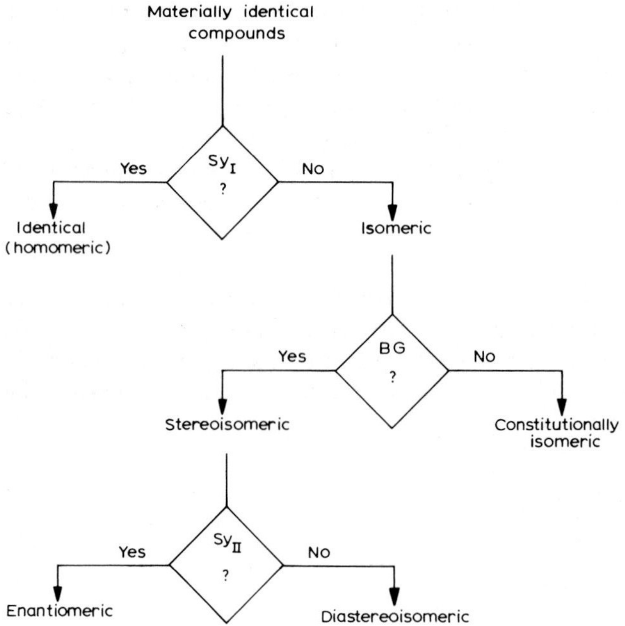

Scheme 1. Comparison of whole molecules to determine isomeric relationships. The question marks signify "Superposition ? yes or no". The three tests are: Sy_I—comparison by symmetry operations of the first kind (rotation, torsion); BG—comparison of bonding (connectivity) graphs vertex by vertex; Sy_{II}—comparison by symmetry operations of the second kind (reflection).

examined by three tests; the intelligent enzyme will conclude that the compounds are either identical, or constitutional isomers or stereoisomers (enantiomers or diastereoisomers). Assuming non-identity, the isomeric classes must be discussed.

(a) Constitutional isomers

This classification includes compounds which are isomeric by some simple structural difference, involving the sequence of atoms or linkages between the molecules. Such compounds are distinct chemical species. Familiar examples would be propyl and isopropyl alcohol, and guanine and isoguanine. The term 'regioselective' is used (particularly by chemists; the 'chemical' definitions of selectivity used here are those given in [52]) to describe a reaction in which one structural (or positional) isomer is favored over another; the term has also been applied to enzyme reactions. Thus, a purified glutathione transferase from the little skate catalyzes the reaction of glutathione with benzo[*a*]pyrene 4,5-oxide, 17. In discussing the relative proportions of the positional isomers, 4,5-dihydro-4-hydroxy-5-glutathionylbenzo[*a*]pyrene, 18, and 4,5-dihydro-4-glutathionyl-5-hydroxybenzo[*a*]pyrene, 19, the authors referred to the 'regiospecificity' of the enzyme; it has other selectivities as well [53]. (A regiospecific reaction is one in which one specific structural or positional isomer is formed when other isomers are possible. It is said to be used synonymously with regioselective [52].)

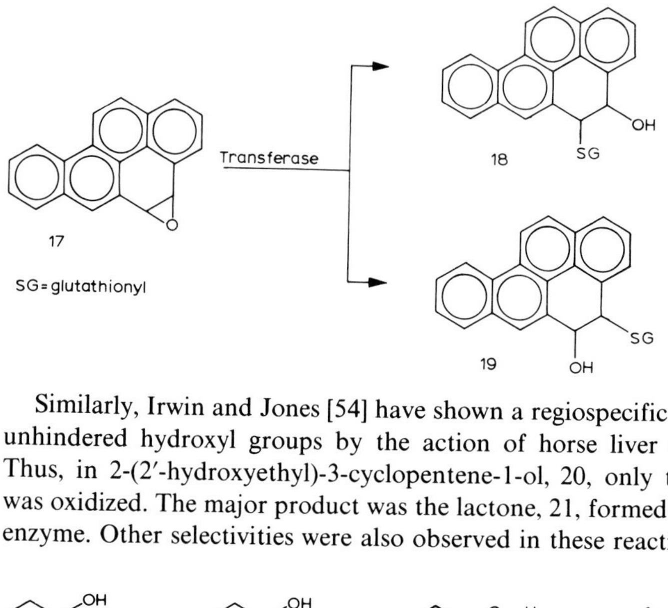

SG = glutathionyl

Similarly, Irwin and Jones [54] have shown a regiospecific oxidation of one of two unhindered hydroxyl groups by the action of horse liver alcohol dehydrogenase. Thus, in 2-(2′-hydroxyethyl)-3-cyclopentene-1-ol, 20, only the hydroxyethyl group was oxidized. The major product was the lactone, 21, formed by further action of the enzyme. Other selectivities were also observed in these reactions.

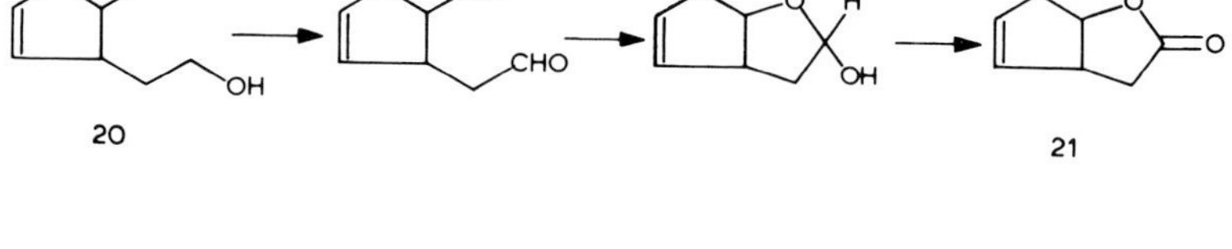

(b) Stereoisomers

While the subdivisions of constitutional isomerism are not well defined, those of stereoisomerism are. Isomers which differ in their three-dimensional spatial organization may be related as object and mirror image, in which case they are termed enantiomers. Stereoisomers not in an object–mirror-image relationship are termed diastereoisomers. Many examples of the ability of enzymes to make these spatial distinctions are known. There are, for example, two separate lactate dehydrogenase enzymes, one active with (−)-D-lactate, 22, and the other with (+)-L-lactate, 23. Pairs of enzymes forming the same product from either the (*R*)- or (*S*)-enantiomer of a substrate have been termed enantiozymes (for a listing of more than 20 such pairs, see [55]). Furthermore, there are many examples to illustrate the selection of one diastereoisomer over another; fumarase does not hydrate maleic acid, and isocitrate dehydrogenase is inactive with the allo-isocitric acids.

COOH, H—C—OH, CH_3 (−)-D 22 → D-lactate dehydrogenase (EC 1.1.1.28) → COOH, CO, CH_3 ← L-lactate dehydrogenase (EC 1.1.1.27) ← COOH, HO—C—H, CH_3 (+)-L 23

In dealing with processes where stereoisomer formation is subject to some kind of control, two terms have been used—stereoselective and stereospecific. These terms have caused some problems, and to some extent have been used synonymously. The meanings attached to these terms by chemists are unambiguous [52]. A stereospecific process is one in which a particular stereoisomer reacts to give one specific stereoisomer (or a racemate) of a product. Thus, two starting materials, differing only in stereoisomerism, must be converted into stereoisomerically different products. A classical example of chemical stereospecificity is the free radical addition of HBr to 2-bromo-2-butene. At −78°C, the (*E*)-olefin, 24, forms *meso*-2,3-dibromobutane, 25, while the (*Z*)-isomer, 26, yields (±)-*threo*-2,3-dibromobutane, 27 plus 28. (Note in addition that this addition is regiospecific—2,2-dibromobutane is not formed.)

24 (E) + HBr —(−78 °C)→ 25, meso: CH_3 (1), Br—C—H (2), Br—C—H (3), CH_3 (4)

26 (Z) + HBr —(−78 °C)→ 27: CH_3, Br—C—H, H—C—Br, CH_3 + 28: CH_3, H—C—Br, Br—C—H, CH_3 (±) racemate

A stereoselective process is one which one stereoisomer is either produced or destroyed more rapidly than another. There will, therefore, be a predominant amount of one stereoisomer in the product (or one stereoisomeric form will have been destroyed more rapidly). Thus, in the alkaline peroxidation of both the (*E*)- and (*Z*)-α-phenylbenzalacetones, 29 and 31, the product is exclusively the racemic (*RS*) + (*SR*) product, 30:

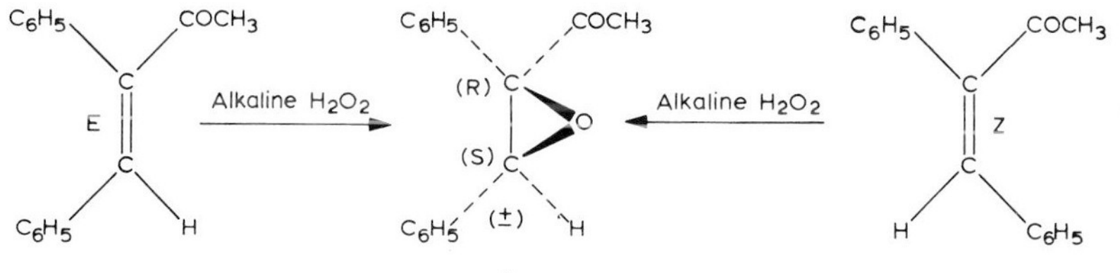

To give another example, in the reduction of the dimethylcyclohexene, 32, a marked selectivity for the *cis*-isomer of the dimethylcyclohexane was observed (product composition: 33, 84%; 34, 16%) [56].

It will be appreciated that, with these definitions, any stereospecific process is also stereoselective; however, not all stereoselective reactions are stereospecific. In some papers, the terms are modified with the prefixes, enantio- or diastereo-.

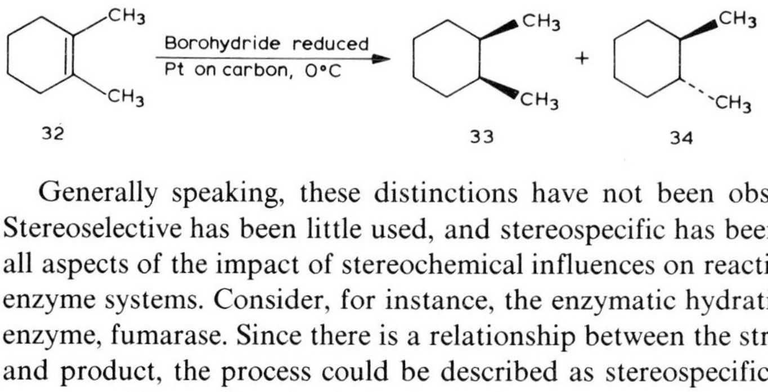

Generally speaking, these distinctions have not been observed by biochemists. Stereoselective has been little used, and stereospecific has been used to cover almost all aspects of the impact of stereochemical influences on reactions in living tissues or enzyme systems. Consider, for instance, the enzymatic hydration of fumarate by the enzyme, fumarase. Since there is a relationship between the structure of the substrate and product, the process could be described as stereospecific. Yet the definition of stereospecific requires that it be shown that the isomer of fumaric acid gives rise to a product which is stereochemically different from L-malate. Since the enzyme, however, does not catalyze any reaction with the (*Z*)-isomer (maleic acid) it is not clear whether stereospecific actually applies.

One case of a stereospecific enzyme reaction (in the chemical sense) has been described [57]. Chloroperoxidase adds the elements of hypochlorous acid across the double bond of propenylphosphonic acid. From the (*Z*)-isomer, 35, the product is the ($\pm$)-*threo* form of 1-chloro-2-hydroxypropyl phosphonate, 36. On the other hand, from the (*E*)-isomer, 38, the product is the ($\pm$)-*erythro* form, 37. Since stereoisomerically different substrates give different stereoisomers as products, this enzymatic reaction is stereospecific in chemical terms. In biochemical terms the enzyme is almost a contradiction since the products are formed as racemates! In a

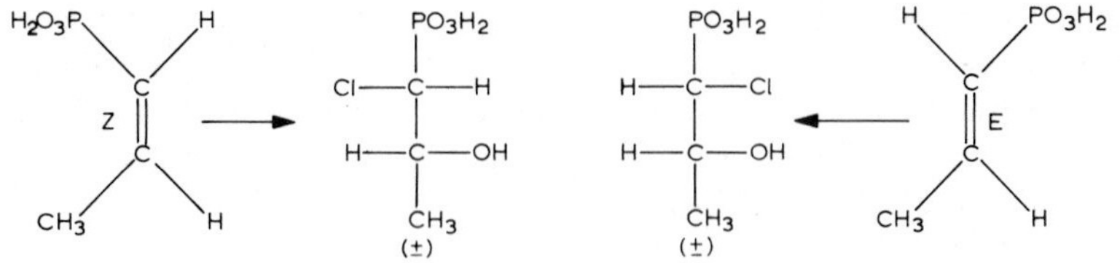

35 → 36 (±) 37 (±) ← 38

discussion of the action of chloroperoxidase on these substrates, the authors stated "Enzymatic synthesis of chiral molecules from achiral substrates proceeds in a stereospecific and asymmetric manner. To our knowledge, there is no exception in the literature to this rule." However, other exceptions are known; highly purified preparations of wheat germ pyruvate decarboxylase convert pyruvate and acetaldehyde to partially racemic acetoin (72% +, 28% −), and some esterases produce partially racemic products.

A further problem is that the two terms stereospecific and stereoselective are also used in pharmacology [58–61]. In this usage, stereoselective means that a pharmacologic action is found predominantly, though not exclusively, in one stereoisomer; stereospecific means that only one single stereoisomer has activity.

In recent years, even chemists have become concerned about terminology to be used for 'asymmetric syntheses' and 'asymmetric reaction processes'. Since catalysis by enzymes represents the ultimate in an asymmetric reaction, it is appropriate to consider briefly a new proposal. Izumi and Tai have proposed that the time has come to abandon the use of stereoselective and stereospecific [62]. They point to two components in the transformation of a substrate to a product. The first resides in chemical structures (e.g., a double bond) rather than in a particular steric structure and the reaction is governed by the nature of the reagent or catalyst (whether the process proceeds with retention or inversion; whether an addition is *syn* or *anti*). In the second component, the reagent or catalyst interacts topologically with the three-dimensional structure of the substrate. This is described as 'stereo-differentiation' and results from the 'stereo-differentiating ability' of the catalyst or reagent.

Thus, in the action of the enzyme aspartase, fumaric acid, 39, is transformed into L-aspartic acid, 40. This process can be considered as a result of three steps:

(a) The enzyme is able to differentiate fumaric acid from other materials (including the geometrical isomer, maleic acid). This ability is not simply that of being able to distinguish the two geometrical isomers; it is a more general ability, a 'substrate specificity'.

(b) The enzyme has the stereo-differentiating ability to recognize the enantiotopic faces of fumaric acid. The NH_2 group is added only at the *Si–Si* face.

(c) The third step, the overall *anti* addition of the elements of ammonia is essentially determined by the reaction character of ammonia.

Thus, to these authors, the important stereochemical consideration is neither the *E* or *Z* nature of the substrate, nor the *syn* or *anti* nature of the addition, but the distinction between the two enantiotopic faces of fumarate.

NH_2 HOOC H E H COOH H 39

Other compounds (including the Z isomer, maleic acid)

NH_2 HOOC H H COOH H 40

On the basis of such considerations, they have classified stereo-differentiating reactions into six types. While the terminology used is readily understandable, some comments on prochiral distinctions are in order. The thinking enzyme discussed earlier can use a sequence of binary congruence tests to examine the environmental relationships of materially identical groups within a molecule (topic analysis). The flow chart resembles that shown earlier for isomeric distinctions (see Scheme 2).

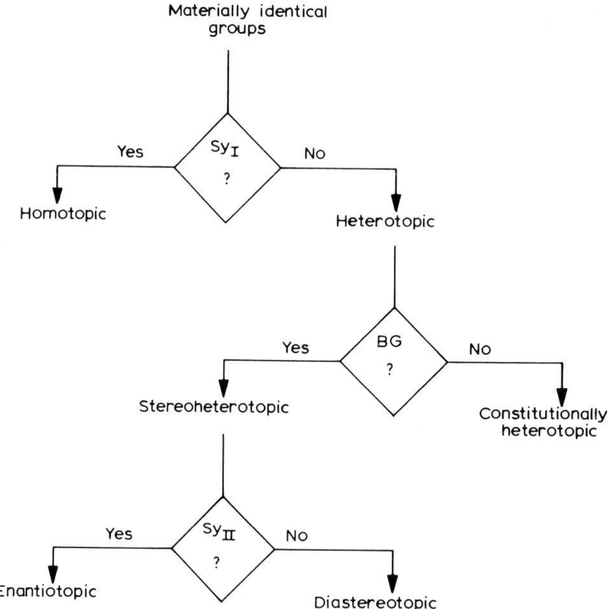

Scheme 2. Comparison of partial structures in intact molecules (or in different molecules). The symbols have the same meaning as those in Scheme 1.

Enantiotopic groups are those which can be interchanged one with the other, by a rotation–reflection operation. If two such groups, a′ and a″ in Ca′a″bc are separately substituted by an achiral group d, the products are the two enantiomers of Cabcd. The central carbon atom is described as prochiral. The two H atoms in ethanol, 41, are enantiotopic and are distinguished as H_R (=*pro-R*) and H_S (=*pro-S*). A location in an enantiotopic relationship has now been termed an 'enantiotopos' by Izumi and Tai.

For an sp^2-bonded carbon, as for example, the carbonyl carbon in acetaldehyde, 42, another kind of prochirality exists, termed sp^2 prochirality.

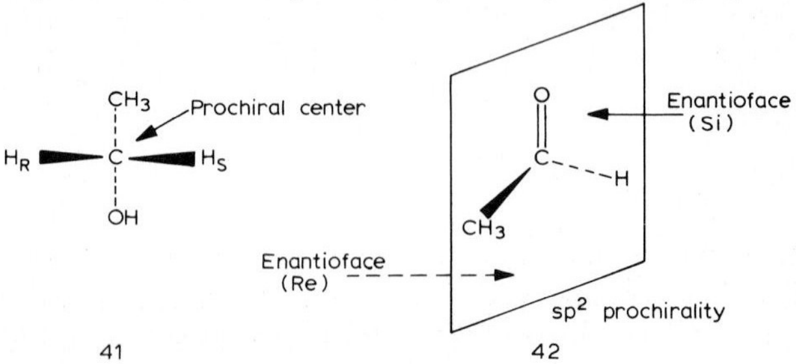

In this situation two 'faces' or 'sides' of a molecule, or part of a molecule, are being distinguished; these faces have been called 'enantiofaces'.

Diastereotopic groups are those which cannot be interchanged by any symmetry operations. Their locations are in a diastereoisomeric relationship and such a location is termed a 'diastereotopos'. The methylene hydrogen atoms of citrate are diastereotopically located, 43.

Faces of molecules can also be diastereotopic; an example is provided by the enediol intermediate, 44, postulated to occur in the glucose-phosphate isomerase reaction:

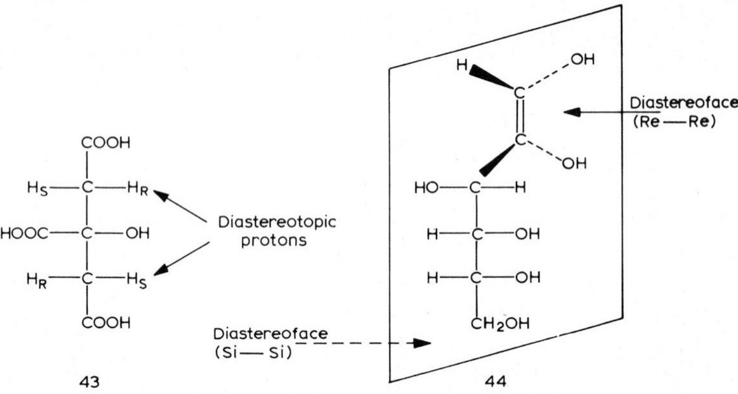

A stereo-differentiation process can fall into one of two main groups, enantio-differentiation or diastereo-differentiation:

(1) The reaction is enantio-differentiating when the chirality participating in the differentiation process occurs in a reagent, catalyst, or reaction medium. The products are typically enantiomers.

(2) The reaction is diastereo-differentiating when the chirality participating in the differentiation is present in the substrate. The products are typically diastereo-isomers.

Since the differentiation can occur at a face (sp^2 prochiral center), prochiral center, or chiral center, there are three subdivisions in each main group:

(1) Enantio-differentiating reactions
  (a) Enantioface differentiation
  (b) Enantiotopos differentiation
  (c) Enantiomer differentiation

(2) Diastereo-differentiating reactions
  (a) Diastereoface differentiation
  (b) Diastereotopos differentiation
  (c) Diastereoisomer differentiation

These categories can be clarified by consideration of specific cases. For more detail, Izumi and Tai's book should be consulted [62]; a comprehensive treatment of asymmetric organic reactions is given by Morrison and Mosher [63].

(i) Enantioface differentiation

An optically active Grignard reagent has the ability to differentiate between the two enantiofaces of a carbonyl compound such as 45 [64]. In the example shown, the (*S*)-enantiomer of the product alcohol, 46, is obtained with a high degree of optical purity (= specific rotation of mixture ÷ specific rotation of one pure enantiomer × 100; for definitions of other terms used in this work, see [65]).

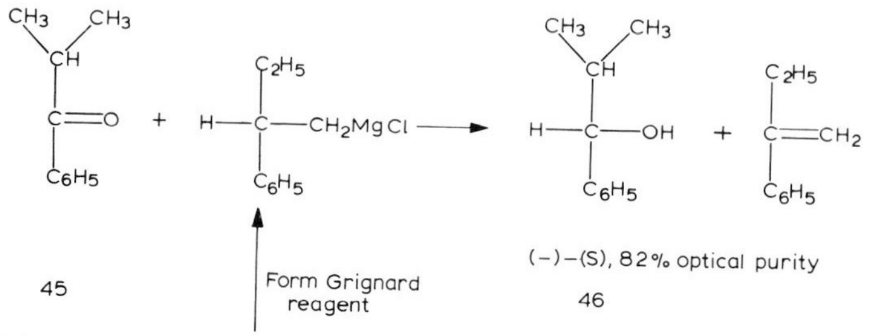

$(CH_3)_2CH{-}C(=O){-}C_6H_5$ + $H{-}C(C_2H_5)(C_6H_5){-}CH_2MgCl$ ⟶ $H{-}C(CH(CH_3)_2)(C_6H_5){-}OH$ + $C_2H_5(C_6H_5)C{=}CH_2$

45

Form Grignard reagent

(+)-1-Chloro-2-phenylbutane

(−)-(S), 82% optical purity

46

The chiral influence can also be present in a catalyst; many asymmetric hydrogenation reactions have been carried out. The example shown (47 → 48) requires differentiation of the enantiofaces of a double bond [66].

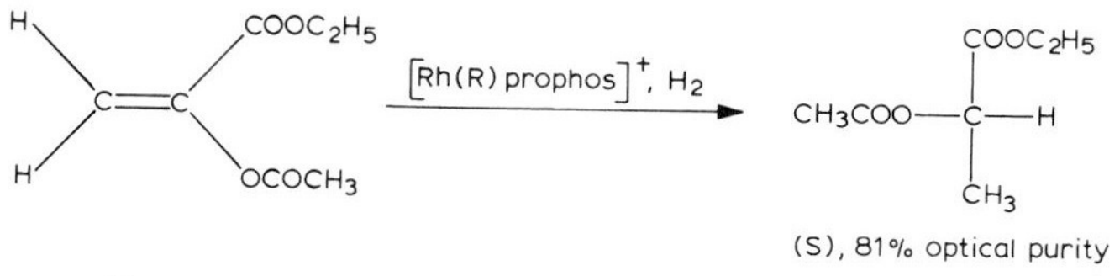

$H_2C{=}C(COOC_2H_5)(OCOCH_3)$ —[Rh(R) prophos]$^+$, H_2⟶ $CH_3COO{-}C(COOC_2H_5)(CH_3){-}H$

47

(S), 81% optical purity

48

(ii) Enantiotopos differentiation

The compound β-phenylglutaric anhydride, 49, contains enantiotopic ligands. On reaction with $(-)$-α-phenylethylamine the two diastereoisomers of the monoamide, 50 and 51, were formed in unequal amounts [67]. In contrast to the earlier statement (the products are usually enantiomers in an enantiodifferentiating process), the products here are diastereoisomers. Of course, if the amine component of the amide were to be removed, the products from the 'substrate' anhydride would be enantiomers. This differentiation between enantiotopic groups was important in the early days of the citrate story. It proved the possibility of differentiation in homogeneous solution, presumably without a three-point attachment.

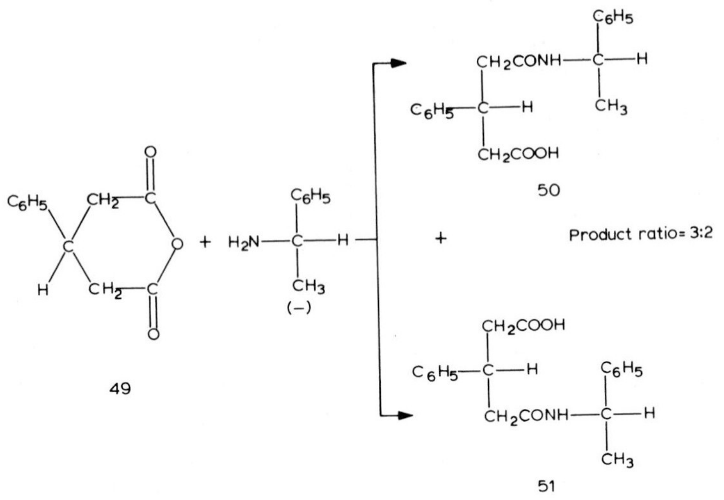

C6H5 CH2 C O C O H CH2 C O + H2N—C—H C6H5 CH3 (−)

49

CH2CONH—C—H C6H5 CH3 C6H5—C—H CH2COOH

50

+ Product ratio= 3:2

CH2COOH C6H5—C—H CH2CONH—C—H C6H5 CH3

51

(iii) Enantiomer differentiation

If racemic *sec* alcohols (e.g., butan-2-ol, 52) undergo partial acylation with an acid anhydride in the presence of an optically active amine, an enantiomer differentiation occurs; in the example shown, the (S)-alcohol reacts preferentially forming but-2-yl acetate, 53, which is levorotatory, and leaving behind unchanged alcohol which has been optically enriched [68]. This is a process under kinetic control. For further examples of kinetic resolutions, see [69].

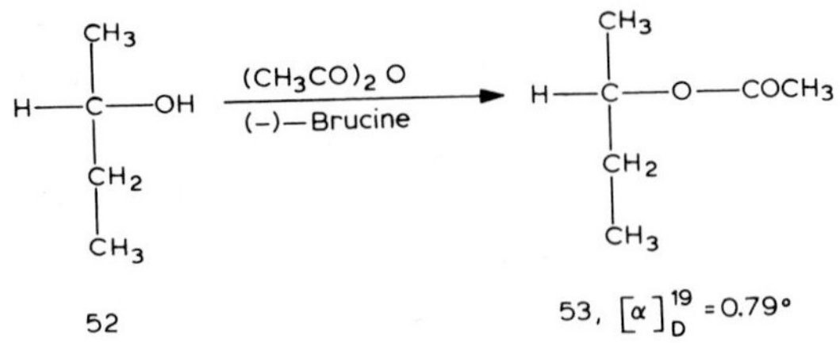

CH3 H—C—OH CH2 CH3 $\xrightarrow[(-)\text{—Brucine}]{(CH_3CO)_2O}$ H—C—O—COCH3 CH3 CH2 CH3

52

53, $[\alpha]_D^{19} = 0.79°$

(iv) Diastereoface differentiation

An example of diastereoface differentiation is provided in an asymmetric synthesis of the dipeptide Ala-Ala [70]. The element of chirality is that of one of the alanine groups, and the faces are those of a Schiff base, >C=N—. The material undergoing a catalytic reduction is the isobutyl ester of the benzylamine Schiff base of *N*-pyruvoyl-(*S*)-alanine, 54. The ratio of *R*:*S* (55) to *S*:*S* dipeptide was 82:18 for an optical purity of 64%.

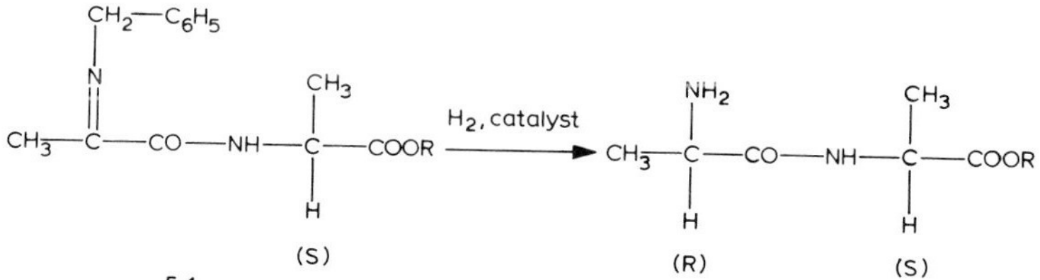

H_2, catalyst

54 (S) → 55 (R) (S)

R = isobutyl

An asymmetric synthesis of alanine itself is provided by the Grignard reaction of the benzylamine Schiff base, 57, of (−)-menthyl glyoxylate, 56. The (*S*)-alanine, 58, was obtained with a 53% optical yield [71]. This yield could be influenced by another

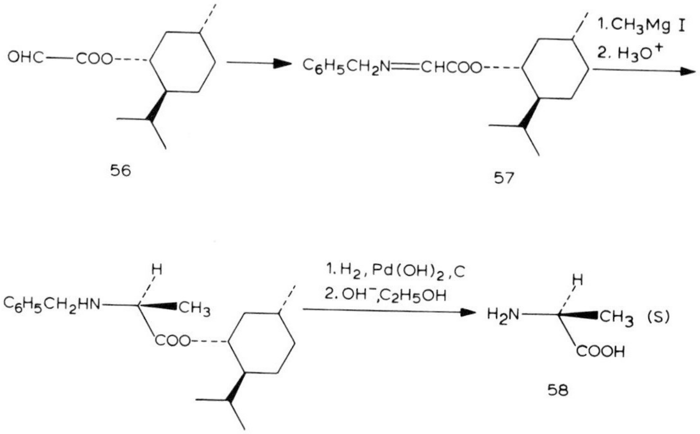

OHC—COO---

$C_6H_5CH_2N$=CHCOO---

1. CH_3MgI
2. H_3O^+

56 57

$C_6H_5CH_2HN$—C(H)(CH_3)COO---

1. H_2, $Pd(OH)_2$, C
2. OH^-, C_2H_5OH

H_2N—C(H)(CH_3)COOH (S)

58

chiral element in the starting material. If the benzyl unit was replaced with (*S*)-phenylethyl the optical yield of the (*S*)-alanine increased by 10.5%; with the (*R*)-phenylethyl unit, the yield of (*S*)-alanine diminished by 7%.

(v) Diastereotopos differentiation

An example of the differentiation of diastereotopic hydrogens occurs in the formation of a β-lactam, 60, from an optically active *N*-methylbenzyl-*N*-chloroacetamide acetonitrile, 59. When the starting chiral compound was (+)-R-α-(1-

naphthyl)ethylamine, 62, the (*S*) form of aspartic acid, 61, was the predominant product [72].

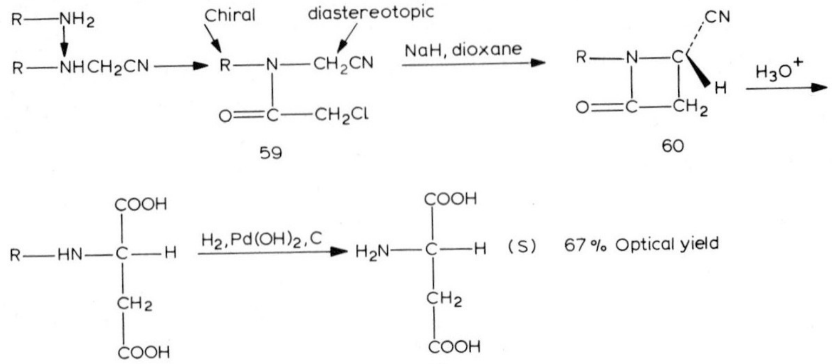

(vi) *Diastereoisomer differentiation*

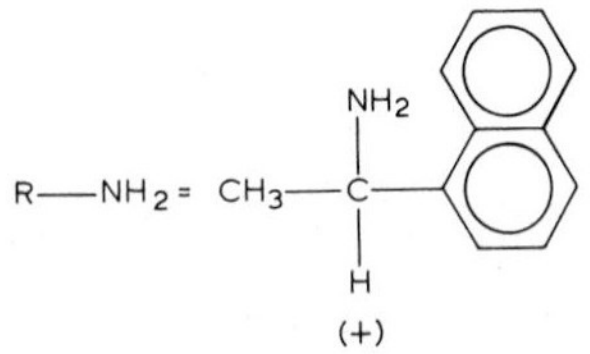

Diastereoisomers have different free energies (internal energies) so reactions of this kind are similar to ordinary competitive reactions between two different substances, and need not be exemplified.

These concepts have been described here since they do offer a way out of the specific–selective dilemma, and they could be applied in biochemistry. For instance alcohol dehydrogenase differentiates between the enantiofaces of acetaldehyde, 63, and, in addition, utilizes H_A of NADH, 64, thereby showing a diastereotopos-differentiating capability as well.

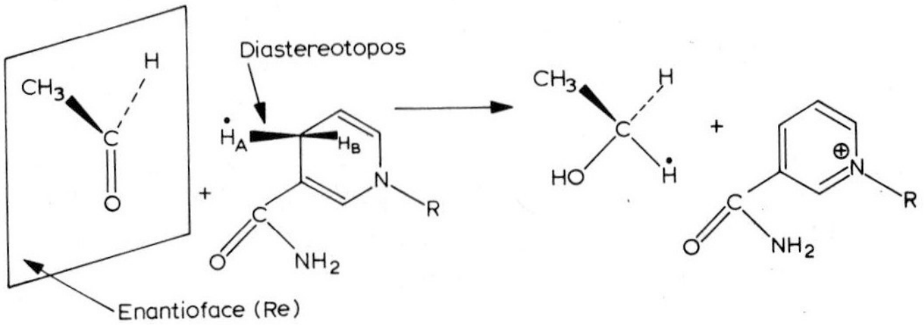

R=adenine–ribose–pyrophosphate–ribose–

Whether these ideas will receive wide acceptance in biochemistry remains to be seen. A difficulty is that many enzymatic processes often show several kinds of differentiation. As just noted in the direction of acetaldehyde reduction, alcohol dehydrogenase would be perhaps described as enantioface (acetaldehyde) differentiating and diastereotopos (NADH) differentiating. However, this steric description does not describe the differentiating ability of the enzyme when the reaction is carried out in the direction of alcohol oxidation. Then the enzyme is enantiotopos (alcohol) differentiating and diastereoface (NAD^+) differentiating.

Another problem seems to arise in enzymatic work where, for example, the addition of a reagent to a double bond is not determined by the 'reaction character' of the reagent (as Izumi and Tai postulate). Thus although the enzymatic addition of the elements of water to a double bond is generally *anti*, several examples of *syn* addition have been well characterized.

It seems likely that biochemists will continue to use a more pragmatic and less comprehensive approach. In biochemical processes, two important features are the structures of substrate and product. The *overall* steric structure of the substrate (and not just the possession of some structural feature such as a double bond) is important in terms of binding to an enzyme or receptor. Since many enzymatic reactions are readily reversible, overall product structure is important for the same reason. Furthermore, since many enzymes make more than one type of stereo-differentiation, the use of the stereo-differentiating terminology of Izumi and Tai would be somewhat cumbersome. The overall steric structure of molecules (as opposed to isolated structural features) is also important in the area of drug–receptor interactions.

Considerations of the importance of substrate and product structures were a feature of the work of Prelog and his colleagues on various ketone reductions by microorganisms (for a summary, see [48]). As early as 1958, Prelog was referring to substrate selectivity (Edukt-Selektivität) and product stereospecificity (Produkt-Stereospecifizität) [73]. This idea was emphasized in Prelog's lecture at the International Symposium of Organic Chemistry (Brussels, 1962) [74]: "In cases when the substrate is racemic the two enantiomers are reduced at different rates; such reductions are called substrate-specific. In cases where the reduction results in a new asymmetric carbon atom, the two theoretically possible stereoisomers form at different rates, and we speak of the product stereospecificity of the reaction."

These terms are finding increased acceptance, with some modifications. First of all, there are good reasons for a very restricted use of "stereo-specific" in biochemical systems. While this may seem heretical, it has rarely been used by biologists in the precisely defined chemical sense, and moreover, the term rarely carries the meaning of 100% specific. Indeed, the question has been raised (for immunological reactions), "how specific is specific?" It is a question that covers the whole range of biological interactions. Boyd has suggested that the word 'virtually' should be applied to all cases of 'specificity', pointing out that complete specificity, like perfect virtue, is seldom, if ever, encountered in this world [75]. For example, fumarase is generally regarded as a 'highly specific' enzyme, but how specific is fumarase? For the chiral center, D-malate is not a substrate, but does function as a competitive inhibitor; it is, therefore, bound at the active site [76]. The dissociation constant is independent of pH over a certain range as is the case for the 'natural' substrates, L-malate and fumarate. This observation is of interest since a well-known text states that "where the stereochemical specificity is absolute the unattached antipode does not usually inhibit the utilization of the substrate" [77]. In another investigation of this enzyme it was found that when the reaction was run in 99.5% ^2H_2O the L-malate contained 0.97 excess atom of ^2H [78]. This was described as "absolute stereospecificity", but is actually somewhat less than 100%. In such cases, the 'discrepancy' may arise from technical problems in the experiment, or may represent a real finding with respect to the enzyme. If selectivity is used, it could be qualified numerically if so desired.

There is increasing acceptance in the biological sciences of the terms substrate selective and product selective with the following meanings defined by Testa [65]: Any biological process is termed stereoselective when one stereoisomer (enantiomer or diastereoisomer) is preferred over the other(s). If the reaction yields material in which one stereoisomer predominates the process is described as *product selective*. If the reaction utilizes stereoisomers at different rates the process is *substrate selective*. As will be seen, some reactions may exhibit both product and substrate selectivity. If it is desired to emphasize an enantiomeric or diastereoisomeric differentiation (to a face, group or atom in a stereoisomeric location, or actual isomer), the selectivity term can be qualified as enantioselective or diastereoselective. This nomenclature has the advantage that it can be widely used in the general sense, it can be quantified, and it can be sufficiently precise to convey the necessary stereochemical information. Substrate and product are defined by the reaction direction as given by the standard enzyme nomenclature [79]; for all enzymes in a given class the direction chosen is the same for all.

Thus, for aspartase (aspartate ammonia-lyase) the reaction direction is L-aspartate → fumarate + NH_3. The enzyme uses L-aspartate but not the D-enantiomer; the product is fumarate not maleate. Hence, this enzyme has strict substrate and product selectivity. For greater detail, this could read: strict substrate enantioselectivity and product diastereoselectivity. If necessary, information concerning prochirality could be conveyed in the same way; for this same enzyme there is substrate diastereoselectivity for the protons at C-3.

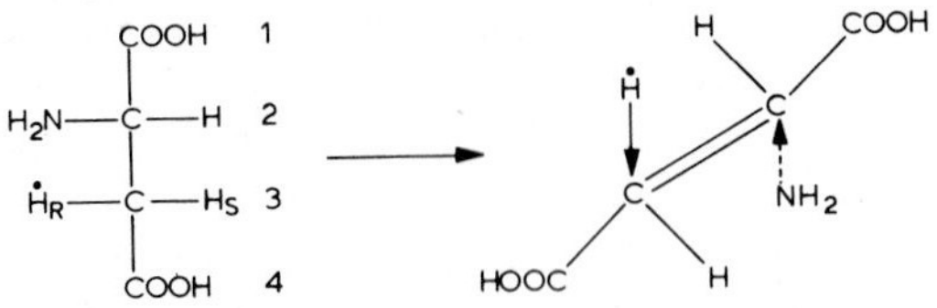

COOH 1
H_2N—C—H 2
H_R—C—H_S 3
COOH 4

H COOH
C
C
HOOC H
NH_2

40 39

The flexibility of this nomenclature may be appreciated from a consideration of the possible results for the reduction of a ketone by a fermentation process. Prelog and his colleagues initiated work in this area by investigating the action of the mold, *Curvularia falcata* on Δ^4-9-methyloctalin-3,8-dione, as the racemic mixture, 65 plus 66.

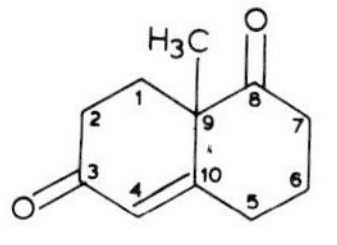

(+)-(9S) 65 (−)-(9R) 66

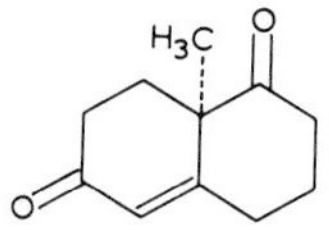

The following situations would be possible if the substrate ketone was initially present as a racemic mixture, 65 plus 66:

(i) Substrate and product enantioselective.
Only 65 attacked and only 67 formed.
Residual ketone is 66 (optically active).

(ii) Substrate enantioselective, product non-enantioselective.
Only 65 attacked and only 67 and 68 formed.
Residual ketone is 66 (optically active).

(iii) Substrate non-enantioselective, product enantioselective.
Both 65 and 66 are attacked and only 67 and 69 formed.

(iv) Substrate and product non-enantioselective.
Both 65 and 66 are attacked and 67 to 70 are formed.

With *C. falcata*, there was a high product selectivity for 67 and 69; with this organism, substrate selectivity was low (case iii) [80]. On the other hand, another mold, *Aspergillus niger*, had little selectivity for either substrate utilization or product formation [81]. All four possible products were obtained (case iv).

Since a great deal of stereochemical information has been obtained for biological processes, and since much work remains to be done, the time seems ripe for some general agreement on the terminology used in connection with the various distinctions made by enzymes and receptors.

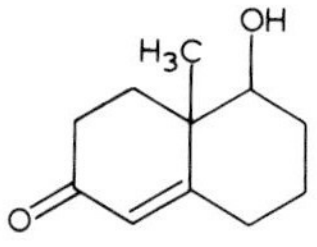

(+)-(8S,9S) 67 (+)-(8R,9S) 68 (−)-(8S,9R) 69 (−)-(8R,9R) 70

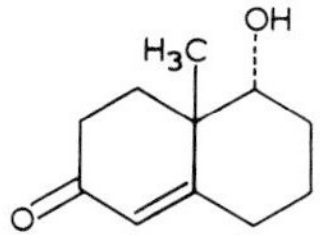

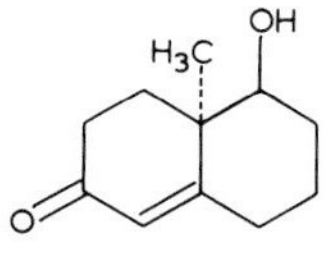

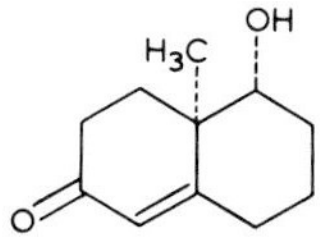

4. The determination of configuration

(a) For chiral elements

Many methods have been used for the determination of configuration of chiral compounds; some of these were reviewed in the first volume of *Comprehensive Biochemistry* [44]. In addition, within the last few years, comprehensive listings of configurations have become available [13–15], and detailed discussions of methodologies have been compiled [82]. In chemical work it is always important to carry out correlations by reactions which do not affect the stereochemical integrity of the chiral grouping, or alternatively to use reactions for which the stereochemical course is known with certainty. Frequently, the selectivity properties of enzymes have been put to good use.

Since Bijvoet in 1951 introduced the anomalous dispersion of X-rays as a method for determining absolute configurations, an increasingly large number of such determinations has been derived. While some calculations suggested that the Bijvoet method was giving 'wrong' answers [83], it now appears that this anomalous dispersion technique has been fully vindicated [84,85]. In many cases, structure determination by X-ray crystallography and the determination of absolute configuration with the anomalous dispersion variation, are carried out simultaneously. In view of the extensive compilations now available, this topic will not be further discussed in this chapter.

(b) For prochiral elements

With the development of the role for prochirality in biology, the need to demonstrate the existence of this phenomenon and to characterize it in precise structural terms became urgent. Along with such developments, chemists were interested in the study of compounds in which isotopic replacement led to chirality which could be expressed as optical activity. In other words, it was necessary to carry out determinations of the configurations of prochiral compounds.

(i) Compounds containing a hydrogen isotope

The possibility that optical activity could be demonstrated in a compound in which deuterium replaced hydrogen had long been considered by chemists. The first successful validations of this expectation were reported in 1949; the historical developments have been reviewed [10,11].

A large number of compounds has now been prepared in which chirality arises as a result of the presence of $=CH^2H$ or $=CH^3H$. A few of them, which have been of particular importance, will be considered here to indicate some of the experimental strategies which have been used. More extensive general information is available [10,11], and Parry [86] has discussed chirally labeled α-amino acids.

In only one case, that of glycolic acid, has a configuration involving $=CH^2H$ been determined on an absolute basis; however, a number of compounds can be related to this standard. In 1965, Johnson et al. applied a neutron diffraction technique similar to the 'anomalous dispersion' technique pioneered by Bijvoet with X-ray diffraction [87]. The anomalous neutron-scattering amplitude of ^6Li, and the markedly different neutron-scattering amplitudes of hydrogen and deuterium, were applied to the lithium salt of glycolic acid, 72. The glycolic acid was obtained by the reduction of [2-^{2}H]glyoxylate, 71, with lactate dehydrogenase. The neutron diffraction technique revealed that it had (*S*) configuration. Consideration of the stereochemistry of the L-lactate dehydrogenase reaction with its 'normal' substrate had earlier led to the same conclusion [88].

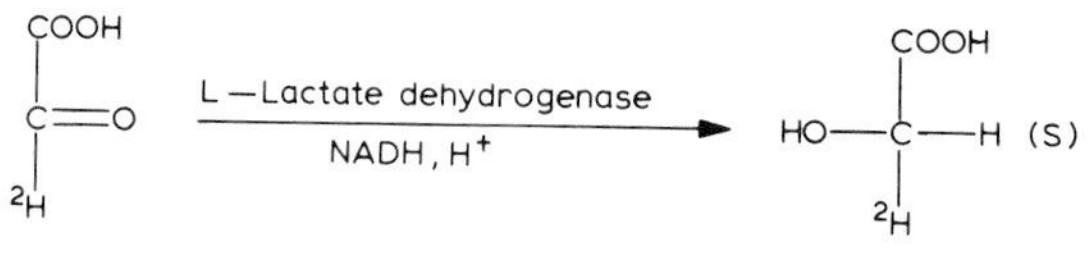

COOH
C=O
^{2}H
71

L-Lactate dehydrogenase
NADH, H^+

COOH
HO—C—H (S)
^{2}H
72

The correlation between glycolate, 72, and ethanol, 73, carried out in Arigoni's laboratory is of considerable importance (see [89]; these results are also reported in [10]).

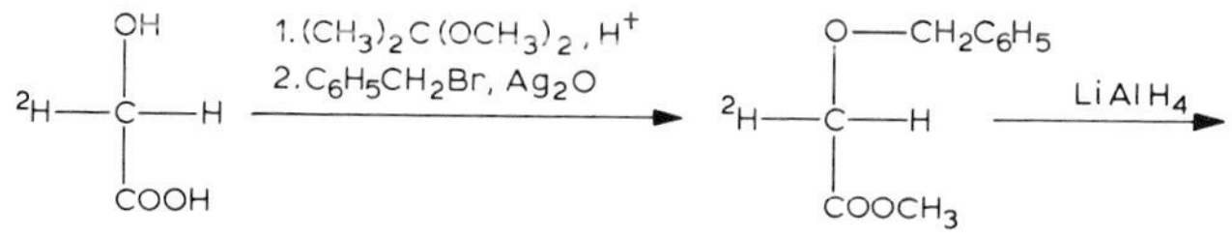

OH
^{2}H—C—H
COOH
(−)-(S)
72

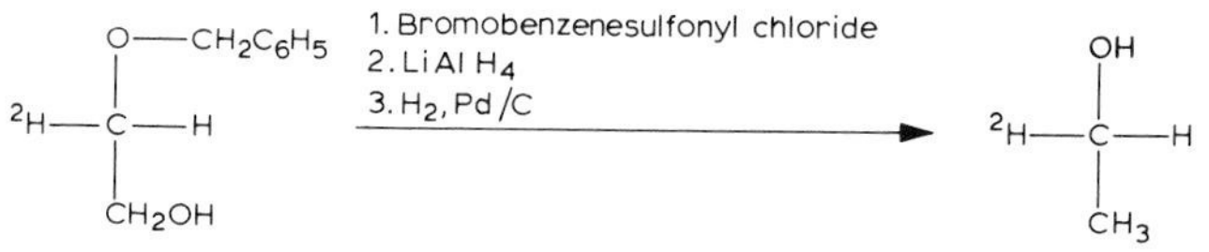

1. $(CH_3)_2C(OCH_3)_2$, H^+
2. $C_6H_5CH_2Br$, Ag_2O

O—$CH_2C_6H_5$
^{2}H—C—H
$COOCH_3$

$LiAlH_4$

O—$CH_2C_6H_5$
^{2}H—C—H
CH_2OH

1. Bromobenzenesulfonyl chloride
2. $LiAlH_4$
3. H_2, Pd/C

OH
^{2}H—C—H
CH_3
(−)-73 (S)

The ethanol obtained by the reactions shown was characterized by its enzymatic behavior, rather than by optical properties. The acetaldehyde obtained on treatment with alcohol dehydrogenase retained ^{2}H. Since the (−)-enantiomer of [2-^{2}H]ethanol behaves in the same way, it follows that (*S*)-[2-^{2}H]ethanol (absolute configuration defined by its derivation from (*S*)-[2-^{2}H]glycolate) is levorotatory.

In addition to this correlation with an absolute configurational standard, the C-2 position in ethanol had earlier [90] been correlated with D-xylose, 74, containing ^{2}H at C-5 (configuration determined by NMR spectroscopy), and also with (−)-*erythro*-

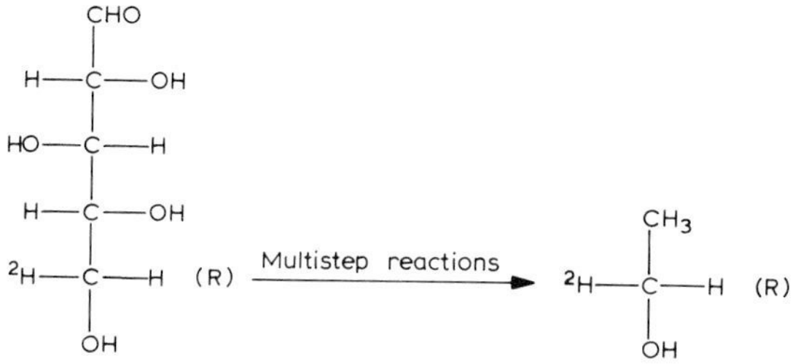

CHO
H—C—OH
HO—C—H
H—C—OH
^{2}H—C—H (R)
OH
74

Multistep reactions

CH_3
^{2}H—C—H (R)
OH
(+)-73

(2*R*,3*S*)-[3-^{2}H]butan-2-ol, 75 [91].

Another compound which can be correlated with the absolute configurational standard is [2-^{2}H]propionic acid. Using (*S*)-[2-^{2}H]ethanol as starting material,

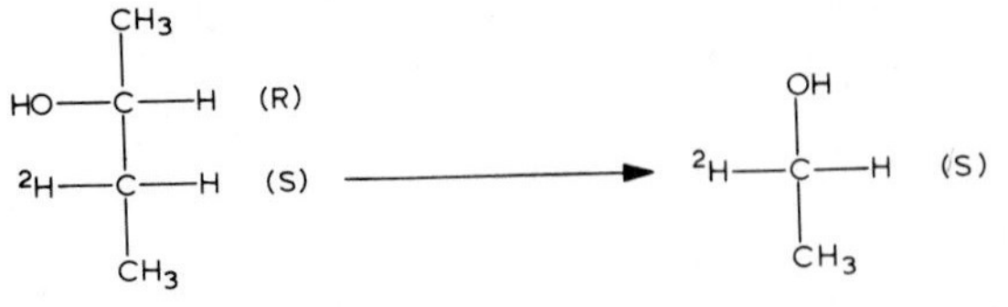

CH_3 / HO—C—H (R) / ^{2}H—C—H (S) / CH_3 → OH / ^{2}H—C—H (S) / CH_3

(−)-75 (−)-73

Zagalak et al. [92] obtained (−)-(R)-[2-^{2}H]propionic acid, 76, as follows:

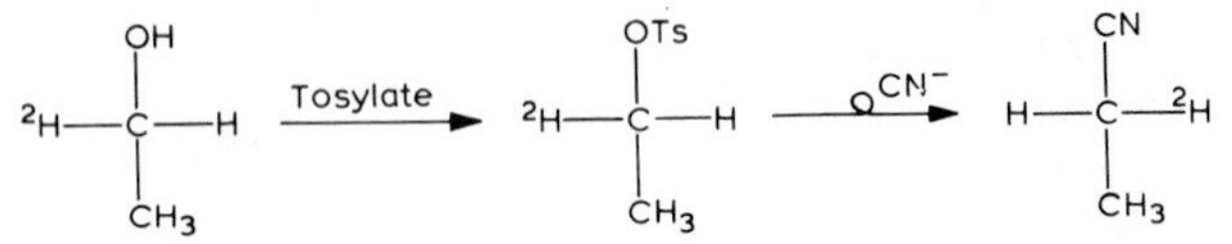

OH / ^{2}H—C—H / CH_3 —Tosylate→ OTs / ^{2}H—C—H / CH_3 —CN^-→ CN / H—C—^{2}H / CH_3

(−)-73

—Lithium aluminum triethoxy hydride→ CHO / H—C—^{2}H / CH_3 —$KMnO_4$→ COOH / H—C—^{2}H (R) / CH_3

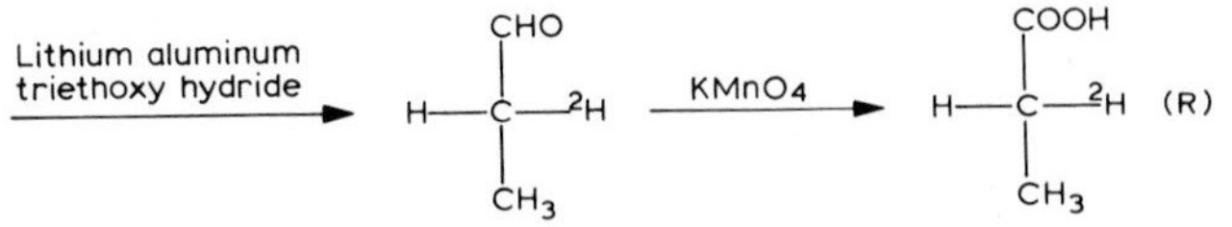

(−)-76

The postulated inversion of configuration during displacement of the tosyl group is unexceptionable. Other individuals [93] have prepared both forms of [2-^{2}H]propionic acid from alanine (shown for D-Ala, 77, to (+)-76).

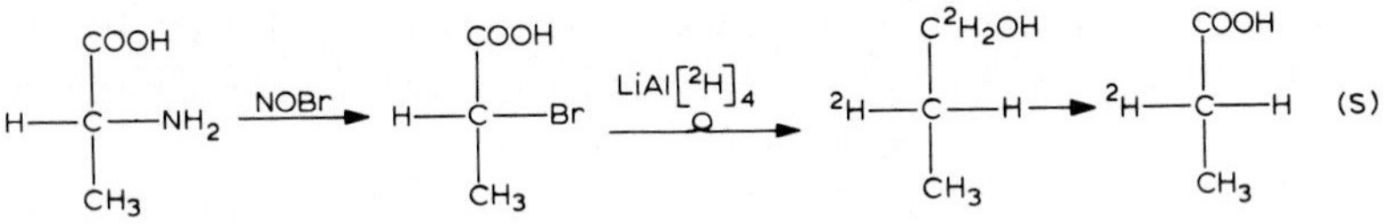

COOH / H—C—NH_2 / CH_3 —NOBr→ COOH / H—C—Br / CH_3 —$LiAl[^2H]_4$→ C^2H_2OH / ^{2}H—C—H / CH_3 → COOH / ^{2}H—C—H (S) / CH_3

77 (+)-76

One compound which has been of unique importance in the stereochemical investigations in biology is malic acid. It was the first 'biochemical' of the type $CabH^2H$ to be prepared. Despite its importance, there has apparently never been an independent confirmation of its configuration at C-3, and it has not been correlated with the glycolic acid standard. The assignment of configuration at C-3 rests on the assumed stereochemistry of chemical syntheses. Before describing these syntheses, the general stereochemistry of malic acid, 78, will be discussed. There is a chiral center at position number 2 and the form of malate produced by the fumarase reaction is known to be (+)-L (or S). At the prochiral center, the two hydrogens are diastereotopic:

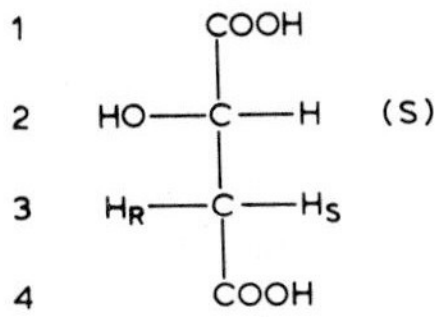

1 COOH
2 HO—C—H (S)
3 H_R—C—H_S
4 COOH

78

If one H at position 3 is replaced by ^{2}H the spatial relationship between the remaining H at C-3 and that at C-2 can be determined from the coupling constant for the two protons determined by NMR spectroscopy.

Gawron and Fondy utilized the stereospecific cleavage of 3,4-epoxy-2,5-dimethoxytetrahydropyran [94]. Formally, this molecule, 79, contains four chiral centers; those at positions 2 and 5, however, suffer loss of chirality during the synthesis. Since the epoxide ring cannot be formed from two *trans* oriented OH groups, the molecular structure must be that shown; the molecule is actually a meso structure. On treatment with $LiAl^2H_4$, an *anti* opening of the epoxy ring was expected. Since the incoming ^{2}H could be placed at either position 3 or 4 of 79, the ring opening gave two products, 80 and 81. On treatment with acid the tetrahydrofuran ring was opened (with loss of chirality at positions 2 and 5) to form a dialdehyde. Oxidation of the latter with HNO_3 gave malic acid as a racemic mixture of the two structures, 82 and 83:

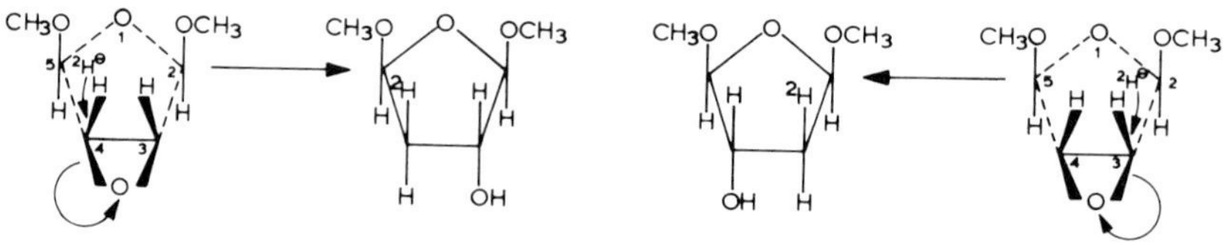

79 80 81 79

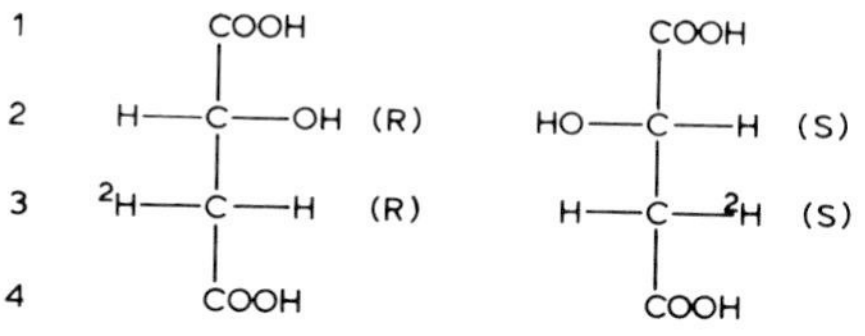

	82		83	
1	COOH		COOH	
2	H—C—OH	(R)	HO—C—H	(S)
3	^{2}H—C—H	(R)	H—C—^{2}H	(S)
4	COOH		COOH	

Note that, although the product is racemic $(2R,3R+2S,3S)$-malic acid, in each enantiomer, the protons at C-2 and C-3 are both in the *threo* relationship. Since the deuteron and OH group are also *threo*, the material may be described as (±)-*threo*-[3-^{2}H]-malic acid.

The validity of this work rests on the expected *anti* opening of the epoxy compound with $LiAl^2H_4$. In earlier work, Trevoy and Brown [95] had provided evidence for this stereochemical course for the reduction 84 → 85:

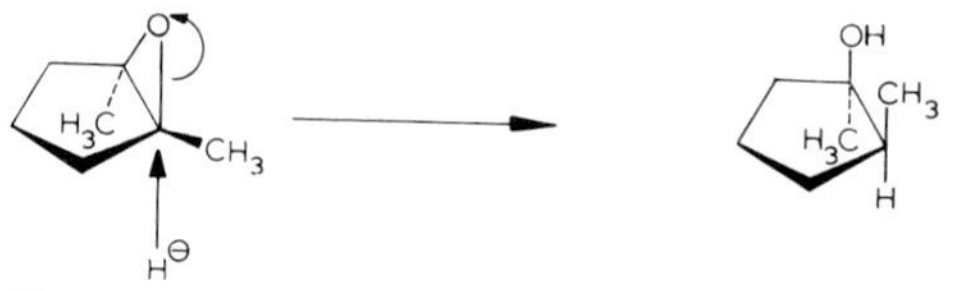

84 85

The epoxy compound was presumed to be *cis*, and evidence indicated the product had the *trans* structure. In another case, 3β-acetoxy-5β,6β-epoxycoprostane, 86, was reduced with $LiAlH_4$ to form 3β,6β-dihydroxycholestane, 87. This ring opening clearly occurred with *anti* stereochemistry [96]. There is thus good reason to postulate the *anti* opening of 3,4-epoxy-2,5-dimethoxytetrahydropyran, and this stereochemical course is generally accepted in many other cases. In a similar synthesis of [3-^{2}H]malic acid from the same starting material, Anet used oxidation with bromine in the presence of calcium carbonate for the final step [97].

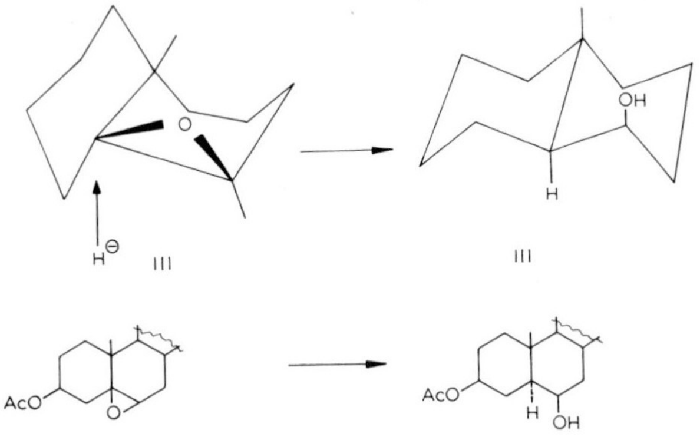

(Only the A and B rings of the steroid system are shown)

86 87

The (±)-*threo*-[3-^{2}H]malic acid showed a coupling constant of 4.4 ± 0.2 Hz for the two non-equivalent protons. This was quite different from the value of $J = 7.1–7.3$ Hz for the [3-^{2}H]malic acid obtained from the action of fumarase in ^2H_2O. Hence the stereochemistry of fumarase was assigned as *anti*, 39 → 88. It should be noted that the coupling constants for the *threo* and *erythro* isomers are in accordance with the generalization, $J_{trans} > J_{gauche}$.

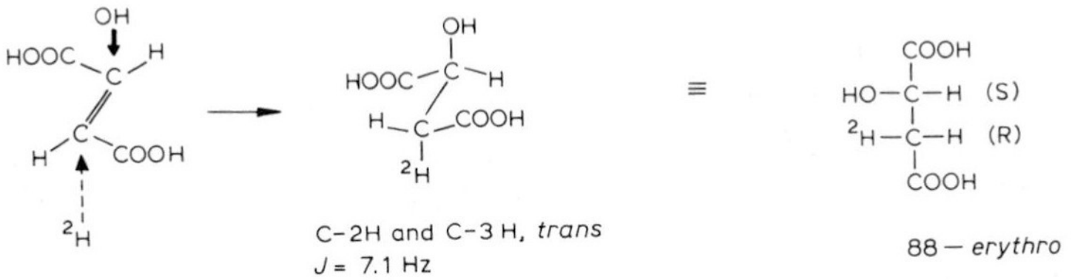

39

C-2H and C-3 H, *trans*
$J = 7.1$ Hz

88 — *erythro*

C-2 H and C-3 H, *gauche*
$J = 4.4$ Hz

83 — *threo*

A combination of enzymatic and chemical methods was used to establish configuration for [2-^{2}H]succinic acid [98]. (2*S*,3*R*)-[3-^{2}H]Malic acid was prepared by use of fumarase, and chemical methods (OH → Cl → H) were used to remove the chirality at position 2.

H COOH
C=C
HOOC H
39

^2H_2O, Fumarase →

COOH
H—C—^{2}H (R)
H—C—OH (S)
COOH
88

1. Form ester
2. $SOCl_2$, pyridine →

COOH
H—C—^{2}H
Cl—C—H
COOH

1. Zn/Cu, HAc
2. H^+ →

COOH
H—C—^{2}H (R)
H—C—H
COOH
(−)-89

In the conversion of malic acid to chlorosuccinic acid, an inversion of configuration at C-2 was presumed. This point is immaterial, however, since chirality is finally lost at this position. The (2*R*)-[2-^{2}H]succinic acid, 89, had a plain negative ORD curve. The configurational assignment rests on the correctness of that for malic acid.

Aspartate ammonia-lyase was also used to provide (2*S*,3*R*)-[3-^{2}H]aspartic acid, and the latter was converted (NH_2 → Br → H) to (2*R*)-[2-^{2}H]succinic acid [99]. Since the stereochemical assignment for aspartate ammonia-lyase also relies on the configuration of malic acid, this synthesis does not provide an unambiguous verification of the succinate configuration.

(ii) The configuration of NADH and NADPH

Following a demonstration that solvent protons were not incorporated into NADH during the action of alcohol dehydrogenase, Fisher et al. provided the first demonstration of prochirality involving hydrogen atoms of a methylene group in 1953 [100]. In this work, two samples of [4-^{2}H]NADH were prepared, one chemically and the other enzymatically; in subsequent reactions with alcohol dehydrogenase, they were shown to behave differently. The enzymatic preparation of [4-^{2}H]NADH, 92, was carried out by the action of yeast alcohol dehydrogenase on [1-$^2H_2$]ethanol, 90. When the reaction was run in the reverse direction with the same enzyme, the NAD^+ obtained was without ^{2}H (the configuration shown for [4-^{2}H]NADH, 92, was not actually determined until later). When the sample of [4-^{2}H]NADH was prepared by a chemical reduction, the NAD^+ obtained in the alcohol dehydrogenase reaction contained 0.44 atom of ^{2}H per molecule (91 + 94). Clearly the chemically prepared [4-^{2}H]NADH was a mixture of two diastereoisomeric forms, 92 + 93 (formed in unequal amounts since the reduction is subject to chiral influences from the ribose components).

It was shown later (in 1955) that some enzymes behave differently from alcohol dehydrogenase; these enzymes (e.g., *β*-hydroxysteroid dehydrogenase), in fact, utilize the other hydrogen of the diastereotopic methylene at C-4. The two hydrogens were distinguished as H_A and H_B; enzymes (such as alcohol dehydrogenase) utilizing H_A

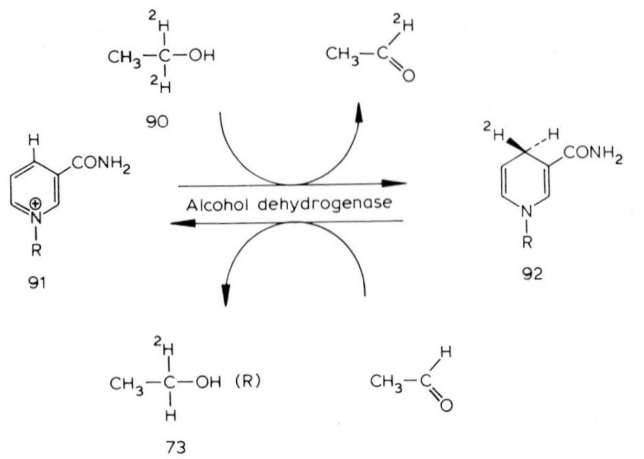

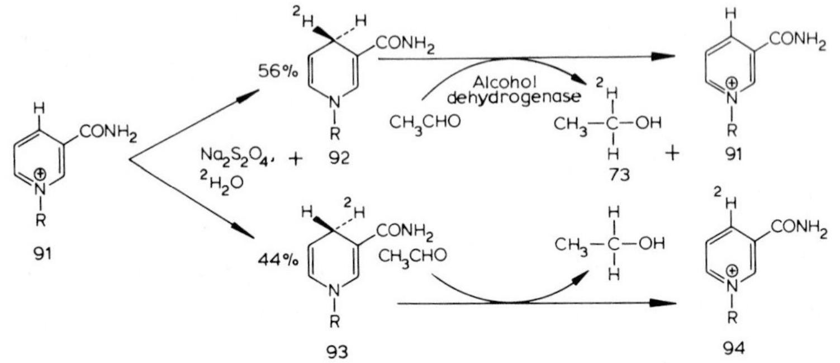

are said to have A selectivity, those utilizing H_B (such as β-hydroxysteroid dehydrogenase) have B stereoselectivity. Many enzymes utilizing the pyridine nucleotides have been examined for this selectivity; a 1978 compilation listed 127 examples [101].

The configuration of NADH at the 4 position was determined by a chemical transformation to succinic acid [98]. A sample of NADH containing ^2H in the 'A' position, 92, was converted to a 6-methoxy derivative, 95; succinic acid was obtained from this derivative by treatment with ozone and then peroxyacetic acid.

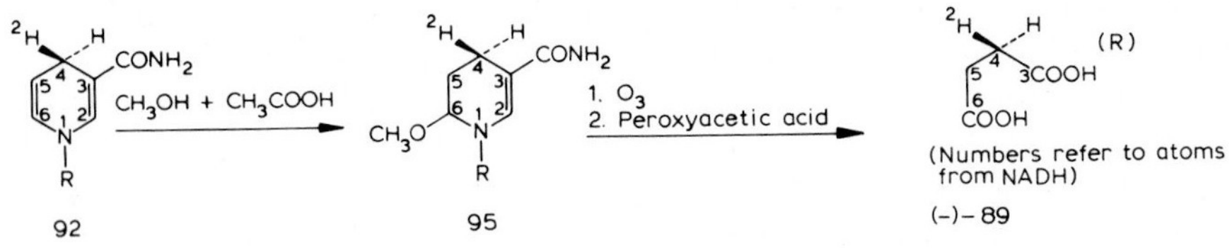

The succinic acid has a (−) ORD curve and was, therefore, (*R*). In the same way a sample of 'B'-labeled NADH was converted to (+)-(*S*)-succinic acid. In other words the 'A' hydrogen of NADH is the *pro-R* hydrogen, the 'B' hydrogen, the *pro-S* (see 96).

$H_R = H_A$, $H_B = H_S$

96

The conclusion with respect to NADH can be extended to NADPH by use of glutamate dehydrogenase, an enzyme with the ability to use both NAD and NADP [102]. A sample of [4-^{2}H]NADPH, 98, was prepared from [4-^{2}H]NADP$^+$, 97, by the action of isocitrate dehydrogenase. On reaction of the [4-^{2}H]NADPH with glutamate dehydrogenase, of known B stereoselectivity, NADP$^+$, 99, was obtained without ^{2}H, while glutamate was labeled. This experiment establishes A stereoselectivity for isocitrate dehydrogenase. Furthermore, the [4-^{2}H]NADPH was converted to [4-^{2}H]NADH, 93, by treatment with intestinal phosphatase to remove the extra phosphate group at the 2′ position of ribose. When this sample of [4-^{2}H]NADH was treated with glutamate dehydrogenase, NAD$^+$ was obtained without ^{2}H. Hence it can be concluded that H_A and H_B of NADPH correspond directly with H_A and H_B of NADH.

97: Isocitrate dehydrogenase (isocitrate → 2-ketoglutarate + CO_2) → 98: Glutamate dehydrogenase (2-ketoglutarate → [2-^{2}H]glutamate + NH_4^+) → 99

98: Intestinal phosphatase (P_i) → 93: Glutamate dehydrogenase (2-ketoglutarate → [2-^{2}H]glutamate + NH_4^+) → 91

Determination of selectivity for a particular enzyme is carried out with the aid of an enzyme of known properties. This is approached by one of two ways:

(1) Either the [4-^{2}H]NADH produced from [4-^{2}H]NAD$^+$ by the enzyme of unknown specificity is examined with an A- or B-selective enzyme (^{3}H can, of course, be used instead of ^{2}H). The NAD$^+$ produced is examined for the presence of either ^{2}H or ^{3}H, depending on which isotope was used initially.

(2) Alternatively, an A- or B-selective enzyme is used for the preparation of A or B labeled NADH. The action of the enzyme of unknown selectivity on this material is then examined. Viola et al. have described improved stereoselective methods for the preparation of the [4-^{2}H]NADPH isomers [103]. For the A-labeled material,

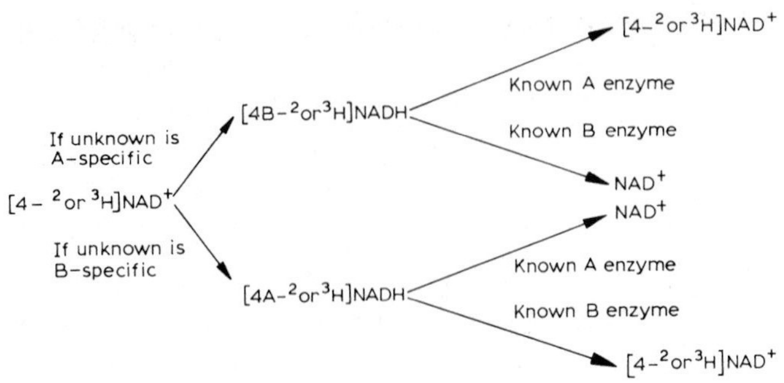

alcohol dehydrogenase is used; the ^2H is introduced from $[^2H_6]$ethanol (which is less expensive than $[1,1\text{-}^2H_2]$ethanol) and aldehyde dehydrogenase is present to drive the reaction to completion:

$$C^2H_3C^2H_2O^2H + NAD^+ \xrightarrow{\text{Alcohol dehydrogenase}} [4A\text{-}^2H]NADH + C^2H_3C^2HO$$

$$C^2H_3C^2HO \xrightarrow{\text{Aldehyde dehydrogenase}} C^2H_3COOH$$

For the B-labeled NADH, β-[1-^{2}H]glucose, 100, was used with glucose-6-phosphate dehydrogenase in the presence of 40% dimethylsulfoxide (this enhances the glucose dehydrogenase activity of the glucose-6-phosphate dehydrogenase). The reaction is essentially irreversible and goes to completion:

100 + NAD⁺ → [4B-²H]NADH + (lactone) → Gluconic acid

(Substituents omitted at atoms other than C-1)

Although the determination of H_A or H_B selectivity is relatively straightforward the techniques for isolation of pyridine nucleotides from the reaction mixtures are tedious and time consuming. Two more recent techniques use either proton magnetic resonance or electron impact and field desorption mass spectrometry. The technique of Kaplan and colleagues requires a 220 MHz nuclear magnetic resonance spectrometer interfaced with a Fourier transform system [104]. It allows the elimination of extensive purification of the pyridine nucleotide, is able to monitor the precise oxidoreduction site at position 4, can be used with crude extracts, and can be scaled down to μmole quantities of coenzyme. The method can distinguish between [4-^{2}H]NAD$^+$ (no resonance at 8.95 δ) and NAD$^+$ (resonance at 8.95—which is preferred) or between [4A-^{2}H]NADH (resonance at 2.67 δ, $J_{5\text{-}4B} = 3.8$ Hz) and [4B-^{2}H]NADH (resonance at 2.77 δ, $J_{5\text{-}4A} = 3.1$ Hz).

The technique of Ehmke et al. also avoids any purification, and is sensitive in the μg range [105]. Both low-resolution electron impact mass spectrometry and low-resolution field desorption mass spectrometry were used as the final analytical tool. These workers prepare [4B-^{2}H]NADH for examination by the enzyme of unknown selectivity. An A-selective enzyme yields [4-^{2}H]NAD$^+$, and B-selective enzyme yields NAD$^+$ without ^{2}H. These samples are analyzed by the mass spectrometer. In electron impact mass spectrometry, NAD$^+$ shows the m/z sequence:

$$122 \xrightarrow{-NH_2} 106 \xrightarrow{-CO} 78 \xrightarrow{-HCN} 51$$

The base peak at m/z 122 = [nicotinamide]$^{+\cdot}$

For [4-^{2}H]NAD$^+$ the characteristic fragmentation sequence is 123 → 107 → 79 → 52. With field desorption mass spectrometry, molecular ions could be observed for the reduced substrate (but not the pyridine nucleotide). Thus, with a glutamate dehydrogenase from duckweed (*Lemma minor*), the [M + H]$^+$ ion of glutamate was present at m/z 149 indicating the presence of [2-^{2}H]glutamic acid. In addition, in the mass range of nicotinamide only one intense signal at m/z 122 was seen (for the base peak of NAD$^+$). These results clearly show transfer of ^{2}H, from the B position of [4B-^{2}H]NADH.

(iii) The configuration of citric acid

The prochiral center in citric acid is attached to two CH_2COOH groups. The configurational question can, therefore, be phrased as the question, does acetyl-CoA contribute the *pro-R* or *pro-S* CH_2COOH groups? To provide an answer proved more difficult than for the relatively simple CabH^{2}H cases so far discussed. The 'proof' requires knowledge about several other enzymes and, as well, information about the chirality of compounds related to shikimic acid.

The first attempt to determine the configuration of citrate was that of Martius and Schorre [106,107]. At the time of their work (1950) it was reasonably clear that the action of aconitase was on that portion of the citrate molecule derived from oxaloacetate. This follows from the following observations.

(a) When radioactive CO_2 was added to systems such as minced pigeon liver, radioactive 2-ketoglutarate could be isolated; chemical degradation of the 2-ketoglutarate revealed that radioactivity was only present in C-1 [108–110]. The CO_2 was believed to form C-4 of oxaloacetate.

(b) $HOO^{14}CCH_2COCOOH$ (the expected product of the CO_2 fixation, 101) was incubated with an acetone powder of pigeon liver and citrate was isolated. This purified citrate was converted enzymatically to 2-ketoglutarate, 103, and the latter was subjected to chemical degradation. The label was again present only in C-1 of 2-ketoglutarate [111].

These experiments can be interpreted in two ways, depending on the configuration of citrate. However, in each case, the action of aconitase must be such as to lead to introduction of the double bond of *cis*-aconitic acid between the carbons originally derived from oxaloacetate. This follows from the fact that the carbonyl

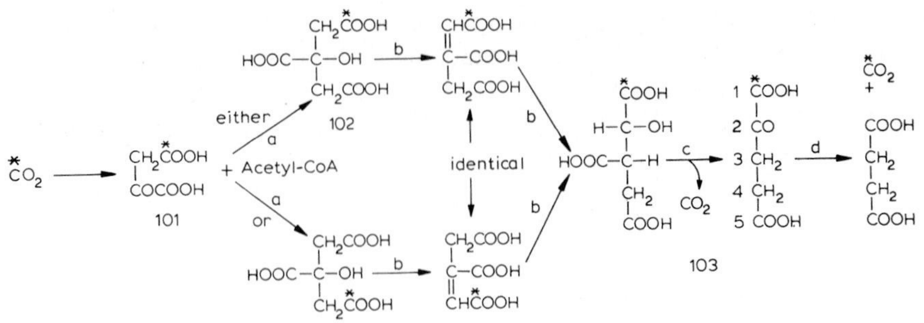

a = citrate synthase; b = aconitase; c = isocitrate dehydrogenase;
d = $KMnO_4$; * = ^{14}C

group of 2-ketoglutarate must be derived by the action of aconitase and isocitrate dehydrogenase. Since the adjacent carboxyl at C-1 is radioactive, and since the radioactivity was introduced by way of oxaloacetate, the action of aconitase is on the carbon atoms derived from oxaloacetate.

Martius and Schorre next prepared two enantiomeric forms of $[^2H_2]$citrate as follows. The aldol reaction of pyruvate and oxaloacetate under weakly alkaline conditions gave the lactone of oxalocitramalic acid, 104 and 105. This was resolved by way of the brucine salt into (+) and (−) forms. The enolization of the CH_2 adjacent to the CO group was used to introduce ^2H and, on oxidation with $^2H_2O_2$, dideuterocitric acids were obtained. From the (−)-oxalocitramalic lactone, 105, a (−)-dideuterocitrate, 106, resulted, and from the (+)-lactone, a (+)-dideuterocitrate.

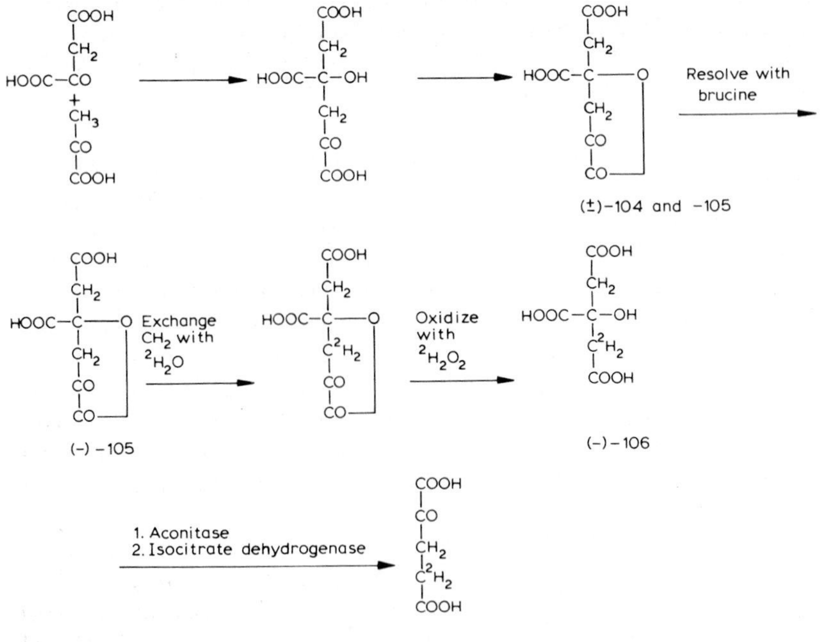

Resolve with brucine

(±)-104 and -105

Exchange CH_2 with ^2H_2O

Oxidize with $^2H_2O_2$

(−)-105

(−)-106

1. Aconitase
2. Isocitrate dehydrogenase

107

In enzymatic experiments, a mince of pigeon breast muscle was used as a source of aconitase and isocitrate dehydrogenase; 2-ketoglutarate, 107, was isolated as a dinitrophenylhydrazone and this derivative was analyzed for ^2H. The (−)-citrate yielded ketoglutarate in which all of the ^2H was retained, and the (+)-citrate yielded ketoglutarate without ^2H. When a ^2H citrate was prepared from the racemic (±)-lactone, the ketoglutarate contained 50% of the ^2H present in the citrate. Since the action of aconitase and isocitrate dehydrogenase converts a CH_2 group to CO, and since this original CH_2 is associated with oxaloacetic acid, these experiments would lead to the configuration of citrate if one further piece of information was available—that is, the configuration of the oxalocitramalic acid precursor. For this purpose, Martius and Schorre applied the Hudson Lactone Rule; for the aldonic acids, in which the CO of the lactone group is at the top in a Fischer projection formula, the (+) form has the OH group to the right [112].

CO	CO	CO	CO
H—C—OH	CO	CO	H—C—OH
HO—C—H	CH_2	CH_2	HO—C—H
H—C——O	HOOC—C——O	O——C—COOH	O——C—H
H—C—OH	CH_2	CH_2	H—C—OH
CH_2OH	COOH	COOH	CH_2COOH
(+)	(+)	(−)	(−)
γ-lactone of D-gluconic acid	104	105	γ-lactone of D-galactonic acid

Using this empirical rule, the configurations shown as 104 and 105, were assigned to the oxalocitramalate lactone. Since the 'aconitase active' portion of the molecule derives from oxaloacetic acid, and since the ^2H citrate, 106, from the (−)-lactone retains deuterium, the two CH_2COOH groups can be identified.

Generalizations such as the Hudson Lactone Rule are not always completely reliable and so a more direct proof was desirable. It did turn out, however, that the conclusions of Martius and Schorre were correct. The new proof also required the synthesis of a labeled citric acid of defined structure. For this purpose, quinic acid of known configuration was used to provide citrate labeled with ^3H in the *pro-R* CH_2COOH [113]. On treatment with aconitase, all of the ^3H of this material was lost to water. Since it was rigorously established that the hydrogen exchanged by aconitase is derived from oxaloacetic acid, it was concluded that this oxaloacetate precursor of citric acid contributes the *pro-R* CH_2COOH group.

Since the details of this proof are somewhat involved, it is worthwhile to review the procedure. The work provides a classic example of the interplay between organic chemistry, biochemistry, enzymology and microbiology.

(1) Configurations of shikimic acid and related compounds: Beginning in 1934, H.O.L. Fischer and G. Dangschat [114] established configurations for shikimic acid, 109, quinic acid, 108, and dehydroquinic acid, 111. By a multistep procedure (which did not change configurations at C-3, -4 and -5), shikimic acid was converted to the lactone of 2-deoxy-D-gluconic acid, 110. Since this compound could be derived from D-glucose itself by another multistep procedure (D-glucose → 2,3,4,6-tetra-*O*-acetyl-D

-glucopyranosyl bromide → D-glucal → 2-deoxy-D-glucose), the configuration of the naturally occurring (−)-shikimic acid was established. Furthermore, (−)-quinic acid has been synthesized from D-arabinose, retaining the configuration of the three

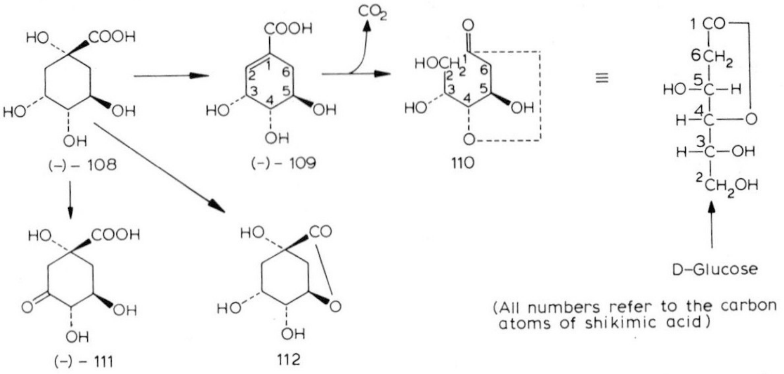

(−)-108 (−)-109 110 D-Glucose (−)-111 112

(All numbers refer to the carbon atoms of shikimic acid)

chiral centers [115]; for (−)-quinic acid, the configurations at the three hydroxyl bearing carbon atoms were earlier established by the conversion of (−)-quinic acid to (−)-shikimic acid and by the reverse synthesis of quinic acid from shikimic acid. The chirality at the fourth center in (−)-quinic acid follows from the fact of lactone formation between the carboxyl group and the hydroxyl which is β to it (forms quinide, 112). For this to occur, both of these groups must be *trans* to form the chair conformation.

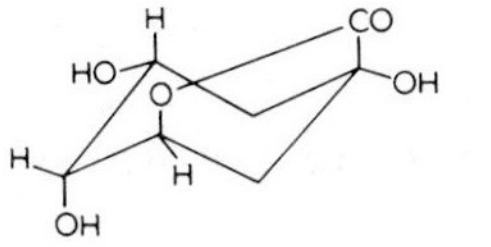

112

(Note that this is not the preferred conformation of quinic acid itself)

(2) The basic chemistry of shikimic and quinic acids was elucidated long before they were realized to be important biosynthetic intermediates. The history of the developments relating to the biosynthesis of aromatic amino acids has been reviewed recently [116]; it is of interest that quinic acid was first isolated almost two centuries ago—in 1790. For the present purpose, two enzymes of the shikimate pathway are of concern: 3-dehydroquinate dehydratase and quinate dehydrogenase (for a general review, see [117]). By using enzyme extracts from *Aerobacter aerogenes*, dehydroshikimic acid, 113, was converted, in ^3H_2O, to quinic acid containing tritium at carbon number 6 (IUPAC—IUB nomenclature) [113].

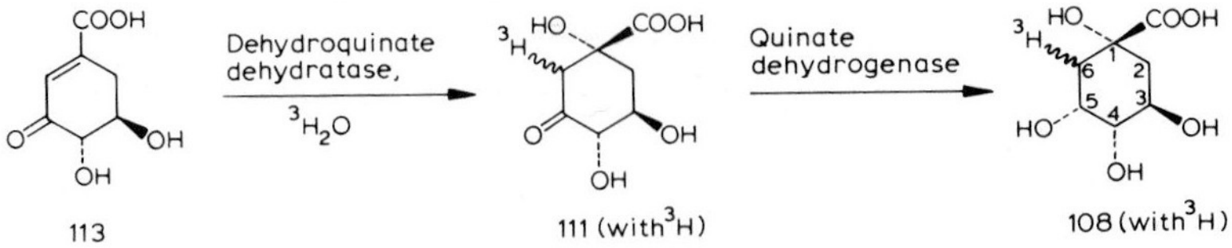

Dehydroquinate dehydratase, ^3H_2O → Quinate dehydrogenase →

113 111 (with ^3H) 108 (with ^3H)

The quinic acid, of necessity, contains ^3H as shown. This follows from the regioselective hydration of the double bond—the OH group is added by the dehydratase at the carbon already carrying a carboxyl. For the purposes of the citric acid proof, it is not necessary to know the actual configuration at carbon atom 6 (this point will be discussed later).

(3) A final contribution to the citric acid proof came from studies of aconitase and other enzymes of the tricarboxylic acid cycle. It is necessary to know that (a) aconitase 'exchanges' one of the hydrogen atoms of citrate; and (b) that the exchanged hydrogen is derived biosynthetically from the oxaloacetate portion of the molecule. This could be inferred from the earlier work but a more rigorous demonstration was desirable. It is typical of work in this area, that information regarding two other enzymes, fumarase and isocitrate dehydrogenase, is also needed.

(4) Isocitrate dehydrogenase and aconitase: Isocitrate dehydrogenase was shown in 1957 to catalyze a direct transfer of hydrogen from isocitrate to NADP [118]. That is to say, if the isocitrate dehydrogenase reaction is carried out in ^2H_2O, the NADPH obtained does not contain ^2H. Although it seemed highly likely that the transferred hydrogen was that associated with the α carbon atom (see 114 → 115), the alternate possibility was eliminated by the following evidence [119]:

COOH–C$_\alpha$H*(OH)–CH(COOH)–CH$_2$–COOH (114) —Isocitrate dehydrogenase, NADP$^+$ → NADPH*→ COOH–CO–CH(COOH)–CH$_2$–COOH (115) —(−CO$_2$)→ COOH–CO–CH$_2$–CH$_2$–COOH (103)

If aconitase is incubated with *cis*-aconitic acid (or either of the citrates) there is formed an equilibrium mixture containing 88.4% of citrate, 7.5% isocitrate and 4.1% of *cis*-aconitate. It is possible that in the hydration–dehydration process

Citrate —($-H_2O$)→ *cis*-Aconitate —($+H_2O$)→ Isocitrate

cis-Aconitate —($+H_2O$)→ Citrate

the hydrogen removed in dehydration is not that added in the hydration (i.e., the process is not selective). In this event, at equilibrium, citrate would contain 2 atoms of ^2H ($\alpha\alpha'$), and isocitrate similarly 2 atoms of ^2H ($\alpha\beta$). On the other hand, if

CHCOOH=C–COOH, CH$_2$COOH (*cis*-Aconitate) —($+^2H_2O$)→ CH^{2}HCOOH–C(OH)COOH–CH$_2$COOH (Citrate) —($-H_2O$)→ C^{2}HCOOH=C–COOH, CH$_2$COOH (*cis*-Aconitate) —($+^2H_2O$)→ α C^{2}H$_2$COOH–β C(OH)COOH–CH$_2$COOH (Citrate); —($+^2H_2O$)→ α C^{2}H(OH)COOH–β C^{2}HCOOH–CH$_2$COOH (Isocitrate, 116)

Aconitase–non–selective mechanism

aconitase operated selectively on one hydrogen in both hydration and rehydration, the equilibrium citrate would contain one $^2H(\alpha)$ and the isocitrate one ^2H (β).

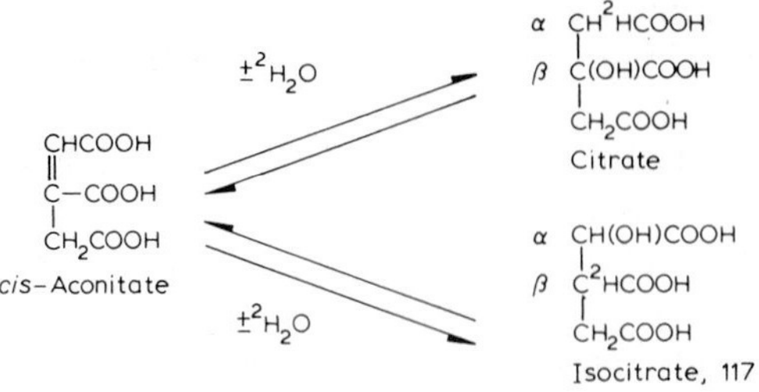

Aconitase-selective mechanism

Depending on the properties of aconitase, isocitrate 116 or 117 would be produced. If these two isocitrates are now examined as substrates for isocitrate dehydrogenase, the labeling in NADH for removal of either the α or β hydrogen may be derived.

In actual experiments, Englard and Colowick incubated citrate with aconitate hydratase in 85% ^2H_2O for varying periods of time [118]. Isocitrate dehydrogenase and $NADP^+$ were then added; finally the reduced NADPH and citrate were isolated and examined for their content of ^2H. The reduced NADPH contained a low level of ^2H which did not increase with time. The citrate, however, accumulated an increasing level of ^2H to a maximum value of 0.55 atoms ^2H per molecule at 8 hours (see Table 4). Since the ^2H content of NADPH was low and did not increase with incubation time, it was concluded that 'Alternative C' represented the pathway actually used. Hence, the action of aconitase was selective, the hydrogen added in hydration being that removed in dehydration. In support of this conclusion was the level of ^2H in citrate. Since the reaction was carried out in 85% ^2H_2O, a non-selective process would have incorporated $2 \times 0.85 = 1.7$ atoms ^2H per molecule of citrate. The observed value after 8 hours of incubation was only 0.55—approaching the value expected for 1 atom of ^2H per molecule.

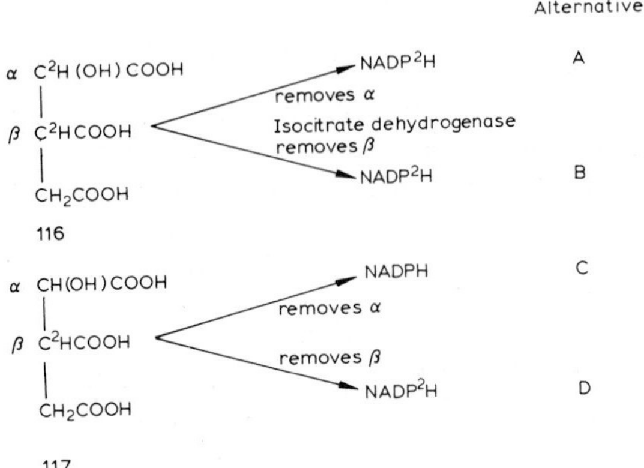

These experiments with aconitase and isocitrate dehydrogenase established a selectivity for aconitase. When citrate is incubated with this enzyme in ^2H_2O, one of the methylene hydrogens of citrate is replaced by ^2H ('exchanged'). The other

TABLE 4
Labeling of NADPH and citric acid from incubation of citric acid, aconitase, isocitrate dehydrogenase and $NADP^+$ in 85% ^2H_2O [a]

Time (min)	Atoms ^2H per molecule	
	NADPH	Citrate
20	0.03	0.09
210	0.04	0.43
480	0.03	0.55

[a] Other data points omitted.

hydrogen of the methylene pair is retained on the α carbon of *cis*-aconitate and isocitrate; it is removed (to $NADP^+$) by the action of isocitrate dehydrogenase. In other words, the citrate methylene behaves as a typical prochiral center. For the proof of the citrate configuration it is not necessary to know the actual configuration at the α carbon atom. The sequence is shown below with the actual (known) configuration indicated for convenience:

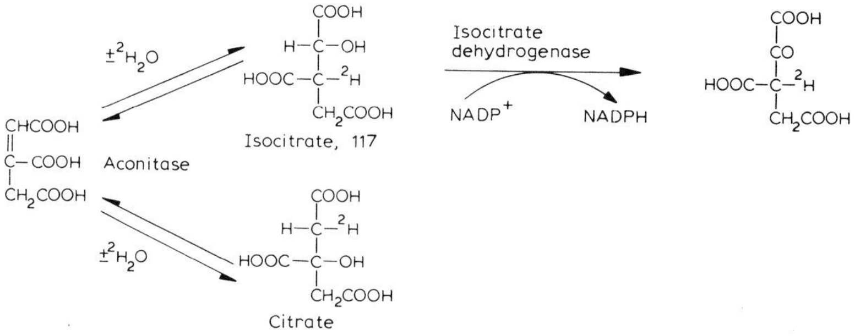

CHCOOH=C(COOH)–CH_2COOH ⇌ (Aconitase, $\pm{}^2H_2O$) Isocitrate, 117: COOH–CH(OH)–C(^{2}H)(COOH)–CH_2COOH → (Isocitrate dehydrogenase; $NADP^+$ → NADPH) COOH–CO–C(^{2}H)(COOH)–CH_2COOH

CHCOOH=C(COOH)–CH_2COOH ⇌ (Aconitase, $\pm{}^2H_2O$) Citrate: COOH–CH(^{2}H)–C(OH)(COOH)–CH_2COOH

(5) It now remains to determine whether the hydrogen exchanged by aconitase has its biosynthetic origin in oxaloacetate or acetyl-CoA. For this purpose, Englard [119] used the fumarase reaction to prepare two samples of ^{2}H-labeled L-malate, and the subsequent action of malate dehydrogenase to yield oxaloacetate. The two samples of oxaloacetate must have had enantiomeric configurations for the methylene group; this follows from the method of preparation (shown in Table 5 with the actual, known configurations). In turn, these samples were subjected to the action of citrate synthase, and labeled samples of citrate were isolated and analyzed. (To prevent ^{2}H loss by enolization of oxaloacetate, limiting amounts of malate dehydrogenase and excess amounts of citrate synthase and acetyl-CoA were used.) The ^{2}H contents are recorded in Table 5. The citrate samples were next equilibrated with aconitase in H_2O and the isocitrate was treated with isocitrate dehydrogenase and NADP. The reduced NADP obtained in this reaction was analyzed for ^{2}H as previously described. When the starting material was ordinary fumarate and when the fumarase reaction was conducted in ^2H_2O, the NADPH contained no ^{2}H. In the complementary reaction from [2,3-$^2H_2$]fumarate in H_2O, the NADPH contained

TABLE 5

Deuterium content of tricarboxylic acid materials obtained from either labeled or non-labeled fumarate [a]

Procedure in fumarase reaction	Atom ^2H/molecule		
	Malic acid	Citric acid	NADPH
Fumarate in ^2H_2O (99.8%)	1.00	0.82	0.0
[2,3-$^2H_2$]Fumarate in H_2O	1.73	0.85	0.67

H–C–COOH ‖ HOOC–C–H $\xrightarrow{+^2H_2O}$ COOH, H–C–^{2}H, H–C–OH, COOH → COOH, H–C–^{2}H, CO, COOH → Citric acid, 118 or 121

Fumarase   Malate dehydrogenase

^{2}H–C–COOH ‖ HOOC–C–^{2}H $\xrightarrow{+H_2O}$ COOH, ^{2}H–C–H, ^{2}H–C–OH, COOH → COOH, ^{2}H–C–H, CO, COOH → Citric acid, 119 or 120

Malic acid   Oxaloacetic acid

[a] The NADPH was isolated after the subsequent action of aconitase and isocitrate dehydrogenase on the two citrate samples.

slightly less than one atom of ^2H per molecule (see Table 5).

These results may be analyzed as follows without knowledge of actual configurations. The two citrate samples must, from the mode of synthesis, be labeled at the same methylene (either *pro-R* or *pro-S*); one of them must contain ^2H as H_R, the other as H_S of this methylene. They can be represented as follows:

118: COOH (1), H_S–C–H_R● (2), HOOC–C–OH (3), H_R–C–H_S (4), COOH (5)

119: COOH (1), ●H_S–C–H_R (2), HOOC–C–OH (3), H_R–C–H_S (4), COOH (5)

Two citrate samples if acetyl-CoA forms C-4 and C-5 (i.e., *pro*-S)

120: COOH (1), H_S–C–H_R (2), HOOC–C–OH (3), H_R–C–H_S● (4), COOH (5)

121: COOH (1), H_S–C–H_R (2), HOOC–C–OH (3), ●H_R–C–H_S (4), COOH (5)

Two citrate samples if acetyl-CoA forms C-1 and C-2 (i.e., *pro*-R)

In the reaction with aconitase, one citrate methylene was shown to behave as follows: one of its prochiral hydrogens is lost to water, while the other becomes the hydrogen on the α-carbon of isocitrate and is removed to form NADPH by isocitrate dehydrogenase. With the two citrate samples, 118 and 119, the observed result shows that the 'active' methylene cannot be at C-4 (both hydrogens are ^1H) and is, therefore, at C-2. Similarly, with the two citrate samples 120 and 121, the 'active' methylene cannot be at C-2, but must be at C-4. The results are only consistent with the conclusion that the methylene acted on by aconitase was contributed as the methylene present in oxaloacetate. It is not necessary to know the actual configurations at the prochiral methylenes—all that is required is the knowledge that aconitase does behave in a selective fashion and that isocitrate dehydrogenase removes the hydrogen from the α carbon of its substrate.

The final component was provided by Hanson and Rose [113]. These workers converted the ^{3}H-labeled quinic acid, 108, to citric acid, 122, by the following chemical steps (ca. 50% yield):

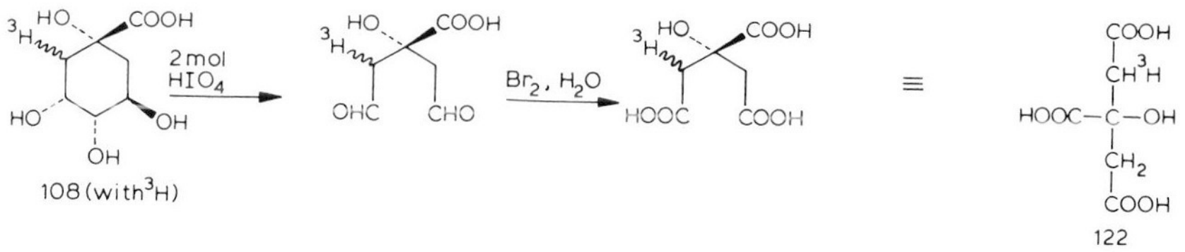

From the regioselectivity of the initial enzymatic hydration of dehydroshikimate, the ^{3}H label had to be associated with one of the methylene protons at C-2 (i.e., the *pro-R* CH_2COOH group of citrate). When this citrate sample was brought to equilibrium with aconitase, essentially all (95–97%) of the ^{3}H was exchanged into the solvent water. Since the previous work establishes that the exchangeable proton has its origin in oxaloacetate, it follows that the *pro-R* CH_2COOH group of citrate derives from oxaloacetate.

The 'proof' just given made no assumptions as to actual configurations. It can also be stated starting from the knowledge that fumarase hydrates fumaric acid to L-malic acid by *anti* addition on the *Si–Si* face. Hence, if the reaction is carried out in ^2H_2O, the product is *erythro*-L-[3-^{2}H]malic acid with (*S*) configuration at C-2, and (*R*) at C-3, 88. Hence, the sequence of reactions just discussed can be represented as follows:

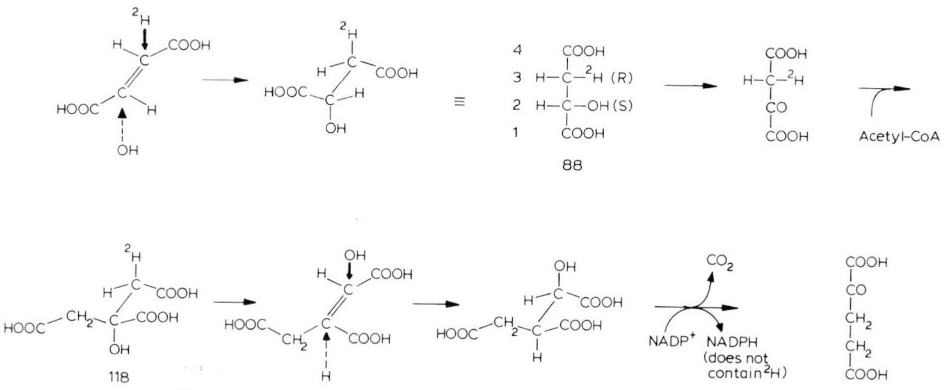

If [2,3-$^2H_2$]fumarate is used as substrate for the fumarase reaction in H_2O, the malic acid has (*S*) configuration at C-3, 123; in this case, the proton lost by action of aconitase is H not ^{2}H, and ^{2}H is finally removed by action of isocitrate dehydrogenase. Using normal projection formulae, this sequence is as follows:

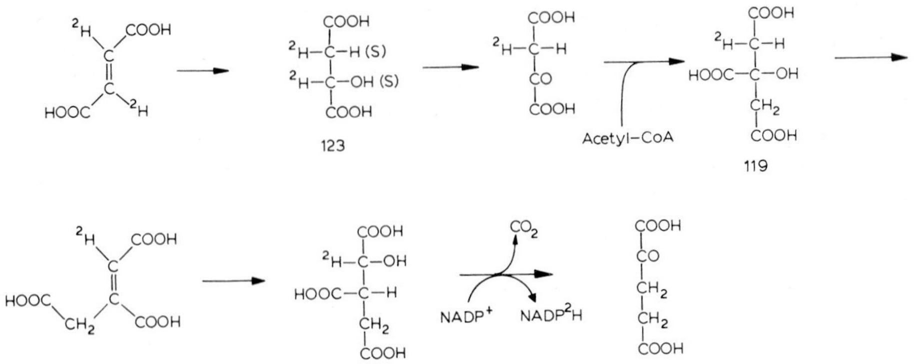

From these refinements, it is also possible to deduce the stereochemical course of the hydration of 5-dehydroshikimic acid; this proves to be a case of the unusual *syn* addition.

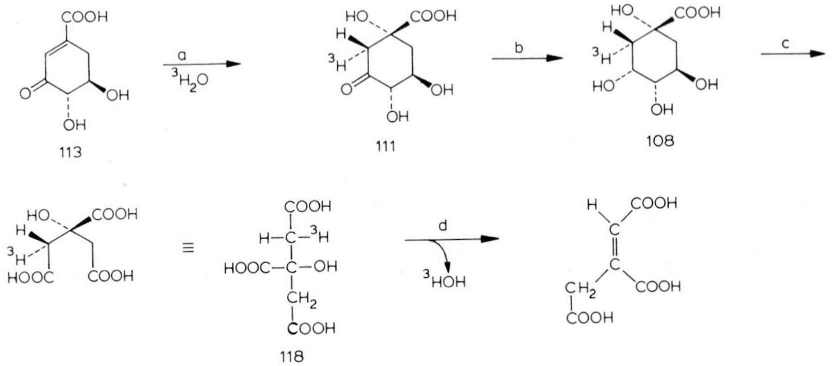

a = dehydroquinate dehydratase; b = quinate dehydrogenase; c = HIO_4, then Br_2/H_2O; d = aconitase

An independent confirmation of the citric acid configuration has been provided by work in Arigoni's laboratory [89]. It is based on a synthesis of (S)-[*sn*-5-^{14}C]citric acid, 125, from a derivative of chlorocitramalic acid, 124, and a correlation of the latter compound with the chiral center of citramalic acid, 126. When the synthetic citric acid, 125, was converted to 2-ketoglutarate, 127, by the tricarboxylic acid cycle enzymes, all of the radioactivity was located at C-5 of the latter compound (as shown by the usual chemical degradation). It follows from this result, that the aconitase active methylene is that in the *pro-R* CH_2COOH group. Since this result is different from that when [4-^{14}C]oxaloacetate is used with the tricarboxylic acid cycle enzymes, it also follows that oxaloacetate provides the *pro-R* CH_2COOH group. An interesting corollary to this work is that it provides an independent confirmation of the correctness of the stereochemistry assigned to fumarase.

In all of the foregoing, it has been assumed that the citrate synthase is that usually encountered, and typified by the pig heart enzyme. It has been found that in a small number of anaerobic bacteria (e.g., *Clostridium kluyveri*) the citrate synthase behaves in a different fashion—the acetyl group contributes the *pro-R* CH_2COOH group. When it is necessary to distinguish these two enzyme types, the usual mammalian enzyme is termed (*Si*)-citrate synthase since the addition is to the *Si* face of the oxaloacetate carbonyl.

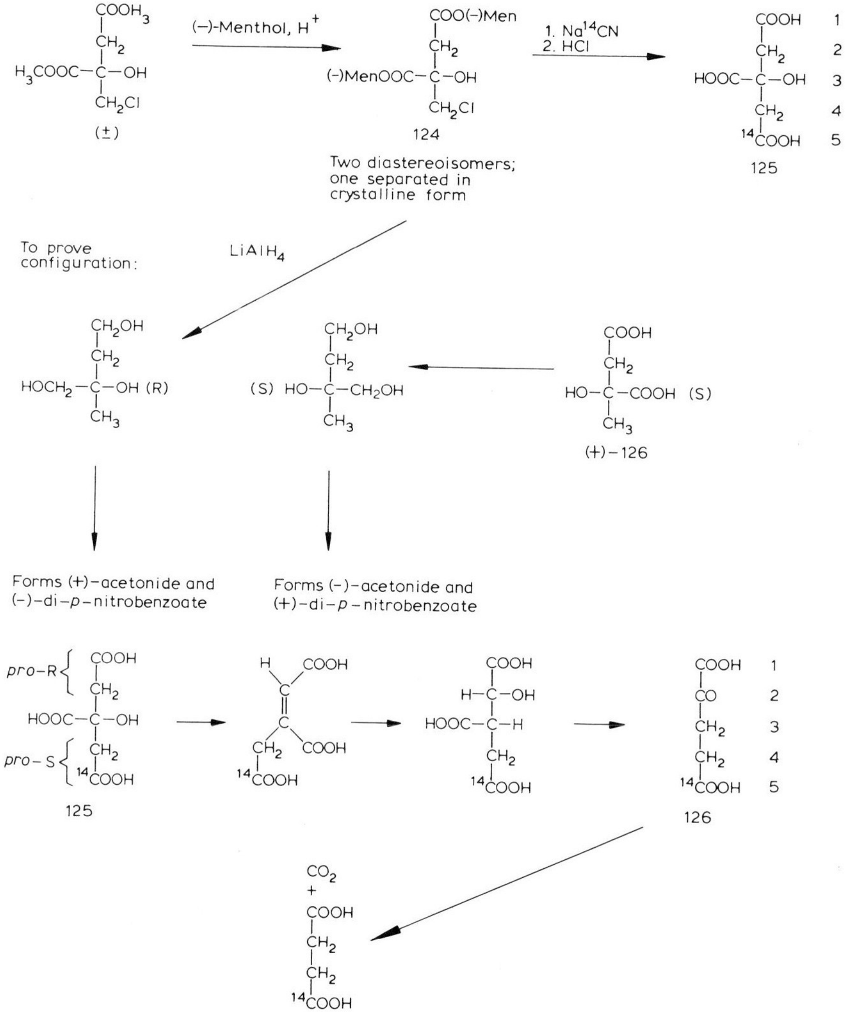

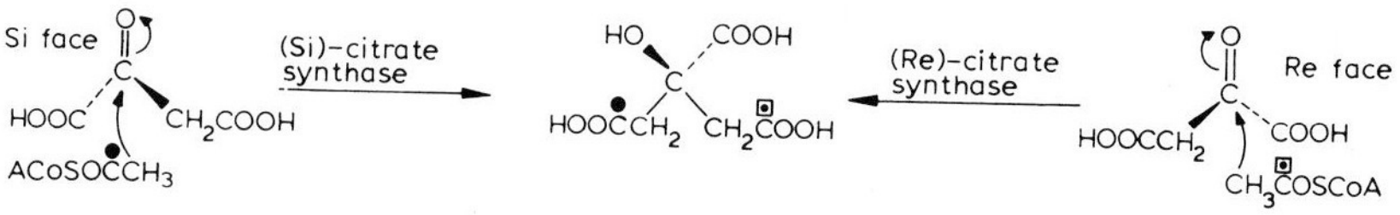

5. The study of chiral methyl groups

Chemical methods have been very important in investigations of 'chiral methyl groups'. In 1962, Levy et al. [120] pointed out that "the carbon atom C(R,H,D,T), where H, D and T represent hydrogen and its isotopes and R is any dissimilar group, is truly asymmetric". They noted further that although two enantiomeric forms of such a molecule must exist, no satisfactory methods were then available for their preparation. Satisfactory methods for the preparation of chiral acetate samples were developed independently in 1969 by two groups of investigators. Cornforth and his group [121] used a purely chemical approach, whereas Arigoni and his colleagues [122] used two enzymes of opposed stereoselectivity and additional chemical manipulations.

Cornforth et al. [121] first prepared *cis*-[2-^{2}H]vinylbenzene, 129, by diimide reduction of [2-^{2}H]phenylacetylene, 128; the assigned structure of the vinylbenzene and its isotopic homogeneity were carefully checked by NMR. Reaction of the olefin with *m*-chloroperoxybenzoic acid formed an epoxide, 130 and 131; the process is known to be a *syn* addition and leads to a racemic mixture. The racemate was reduced with LiB^3H_4 (process known to involve inversion of configuration as discussed previously) to yield a racemic mixture (132 and 133) of phenylethanols. A key step was then the resolution of this racemate by means of the brucine phthalate derivatives. The resolution of the conventional chiral center (C-1) also served to resolve the carbon carrying H, ^{2}H and ^{3}H (C-2). The resolved alcohols were converted to acetic acid by the reactions shown. It can be seen that the configurations assigned to the acetate samples, 134 and 135, depend on the following stereochemical elements: (1) a *syn* addition to a double bond of defined *cis* structure; and (2) a ring opening of an epoxide in an *anti* sense. In a modification of this synthesis, Cornforth and his colleagues started with *trans*-2-bromostyrene; the bromine atom was replaced with ^{3}H by treatment with butyllithium in ^3H_2O [123]. As in the first synthesis, *m*-chloroperoxybenzoic acid was used to prepare a racemic epoxide (similar to 130 and 131 but containing H and ^{3}H rather than H and ^{2}H). Reduction with $LiAl^2H_4$ gave the same racemic mixture of phenylethanols as was obtained in the first synthesis; the resolution and subsequent steps were carried out as before.

Arigoni and his colleagues [122] used the opposed specificities of lactate dehydrogenase and glyoxylate reductase to reduce [^{3}H]glyoxylic acid, 136, to either (*S*)- or (*R*)-[2-^{3}H]glycolate. As indicated for the synthesis using lactate dehydrogenase, chemical transformations were used to convert the glycolate, 137, to acetate, 134, by way of ethylene glycol, 138. When the monobrosylate derivative of this compound was prepared, 50% of the molecules contained a $-CH_2OBrs$ group and the other 50%, a $-CH^3HOBrs$ group (139). The final acetic acid was hence obtained as a mixture of achiral $CH_2{}^2HCOOH$ and chiral CH^2H^3HCOOH. With lactate dehydrogenase as initial enzyme for reduction of glyoxylic acid, (*R*)-acetate was obtained. The (*S*)-acetate was prepared by similar reactions, but using glyoxylate reductase and NADH to reduce $O^3HCCOOH$ to (*R*)-$HOCH^3HCOOH$.

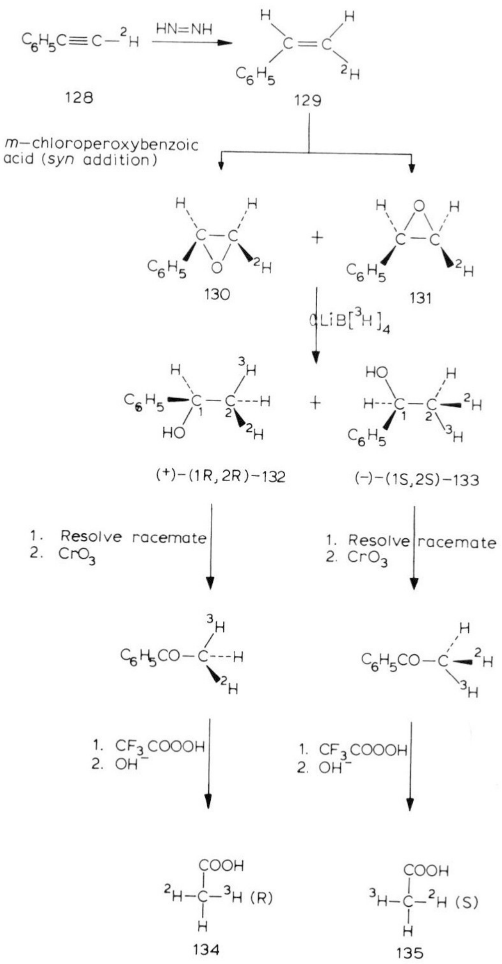

$C_6H_5C{\equiv}C{-}^2H$ $\xrightarrow{HN=NH}$ (H)(C₆H₅)C=C(H)(²H)

128 129

m-chloroperoxybenzoic acid (*syn* addition)

130 + 131

$LiB[^3H]_4$

(+)-(1R,2R)-132 + (−)-(1S,2S)-133

1. Resolve racemate
2. CrO_3

1. Resolve racemate
2. CrO_3

$C_6H_5CO{-}C(^3H)(^2H)H$ $C_6H_5CO{-}C(H)(^2H)(^3H)$

1. CF_3COOOH
2. OH^-

1. CF_3COOOH
2. OH^-

COOH, $^2H{-}C{-}^3H$ (R), H — 134

COOH, $^3H{-}C{-}^2H$ (S), H — 135

A later synthesis from Arigoni's laboratory uses a very ingenious, completely chemical process [124]. There are two crucial steps in which the geometries of two transition states in a pyrolytic reaction determine the overall outcome. In the first of these (which is an 'ene reaction') a triple bond, $-C{\equiv}C{-}^3H$ acquires H from one direction only, forming

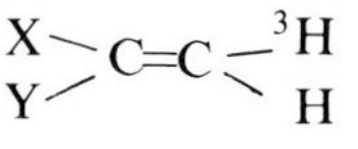

$$\begin{matrix}X\\Y\end{matrix}{>}C{=}C{<}\begin{matrix}^3H\\H\end{matrix} \quad \text{and not} \quad \begin{matrix}X\\Y\end{matrix}{>}C{=}C{<}\begin{matrix}H\\^3H\end{matrix}$$

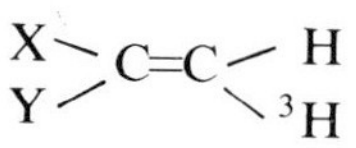

In the second crucial step, the double bond

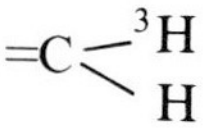

$$=C{<}\begin{matrix}^3H\\H\end{matrix}$$

acquires ^2H from a $>C^2H_2$ group present in the same molecule by pyrolysis. The C^2H_2 group is 'poised' either above or below the plane of the olefin bond depending on whether an alcohol is (*R*) or (*S*). As shown, a propargylic alcohol is first

*Inversion presumably takes place, but is immaterial

resolved, into its enantiomeric forms, 141 and 142. The alcohol, for example, 141, was converted to the propargylic ether, 143. Configuration at the chiral center in the propargylic alcohol was established by conversion of the (+) form, 142, to the (−)-diamide, 144, of known configuration [125] (diamide of 2-hydroxyglutaric acid). Pyrolysis of the propargylic ether proceeded with the geometry indicated by 143 → 145 → 146. Following the pyrolytic reaction the doubly-unsaturated bicyclic hydrocarbon, 146, was subjected to Kuhn–Roth oxidation (negligible exchange) leading to (*R*)-acetate. The (*S*)-acetate could be prepared analogously from the (+)-propargylic alcohol.

In another chemical synthesis, (2*R*)-[^2H]glycine, 147, has been converted to (*R*)-acetate in excellent ('optical') yield [126] (earlier methods were less efficient as noted by Floss and Tsai [127]). The conversion of bromoacetic acid, 148, to ethanol occurred with inversion of configuration. Other, enzymatic, syntheses of chiral acetates have been reported and are summarized by Floss and Tsai [127].

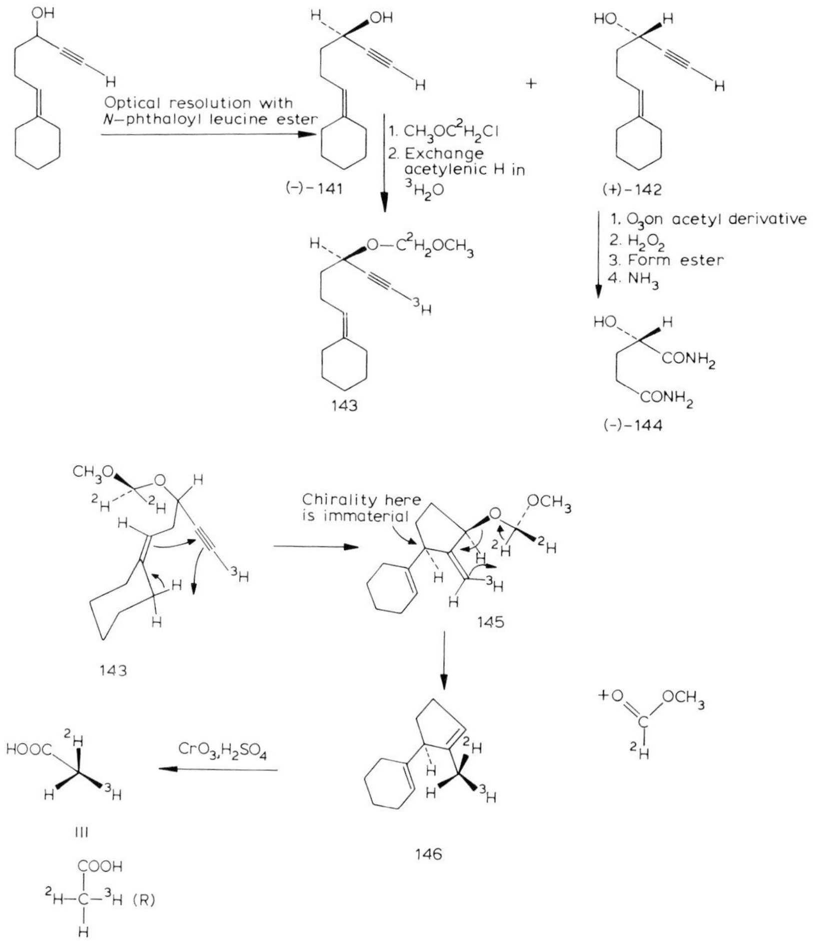

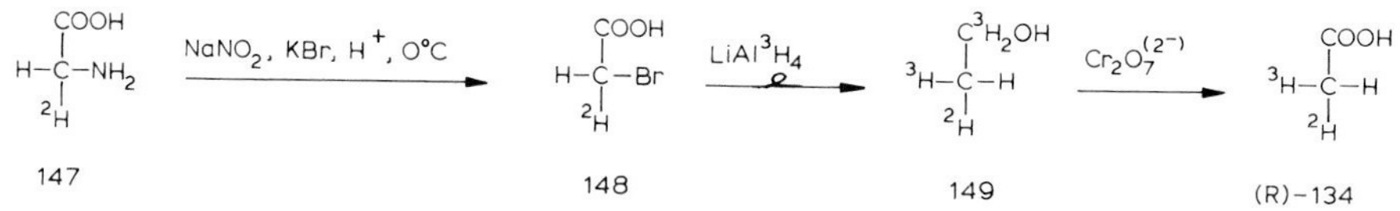

Optical resolution with *N*-phthaloyl leucine ester

(−)-141 + (+)-142

1. $CH_3OC^2H_2Cl$
2. Exchange acetylenic H in ^3H_2O

143

1. O_3 on acetyl derivative
2. H_2O_2
3. Form ester
4. NH_3

(−)-144

Chirality here is immaterial

143 → 145 → 146

CrO_3, H_2SO_4

(R)

147 $\xrightarrow{NaNO_2,\ KBr,\ H^+,\ 0°C}$ 148 $\xrightarrow{LiAl^3H_4}$ 149 $\xrightarrow{Cr_2O_7^{(2-)}}$ (R)-134

While the syntheses described here should produce chiral acetate samples, methods are needed to prove the existence of chirality and to determine its extent. The first method was devised independently by the Cornforth and Arigoni groups and is still widely used [122,128,129]. It depends on the use of two enzymes, malate synthase and fumarase. Thus, overall, the sample of acetic acid is first converted to its CoA derivative (either with acetate kinase and phosphotransacetylase or chemically with $ClCOOC_2H_5$ and CoASH) and then to malate and fumarate:

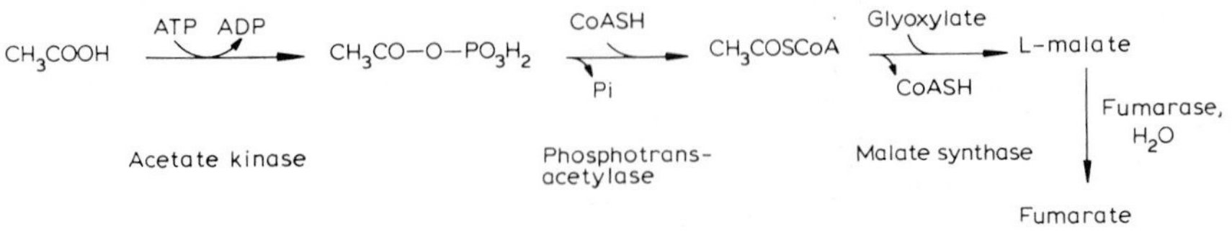

$CH_3COOH \xrightarrow[\text{Acetate kinase}]{\text{ATP ADP}} CH_3CO{-}O{-}PO_3H_2 \xrightarrow[\text{Pi; Phosphotrans-acetylase}]{\text{CoASH}} CH_3COSCoA \xrightarrow[\text{CoASH; Malate synthase}]{\text{Glyoxylate}}$ L-malate $\xrightarrow{\text{Fumarase, } H_2O}$ Fumarate

The method relies on an isotope effect so that H is removed in preference to ^{2}H and ^{3}H in the malate synthase reaction. In other words, the hydrogen isotope which is eliminated (in formation of the methylene group of malate) is determined by an intramolecular isotope effect; such effects normally follow the sequence $k_H > k^2{}_H > k^3{}_H$. Whether ^{3}H is located in the *pro-R* or *pro-S* position of malate is then determined from the known selectivity of fumarase.

In the conversion of acetate (as its CoA derivative) to malate, the methyl group is converted to methylene. The incoming group (from glyoxylate) could be introduced with inversion of configuration or with retention of configuration. For the process with inversion of configuration, three molecular species could be derived depending on whether the eliminated atom was H, ^{2}H or ^{3}H:

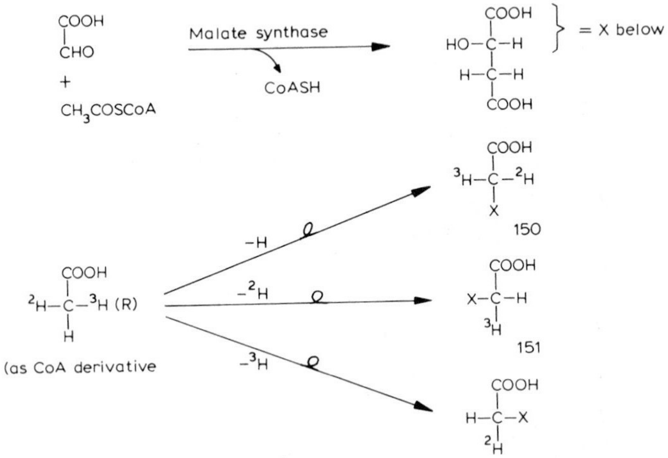

COOH–CHO + $CH_3COSCoA$ → (Malate synthase; CoASH) → COOH–C(HO)(H)–C(H)(H)–COOH (= X below)

^{2}H–C(^{3}H)(H)–COOH (R) (as CoA derivative):
- −H → COOH–C(^{3}H)(^{2}H)–X, 150
- −^{2}H → COOH–C(X)(H)(^{3}H), 151
- −^{3}H → COOH–C(H)(X)(^{2}H), 152

The species, 152, derived by elimination of ^{3}H is of no practical interest. Tritium is only present at tracer levels so there are large numbers of $CH_2{}^2HCOOH$ molecules also present; these will give rise to 152 by elimination of H. As for 150 and 151, if the malate synthase shows a normal isotope effect ($k_H > k^2{}_H$), 150 will predominate over 151.

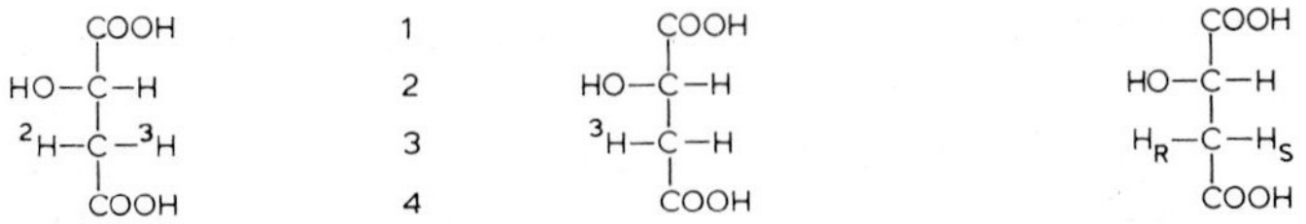

150		151	
COOH	1	COOH	COOH
HO–C–H	2	HO–C–H	HO–C–H
^{2}H–C–^{3}H	3	^{3}H–C–H	H_R–C–H_S
COOH	4	COOH	COOH

Thus, one expects more ^3H in the *pro*-3*S* position of malate. The amount of ^3H in the individual diastereotopic protons at the C-3 position can be determined using the known *anti* elimination catalyzed by fumarase.

After equilibration with fumarase, the mixture of fumarate and malate is analyzed for ^3H. If tritium is originally present in the *pro*-3*S* position, the equilibration will not remove it (150 plus 152). On the other hand, tritium in the *pro*-3*R* position will be lost to the water by way of 153. To facilitate the analysis, [^{14}C]acetate is added initially and the ^3H:^{14}C ratio is determined on the malate produced by malate synthase, before and after incubation with fumarase. The tritium content of the incubation water is also determined. The % retention of ^3H in the fumarase reaction is given the symbol, *F* [127]. In detailed analyses, it has been shown that chirally pure (*R*)-acetate is characterized by $F = 79$ and (*S*)-acetate by $F = 21$ [129]. The *F* value actually depends on four factors:

(a) the chiral purity of the acetate sample used;
(b) the chirality of the methyl group;
(c) the nature and magnitude of the isotope effect in the malate synthase reaction;
(d) the steric course of the malate synthase reaction.

Malate synthase does not incorporate hydrogen from water into malate so there is no enolization of acetyl-CoA; furthermore, for this enzyme, a normal isotope effect has been determined: $k_H > k_{^2H} = 3.8 \pm 0.1$ [129].

By determining the *F* factor it is possible to show whether a given sample of acetate is (*R*) or (*S*). This determination is unequivocal (given the validity of the syntheses just described). Hence for reactions in which a methyl group is formed (and assuming it can be converted to acetate) the stereochemical course can be determined: $Y{-}CH_2{-}X \rightarrow CH_3{-}X \rightarrow CH_3COOH$. For the conversion of chiral methyl to methylene, $CH_3{-}X \rightarrow Y{-}CH_2{-}X$, an assumption is necessary regarding an isotope effect before stereochemistry can be assigned. For example, in the malate synthase step used for assay of chiral methyl groups, it can be said that the reaction proceeds with inversion of configuration if a normal isotope effect operates. A listing of more than 50 enzymes which have been studied from the point of view of chiral methyl groups is available [127].

Rather than reviewing a number of enzymatic reactions, it is more appropriate in this chapter to focus on some other chemical methods which have been used in the study of chiral methyl groups.

Many of the methods that have been so far described in this chapter have used the defined stereochemical course of a chemical reaction to achieve a desired end—for example, the *anti* opening of epoxide rings or the *syn* or *anti* addition of a reagent to a double bond (e.g., in the Cornforth synthesis of chiral acetate, *syn* epoxidation of the *cis* double bond of vinylbenzene gave a racemic mixture of enantiomers). It is possible, however, to carry out a chemical synthesis and obtain a marked selection for one enantiomeric form. Such processes are termed 'asymmetric organic reactions', and were discussed briefly in connection with the nomenclature of stereoselective processes. Consider, for instance, one of the classical cases: An achiral keto acid, with a prochiral carbonyl, is reacted with one enantiomer of a chiral alcohol to form an ester; the ester is subjected to a Grignard reaction at the keto group, and the so formed hydroxy acid ester is hydrolyzed to the hydroxy acid. In this acid, one enantiomer will usually predominate. In the example shown here, the chiral influence of the (−)-menthol, 154, component (which molecule was ultimately removed) had led to an overall asymmetry in the synthesis of a chiral center, and the (−) form of atrolactic acid, 157, is synthesized preferentially. The reason, of course, is that of the two diastereoisomeric intermediates, 155 and 156, 156 has been formed in a larger amount. The situation arises from the frequency of actual conformations in the ketoester and from differences in steric hindrance when the Grignard reagent approaches the carbonyl. A detailed description is beyond the scope of this article—the original papers of Prelog [130,131] should be consulted, or the review [132] (in English).

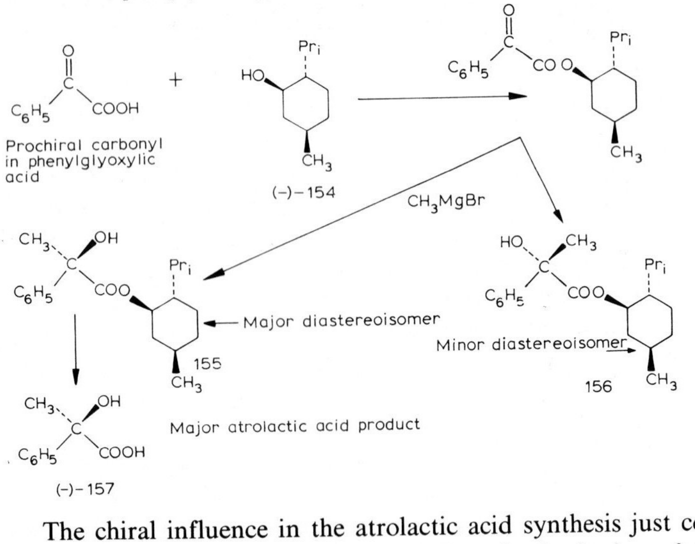

Prochiral carbonyl in phenylglyoxylic acid

(−)-154

CH_3MgBr

Major diastereoisomer

155

Minor diastereoisomer

156

Major atrolactic acid product

(−)-157

The chiral influence in the atrolactic acid synthesis just considered is a reagent which actually forms a chemical bond at the beginning of the process; later this bond is broken. The same overall result can be obtained by use of a physical chiral influence (e.g., circularly polarized light) or a catalyst. Biochemists, of course, are very familiar with the latter situation since enzymes are chiral catalysts—many enzymatic reactions produce a pure, single enantiomer from a prochiral substrate. However, simpler catalysts showing a high degree of asymmetric synthesis are available and can be put to use by chemists.

This technique played a big role in the synthesis of all possible isomeric forms of lactic acid in which the methyl contains all three hydrogen isotopes and the hydrogen at C-2 can be either H, ^{2}H or ^{3}H. Each of the three structures, $CH_3CHOHCOOH$, $CH_3C^2HOHCOOH$ and $CH_3C^3HOHCOOH$, contains two chiral centers when the methyl is CH^2H^3H. Each structure thus exists in four forms, for a total of 12 isomers. They can all be prepared in substantial amounts and with high optical purity [133].

The asymmetric synthesis step is actually a case of homogeneous asymmetric hydrogenation of a C=C bond with a chiral catalyst; it represents an example of enantio-differentiation. Two rhodium(I) catalysts incorporating as ligand either '(*S*,*S*)-chiraphos' {=(*S*,*S*)-(2,3)-bis[diphenylphosphino]butane}, 158, or '(*R*)-prophos' {=(*R*)-1,2-bis[diphenylphosphino]propane}, 159, were used.

CH_3
H—C_2—$P(C_6H_5)_2$ (S)
$(C_6H_5)_2P$—C_3—H (S)
CH_3

'(S,S)-chiraphos,' 158

CH_3
$(C_6H_5)_2P$—C—H (R)
H—C—H
$P(C_6H_5)_2$

'(R)-prophos,' 159

It was earlier shown that ethyl 2-acetoxyacrylate was reduced, in the presence of the (*S*,*S*)-chiraphos catalyst to (*R*)-ethyl *O*-acetyllactate in 84% optical yield, the (*R*)-prophos catalyst similarly gave the (*S*)-enantiomer.

Consider now the application of this hydrogenation method to 2-[3-^{3}H]ethyl 2-acetoxyacrylate, 160, with the introduction of two ^{2}H atoms by the [Rh(*R*)-prophos]$^+$ catalyst. Since there is considerable evidence that these catalytic hydrogenations proceed in a *syn* fashion, there is a direct relationship between the newly developed configurations at both C-2 and the C-3 methyl group in 161. Moreover, if the absolute configuration at C-2 is known (and this follows from measurement of optical rotation) the absolute configuration of the chiral methyl group also follows.

H, ^{3}H (Z) C=C $COOC_2H_5$, $OCOCH_3$ — 160

$^2H_2$, [Rh(R)prophos]$^+$ →

H, ^{3}H, ^{2}H —C—C— $COOC_2H_5$, $OCOCH_3$, ^{2}H

↓

$COOC_2H_5$ 1
CH_3COO—C—^{2}H 2
^{3}H—C—^{2}H 3
H

161

1. NaOH
2. HCl ←

COOH
HO—C—H (S)
^{3}H—C—^{2}H (S)
H

162

The use of these asymmetric hydrogenation catalysts gives the C-2 chiral center in about 80% optical purity. The same value would apply also to the chiral methyl. For further purification, a crystallization process was used. The optically impure lactic acid (an oil) was dissolved in an approximately equal volume of boiling diethylether: diisopropyl ether, 1 : 1; on standing at 5°C large, colorless, crystals of optically pure chiral methyl chiral lactic acid, 162, were deposited. The recovery of the purified material was 60%. Because of the inherent relationship between the two chiral centers, optical purity at C-2 guarantees optical purity at C-3.

Methods were worked out so that the required 2-acetoxyacrylates can be prepared from the bromine derivatives; Table 6 shows the derivation of six of the possible isomers using the [Rh(*R*)-prophos]$^+$ catalyst.

The use of [Rh(*S*,*S*)-chiraphos]$^+$ for the catalytic hydrogenation step would, of course, lead to the formation of the 6-isomers with (*R*) configuration at C-2.

Another interesting development is the recent use of tritium NMR spectroscopy to study the removal of hydrogen from a chiral methyl group. In the degradation of L-valine, isobutyryl-CoA, 163, is an intermediate; this metabolite is then oxidized to metacrylyl-CoA, 164. In the oxidation, one hydrogen is removed from C-2, and one from the *pro-S* methyl group on the same carbon. To determine the stereochemistry

TABLE 6
The preparation of six isomers of chiral methyl chiral lactic acid using [Rh(*R*)-prophos]$^+$ catalyst

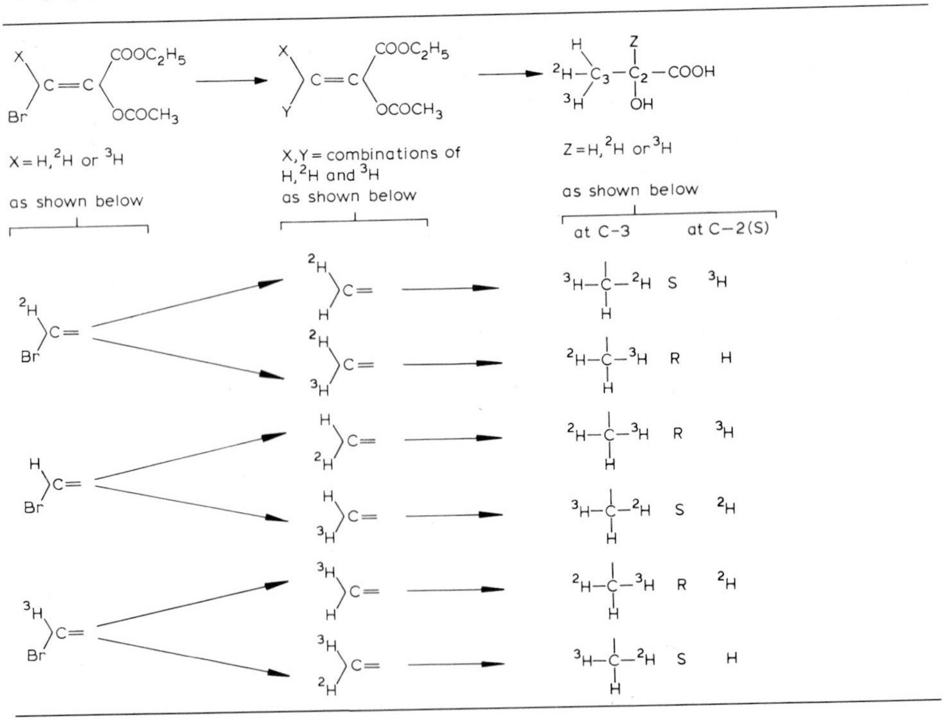

X(Br)C=C(COOC$_2$H$_5$)(OCOCH$_3$) → X(Y)C=C(COOC$_2$H$_5$)(OCOCH$_3$) → H(^{2}H)(^{3}H)C$_3$–C$_2$(Z)(OH)–COOH

X = H, ^{2}H or ^{3}H as shown below | X,Y = combinations of H, ^{2}H and ^{3}H as shown below | Z = H, ^{2}H or ^{3}H as shown below

X (with Br)	X, Y	at C-3		at C-2(S)
^{2}H	^{2}H, H	^{3}H–C(H)–^{2}H	S	^{3}H
^{2}H	^{2}H, ^{3}H	^{2}H–C(H)–^{3}H	R	H
H	H, ^{2}H	^{2}H–C(H)–^{3}H	R	^{3}H
H	H, ^{3}H	^{3}H–C(H)–^{2}H	S	^{2}H
^{3}H	^{3}H, H	^{2}H–C(H)–^{3}H	R	^{2}H
^{3}H	^{3}H, ^{2}H	^{3}H–C(H)–^{2}H	S	H

of the dehydrogenation, Aberhart and Tann prepared chiral methyl isobutyrates and subjected them to the action of *Pseudomonas putida*. This led to accumulation of β-hydroxyisobutyric acid produced by the *syn* addition of water to the *Re–Re* face of the double bond of metacrylyl-CoA:

L-Valine → (*pro*-S) CH_3–C(COSCoA)(H)–CH_3 (*pro*-R) → 164 →

163

164

Deacylate → β-Hydroxyisobutyric acid

Form derivative for analysis

165

Since the C-3 proton signals in the Eu(fod)$_3$ shifted ^{1}H NMR spectrum of the benzyl ester, 165, had been assigned earlier the location of ^{3}H (derived from the chiral methyl isobutyrate substrates) could be determined.

The chiral methyl isobutyrates were prepared by a *syn* specific catalytic hydrogenation of metacrylic acid derivatives, 166. The catalyst used was not chiral (as in the work on chiral methyl chiral lactic acid) so a diastereoisomeric mixture of 167 and 168 was obtained. Since they contain ^{3}H on the C-2 carbon atom, and assuming

166

$(C_6H_5P)_3RhCl$ + 5 Ci of carrier-free $^3H_2$

167 + 168

Diastereoisomer mixture

1. KOH, C_2H_5OH
2. *P. putida* to form methacrylyl CoA and hydroxyisobutyric acid
3. Derivatization

Derivative as 165, but with chiral methyl (see also 169 and 170) a normal isotope effect, $k_H > k_{^2H} > k_{^3H}$, the abstracted species in the formation of the metacrylyl-CoA will be H^3H. The predicted result for an *anti* elimination is shown below:

HOOC CH3 H 3H 167 2H pro-S 3H

HOOC H 2H 3H H 3H 168 H pro-S H

−H3H (*anti* elimination)

−H3H (*anti* elimination)

HOOC CH3 H 2H 3H OH

HOOC H 2H 3H H H H OH

+ H2O (*syn* addition)

+ H2O (*syn* addition)

HOOC CH3 2H pro-S 3H H pro-R OH

169

HOOC H 2H 3H H H H OH

170

In these structures, the CoA group has been omitted for the isobutyrates and metacrylates; similarly, the groups used to derivatize the hydroxyisobutyrates have been omitted.

Of the two diastereoisomers, 168 contains the chiral methyl group in the *pro-R* position and will lead to a $-CH_2OH$ group in the hydroxyisobutyrate, 170, which contains only ^1H. Hence, it remains to analyze 169, to determine whether ^3H is in the *pro-R* or *pro-S* position. In the ^1H-decoupled ^3H NMR spectrum the product (169 plus 170) obtained from 167 and 168, showed the following signals which had been assigned in the earlier work:

Chemical shift

δ	5.80	*pro*-3*S* proton
δ	5.96	*pro*-3*R* proton
δ	1.74	*pro*-2*R* chiral methyl

Of the relevant signals for the C-3 position, that at δ 5.80 was much larger than that at δ 5.96 indicating a predominant presence of ^3H in *pro-S* position. This is in accordance with the *anti* elimination mechanism illustrated as 167 → 169. When the chiral methyl isobutyrate was synthesized from the (*E*)-isomer of [3-$^2H_1$]metacrylic acid and submitted to the operations just described, the ^3H was present in higher amounts in the *pro*-3*R* position.

6. Epilogue

Although the title of this chapter is 'Chemical Methods for the Investigation of Stereochemical Problems in Biology', it has hopefully become clear that many methods require a combination of both chemical and biochemical techniques. Nevertheless, it will be apparent from later chapters that rigorous chemical methods have provided a sure foundation for the enormous developments which continue to take place.

References

1 Cornforth, J.W. (1970) Chem. Brit. 6, 431–436.
2 Mautner, H.G. (1967) Pharmacol. Rev. 19, 107–144.
3 Fischer, E. (1894) Berichte 27, 2985–2993.
4 Pasteur, L. (1860) Researches on the Molecular Asymmetry of Natural Organic Products, Alembic Club Reprint No. 14 (reissue edition 1948), Livingstone Ltd., Edinburgh, pp. 1–46.
5 Freudenberg, K. (1933) Stereochemie: Eine Zusammenfassung der Ergebnisse, Grundlagen und Probleme, Franz Deuticke, Leipzig.
6 Ogston, A.G. (1948) Nature (London) 162, 963.
7 Bentley, R. (1978) Nature (London) 276, 673–676.
8 Hanson, K.R. (1966) J. Am. Chem. Soc. 88, 2731–2742.
9 Various authors (1976) in J.B. Jones, C.J. Sih and D. Perlman (Eds.), Applications of Biochemical Systems in Organic Chemistry, Parts 1 and 2, Wiley-Interscience, New York.
10 Arigoni, D. and Eliel, E.L. (1969) in E.L. Eliel and N.L. Allinger (Eds.), Topics in Stereochemistry, Vol. 4, Wiley-Interscience, New York, pp. 127–243.
11 Bentley, R. (1969) Molecular Asymmetry in Biology, Vol. I, Academic Press, New York, pp. 190–221.
12 For references, see Andersen, K.K., Colonna, S. and Stirling, C.J.M. (1973) J. Chem. Soc. Chem. Commun. 645–646.
13 Bentley, R. (1970) Molecular Asymmetry in Biology, Vol. II, Academic Press, New York, pp. 345–526.
14 Jacques, J., Gros, C. and Bourcier, S. (1977) in H.B. Kagan (Ed.), Absolute Configurations of 6000 Selected Compounds with One Asymmetric Carbon Atom, Stereochemistry—Fundamentals and Methods, Vol. 4, Georg Thieme, Stuttgart.
15 Klyne, W. and Buckingham, J. (1978) Atlas of Stereochemistry, 2nd Edn., Vols. 1 and 2, Oxford University Press, New York.
16 Ref. 4, p. 45.
17 Frankland, P. (1897) Pasteur Memorial Lecture, J. Chem. Soc. Trans. 71, 683–743.
18 Le Bel, J.-A. (1891) Compt. Rend. 112, 724–726.
19 Pope, W.J. and Read, J. (1912) J. Chem. Soc. 101, 519–529.
20 Piutti, M.A. (1886) Compt. Rend. 103, 134–137.
21 Pasteur, L. (1886) Compt. Rend. 103, 138.

22 Stewart, A.W. (1919) Stereochemistry, 2nd Edn., Longmans Green and Co., London.
23 Cushny, A.R. (1926) Biological Relations of Optically Isomeric Substances, Williams and Wilkins, Baltimore.
24 Parascandola, J. (1975) in O.B. Ramsay (Ed.), van't Hoff-Le Bel Centennial American Chemical Society Symposium Series No. 12, American Chemical Society, Washington, D.C., pp. 143–158.
25 Easson, L. and Stedman, E. (1933) Biochem. J. 27, 1257–1266.
26 Ogston, A.G. (1979) Letter to R. Bentley, dated January 24.
27 Bergmann, M. and Fruton, J.S. (1937) J. Biol. Chem. 117, 189–202.
28 Bergmann, M., Zervas, L., Fruton, J.S., Schneider, F. and Schleich, H. (1935) J. Biol. Chem. 109, 325–346.
29 Bergmann, M., Zervas, L. and Fruton, J.S. (1936) J. Biol. Chem. 115, 593–611.
30 Wilen, S.H., Collet, A. and Jacques, J. (1977) Tetrahedron 33, 2725–2736.
31 Arnett, E.M. and Zingg, S.P. (1981) J. Am. Chem. Soc. 103, 1221–1222.
32 Walsh, C. (1979) Enzymatic Reaction Mechanisms, Freeman and Co., San Francisco, p. 41.
33 Hartsuck, J.A. and Lipscomb, W.N. (1971) in P.D. Boyer (Ed.), The Enzymes, 3rd Edn., Vol. III, Academic Press, New York, pp. 1–79.
34 Bentley, R. (1969) Molecular Asymmetry in Biology, Vol. I, Academic Press, New York, pp. 271–280.
35 Rosso, R.G. and Adams, E. (1967) J. Biol. Chem. 242, 5524–5534.
36 Dekker, E.E. (1977) in E.E. van Tamelen (Ed.), Bioorganic Chemistry, Vol. 1, Academic Press, New York, pp. 59–77.
37 Lambert, M.P. and Neuhaus, F.C. (1972) J. Bacteriol. 110, 978–987.
38 Kenyon, G.L. and Hegeman, G.D. (1979) in A. Meister (Ed.), Advances in Enzymology and Related Areas of Molecular Biology, Vol. 50, Wiley and Sons, New York, pp. 325–360.
39 Cornforth, J.W. (1976) Science 193, 121–125.
40 Barton, D.H.R. in interview with Farago, P. (1973) J. Chem. Educ. 50, 234–237.
41 Vennesland, B. (1974) Topics in Current Chemistry 48, 39–65.
42 Ogston, A.G. (1978) Nature (London) 276, 676.
43 Hirschmann, H. (1956) in S. Graff (Ed.), Essays in Biochemistry, Wiley, New York, pp. 156–180.
44 Mislow, K. (1962) in M. Florkin and E.H. Stotz (Eds.), Comprehensive Biochemistry, Vol. 1, Elsevier, Amsterdam, pp. 192–243.
45 Prelog, V. (1959) in G.E.W. Wolstenholme and C.M. O'Connor (Eds.), Steric Course of Microbiological Reductions, Ciba Foundation Study Group No. 2, Little, Brown and Co., Boston, pp. 79–90.
46 Westheimer, F.H. (1959) in G.E.W. Wolstenholme and C.M. O'Connor (Eds.), Steric Course of Microbiological Reductions, Ciba Foundation Study Group No. 2, Little, Brown and Co., Boston, pp. 91–92.
47 Ogston, A.G. (1958) Nature (London) 181, 1462.
48 Bentley, R. (1970) Molecular Asymmetry in Biology, Vol. II, Academic Press, New York, pp. 22–41.
49 Jones, J.B. and Beck, J.F. (1976) in J.B. Jones, C.J. Sih and D. Perlman (Eds.), Applications of Biochemical Systems in Organic Chemistry, Part I, Wiley-Interscience, New York, pp. 107–403.
50 Brändén, C.-I., Jörnvall, H., Eklund, H. and Furugren, B. (1975) in P.D. Boyer (Ed.), The Enzymes, Vol. XI, Part A, 3rd Edn., Academic Press, New York, pp. 103–190.
51 Dutler, H. and Brändén, C.-I. (1981) Bioorg. Chem. 10, 1–13.
52 Orchin, M., Kaplan, F., Macomber, R.S., Wilson, R.M. and Zimmer, M. (1980) The Vocabulary of Organic Chemistry, Wiley and Sons, New York.
53 Hernandez, O., Walker, M., Cox, R.H., Foureman, G.L., Smith, B.R. and Bend, J.R. (1980) Biochem. Biophys. Res. Commun. 96, 1494–1502.
54 Irwin, A.J. and Jones, J.B. (1977) J. Am. Chem. Soc. 99, 1625–1630.
55 Bentley, R. (1976) in J.B. Jones, C.J. Sih and D. Perlman (Eds.), Applications of Biochemical Systems in Organic Chemistry, Part 1, Wiley-Interscience, New York, pp. 403–477.
56 Brown, C.A. (1969) J. Am. Chem. Soc. 91, 5901–5902.
57 Kollonitsch, J., Marburg, S. and Perkins, L.M. (1970) J. Am. Chem. Soc. 98, 4489–4490.
58 Jenner, P. and Testa, B. (1973) Drug Metab. Rev. 2, 117–184.

59 Testa, B. and Jenner, P. (1978) in E.R. Garrett and J.L. Hirtz (Eds.), Drug Fate and Metabolism, Vol. 2, Marcel Dekker, New York, pp. 143–193.
60 Lehmann, F., P.A. (1976) Prog. Drug Res. 20, 101–142.
61 Lehman, F., P.A. (1978) in P. Cuatrecasas and M.F. Greaves (Eds.), Receptors and Recognition, Ser. A, Vol. 5, Chapman and Hall, London, pp. 1–77.
62 Izumi, U. and Tai, A. (1977) Stereo-differentiating Reactions, Academic Press, New York.
63 Morrison, J.D. and Mosher, H.S. (1976) Asymmetric Organic Reactions, Second Printing, American Chemical Society, Washington, D.C.
64 Birtwistle, J.S., Lee, K., Morrison, J.D., Sanderson, W.A. and Mosher, H.S. (1964) J. Org. Chem. 29, 37–40.
65 Testa, B. (1979) Principles of Organic Stereochemistry, Marcel Dekker, New York, pp. 156–215.
66 Fryzuk, M.D. and Bosnich, B. (1978) J. Am. Chem. Soc. 100, 5491–5494.
67 Schwartz, P. and Carter, H.E. (1954) Proc. Natl. Acad. Sci. U.S.A. 40, 499–508.
68 Bird, C.W. (1962) Tetrahedron 18, 1–5.
69 Wilen, S.H. (1971) in N.L. Allinger and E.L. Eliel (Eds.), Topics in Stereochemistry, Vol. 6, Wiley-Interscience, New York, pp. 107–176.
70 Harada, K. and Matsumoto, K. (1971) Bull. Chem. Soc. Japan 44, 1068–1071.
71 Fiaud, J.-C. and Kagan, H.B. (1970) Tetrahedron Lett. 1813–1816.
72 Okawara, T. and Harada, K. (1972) J. Org. Chem. 37, 3286–3289.
73 Acklin, W., Prelog, V. and Prieto, A.P. (1958) Helv. Chim. Acta 41, 1416–1424.
74 Prelog, V. (1962) Ind. Chim. Belge 27, 1309–1318.
75 Boyd, W.C. (1974) Science 186, 846.
76 Massey, V. (1953) Biochem. J. 55, 172–177.
77 Dixon, M., Webb, E.C., Thorne, C.J.R. and Tipton, K.F. (1979) Enzymes, 3rd Edn., Academic Press, New York, p. 236.
78 Fisher, H.F., Frieden, C., McKee, J.S.M. and Alberty, R.A. (1955) J. Am. Chem. Soc. 77, 4436.
79 IUPAC and IUB (1972) Enzyme Nomenclature: Recommendations of the Commission on Biochemical Nomenclature on the Nomenclature and Classification of Enzymes together with their Units and the Symbols of Enzyme Kinetics. Elsevier, Amsterdam.
80 Prelog, V. and Acklin, W. (1956) Helv. Chim. Acta 39, 748–757.
81 Acklin, W., Dütting, D. and Prelog, V. (1958) Helv. Chim. Acta 41, 1424–1427.
82 Various authors (1977) in Stereochemistry—Fundamentals and Methods, H.B. Kagan (Ed.), Vols. 1, 2 and 3, Georg Thieme, Stuttgart.
83 Tanaka, J., Katayama, C., Ogura, F., Tatemitsu, H. and Nakagawa, M. (1973) J. Chem. Soc. Chem. Commun., 21–22.
84 Mason, S.F. (1973) J. Chem. Soc. Chem. Commun., 239–241.
85 Hezemans, A.M.F. and Groenewege, M.P. (1973) Tetrahedron 29, 1223–1226.
86 Parry, R.J. (1978) in E.E. van Tamelen (Ed.), Bioorganic Chemistry, Vol. II, Academic Press, New York, pp. 247–272.
87 Johnson, C.K., Gabe, E.J., Taylor, M.R. and Rose, I.A. (1965) J. Am. Chem. Soc. 87, 1802–1804.
88 Rose, I.A. (1958) J. Am. Chem. Soc. 80, 5835–5836.
89 Weber, H. (1965) Dissertation No. 3591, Eidgenossische Technische Hochschule, Zurich.
90 Lemieux, R.U. and Howard, J. (1963) Can. J. Chem. 41, 308–316.
91 Weber, H., Seibl, J. and Arigoni, D. (1966) Helv. Chim. Acta 49, 741–748.
92 Zagalak, B., Frey, P.A., Karabatsos, G.L. and Abeles, R.H. (1966) J. Biol. Chem. 241, 3028–3035.
93 Prescott, D.J. and Rabinowitz, J.L. (1968) J. Biol. Chem. 243, 1551–1557.
94 Gawron, O. and Fondy, T.P. (1959) J. Am. Chem. Soc. 81, 6333–6334.
95 Trevoy, L.W. and Brown, W.G. (1949) J. Am. Chem. Soc. 71, 1675–1678.
96 Plattner, Pl.A., Heusser, H. and Feurer, M. (1949) Helv. Chim. Acta 32, 587–591.
97 Anet, F.A.L. (1960) J. Am. Chem. Soc. 82, 994.
98 Cornforth, J.W., Ryback, G., Popják, G., Donninger, C. and Schroepfer, G. (1962) Biochem. Biophys. Res. Commun. 9, 371–375.
99 Sprecher, M., Switzer, R.L. and Sprinson, D.B. (1966) J. Biol. Chem. 241, 864–867.

100 Fisher, H.F., Conn, E.E., Vennesland, B. and Westheimer, F.H. (1953) J. Biol. Chem. 202, 687–697.
101 You, K., Arnold, L.J., Allison, W.S., and Kaplan, N.O. (1978) Trends Biochem. Sci. 3, 265–268. See expanded data in TIBS Databank No. 2, IUB Publications, P.O. Box 016129, Miami, Florida 33101, U.S.A.
102 Nakamoto, T. and Vennesland, B. (1960) J. Biol. Chem. 235, 202–204.
103 Viola, R.E., Cook, P.F. and Cleland, W.W. (1979) Anal. Biochem. 96, 334–340.
104 Arnold, L.J., You, K., Allison, W.S. and Kaplan, N.O. (1976) Biochemistry 15, 4844–4845.
105 Ehmke, A., Flossdorf, J., Habicht, W., Schiebel, H.M. and Schulten, H.-R. (1980) Anal. Biochem. 101, 413–420.
106 Martius, C. and Schorre, G. (1950) Liebig's Annal. Chem. 570, 140–143.
107 Martius, C. and Schorre, G. (1950) Liebig's Annal. Chem. 570, 143–147.
108 Evans, E.A. and Slotin, L. (1941) J. Biol. Chem. 141, 439–450.
109 Wood, H.G., Werkman, C.H., Hemingway, A. and Nier, A.O. (1942) J. Biol. Chem. 142, 31–45.
110 Potter, Van R. and Heidelberger, C. (1949) Nature (London) 164, 180–181.
111 Lorber, V., Utter, M.F., Rudney, H. and Cook, M. (1950) J. Biol. Chem. 185, 689–699.
112 Bates, F.J. (1942) Polarimetry, Saccharimetry and the Sugars, U.S. Government Printing Office, Washington, D.C., p. 434.
113 Hanson, K.R. and Rose, I.A. (1963) Proc. Natl. Acad. Sci. U.S.A. 50, 981–988.
114 Fischer, H.O.L. and Dangschat, G. (1937) Helv. Chim. Acta 20, 705–716.
115 Bestmann, H.J. and Heid, H.A. (1971) Angew. Chem. 83, 329–331.
116 Florkin, M. (1979) in M. Florkin and E.H. Stotz (Eds.), Comprehensive Biochemistry, Elsevier, Amsterdam, 33 B, pp. 167–206.
117 Haslam, E. (1974) The Shikimate Pathway, Wiley and Sons, New York.
118 Englard, S. and Colowick, S.P. (1957) J. Biol. Chem. 226, 1047–1058.
119 Englard, S. (1960) J. Biol. Chem. 235, 1510–1516.
120 Levy, H.R., Talalay, P. and Vennesland, B. (1962) in P.B.D. de la Mare and W. Klyne (Eds.), Progress in Stereochemistry, Vol. 3, Butterworths, London, p. 301.
121 Cornforth, J.W., Redmond, J.W., Eggerer, H., Buckel, W. and Gutschow, C. (1969) Nature (London) 221, 1212–1213.
122 Lüthy, J., Rétey, J. and Arigoni, D. (1969) Nature (London) 221, 1213–1215.
123 Eggerer, H., Buckel, W., Lenz, H., Wunderwald, P., Gottschalk, G., Cornforth, J.W., Donninger, C., Mallaby, R. and Redmond, J.W. (1970) Nature (London) 226, 517–519.
124 Townsend, C.A, Scholl, T. and Arigoni, D. (1975) J. Chem. Soc. Chem. Commun., 921–922.
125 Lindstedt, S. and Lindstedt, G. (1963) J. Org. Chem. 28, 251–252.
126 Kajiwara, M., Lee, S.-F., Scott, A.I., Akhtar, M., Jones, C.R. and Jordan, P.M. (1978) J. Chem. Soc. Chem. Commun., 967–968.
127 Floss, H.G. and Tsai, M.-D. (1979) in A. Meister (Ed.), Advances in Enzymology and Related Areas of Molecular Biology, Vol. 50, Wiley and Sons, New York, pp. 243–302.
128 Cornforth, J.W., Redmond, J.W., Eggerer, H., Buckel, W. and Gutschow, C. (1970) Eur. J. Biochem. 14, 1–13.
129 Lenz, H. and Eggerer, H. (1976) Eur. J. Biochem. 65, 237–246.
130 Prelog, V. (1953) Helv. Chim. Acta 36, 308–319.
131 Prelog, V. and Meier, H.L. (1953) Helv. Chim. Acta 36, 320–325.
132 Bentley, R. (1969) Molecular Asymmetry in Biology, Vol. I, Academic Press, New York, pp. 100–120.
133 Fryzuk, M.D. and Bosnich, B. (1979) J. Am. Chem. Soc. 101, 3043–3049.
134 Aberhart, D.J. and Tann, C.-H. (1980) J. Am. Chem. Soc. 102, 6377–6380.
135 Aberhart, D.J. and Tann, C.-H. (1979) J. Chem. Soc. Perkin 1, 939–942.

Stereochemistry of dehydrogenases

JONATHAN JEFFERY

Department of Biochemistry, University of Aberdeen,
Marischal College, Aberdeen AB9 1AS, U.K.

1. The enzymes and what they do

(a) Introduction

Dehydrogenases (or oxidoreductases) constitute the first of six main divisions in the Enzyme Commission classification [1]. About 300 dehydrogenases that utilize nicotinamide coenzymes are known, and this chapter deals with some of them. The recommended name and EC number designate "not a single enzyme protein, but a group of proteins with the same catalytic property" [1].

Proteins with the same catalytic property do not always have closely similar structures. Alcohol dehydrogenases (EC 1.1.1.1) from yeast [2] and fruit-flies [3,4], for example, have strikingly different primary structures.

It is also the case that enzymes showing sequence similarities do not necessarily catalyse the same reactions. Sheep liver sorbitol dehydrogenase (EC 1.1.1.14) does not utilize ethanol, though in primary structure it resembles both yeast and horse liver alcohol dehydrogenases [5,6].

The EE and SS isozymes of horse liver alcohol dehydrogenase have very similar primary structures [7,8]. Ethanol is a substrate for both, but both have a wide specificity, and ethanol is not the best substrate for either. The SS isozyme has 3β-hydroxysteroid dehydrogenase activity [9,10], which the EE isozyme does not have, and which is thought to depend upon a single amino acid replacement [11]. It remains to be established whether these different isozymes have different roles in vivo.

What is meant by "the same catalytic activity"? Reduction of pyruvate to L-lactate (Fig. 1a, $X = CH_3$) is different from reduction of pyruvate to D-lactate (Fig. 1b, $X = CH_3$). The enzymes are designated lactate dehydrogenase (EC 1.1.1.27) and D-lactate dehydrogenase (EC 1.1.1.28). When $X = H$, the reactions shown in Figs. 1a and b still differ in exactly the same way (*Re* attack and *Si* attack at the carbonyl carbon), but both reactions lead to glycollate. Enzymes that convert the same starting material (glyoxylate) into the same product (glycollate) have "the same catalytic property" (in this case classified as EC 1.1.1.26).

Tamm (ed.) Stereochemistry

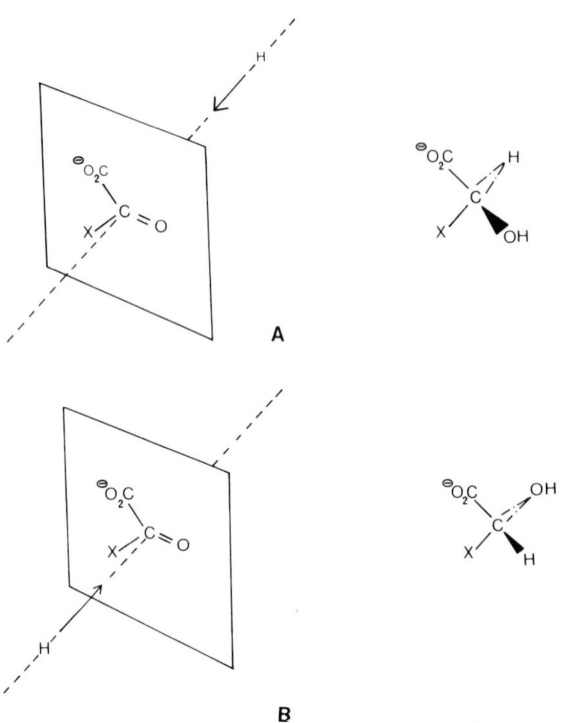

Fig. 1. Reduction of pyruvate to lactate. (a) *Re* attack giving L-lactate ((*S*)-lactate). X is CH_3. (b) *Si* attack giving D-lactate ((*R*)-lactate). X is CH_3. When X is H, *Re* attack (a) and *Si* attack (b) both give glycollate.

(b) General characteristics

(i) Flavin involvement

Some variety in mechanism is found among reactions involving flavin coenzymes. Usually, the flavin coenzyme is bound with high affinity, so that its reduction and re-oxidation occur in situ. Orotate reductase (EC 1.3.1.14), for example, from *Clostridium oroticum* is thought to catalyse hydride transfer from the 4-*pro-R* position of NADH to N-5 of the bound FAD (Fig. 2a) [12–14]. The hydrogen atom transferred therefore becomes the hydrogen of a secondary amine (Fig. 3a), and can undergo proton exchange with water. At the next step of the reaction (believed to be hydrogen transfer from N-5 of the FAD to C-5 or C-6 of bound orotate), the hydrogen atoms added to the orotate are not distinguished from those of the water (the hydrogen from the 4-*pro-R* position of the NADH being accounted for in the water).

In bacterial NADH:FMN oxidoreductase [15], the flavin coenzyme is bound relatively weakly. This enzyme serves to provide reduced FMN to luciferase in

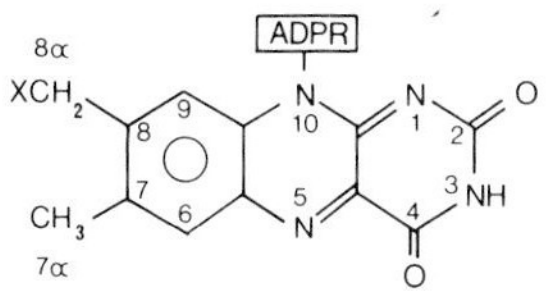

Fig. 2. Flavin adenine dinucleotide, oxidised form. (a) X is H in the ordinary free coenzyme. (b) In cases of covalent attachment to the enzyme, X is the imidazole N (1) or N (3) of a histidine residue, or the S of a cysteine residue.

photobacteria [16,17], but it is present also in non-luminescent bacteria (aerobic and anaerobic), such as *E. coli*, *Azotobacter vinelandii* and *Clostridium perfrigens* [18]. Studies with the enzyme from *Beneckea harveyi* showed hydrogen transfer to occur from the 4-*pro-R* position of NADH, but it was not certain that stereospecificity was complete [19].

At the active site of beef heart mitochondrial succinate dehydrogenase (EC 1.3.99.1), the FAD is covalently linked by C-8α to nitrogen of a histidine residue (Fig. 2b) [20]. In the catalytic reaction, removal of a proton from C-3 of the succinate may be followed by attack of the 3-carbanion on N-5 of the FAD to form an intermediate adduct, which breaks down with loss of a proton from C-2 of the succinate, giving fumarate and a reduced FAD moiety [21]. This mechanism is not certain, but it is established that the succinate loses two non-equivalent hydrogen atoms by a *trans* elimination (Fig. 4) [22]. In other enzymes, different types of covalent attachment of the FAD are known [23].

The flavodoxins, which resemble the larger flavoprotein dehydrogenases [24], are well-characterized flavoproteins [25] in which oxidation involves formation of a semiquinone radical (Fig. 3b) [26].

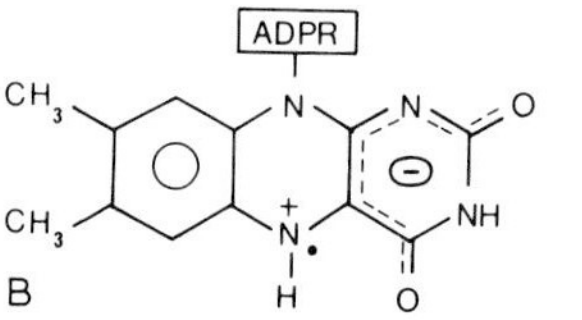

Fig. 3. Reduced forms of flavin adenine dinucleotide. (a) $FADH_2$. (b) Neutral semiquinone radical.

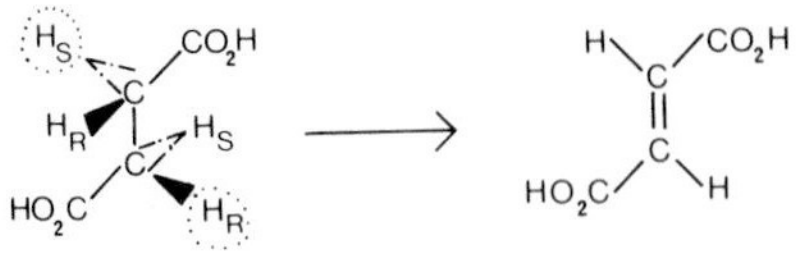

Fig. 4. In the succinate dehydrogenase reaction, two non-equivalent hydrogen atoms are removed by *trans* elimination.

(ii) Solely nicotinamide coenzymes

Reactions not involving flavin coenzymes show a greater uniformity of mechanism. The direct nature of the hydrogen transfer was first demonstrated for [1-$^2H_2$]ethanol and NAD^+, using yeast alcohol dehydrogenase [27]. Discovery that the transfer was stereospecific with respect to both substrate and coenzyme followed [28]. The position involved in the nicotinamide ring was shown to be C-4 [29], and some years later the absolute configuration was established [30,31]. Thus, the side of NAD^+ to which yeast alcohol dehydrogenase transferred hydrogen (and which had arbitrarily been called the A-side) was the *Re* side; in the NADH formed, the hydrogen transferred was 4-*pro-R*. (The system of stereochemical nomenclature is explained in [32] and [33].) This absolute assignment rests upon the belief that laevorotatory [2-$^2H_1$]succinic acid (Fig. 5) has the (2*R*) configuration [30,31,34,35]. The hydrogen transferred from the ethanol was shown to be 1-*pro-R* [36] on the basis that laevorotatory [1-$^2H_1$]ethanol is (1*S*) [37].

Both NAD-dependent and NADP-dependent dehydrogenases were found to be stereospecific. Correlation of the sides of NAD and NADP [38,39] established the absolute configuration of NADP.

It is now generally held that all dehydrogenases requiring nicotinamide coenzymes utilize only the 4-position of the nicotinamide ring, that the specificity for 4-*pro-R* or 4-*pro-S* is essentially complete, and that the mechanism is equivalent to hydride transfer.

(−) R-[2-$^2H_1$]succinic acid

4 R-[4-$^2H_1$]NADH

Fig. 5. The absolute configuration at C-4 of nicotinamide coenzymes followed from the (*R*) configuration of laevorotatory [2-$^2H_1$]succinic acid.

(c) Chemical comparisons

In non-enzymic reactions, hydride transfer between carbon atoms is unusual. Examples are the Cannizzaro reaction [40], and the Meerwein–Ponndorf–Verley–Oppenhauer reaction [41], in which a preferred steric course is discernible in some cases [42].

Reductions of carbonyl groups with lithium aluminium hydride or sodium borohydride occur by hydride transfer to carbon from aluminium or boron, respectively. The course of reaction is subject to 'steric approach control' and 'product development control' [43–45]. Enzymic reactions may or may not form the epimer favoured in the chemical reduction. This has been discussed elsewhere [46]. It is quite clear that the steric course of a dehydrogenase reaction is determined by the structure of the enzyme.

(d) Definitive descriptions of stereospecificity

The stereospecificity of a dehydrogenase can be represented by a characteristic diamond lattice section (Fig. 6) [47,48]. Horse liver alcohol dehydrogenase, for example, catalyses attack by the 4-*pro-R* hydrogen of NADH on the *Re* side of the acetaldehyde carbonyl [49]. In Fig. 6, both the upper right and lower right diagrams represent *Re* attack. Studies with substrates other than acetaldehyde indicate that the upper figure better represents the specificity of this enzyme [50] (though *trans*-1-decalones are not substrates in this case).

Because of the nature of the priority rules [32], substrates sometimes differ in formal chirality, but nevertheless have essentially the same handedness [51,52]. Such reactions are *homofacial* rather than *homochiral.* The problem is of nomenclature interpretation for the enzymologist, not of stereospecific recognition by the enzyme.

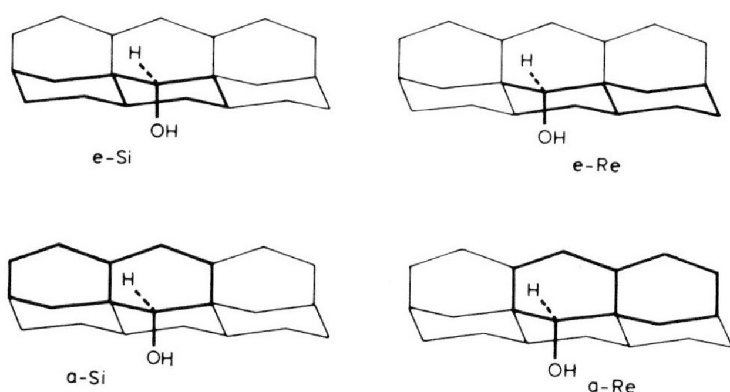

Fig. 6. Characteristic diamond lattice sections representing different stereospecificities of dehydrogenases [48].

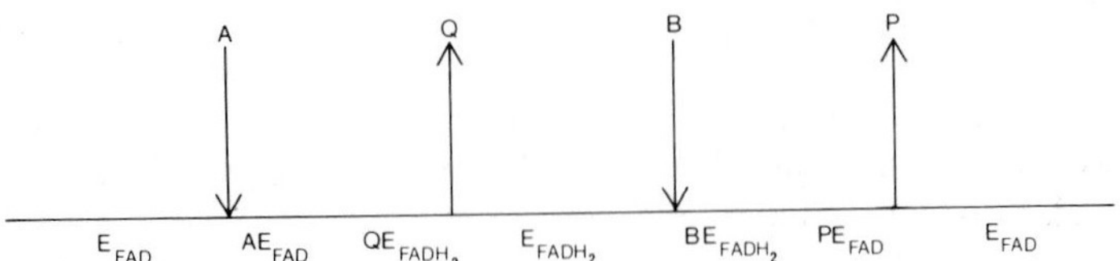

Fig. 7. The stereochemistry of the oxidation of A and the reduction of B is determined by the arrangement in the complexes A–E_{FAD} and B–E_{FADH_2}, respectively. The tightly-bound FAD undergoes reduction and reoxidation in situ.

(e) Dehydrogenase reaction mechanisms

The kinetic mechanism of dehydrogenases that contain a tightly bound flavin coenzyme, which undergoes reduction and re-oxidation in situ, can be represented by the simplified scheme shown in Fig. 7. The steric course for hydrogen removal from A is determined by the arrangement of bound A and bound FAD in the binary complex A–E_{FAD}, whereas the steric course of the addition of hydrogen to B is determined by the arrangement of bound B and bound $FADH_2$ in the binary complex B–E_{FADH_2}.

Numerous dehydrogenase reactions that proceed without the intermediacy of a flavin coenzyme (e.g., hydroxy–oxo interconversions) correspond to Fig. 8. This represents alternative routes to the productive ternary complex EAB/EPQ, and shows the formation of nonproductive ternary complexes EAP and EBQ. Certain of the alternative pathways often predominate, allowing schemes simpler than Fig. 8 to be used. In some cases, however, there is accompanying phosphorylation (e.g., glyceraldehyde-3-phosphate, EC 1.2.1.12) deamination (e.g., glutamate dehydrogenase, EC 1.4.1.2), decarboxylation (e.g., isocitrate dehydrogenase, EC 1.1.1.41) or hydration (e.g., octopine dehydrogenase, EC 1.5.1.11). Hydride transfer occurs in a central complex, such as the ternary complex EAB/EPQ in Fig. 8, and it is the arrangement of the bound coenzyme and the bound substrate in this complex that determines the steric course of the reaction.

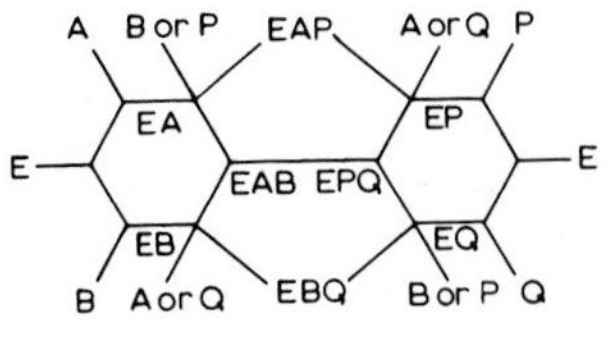

A B or P EAP A or Q P
EA EP
E EAB EPQ E
EB EQ
B A or Q EBQ B or P Q

Fig. 8. The stereochemistry of direct hydrogen transfer from NAD(P)H (A) to substrate (B), to give $NAD(P)^+$ (Q) and product (P) is determined by the arrangement in the central productive ternary complex, EAB/EPQ.

2. *How the stereospecificity arises*

(a) Reactions involving flavin coenzymes

(i) Glutathione reductase (EC 1.6.4.2)

G-S-S-G + NADPH + H^+ → 2 GSH + $NADP^+$

The structure of the enzyme from human erythrocytes has been solved to 3 Å resolution [53]. The chain trace of one subunit (Fig. 9) is represented schematically in Fig. 10. Both the FAD-domain and the NADP-domain (Fig. 10, hatched) contain a structural arrangement similar to that at the nucleotide binding sites in several enzymes [54,55], and often called a Rossmann fold. These two domains are evidently the result in this enzyme of a primordial gene duplication and gene splicing [56].

The native enzyme is dimeric (Fig. 11), and oxidised glutathione binds between the subunits. The suggested mechanism [53,57,58] involves transfer of the 4-*pro-S* hydrogen from NADPH to N-1 of the isoalloxazine, and formation of a transient adduct between C-4a of the isoalloxazine and Cys-46. Nucleophilic attack on the S-atom of glutathione I by Cys-41 (Fig. 11) releases glutathione II, and glutathione I is released by dissociation of the adduct, with nucleophilic attack by Cys-46 on Cys-41 [57]. Direct hydride transfer from NADPH to GSSG is not possible because of the location of the isoalloxazine. Lys-49 on one side, and His-450 on the other side, participate in the catalysis [53,57]. His-450 is from the other subunit. Specificity for the 4-*pro-S* hydrogen is determined by the residues that locate the nicotinamide relative to the isoalloxazine. Details are expected to be available soon (G. Schulz, personal communication).

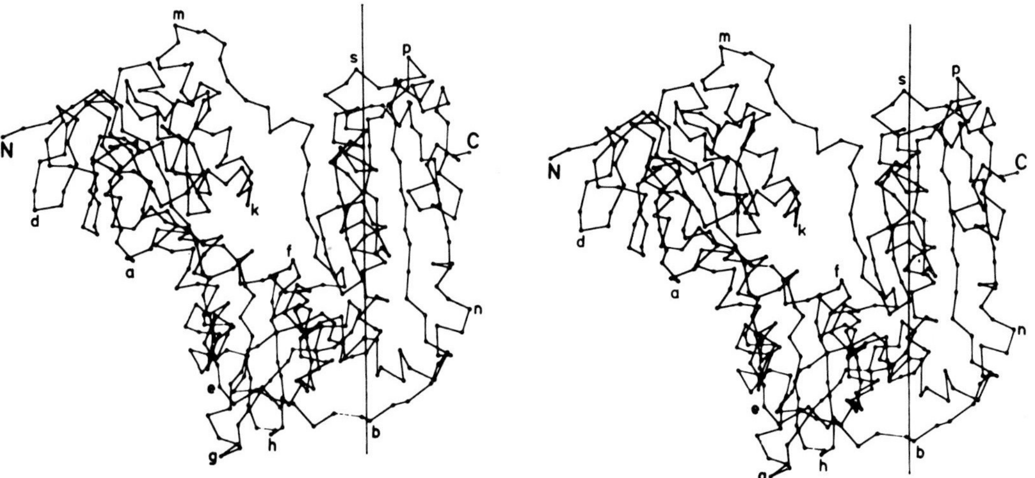

Fig. 9. Chain trace of one subunit of glutathione reductase. Stereo drawing from the work of Schulz and colleagues [53].

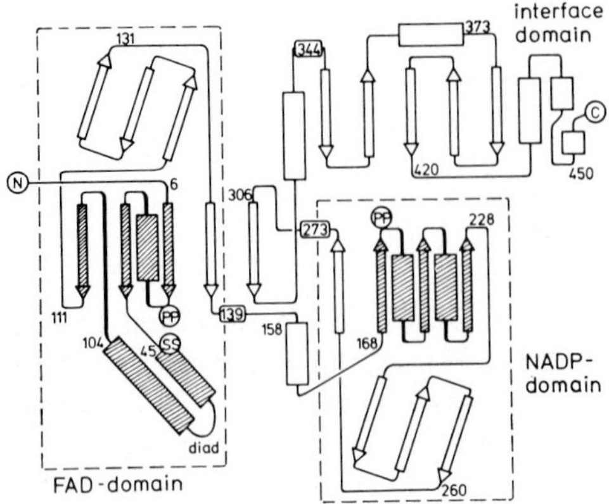

Fig. 10. Schematic representation of a glutathione reductase subunit, from the work of Schulz and colleagues [53]. Rectangles are α-helix, arrows are strands of β-sheet. Rossmann folds are hatched.

(ii) p-*Hydroxybenzoate hydroxylase (EC 1.14.13.2)*

$$H^+ + NADPH + O_2 + HO\text{-}\bigcirc\text{-}CO_2H \rightarrow NADP^+$$

$$+ HO\text{-}\underset{HO}{\bigcirc}\text{-}CO_2H + H_2O$$

This is not strictly a dehydrogenase reaction, but it involves oxidation of NADPH. The enzyme requires FAD and no other prosthetic group. A reaction mechanism has been proposed [59]. The enzyme from *Pseudomonas fluorescens* has been studied by X-ray crystallography [60]. From a chain trace of the subunit (Fig. 12), three domains are discerned (Fig. 13). The topology of the five strands of sheet A (A6, A4, A1, A2, A3) (Fig. 13) corresponds to the arrangement in lactate dehydrogenase [54,61], and the FAD interacts with a $\beta\alpha\beta$-unit (A1, H1, A2), much as in glutathione reductase [53]. Some sequence similarity to the dehydrogenases has been noted in this $\beta\alpha\beta$-unit [62]. Sequence comparison also with pig kidney D-amino acid oxidase (EC 1.4.3.3) [63] allowed the suggestion that an FAD-binding $\beta\alpha\beta$-unit may be located in the N-terminal region of that enzyme too [61].

A groove (Fig. 13) divides the subunit into two lobes. In a derivative made for crystallographic purposes using $NaAu(CN)_2$, the gold was in a pocket close to the isoalloxazine (Fig. 13). The nicotinamide ring of NADPH is thought to be located in this pocket [60]. The *p*-hydroxybenzoate binds (hydroxyl inwards, carboxyl towards

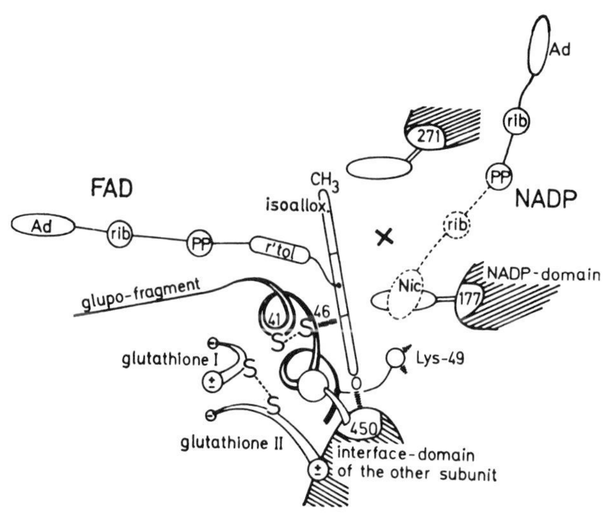

Fig. 11. Sketch of one active site of the dimeric enzyme glutathione reductase, from the work of Schulz and colleagues [53]. Oxidised glutathione binds between subunits. Hydrogen is transferred via the FAD.

the surface) deep inside the subunit in a region formed by the chain reversal between D3 and H5, by sheet B and by the isoalloxazine [60]. The NADPH site is accessible via the groove, and the substrate site via a cleft between domains I and II. Bound substrate and NADPH are thus separated by the isoalloxazine ring. Conformational changes associated with binding/dissociation probably involve movement of domain II relative to domain I (narrowing or widening the cleft). $NADP^+$ leaves before hydroxylation by O_2 occurs. Some details of the FAD binding are known [62]. Most probably the 4-*pro-R* hydrogen of the NADPH is transferred (presumably directly to the isoalloxazine), but the binding of NADPH is not yet known in sufficient detail to explain how this is determined.

(b) *Reactions with direct transfer of hydrogen between nicotinamide coenzyme and substrate*

(i) Dihydrofolate reductase (EC 1.5.1.3)

This is a small monomeric enzyme (molecular weight around 20000) that catalyses reduction of a double bond between carbon and nitrogen, converting 7,8-dihydrofolate into 5,6,7,8-tetrahydrofolate (Fig. 14). Hydride is transferred from the 4-*pro-R* position of NADPH to C-6, N-5 acquiring a proton [64–66].

Crystallographic studies at 2.5 Å resolution have been carried out on the enzyme from *E.coli* [67] (Fig. 15) and *Lactobacillus casei* [68,69] (Fig. 16). Residue numbers in what follows refer to the *L. casei* enzyme sequence [70]. The binding sites for the substrate and coenzyme are not located in separate domains, but are composed of overlapping portions involving mainly the N-terminal two-thirds of the polypeptide

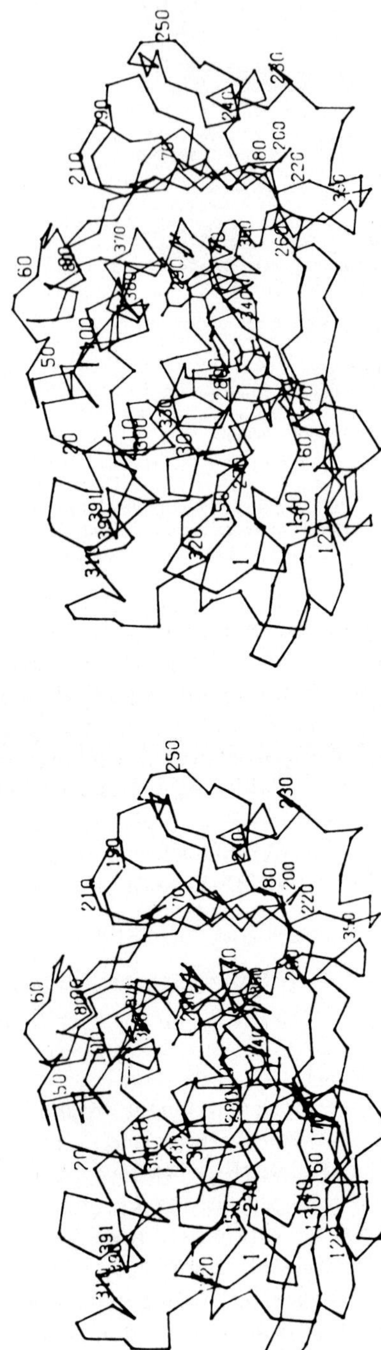

Fig. 12. Chain trace of p-hydroxybenzoate hydroxylase. Stereo drawing from the work of Drenth and colleagues [60].

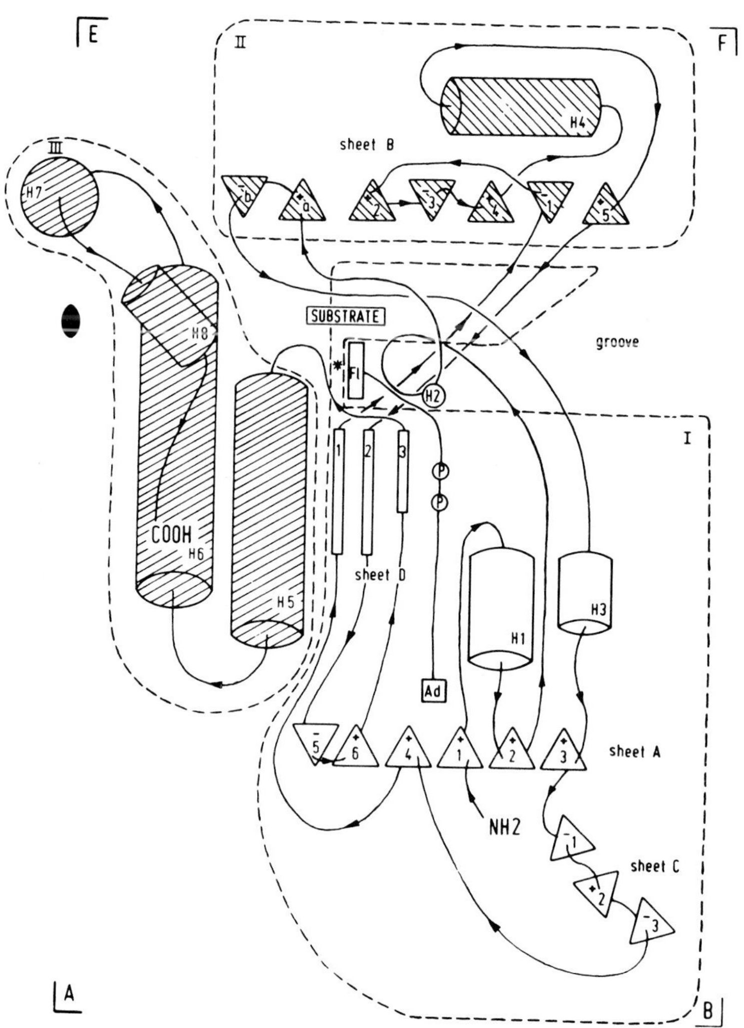

Fig. 13. Schematic representation of *p*-hydroxybenzoate hydroxylase. From the work of Drenth and colleagues [60].

Fig. 14. Dihydrofolic acid. In the dihydrofolate reductase reaction, the double bond between N-5 and C-6 is reduced by hydride transfer from the 4-*pro-R* position of NADPH to C-6, and addition of a proton at N-5.

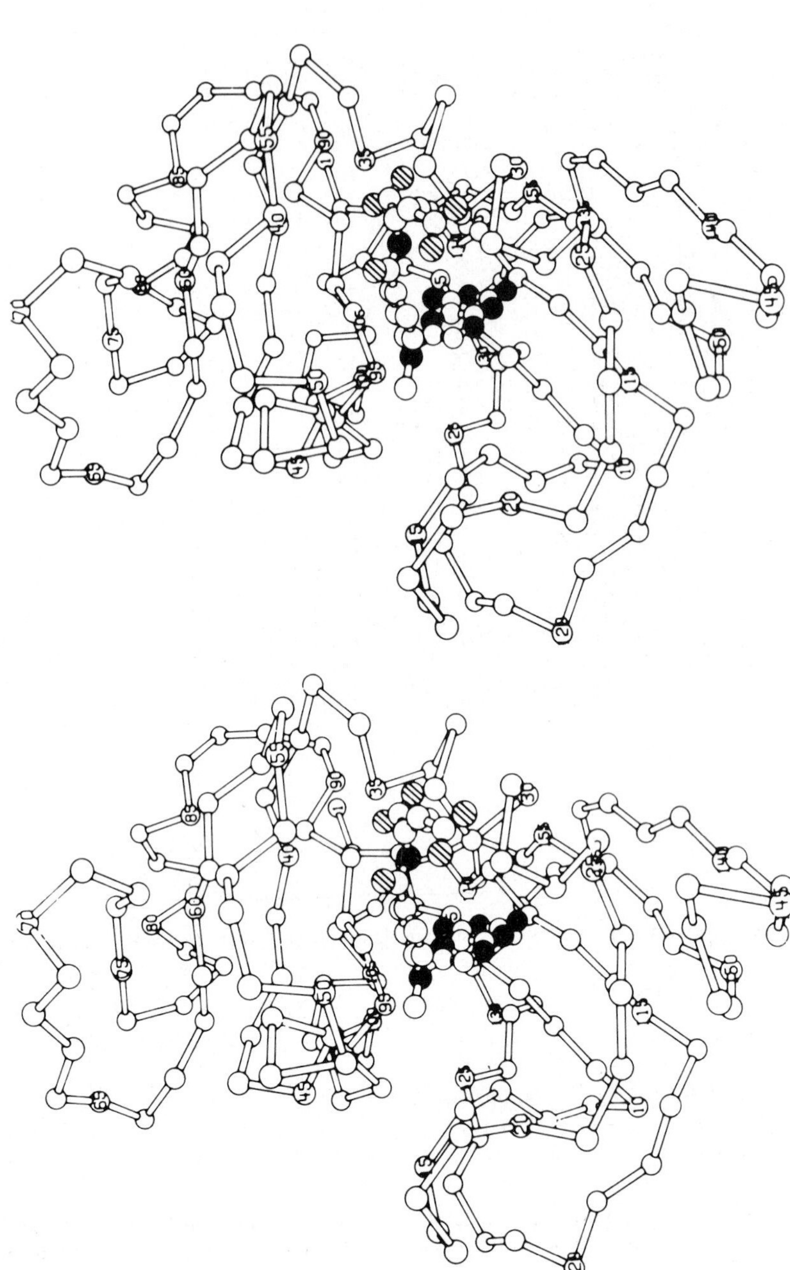

Fig. 15. Chain trace of *E.coli* dihydrofolate reductase. Bound inhibitor (methotrexate) is also shown, with nitrogen (black) and oxygen (shading) atoms indicated. Stereo drawing from the work of Kraut and colleagues [68].

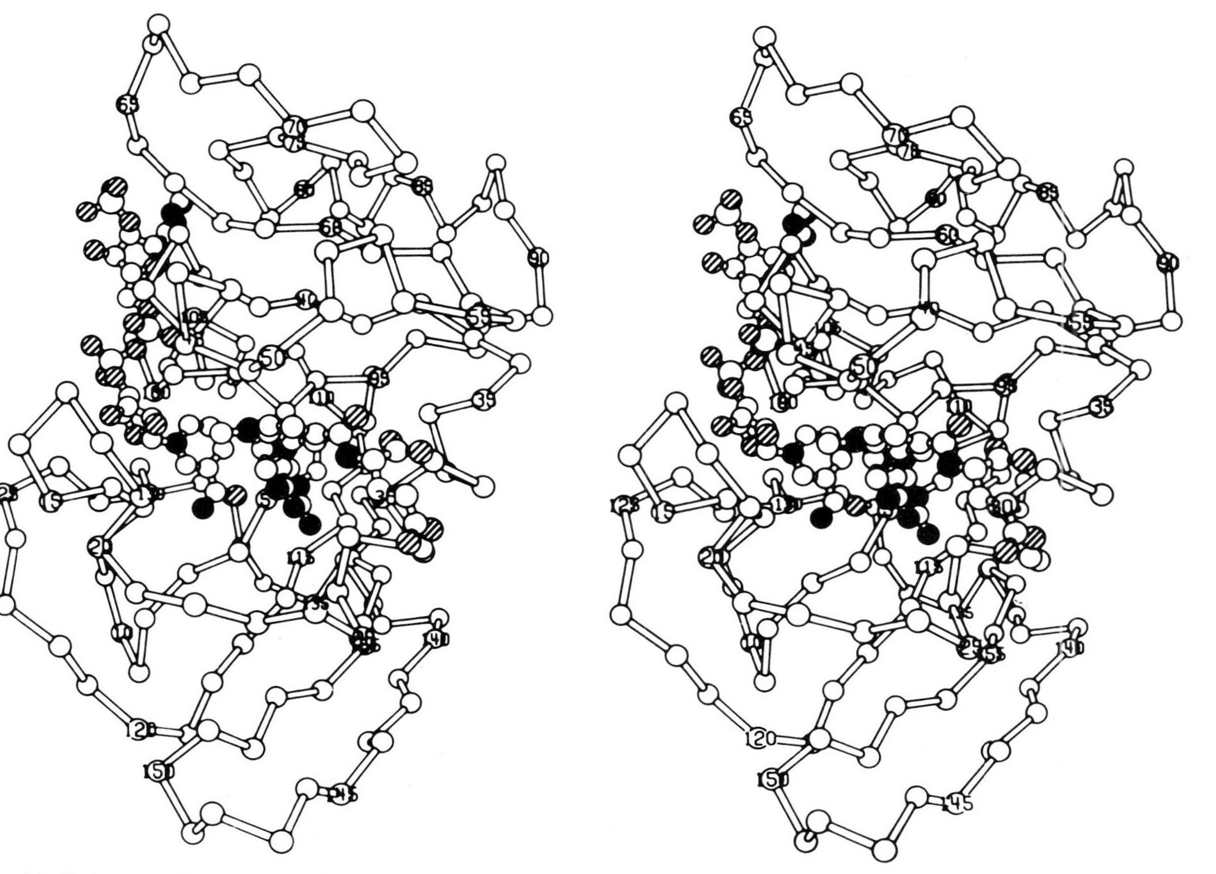

Fig. 16. Chain trace of *L. casei* dihydrofolate reductase. Bound NADPH and inhibitor (methotrexate) are also shown, with nitrogen (black) and oxygen (shading) atoms indicated. Stereo drawing from the work of Kraut and colleagues [68].

chain [68]. Despite the uniform finding that the coenzyme binds at the carboxyl edge of a region of parallel β-sheet flanked by a pair of α-helices, the dinucleotide binding domains of dihydrofolate reductase [69] and various NAD-dependent dehydrogenases [55] do not show overall structural homology in the view of Kraut and coworkers [69]. Others were impressed by the general similarity [71,72].

In dihydrofolate reductase, the adenine ring lies in a hydrophobic cleft, evidently without specific interactions, and the nicotinamide ring is bound in a cavity close to the pyrazine ring of the substrate. Hydrogen bonds between the amido hydrogen of Ala-7 and the amide oxygen of the nicotinamide, as well as those formed by the carbonyls of Ala-6 and Ile-13 with the nicotinamide NH_2 orient the nicotinamide ring so that the 4-*pro-R* hydrogen is positioned for hydride transfer to C-6 of the dihydrofolate [69]. Asp-26 is thought to provide the proton to N-5 [68]. Crystallographic studies were performed using an inhibitory substrate analogue (methotrexate), and it seems likely that the substrate (dihydrofolate), the position of which is known only by inference, binds with the pteridine ring oriented differently, namely, rotated by 180° about the C-6–C-9 bond, and 30° about the C-9–N-10 bond [68]. The tetrahydrofolate formed would then have the (S) configuration at C-6.

(ii) 6-Phosphogluconate dehydrogenase (EC 1.1.1.44)

$$\begin{array}{c} CO_2^- \\ \text{—}\!\!\!\!\!\!\mid\text{—OH} \\ \text{HO—}\!\!\!\mid\text{—} \\ \text{—}\!\!\!\!\!\!\mid\text{—OH} \\ \text{—}\!\!\!\!\!\!\mid\text{—OH} \\ CH_2OPO_3^{2-} \end{array} + NADP^+ \rightarrow \begin{array}{c} CH_2OH \\ \mid\!\!=O \\ \text{—}\!\!\!\!\!\!\mid\text{—OH} \\ \text{—}\!\!\!\!\!\!\mid\text{—OH} \\ CH_2OPO_3^{2-} \end{array} + CO_2 + NADPH + H^+$$

This is a dimeric enzyme (subunit molecular weight 47000). Crystallographic studies of the sheep liver enzyme [71,73–75] show the subunit (Fig. 17) to consist of a small domain (containing 4 helices and a small parallel sheet), and a large domain (containing the eleven remaining helices). The $NADP^+$ binds with the adenine at the C-terminal end of two sheet strands, the ribose 2′-phosphate close to the N-terminus of a helix, and the nicotinamide at the junction of the large and small domains (Fig. 17). Enzymes investigated from other sources transferred hydride from C-3 of the substrate to the 4-*pro-S* position of $NADP^+$ [76], and β-decarboxylation occurred with inversion to give ribulose 5-phosphate [77]. It is likely that this is the course of reaction also for the sheep liver enzyme. Present knowledge of coenzyme and substrate binding [75] does not reveal how this is achieved.

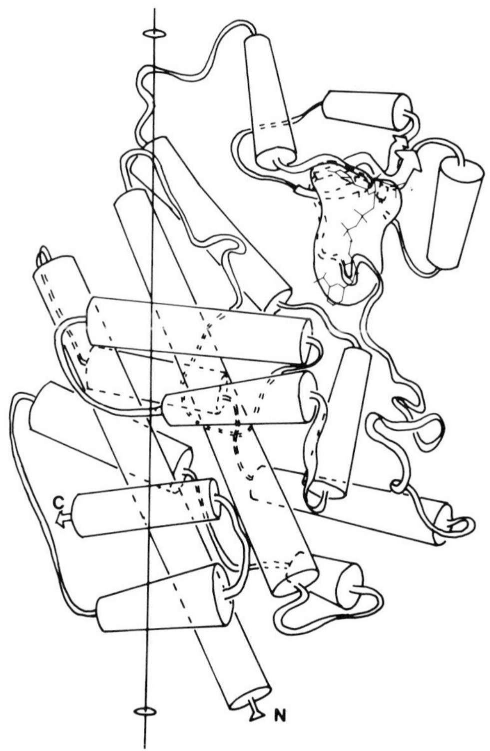

Fig. 17. Chain trace of sheep liver 6-phosphogluconate dehydrogenase subunit based on the work of Adams and colleagues [71]. The $NADP^+$ binding region is indicated.

(iii) Lactate dehydrogenase (EC 1.1.1.27)

$$\begin{array}{l} CH_3 \\ | \\ C{=}O + NADH + H^+ \rightleftharpoons \\ | \\ CO_2^- \end{array} \begin{array}{l} CH_3 \\ | \\ CHOH + NAD^+ \\ | \\ CO_2^- \end{array}$$

This tetrameric enzyme (subunit 36000) has been the subject of several crystallographic studies (summarized in [55]), and information on the dogfish M_4 isozyme has been obtained at high resolution [78]. The subunit structure [79] is illustrated in Fig. 18. On the basis of crystallographic findings, a numbering system for the amino acid chain was introduced [80], and has been widely used [55,79,81]. What were already known as Ser-163, Arg-171 and His-195 would simply become Ser-161, Arg-169 and His-193 in the consecutively numbered complete sequence [82]. However, in other parts of the chain (notably around residue 33, residue 187, and the regions 135–147 and 236–252) there were more extensive differences. For simplicity, the well-known older numbering system [80] is used again here.

The coenzyme is bound mainly as a result of hydrophobic interactions and hydrogen bonds. A lysine and an arginine residue make charge interactions with the pyrophosphate moiety [55]. Hydrogen bonds between the carboxamide group of the coenzyme, Ser-163, and the main-chain carbonyl of residue 139 anchor the nicotinamide ring [55]. The arrangement at the catalytic site is shown in Fig. 19A. Arg-171 forms a salt bridge with the substrate carboxyl, and His-195 forms a hydrogen bond with the oxygen atom bound to the reacting carbon of the substrate. Hydrogen transfer is therefore specific for L-lactate and the 4-*pro-R* position of the coenzyme.

(iv) Malate dehydrogenase (EC 1.1.1.37)

$$\begin{array}{l} CO_2^- \\ | \\ CHOH \\ | \\ CH_2 \\ | \\ CO_2^- \end{array} + NAD^+ \rightleftharpoons \begin{array}{l} CO_2^- \\ | \\ C=O \\ | \\ CH_2 \\ | \\ CO_2^- \end{array} + NADH + H^+$$

This dimeric enzyme (subunit 35000) catalyses a reaction similar to the lactate dehydrogenase reaction, and the subunit structures of the enzymes are strikingly similar [83–85] (Fig. 20). Crystallographic [85] and other [86] evidence suggests that the reaction mechanisms are similar. The 4-*pro-R* hydrogen of NADH is transferred to the *Re* side of the oxaloacetate to give L-malate [87].

(v) Glyceraldehyde-3-phosphate dehydrogenase (EC 1.2.1.12)

$$\begin{array}{l} CHO \\ | \\ CHOH \\ | \\ CH_2OPO_3^{2-} \end{array} + HPO_4^{2-} + NAD^+ \rightarrow \begin{array}{l} CO \cdot OPO_3^{2-} \\ | \\ CHOH \\ | \\ CH_2OPO_3^{2-} \end{array} + NADH + H^+$$

The reaction involves phosphorylation as well as hydride transfer. The enzyme is tetrameric (subunit 36000), and the subunit structure (Fig. 21) resembles that of lactate dehydrogenase (Fig. 18) and malate dehydrogenase (Fig. 20). Coenzyme binding (Fig. 22) is quite similar for all three enzymes, but in glyceraldehyde-3-phosphate dehydrogenase the nicotinamide ring is aligned so that hydrogen is transferred to the 4-*pro-S* position [88]. This results from a hydrogen bond between the nicotinamide carboxamide and Asn-313, and from the positions of Ala-120 and Pro-121, which would prevent the carboxamide from occupying the position neces-

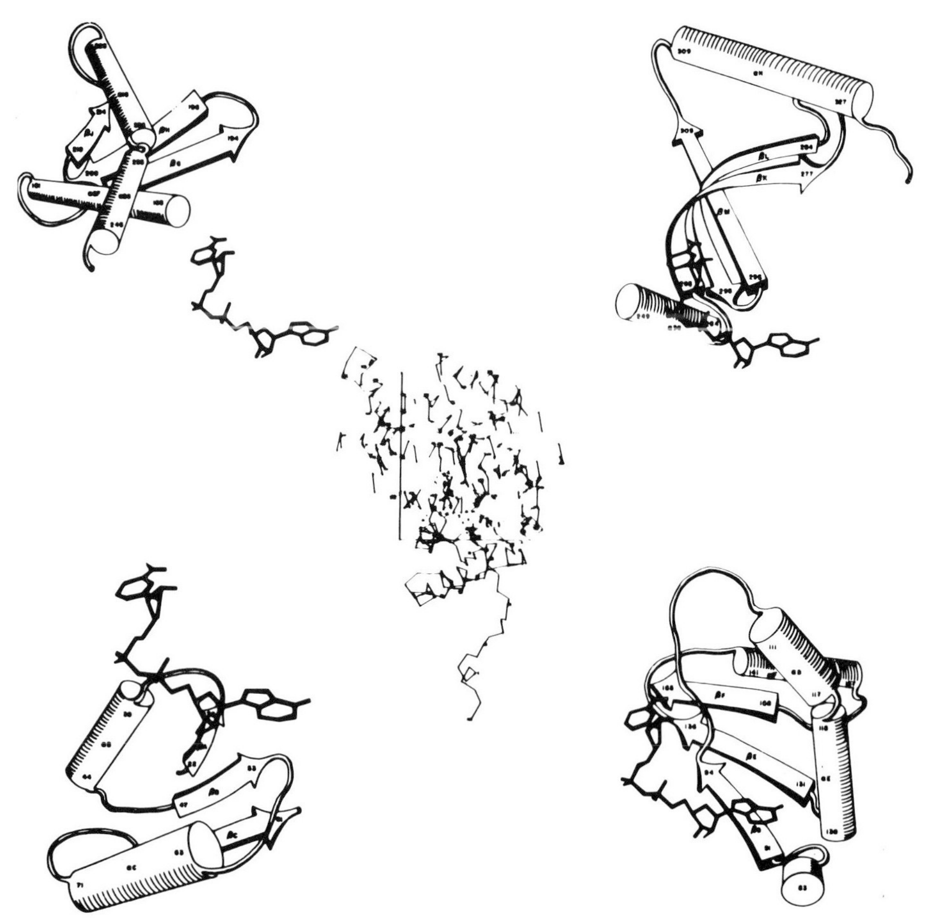

Fig. 18. Chain trace (centre) of a subunit of dogfish muscle lactate dehydrogenase. Diagrammatic representations (corners) show constituent parts of the structure, and their relationship to the bound coenzyme. From the work of Rossmann and colleagues [79].

sary for hydrogen transfer to the 4-*pro-R* position [89]. Asn-313, Ala-120 and Pro-121 are highly conserved in the enzyme from various sources such as lobster muscle [90], pig muscle [91] and *B. stearothermophilus* [92].

Reaction evidently involves formation of the *S*-thiohemiacetal with the super-reactive Cys-149. His-176 is thought to activate Cys-149 and to facilitate proton removal (the oxidation step being transfer of hydride to the coenzyme, with formation of the thioacyl intermediate). Thr-179 and residue 181 (e.g., Thr or Asn) probably interact with the 3-phosphate, while Ser-148 can possibly form hydrogen bonds with the C-2 hydroxyl and with the inorganic phosphate (Fig. 19B) [52]. Thr-208 or Arg-231 may also bind the inorganic phosphate. Nucleophilic attack by the phosphate on C-1, with rupture of the C–S bond, gives 1,3-diphosphoglycerate.

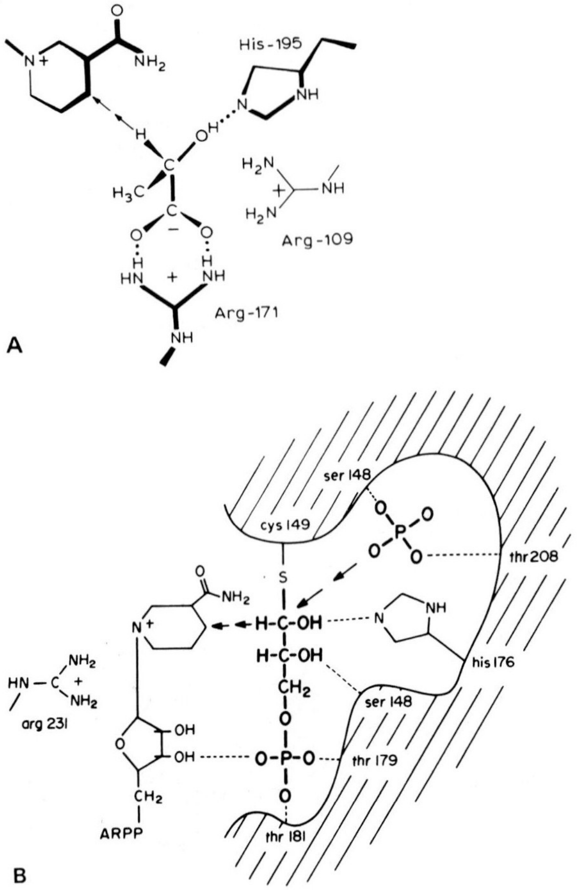

A

B

Fig. 19. The active sites of lactate dehydrogenase (A) and glyceraldehyde-3-phosphate dehydrogenase (B). From the work of Rossmann and colleagues [52].

(vi) Glycerol-3-phosphate dehydrogenase (EC 1.1.1.8)

$$\begin{array}{l} CH_2OH \\ | \\ CHOH \\ | \\ CH_2O \cdot PO_3^{2-} \end{array} + NAD^+ \rightleftharpoons \begin{array}{l} CH_2OH \\ | \\ C=O \\ | \\ CH_2O \cdot PO_3^{2-} \end{array} + NADH + H^+$$

This enzyme utilizes *sn*-glycerol 3-phosphate (L-glycerol 3-phosphate) and the 4-*pro-S* position of the coenzyme [38]. The primary structure of the rabbit muscle enzyme

Fig. 20. Chain trace of pig heart soluble malate dehydrogenase subunit. Stereo drawing from the work of Banaszak and colleagues [83].

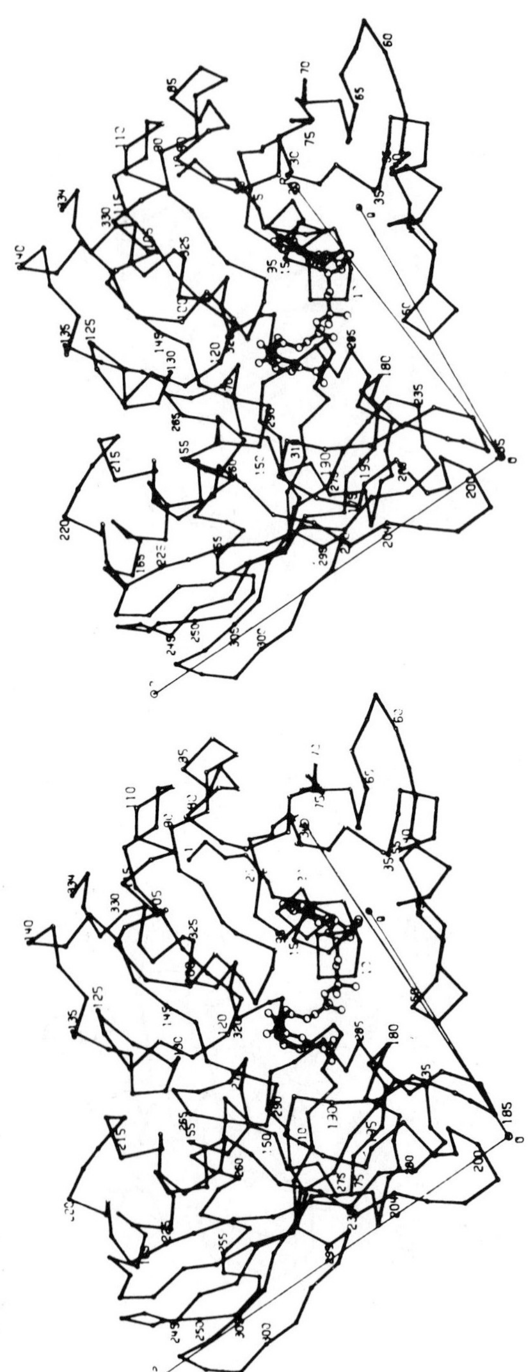

Fig. 21. Chain trace of lobster muscle glyceraldehyde-3-phosphate dehydrogenase subunit. Stereo drawing from the work of Rossmann and colleagues [166]. Bound NAD is also shown.

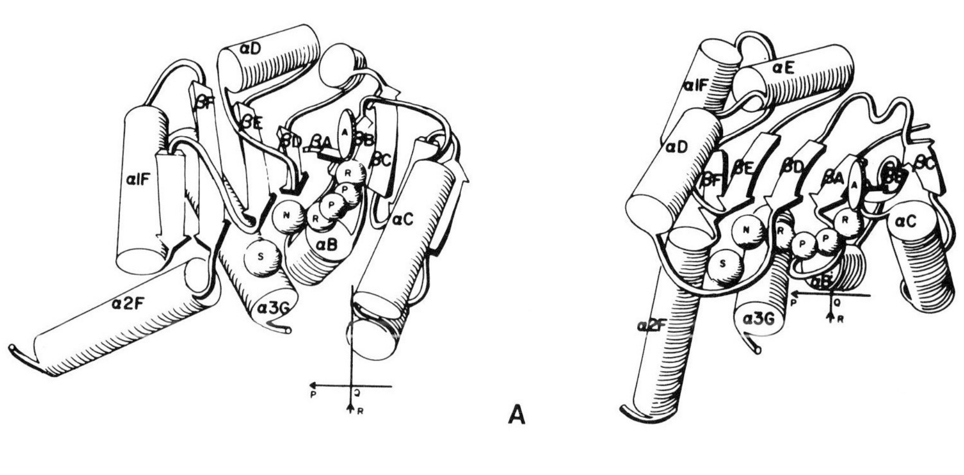

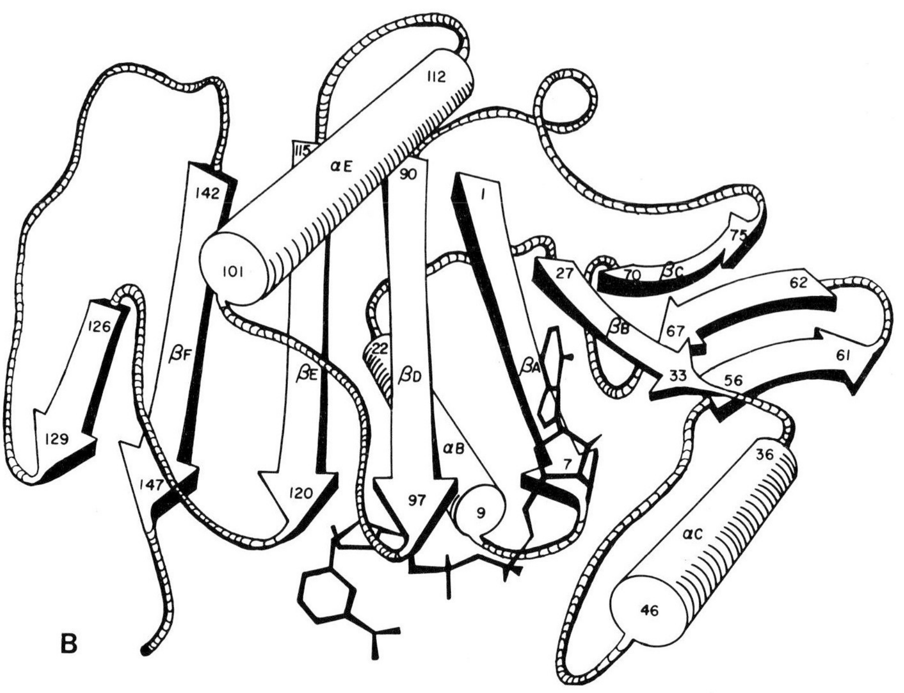

Fig. 22. Sketches depicting coenzyme binding. (A) Lactate dehydrogenase (right) and glyceraldehyde-3-phosphate dehydrogenase (left), allowing general comparison of the binding structures. From the work of Rossmann and colleagues [167]. (B) An alternative representation of coenzyme binding to glyceraldehyde-3-phosphate dehydrogenase, showing the Rossmann fold of the enzyme, the coenzyme lying in a Brändén crevice, and the pyrophosphate moiety of the coenzyme located in a Hol zone (for details see text). Sketch from the work of Rossmann and colleagues [168].

[93] was used to predict the secondary structure, and comparison with the known structures of other NAD-dependent dehydrogenases allowed an arrangement for glycerol-3-phosphate dehydrogenase to be proposed (Fig. 23) [94].

(vii) Glutamate dehydrogenase (EC 1.4.1.2–4)

$$\begin{array}{l} CO_2H \\ | \\ CH \cdot NH_2 + H_2O + NAD(P)^+ \\ | \\ CH_2 \\ | \\ CH_2 \\ | \\ CO_2H \end{array} \rightleftharpoons \begin{array}{l} CO_2H \\ | \\ C=O + NH_4^+ + NAD(P)H \\ | \\ CH_2 \\ | \\ CH_2 \\ | \\ CO_2H \end{array}$$

These enzymes utilize L-glutamate and the 4-*pro-S* position of the coenzyme [38,39]. Crystals have been obtained in many cases (summarized in [95]). Those from mammalian liver were not suitable for structure determination. For example, crystals of the rat liver enzyme had a large unit cell, and were unstable in the X-ray beam [96]. The NADP-dependent enzyme from *Neurospora crassa* is smaller than the mammalian enzyme, and less complicated with respect to allosteric and aggregation effects. Nevertheless, carefully grown crystals did not provide X-ray diffraction patterns favourable for structure determination (A.C.T. North, personal communication). At present, best hopes reside in crystals of the yellowfin tuna liver enzyme [97].

In primary structure, the NADP-dependent *N. crassa* enzyme shows clear but limited homology with the vertebrate glutamate dehydrogenases in the N-terminal two-thirds of the chain [98]. Interestingly, the NAD-dependent enzyme of *N. crassa* has a much larger subunit, with an apparently dissimilar sequence [99].

(viii) Alanine dehydrogenase (EC 1.4.1.1)

$$\begin{array}{l} CO_2H \\ | \\ CH \cdot NH_2 + H_2O + NAD^+ \\ | \\ CH_3 \end{array} \rightleftharpoons \begin{array}{l} CO_2H \\ | \\ C=O + NH_4^+ + NADH \\ | \\ CH_3 \end{array}$$

An enzyme from *Bacillus sphaericus* [100] catalyses this reaction. Like the well-known glutamate dehydrogenases it has a hexameric subunit arrangement, and is specific

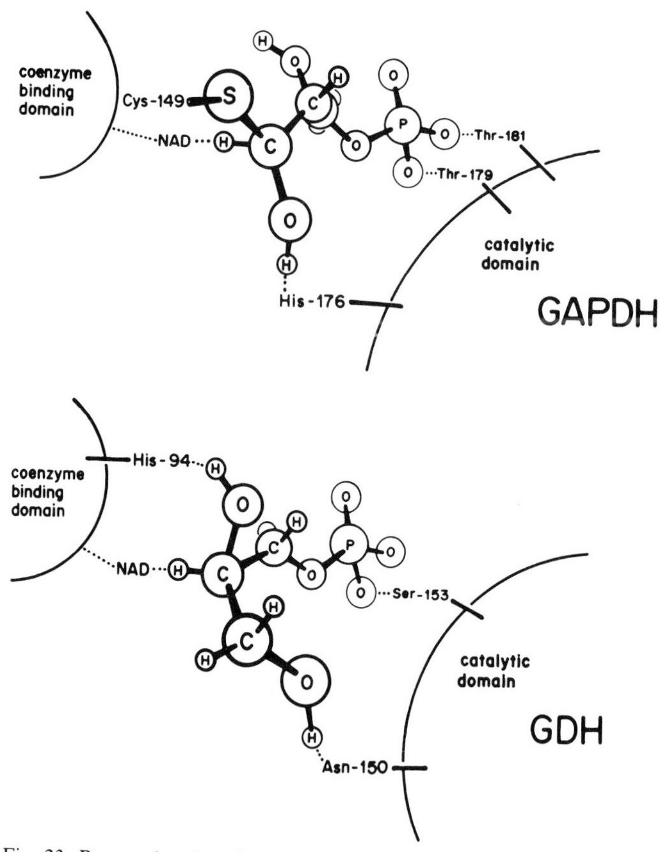

Fig. 23. Proposed active site arrangement of *sn*-glycerol-3-phosphate dehydrogenase (below), based on secondary structure predicted from the known primary structure, and on comparison with the known tertiary structure of glyceraldehyde-3-phosphate dehydrogenase (above). From the work of Rossmann and colleagues [94].

for the L-amino acid. Glutamate dehydrogenases often have alanine dehydrogenase activity [101]. However, this bacterial alanine dehydrogenase has a smaller subunit (38000) than either bovine liver glutamate dehydrogenase (55000) or the NADP-dependent glutamate dehydrogenase of *N. crassa* (48000), and unlike these enzymes it does not utilize NADP, and it is 4-*pro-R* specific with respect to NAD. Despite some similarities, therefore, it could be substantially different from the glutamate dehydrogenases. Crystals have been obtained [100] and it is hoped that structural information will be available in due course.

(ix) Saccharopine dehydrogenase (EC 1.5.1.7)

$$\begin{array}{l} CO_2H \\ | \\ CH\cdot NH_2 \\ | \\ (CH_2)_4 \qquad\qquad +NAD^+ + H_2O \\ | \\ NH \\ | \\ CH\text{-}(CH_2)_2CO_2H \\ | \\ CO_2H \end{array}$$

$$\rightarrow \begin{array}{l} CO_2H \\ | \\ CH\cdot NH_2 \\ | \\ (CH_2)_4 \\ | \\ NH_2 \end{array} + \begin{array}{l} CO_2H \\ | \\ (CH_2)_2 \\ | \\ C=O \\ | \\ CO_2H \end{array} + NADH + H^+$$

This is L-lysine forming. The enzyme from *Saccharomyces cerevisiae* (Bakers' yeast) utilizes the (*S*) configuration at C-2 of the glutaryl moiety (L-saccharopine) and the 4-*pro-R* position of the coenzyme [102]. The enzyme is monomeric [103] with size (39000) similar to the subunit of liver alcohol dehydrogenase. The three-dimensional structure is not known, but the reaction evidently involves a ketimine (Schiff base) intermediate [102]. A cysteine (coenzyme binding site) [104], a histidine [105] and a lysine [106] (2-oxoglutarate binding site), and an arginine (catalysis) [107] are important. It is now pointed out that the sequence immediately around the essential cysteine [108] resembles that around the reactive cysteine of sheep liver sorbitol dehydrogenase [5] (Fig. 24), though such short regions of identity are difficult to interpret (see, for example, the "fortuitous" reactive histidine sequence homology described by Otto and coworkers [94], and four selected examples discussed by Jörnvall [109]).

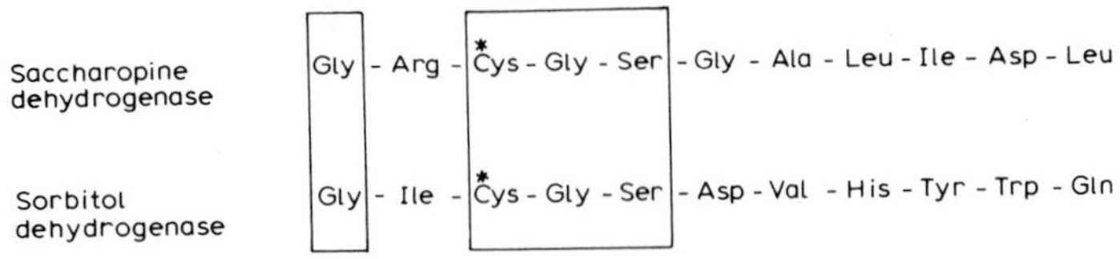

Saccharopine dehydrogenase	Gly	Arg	Cys*	Gly	Ser	Gly	Ala	Leu	Ile	Asp	Leu
Sorbitol dehydrogenase	Gly	Ile	Cys*	Gly	Ser	Asp	Val	His	Tyr	Trp	Gln

Fig. 24. Four out of five amino acid residues around the reactive cysteine are identical in yeast saccharopine dehydrogenase and sheep sorbitol dehydrogenase.

(x) Octopine dehydrogenase (EC 1.5.1.11)

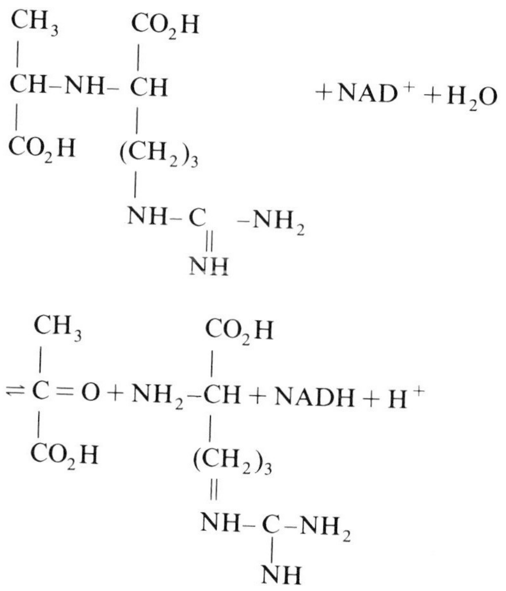

$$
\begin{array}{lllll}
CH_3 & & CO_2H & & \\
| & & | & & \\
CH{-}NH{-} & & CH & & +NAD^+ + H_2O \\
| & & | & & \\
CO_2H & & (CH_2)_3 & & \\
 & & | & & \\
 & & NH{-}C{-}NH_2 & & \\
 & & \quad\;\, \| & & \\
 & & \quad\; NH & &
\end{array}
$$

$$
\begin{array}{lllll}
 & CH_3 & & CO_2H & \\
 & | & & | & \\
\rightleftharpoons & C=O & +\,NH_2{-} & CH & +\,NADH + H^+ \\
 & | & & | & \\
 & CO_2H & & (CH_2)_3 & \\
 & & & \| & \\
 & & & NH{-}C{-}NH_2 & \\
 & & & \quad\;\, | & \\
 & & & \quad\; NH &
\end{array}
$$

In the muscles of cephalopods [110,111] and some bivalves [112,113], octopine dehydrogenase is the terminal enzyme of anaerobic glycolysis, serving these organisms as lactate dehydrogenase serves many others, or as alcohol dehydrogenase serves yeast. Octopine dehydrogenase from the marine mollusc *Pecten maximus L.* is a monomeric protein of molecular weight 38000 [114]. It utilizes D-octopine (systematic name N^2-(*R*1)-carboxyethyl)-(*S*)-arginine) and the 4-*pro-S* position of NAD [115]. The steric course therefore differs with respect to both substrate and coenzyme from those of three analogous enzymes, namely saccharopine dehydrogenase (similar reaction), lactate dehydrogenase and yeast alcohol dehydrogenase (corresponding functions in the organism).

(xi) Alcohol dehydrogenase (EC 1.1.1.1)

$$CH_3CH_2OH + NAD^+ \rightleftharpoons CH_3CHO + NADH + H^+$$

The subunit (size 40000) of horse liver alcohol dehydrogenase (Fig. 25) [116] bears some similarity to lactate dehydrogenase (Fig. 18), malate dehydrogenase (Fig. 20) and glyceraldehyde-3-phosphate dehydrogenase (Fig. 21) subunits, mainly in the coenzyme binding region. Liver alcohol dehydrogenase is dimeric, but the yeast enzyme is tetrameric [117], a situation reminiscent of malate and lactate dehydrogenases, which, though they utilize different substrates, have markedly similar tertiary structures, and have dimeric and tetrameric subunit arrangements, respectively. Liver and yeast alcohol dehydrogenase subunits each contain a zinc atom

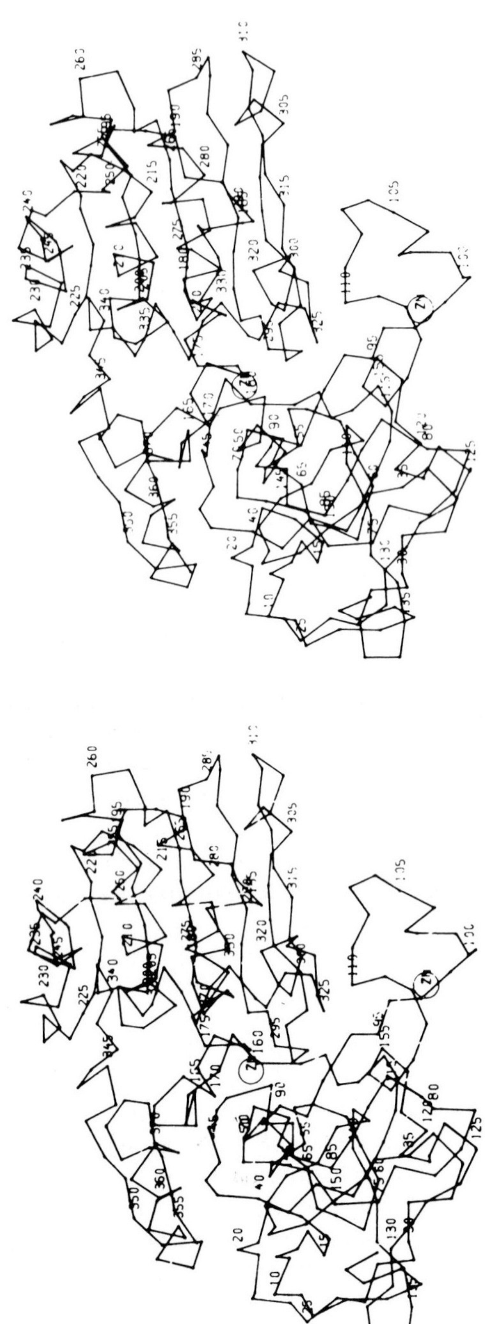

Fig. 25. Chain trace of one subunit of horse liver alcohol dehydrogenase. Stereo drawing from the work of Brändén and colleagues [117]. The catalytic zinc atom is central, the structural zinc atom is at the bottom right.

(Zn^{2+}) at the active site, and the liver enzyme contains a second zinc atom (Zn^{2+}) at a separate site (Fig. 26A) [55]. In the catalytic mechanism, the active-site zinc evidently fulfils a function broadly similar to that of His-195 in lactate dehydrogenase, and His-176 in glyceraldehyde-3-phosphate dehydrogenase [52,118]. This zinc is coordinated in a distorted tetrahedral geometry by two sulphur, one nitrogen and one oxygen atom (Cys-46, Cys-174, His-67, and water or the substrate). It can be replaced by Co^{2+} without total loss of activity [119,120]. Binding of Mn^{2+} at the catalytic site was not detected [121], whereas Cd^{2+} entered faster than Zn^{2+} [121]. Findings by Michael Zeppezauer and his coworkers, obtained in part in collaboration with M.F. Dunn, show that the strength of substrate binding, and the rate of hydride transfer in the ternary complex are inversely proportional to the ionic radius of the metal ion, which suggests that the metal ion coordinates the substrate in a direct way, and has the role of a general Lewis acid (M. Zeppezauer, personal communication). The exception is Cu^{2+}, which is inactive. The Cu^{2+}–enzyme does bind coenzyme, apparently in a similar way to the active enzymes, but the substrate does not then displace the coordinated water efficiently [122].

The rate of hydride transfer should not be confused with the turnover, which is determined by the off-rate of NADH and is not simply related to the size of the catalytic metal ion. However, there is evidence that coenzyme binding is affected by the metal occupying the catalytic site, so altogether this metal influences substrate binding, coenzyme binding, and substrate activation (M. Zeppezauer, personal communication), though the effect on coenzyme binding may not be great.

The tertiary structure of the coenzyme binding domain of horse liver alcohol dehydrogenase resembles that of the other NAD-dependent dehydrogenases described, and a similar combination of hydrophobic interactions and hydrogen bonds positions the coenzyme [55]. Orientation of the nicotinamide moiety with the 4-*pro-R* position aligned for hydrogen transfer results from hydrogen bonds formed with the carboxamide group. Most likely to be involved are the N of Phe-319, and the C=O of Val-292 [55], but not the side chain of Thr-178, despite its proximity [169]. Also, steric hindrance involving Val-203 would possibly prevent the 4-*pro-S* position from attaining this location [117].

Ethanol binding occurs so that the 1-*pro-R* hydrogen is transferred [49]. This results from positioning of the methyl in a region of hydrophobic residues (the 'hydrophobic barrel' [124]). The position occupied by this methyl group could be occupied by a hydrogen atom, and a methyl group could occupy the other position instead of the 1-*pro-S* hydrogen. Thus, in liver alcohol dehydrogenase no specific interaction is known that determines the orientation of the ethanol, but the interactions that cause methanol (both positions H) and propan-2-ol (both positions CH_3) to be poor substrates combine to make hydrogen transfer highly stereospecific in the case of ethanol.

The substrate binding pocket of horse liver alcohol dehydrogenase comprises residues from both subunits (Fig. 26B) [123]. The active site is shown in Fig. 27, with NAD(H) bound, and *p*-bromobenzyl alcohol bound in a non-productive binding mode. The hydrophobic residues (from both subunits) that line the substrate binding

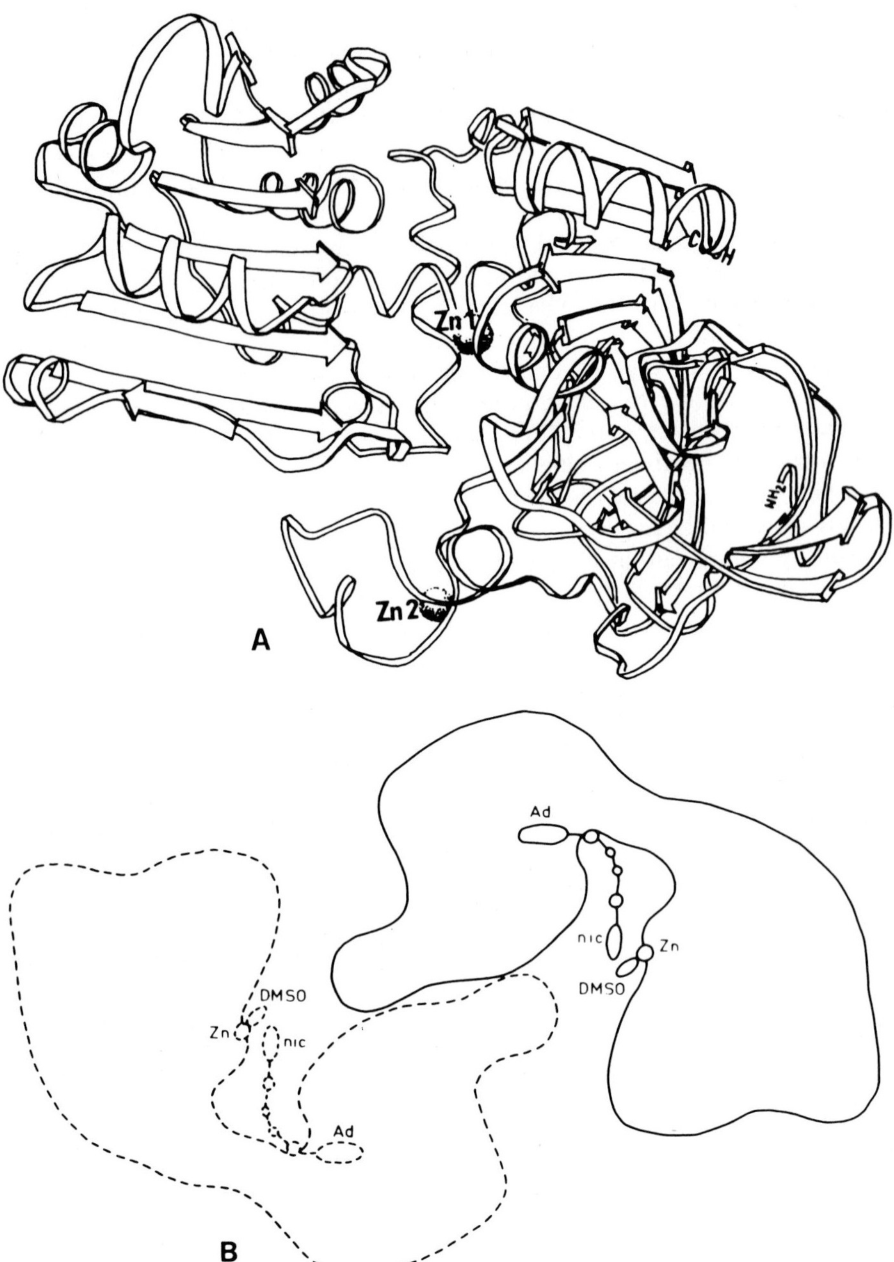

Fig. 26. (A) Schematic diagram of one subunit of horse liver alcohol dehydrogenase. Znl is the active-site zinc. Designed by B. Furugren, from the work of Brändén and colleagues [55]. (B) Schematic diagram of a section through the horse liver alcohol dehydrogenase dimer. The catalytic zinc atoms are shown, with the inhibitory substrate analogue DMSO and coenzyme molecules indicated. The dimer has two active sites, each composed of parts of both subunits. From the work of Brändén and colleagues [123].

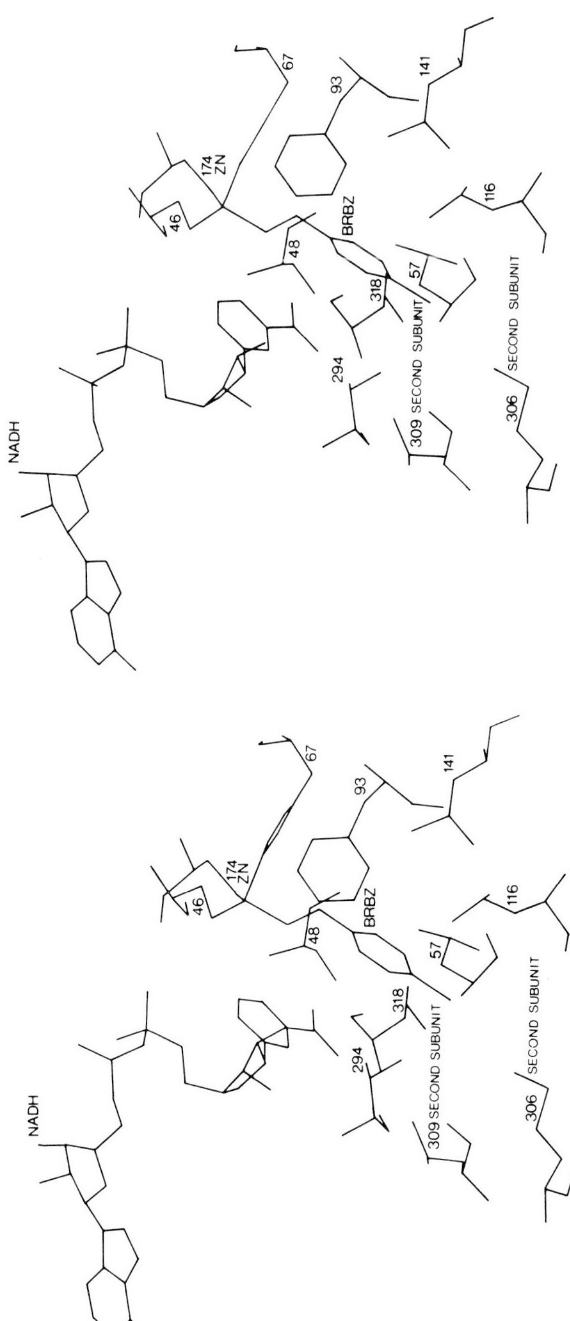

Fig. 27. The active site of horse liver alcohol dehydrogenase, with bound NAD(H) and a non-productive binding mode of *p*-bromobenzyl alcohol. The hydrophobic residues (from both subunits) that line the substrate binding pocket are illustrated. Stereo drawing from the work of Brändén and colleagues.

pocket are indicated. The view into this substrate binding pocket towards the catalytic zinc is illustrated in Fig. 28. The proposed productive position of the alcohol (light line), based on model building, is shown in Fig. 29, together with the observed non-productive complex (dark line).

The inhibitor dimethyl sulphoxide (DMSO) forms, with NADH and alcohol dehydrogenase, a ternary complex in which the sulphoxide is bound directly to the active-site zinc [123]. Fig. 30 shows the substrate binding pocket in the apo-enzyme, and Fig. 31 shows it in the enzyme–NADH–DMSO complex. The structures are compared in Fig. 32. When the complex forms, there is a slight rotation of the catalytic domains of the dimer with respect to the central core [123]. As a result, the clefts between the domains become narrower. The overall effect is to shield the active site zinc and its surroundings from the solution. This is likely to facilitate the catalytic reaction. A similar effect on the active site of lactate dehydrogenase has been described [78], but whereas in alcohol dehydrogenase the residues involved (residues 295–298, 53–57 and 15–18) belong to the catalytic domain, in lactate dehydrogenase the region concerned (residues 103–115) is in the coenzyme binding domain [78], and is a length of polypeptide not present in the alcohol dehydrogenase sequence [123].

(xii) Aldehyde reductase (EC 1.1.1.2) and similar enzymes

Various enzymes that transfer hydrogen from NADPH to a carbonyl compound have broad substrate specificity and, if acetaldehyde will serve, are classifiable as EC 1.1.1.2.

Aldose reductase (EC 1.1.1.21) shows marked similarities to aldehyde reductase [125,126], of which it may be an isozyme [127], though the relationship is not certain [128]. Enzymes previously classified as D-glucuronate reductase (EC 1.1.1.19), mevaldate reductase (EC 1.1.1.33), L-hexonate dehydrogenase and daunorubicin reductase are aldehyde reductases [129–132]. Aldose reductase from human placenta transfers the 4-*pro-R* hydrogen of NADPH to the aldose (e.g., D-glucose), giving the corresponding alditol (e.g., sorbitol) [133]. It is a monomeric enzyme of molecular weight 39000 [134].

Aldehyde reductase from human liver catalyses attack by the 4-*pro-R* hydrogen on the *Re* face of the carbonyl [135]. 4-Nitrobenzaldehyde and 3,4-dihydroxyphenylglycolaldehyde are good substrates [135,136]. This enzyme is monomeric with molecular weight 36000 [137]. Pig kidney [138] and pig liver [139] also contain a monomeric aldehyde reductase that catalyses *Re* attack by the 4-*pro-R* hydrogen, and has molecular weight around 35000. Several other tissues and species contain closely similar enzymes [140–142].

There is, however, a further group of NADPH-dependent carbonyl reductases that transfer the 4-*pro-S* hydrogen [143]. These include human brain aldehyde reductase I [144], liver xenobiotic ketone reductase [145] and prostaglandin 9-ketoreductase [146], all monomeric proteins with molecular weights in the range of 30000–400000.

The synthesis of lignin from phenylalanine in plants involves two successive (and

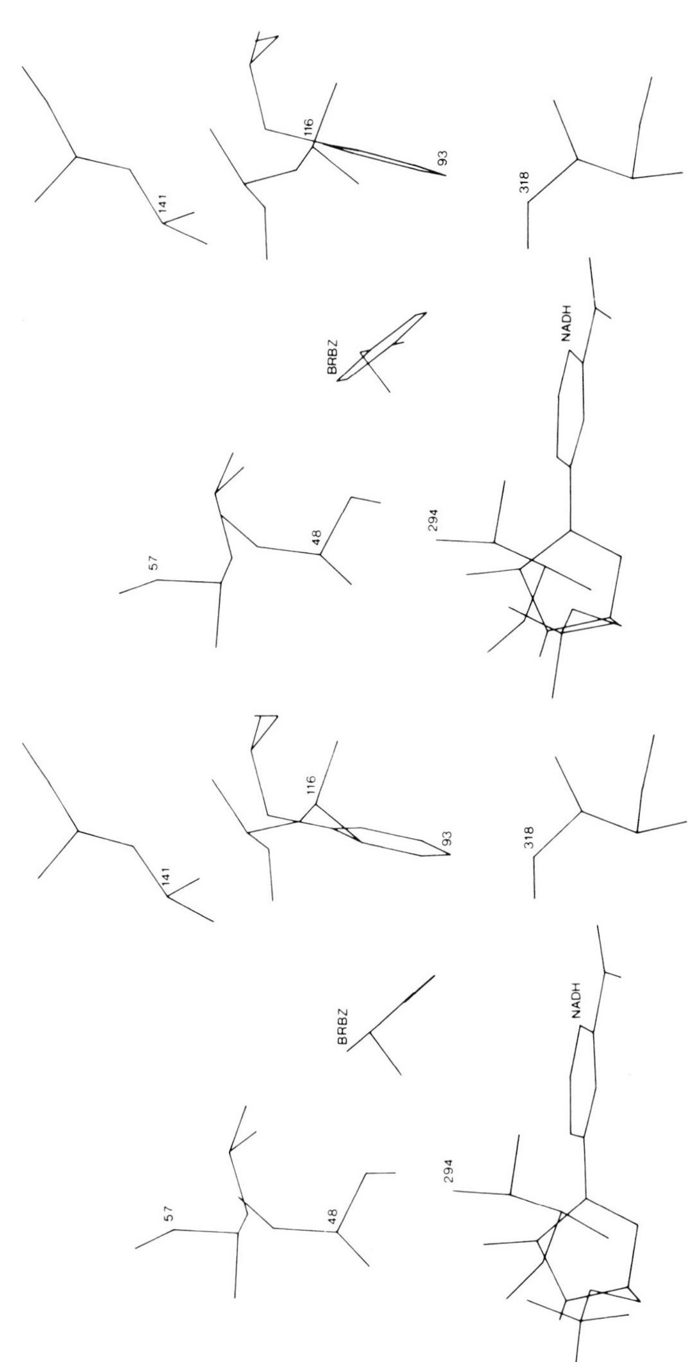

Fig. 28. The substrate binding pocket of horse liver alcohol dehydrogenase, as in Fig. 27, viewed here into the pocket towards the zinc (not itself shown). Stereo drawing from the work of Brändén and colleagues.

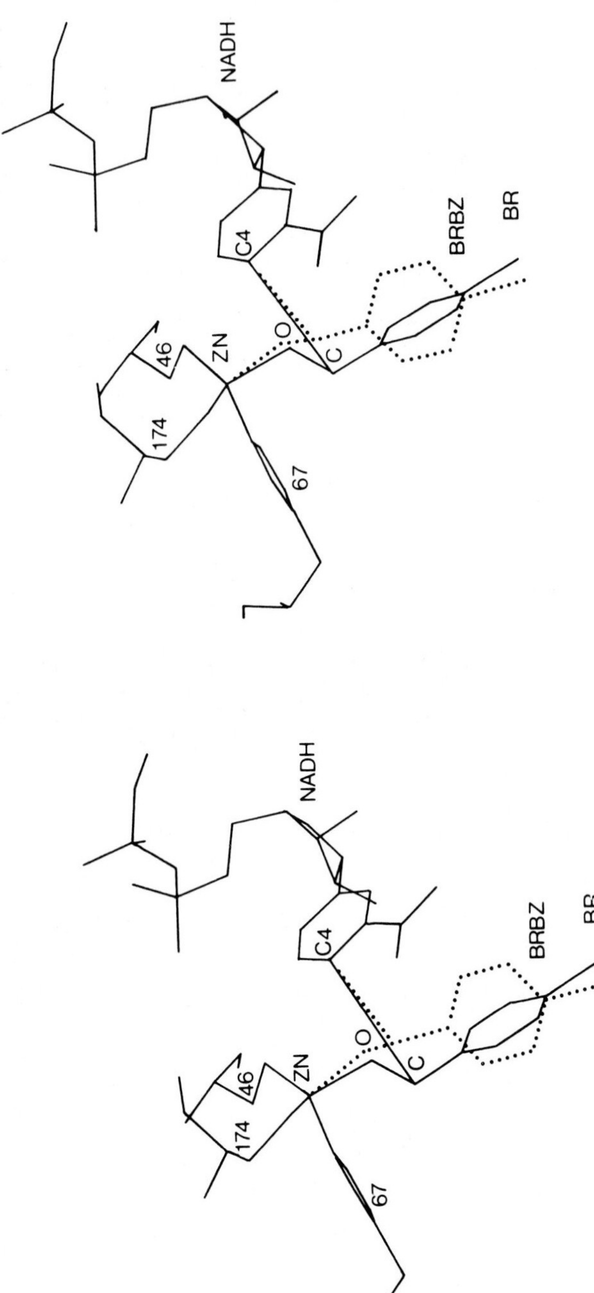

Fig. 29. The observed ternary complex of horse liver alcohol dehydrogenase, NAD(H), and *p*-bromobenzyl alcohol (dark lines) and the proposed productive alcohol position (dotted lines), based on model building. Stereo drawing from the work of Brändén and colleagues.

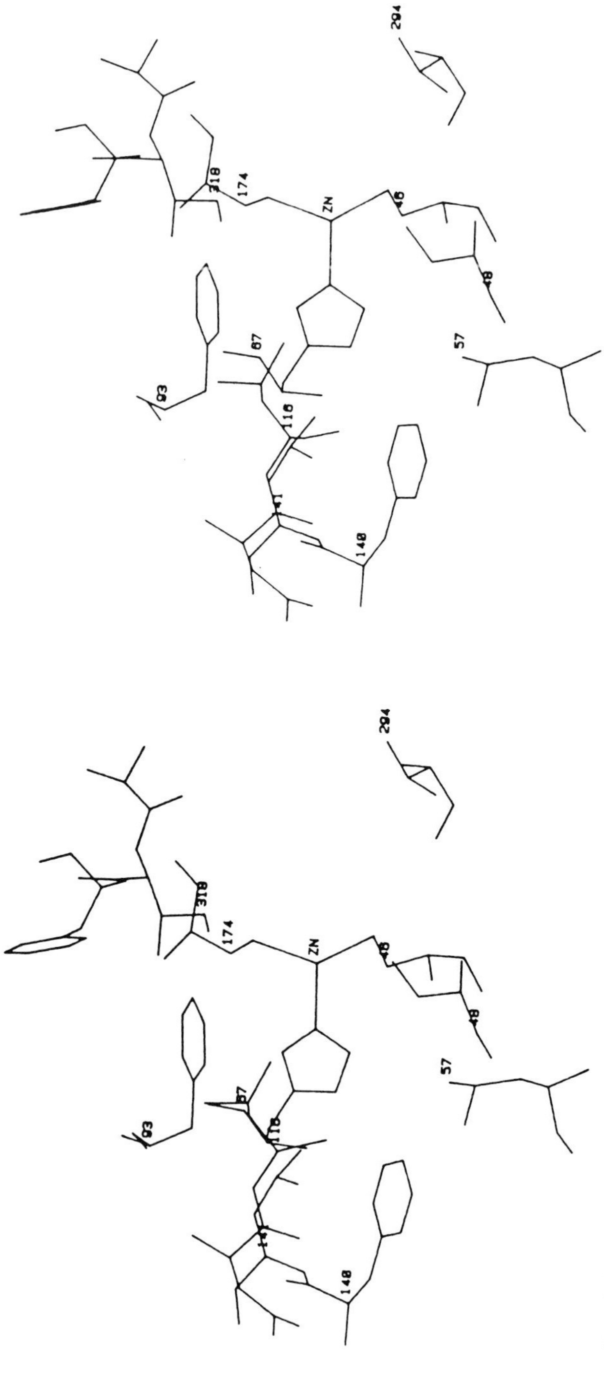

Fig. 30. Horse liver alcohol dehydrogenase substrate binding pocket, unoccupied. Stereo drawing from the work of Brändén and colleagues.

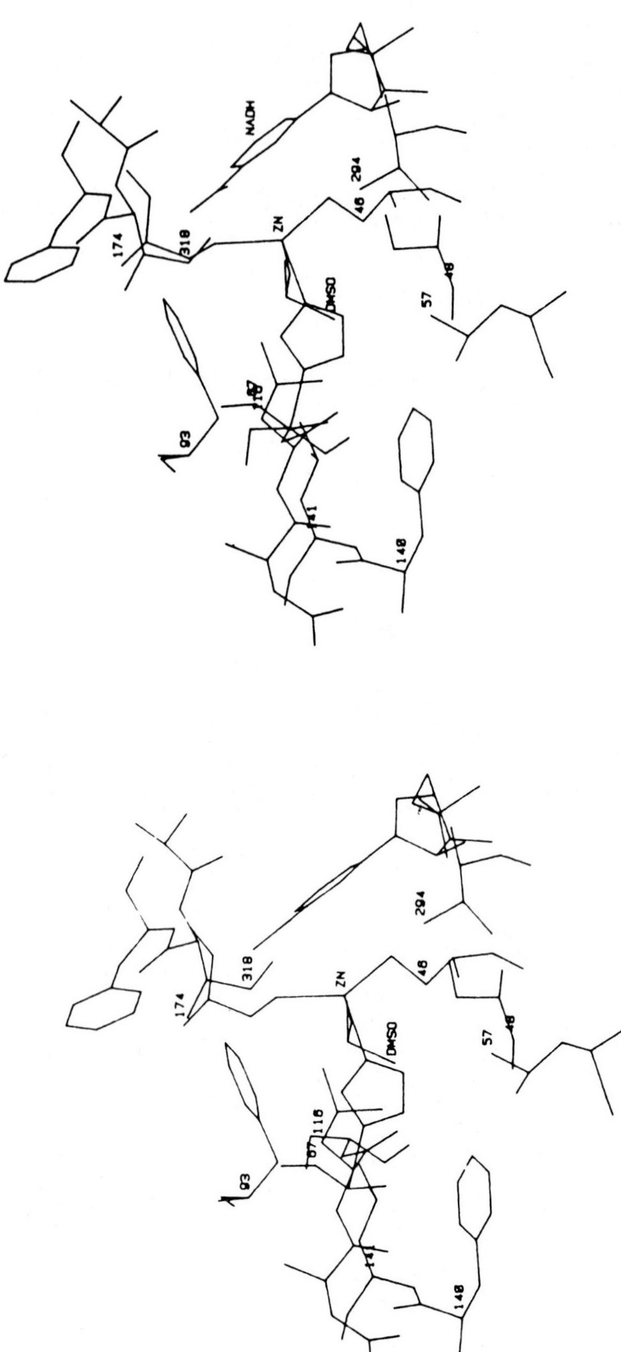

Fig. 31. Horse liver alcohol dehydrogenase ternary complex with NADH and the inhibitory substrate analogue dimethyl sulphoxide. Stereo drawing from the work of Brändén and colleagues.

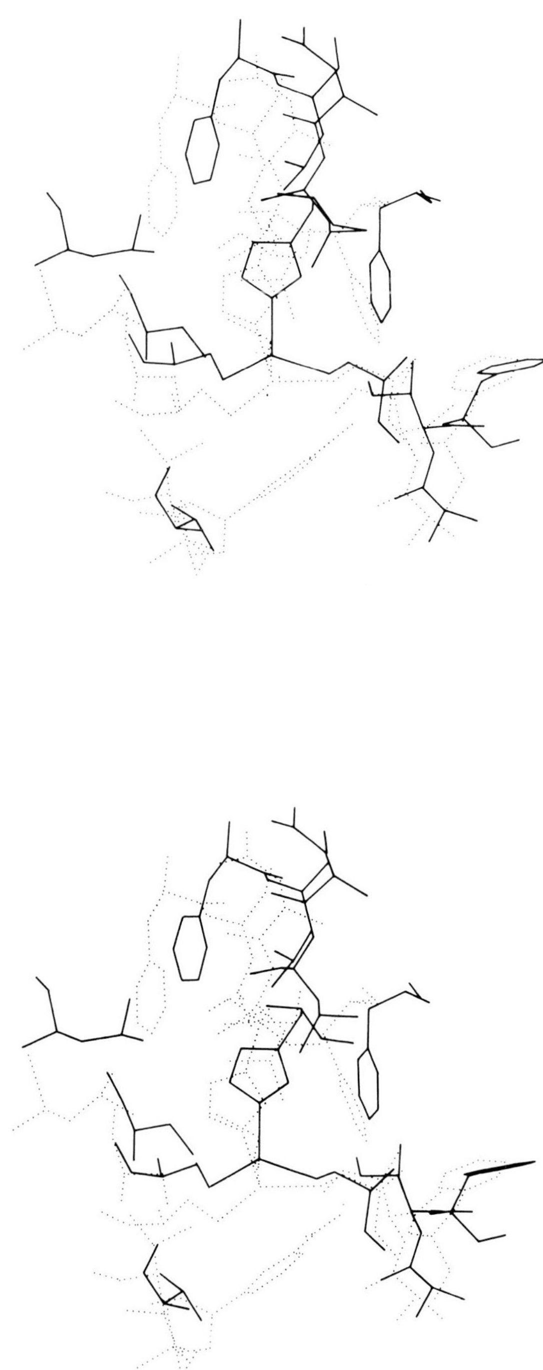

Fig. 32. Horse liver alcohol dehydrogenase. The full line shows positions in the apo-enzyme (Fig. 30) and the broken line shows positions in the ternary complex with NADH and dimethyl sulphoxide (Fig. 31). This figure therefore illustrates the movement of active site residues that occurs when the coenzyme and substrate analogue bind. Stereo drawing from the work of Brändén and colleagues.

rather different) dehydrogenase reactions [147]. In the first of these, cinnamoyl-CoA is converted into cinnamaldehyde:

$$\text{R–CH=CH–CO}\cdot\text{SCoA} + \text{NADPH} + \text{H}^+$$
$$\rightarrow \text{R–CH=CH–CHO} + \text{NADP}^+ + \text{CoASH}$$

The enzyme (studied in soybean, *Glycine max* L.) has molecular weight 38000 [148], and transfers the 4-*pro-S* hydrogen [149].

The second step is the reduction of cinnamaldehyde to cinnamyl alcohol:

$$\text{R–CH=CH–CHO} + \text{NADPH} + \text{H}^+ \rightarrow \text{R–CH=CH–CH}_2\text{OH} + \text{NADP}^+$$

This is catalysed by a zinc-dependent dimeric enzyme of subunit size 40000 [150,151]. The 4-*pro-R* hydrogen is transferred to the *Re* side of the aldehyde [151].

This lignin synthesis pathway is of great importance for trees and woody plants (and therefore, indirectly, life generally) [147]. From the stereochemistry and molecular properties, two interesting points emerge. First, it is an exception to the empirical generalization (discussed in [46]) that successive dehydrogenase steps in metabolic pathways have constant prochiral specificity for coenzyme. Second, this plant cinnamyl alcohol dehydrogenase appears remarkably similar to liver alcohol dehydrogenase (except that it uses NADPH rather than NADH). The similarity contrasts strongly with the difference between liver aldehyde reductase and liver alcohol dehydrogenase. Aldehyde reductases from animal tissues are not zinc-dependent [139,152,153].

3. *Do particular structural features fulfil similar functions in different dehydrogenases?*

The majority of dehydrogenases so far investigated comprise regions of β-sheet connected by α-helices. In this type of protein, a central core often consists of a sheet of strands, most of which are parallel [154]. The connections between the parallel strands frequently contain helices packed on both sides of the sheet in a regular way [155]. These connections are right-handed [156–158]. In a parallel pleated sheet, the amino ends of the strands are at one edge of the sheet, and the carboxyl ends at the other edge. Binding of substrate or cofactor occurs in crevices at the carboxy end of the parallel strands of the sheet. The dipole moment at the amino end of one of the helices adjacent to the carboxy end of the strands could facilitate positioning of negatively charged groups in this region [159]. Because of the righthandedness of the $\beta\alpha\beta$-unit, it is possible to predict from the strand order of the sheet the location of crevices favourable for binding [72]. Such crevices may be expected outside the carboxy end at two adjacent parallel strands that are connected on opposite sides of the sheet [72]. The necessary reversal of strand order occurs in most domains of the α/β type [160]. In the NAD-dependent dehydrogenases, the strand order is 6 5 4 1 2

3 (Fig. 33), giving the required topological features between strands 4 and 1, which is where the pyrophosphate moiety of NAD binds [72].

If the strand order were 1 2 3 4 5 6, all five connections would be on the same side of the sheet, and no crevice would be formed. However, when eight strands are aligned in this way, a closed barrel of twisted strands can be formed in which strand number 8 is adjacent to strand number 1 (Fig. 34) [72]. The helical connections lie outside the barrel, and diverge from the barrel axis, giving geometrical conditions favourable for the formation of crevices at the carboxyl end of the barrel [72]. This arrangement occurs in triose-phosphate isomerase [161], pyruvate kinase [162] and glycollate oxidase [163].

Brändén [72] showed that, in addition to the dehydrogenases already mentioned, the following enzymes had one or more binding sites in the positions predicted from strand order reversals: flavodoxin, adenylate kinase, dihydrofolate reductase, hexokinase, *p*-hydroxybenzoate hydroxylase, phosphoglycerate mutase, carboxypeptidase, subtilisin, phosphoglycerate kinase, glutathione reductase, phosphofructokinase, rhodanase, arabinose binding protein, phosphorylase, glucose-6-phosphate

A

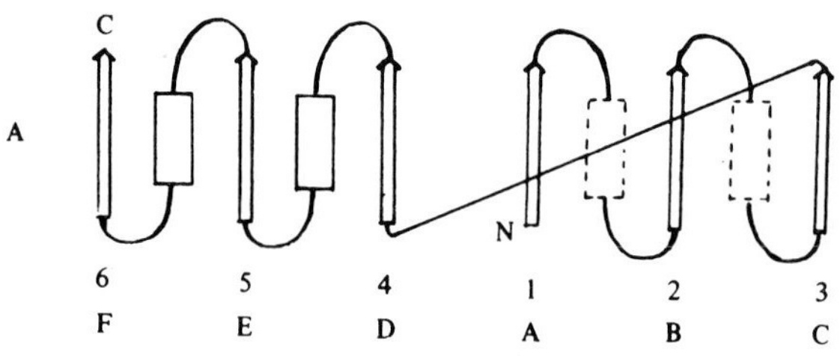

B

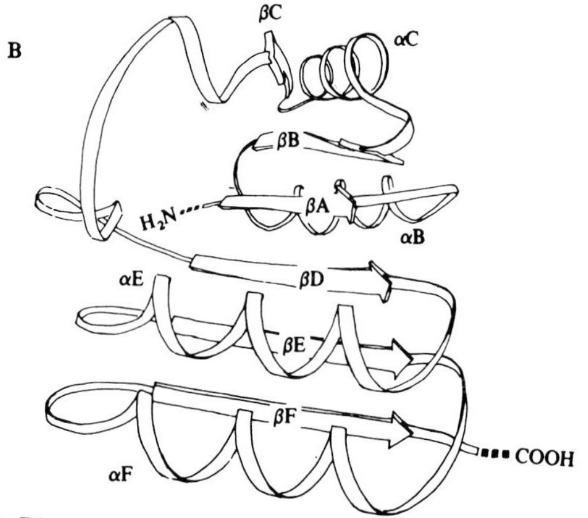

Fig. 33. Diagram illustrating the arrangement of strands and helices in the NAD-binding domain of dehydrogenases. (B designed by B. Furugren.) From the work of Brändén and colleagues [72].

A

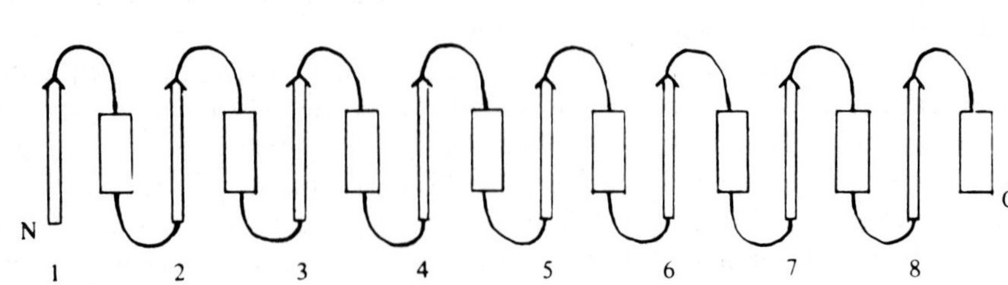

B

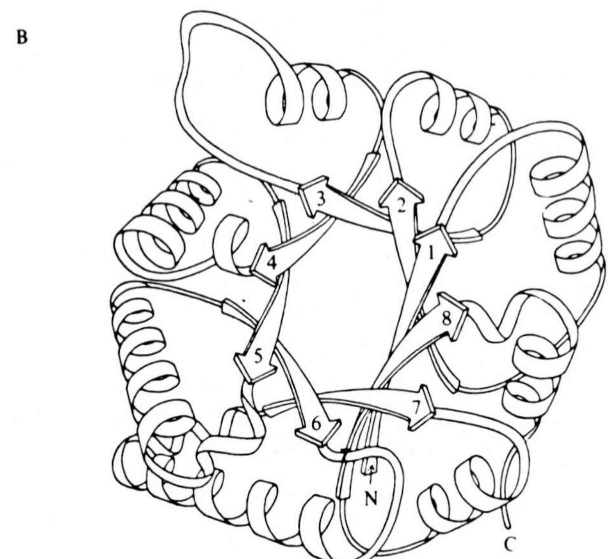

Fig. 34. Diagram illustrating the arrangement of strands and helices in triose-phosphate isomerase. (B designed by J. Richardson). From the work of Brändén and colleagues [72].

isomerase and aspartate carbamoyl transferase. Only aspartate amino-transferase had known binding sites that did not correspond to the predicted location [72]. 6-Phosphogluconate dehydrogenase has an unusual structure (Fig. 17) containing only three strands of β-sheet, which are parallel. They occur in a small domain and have one strand reversal. The coenzyme has been described as binding with the adenine at the C-terminus of two sheet strands, the ribose 2′-phosphate fairly close to the N-terminus of a helix (Fig. 17) and the nicotinamide at the junction of the large and small domains [71].

Do more detailed similarities of structure and mechanism accompany the broad relationships of Brändén's topological analysis [72]? Functionally equivalent groups around the substrate are illustrated in Fig. 35 for lactate dehydrogenase, glyceralde-

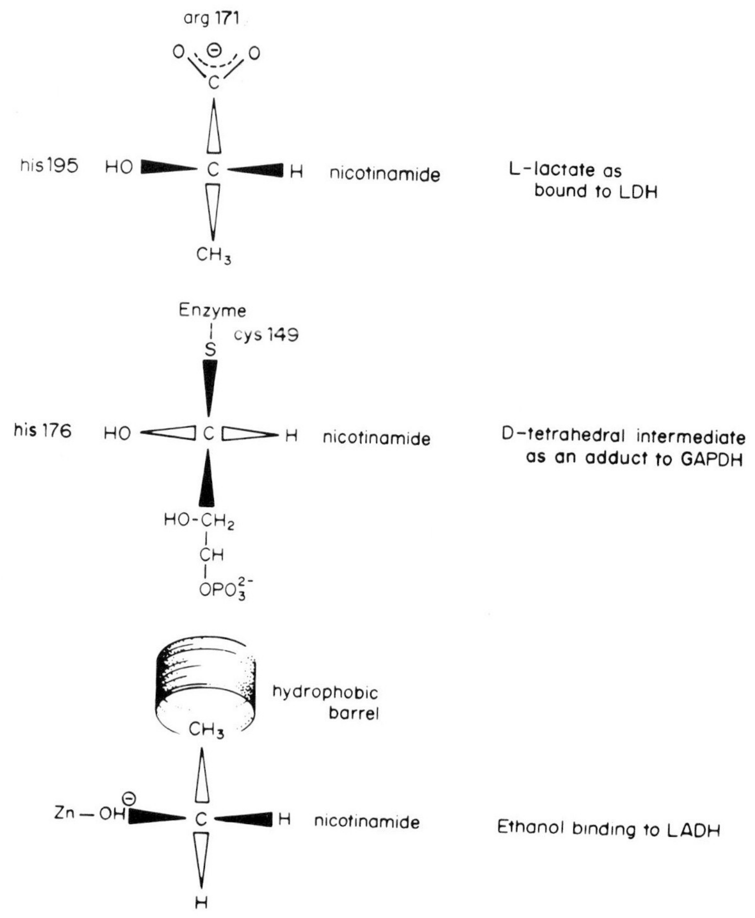

Fig. 35. Diagrammatic representation of functionally equivalent groups around the substrate in lactate dehydrogenase, glyceraldehyde-3-phosphate dehydrogenase and horse liver alcohol dehydrogenase. From the work of Rossmann and colleagues [164].

hyde-3-phosphate dehydrogenase and alcohol dehydrogenase. Fig. 36 compares steps in the reaction catalysed by one of these enzymes (glyceraldehyde-3-phosphate dehydrogenase) with steps in the hydrolytic reactions catalysed by chymotrypsin and papain [52]. Cys-149 of glyceraldehyde-3-phosphate dehydrogenase (GAPDH) is comparable with Cys-25 (papain); His-176 (GAPDH) with His-159 (papain); and Ser-238 (GAPDH) with Asn-175 (papain). These comparable groups can be closely aligned in the crystal structures, and when this is done Gln-19 (papain), which orients the carbonyl of the substrate superimposes on Asn-313 (GAPDH), which orients the nicotinamide ring [52,55].

Chymotrypsin resembles papain quite closely. The tetrahedral intermediates are,

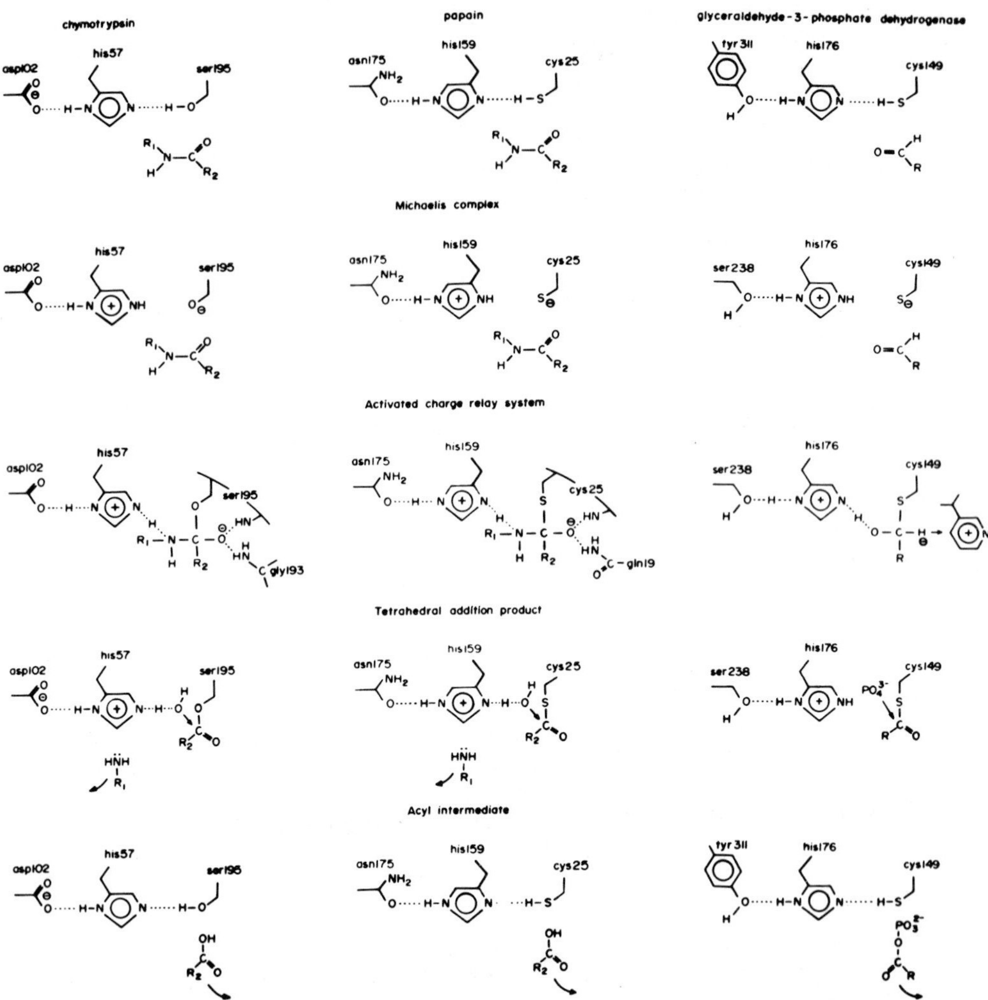

Fig. 36. Comparison of steps in the reaction catalysed by chymotrypsin, papain and glyceraldehyde-3-phosphate dehydrogenase. From the work of Rossmann and colleagues [52].

in fact, of the opposite hand, as also are those of GAPDH and lactate dehydrogenase (or alcohol dehydrogenase) [52,164]. These relationships allow the serine protease family and the NAD-dependent dehydrogenases to be compared.

The presence of zinc at the catalytic site of liver alcohol dehydrogenase suggests comparison with other zinc-dependent enzymes [164], and three are shown in Fig. 37. The hand and geometry of the zinc environment is invariant with respect to the proton abstractor (L4 and Fig. 37), the zinc atom, substrate site, water position, and protein ligand cluster. Ligand L1 is in each case histidine, and the planes of the imidazole rings approximately coincide, though the rotations of the rings within their common plane differ. Other imidazole groups (e.g., L2 or L3) did not necessarily have like orientations. The angle L2–Zn–L3 (98° ± 2°) was always less than the

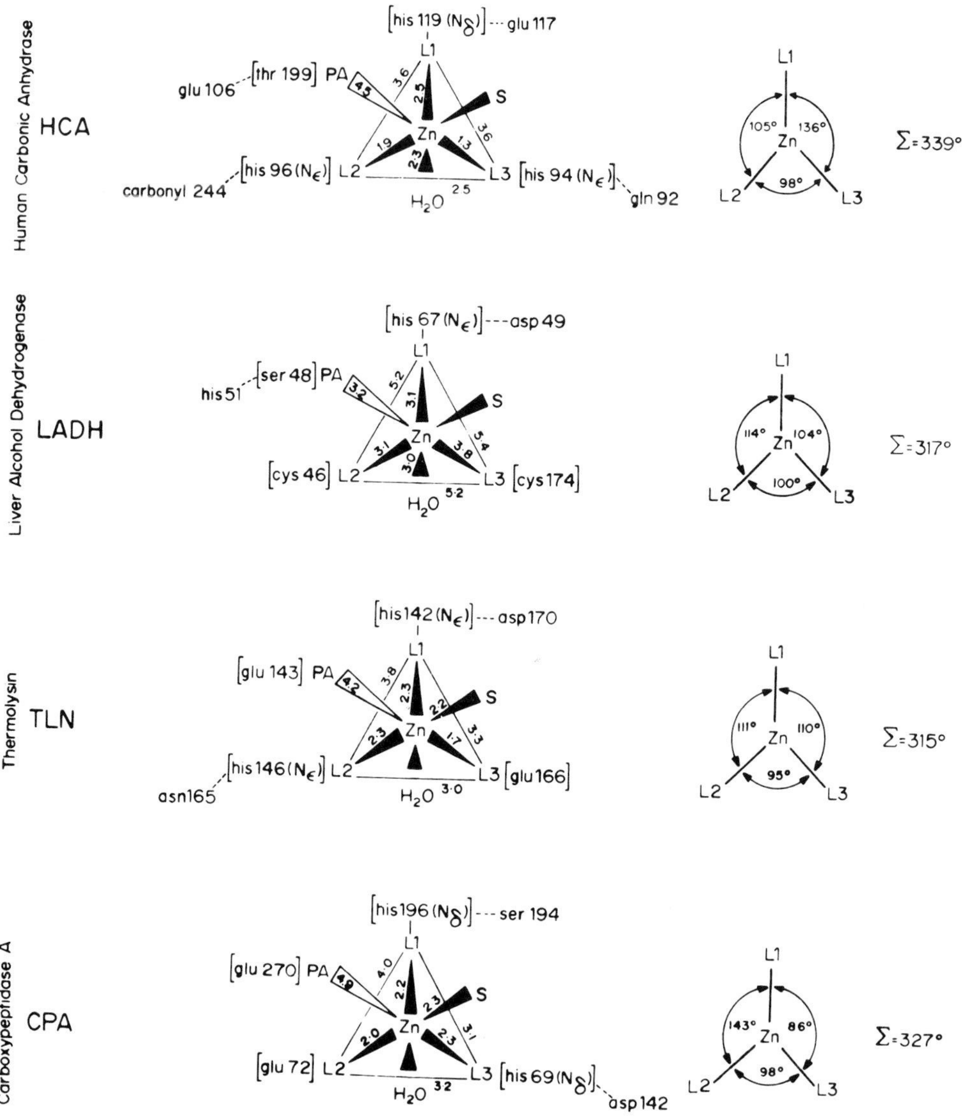

Fig. 37. Diagram representing the environment of the catalytic zinc in horse liver alcohol dehydrogenase, human carbonic anhydrase, thermolysin and carboxypeptidase A. Positions occupied by the substrate (S) and proton-abstracting group (PA) are indicated. The angles subtended by the liganding protein atoms at the zinc atom are shown to the right, and their sum (Σ) is shown extreme right. From the work of Rossmann and colleagues [164].

tetrahedral angle (109°). These findings [164] encouraged the view that in all these enzymes transient pentagonal arrangements may occur, and that zinc-bound water is likely to be an active participant in the reaction, its activation controlled by the proton abstractor.

4. Why are the structures related?

The way in which Brändén crevices [72] are formed at the end of the $\beta\alpha\beta$-units is discussed in Section 3. The opportunities they afford for mutations to affect binding capacity or specificity without making much difference to the stability of the fold could be important in evolution, and could explain why α/β domains of quite similar structure are found in rather different proteins [72]. The functional advantage of an α/β protein may then explain why the dehydrogenases and the other proteins referred to are of this type. It is not necessary to infer that they are genetically related. The structures of many proteins of different origin and different function can be similar simply because the principles of folding limit the number of possible protein topologies [160].

Efficient catalysis requires a specific configuration of substrates and reactive residues, with groups (selected from a small number of suitable candidates) to serve as nucleophiles and electrophiles under physiological conditions. In enzymes catalysing similar reactions, some convergence towards similar spatial arrangements and functional groups at the active site must be expected.

Is it possible that an early dinucleotide binding protein diverged into NAD-binding and NADP-binding proteins, followed by further divergence for substrate specificity? Adams and coworkers [71] concluded that evidence at present available does not support the view that the NADP-dependent enzymes arose in this way. Divergence of the NAD-dependent enzymes from a single nucleotide binding protein,

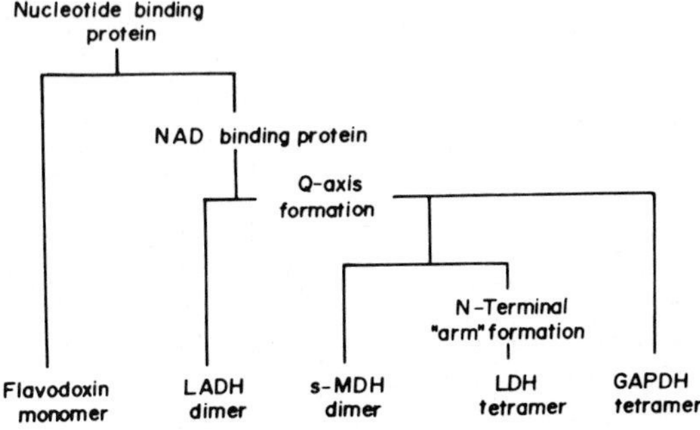

Fig. 38. Evolutionary scheme suggested in 1973 by Rossmann and colleagues [167]. A primordial mononucleotide binding protein is shown diverging to give an FMN-binding unit found in flavodoxin, and an NAD-binding unit found in dehydrogenases. The scheme is based mainly upon similarities of tertiary structure, which are regarded as conserved. It assigns importance to formation of the Q-axis, an event separating liver alcohol dehydrogenase from the other dehydrogenases shown.

with retention of tertiary rather than primary structure, was suggested many years ago [54,165], and is still a tenable hypothesis [71]. One proposed scheme is shown in Fig. 38. The relationships were based on similarities of coenzyme-binding domains, and left open the origin of the catalytic domains. Recent evidence for alcohol dehydrogenases and polyol dehydrogenases suggests an early evolution of different subunit sizes before the evolution of specificity, resulting in different ancestor types from which different specificities, quaternary structures and other enzyme properties evolved [6].

5. Conclusions

General properties of polypeptide chains cause formation of α-helices and β-sheets. The combination of these secondary structures in dehydrogenases leads to tertiary structures that are not uniform, but which show striking similarities, for example in the coenzyme binding domains of NAD-dependent dehydrogenases (Rossmann folds). These are probably special cases of structures favourable for binding, situated outside the carboxy end of parallel β-strands at places where there is a reversal of strand order (Brändén crevices). In regions near the N-terminal end of α-helices (Hol zones), the dipole of the helix could assist the binding of negatively-charged groups, or the polarization of groups for catalysis.

The gross structural characteristics of dehydrogenases therefore seem to derive from general properties of polypeptide chains, and it is found that similar tertiary structures have resulted from considerably different amino acid sequences.

The steric course of reaction is determined by relatively non-specific interactions, such as van der Waal's contacts that would not allow a bulky group to be accommodated the other way round, and by specific interactions such as hydrogen bonds.

The evolutionary pressure on stereospecificity must be related to the advantage it gives the organism. Whether, for example, the glutamate dehydrogenase utilizes D- or L-glutamate is obviously important to the organism. On the other hand, it seems of no importance whether a dehydrogenase is 4-*pro-R* or 4-*pro-S* specific with respect to coenzyme, though the fact is that this property is highly conserved. Possibly this reflects the difficulty of achieving adjustments at the catalytic site.

It might at first seem wholly unimportant whether an aldehyde is reduced by *Re* or *Si* attack; the same primary alcohol will result. However, if a related ketone serves as a poor substrate, the stereospecificity will determine whether the (*R*) or the (*S*) secondary alcohol is formed as an accompanying product. One or other may be advantageous to the organism: for example, glycollate plus L-lactate as opposed to glycollate plus D-lactate.

It is the spirit of the 1980s that all scientific endeavour should be useful. Finally, then, let it be recorded that the stereochemistry of dehydrogenases has important applications in biochemical analysis, in small-scale stereospecific syntheses, and in the design of therapeutic substances.

Acknowledgements

I am most grateful to Margaret Adams, Patrick Argos, Carl Brändén, Jan Drenth, Joseph Kraut, Michael Rossmann and Georg Schulz for kindly providing structural diagrams based on their findings, and allowing me to use them. The cooperation of Academic Press, American Society, American Society of Biological Chemists, Birkhäuser Verlag, Cambridge University Press, Federation of European Biochemical Societies, Macmillan Journals, National Academy of Sciences of the U.S.A., and Oxford University Press is also gratefully acknowledged.

I wish to express my thanks to Margaret Adams, Carl Brändén, Geoffrey Flynn, Motoji Fujioka, Anthony Turner, Bendicht Wermuth and Michael Zeppezauer for allowing me to read some of their papers before publication, and I am grateful to them and to Hans Grisebach, Anna Olomucki, David Parker and John Walker for providing reprints of their published work.

References

1 International Union of Biochemistry (1979) Enzyme Nomenclature, Recommendations (1978) of The Nomenclature Committee, Academic Press, New York, pp. 7, 28–134.
2 Jörnvall, H. (1977) Eur. J. Biochem. 72, 425–442.
3 Schwartz, M. and Jörnvall, H. (1976) Eur. J. Biochem. 68, 159–168.
4 Thatcher, D.R. (1980) Biochem. J. 187, 875–886.
5 Jeffery, J., Cummins, L., Carlquist, M. and Jörnvall, H. (1981) Eur. J. Biochem. 120, 229–234.
6 Jörnvall, H., Persson, M. and Jeffery, J. (1981) Proc. Natl. Acad. Sci. U.S.A. 78, 4226–4239.
7 Jörnvall, H. (1970) Eur. J. Biochem. 16, 25–40.
8 Jörnvall, H. (1970) Eur. J. Biochem. 16, 41–49.
9 Pietruszko, R. and Theorell, H. (1969) Arch. Biochem. Biophys. 131, 288–298.
10 Theorell, H. (1970) in H. Sund (Ed.), Pyridine Nucleotide-Dependent Dehydrogenases, Springer, Berlin, pp. 121–126.
11 Eklund, H., Brändén, C.-I. and Jörnvall, H. (1976) J. Mol. Biol. 102, 61–73.
12 Graves, J.L. and Vennesland, B. (1957) J. Biol. Chem. 226, 307–316.
13 Friedmann, H.C. and Vennesland, B. (1960) J. Biol. Chem. 235, 1526–1532.
14 Blattmann, P. and Rétey, J. (1972) Eur. J. Biochem. 30, 130–137.
15 Duane, W. and Hastings, J.W. (1975) Mol. Cell. Biochem. 6, 53–64.
16 McCapra, F. and Hysert, D.W. (1973) Biochem. Biophys. Res. Commun. 52, 298–304.
17 Dunn, D.K., Michaliszyn, G.A., Bogalki, I.G. and Meighen, E.A. (1973) Biochemistry 12, 4911–4918.
18 Puget, K. and Michelson, A.M. (1972) Biochemie 54, 1197–1204.
19 Fisher, J. and Walsh, C. (1974) J. Am. Chem. Soc. 96, 4345–4346.
20 Walker, W.H., Singer, T.P., Ghisla, S. and Hemmerich, P. (1972) Eur. J. Biochem. 26, 279–289.
21 Alston, T.A., Mela, L. and Bright, H.J. (1977) Proc. Natl. Acad. Sci. U.S.A. 74, 3767–3771.
22 Tchen, T.T. and van Milligan, H. (1960) J. Am. Chem. Soc. 82, 4115–4116.
23 Kenny, W.C. (1980) in K. Yagi and T. Yamano (Eds.), Flavins and Flavoproteins, Japan Scientific Societies Press, Tokyo, pp. 237–243.
24 Massey, V., Müller, F., Feldberg, R., Schuman, M., Sullivan, P.A., Howell, L.G., Mayhew, S.G., Matthews, R.G. and Foust, G.P. (1969) J. Biol. Chem. 244, 3999–4006.
25 Mayhew, S.G. and Ludwig, M.L. (1975) in P.D. Boyer (Ed.), The Enzymes, 3rd Edn., Vol. 12, Academic Press, New York, pp. 57–118.

26 Müller, F., Hemmerich, P., Ehrenberg, A., Palmer, G. and Massey, V. (1970) Eur. J. Biochem. 14, 185–196.
27 Westheimer, F.H., Fischer, H.F., Conn, E.E. and Vennesland, B. (1951) J. Am. Chem. Soc. 73, 2403.
28 Fisher, H.F., Conn, E.C., Vennesland, B. and Westheimer, F.H. (1953) J. Biol. Chem. 202, 687–697.
29 Pullman, M.E., San Pietro, A. and Colowick, S.P. (1954) J. Biol. Chem. 206, 129–141.
30 Cornforth, J.W., Ryback, G., Popják, G., Donninger, C. and Schroepfer, G.L. Jr. (1962) Biochem. Biophys. Res. Commun. 9, 371–375.
31 Cornforth, J.W., Cornforth, R.H., Donninger, C., Popják, G., Ryback, G. and Schroepfer, G.J. Jr. (1966) Proc. Roy. Soc. Lond. Ser. B 163, 436–464.
32 Cahn, R.S., Ingold, C. and Prelog, V. (1966) Angew. Chem. Int. Ed. Engl. 5, 385–415.
33 Hanson, K.R. (1966) J. Am. Chem. Soc. 88, 2731–2742.
34 Englard, S., Britten, J.S. and Listowsky, I. (1967) J. Biol. Chem. 242, 2255–2259.
35 Englard, S. and Hanson, K.R. (1969) Methods Enzymol. 13, 567–601.
36 Levy, H.R., Loewus, F.A. and Vennesland, B. (1957) J. Am. Chem. Soc. 79, 2949–2953.
37 Lemieux, R.U. and Howard, J. (1963) Can. J. Chem. 41, 308–316.
38 Levy, H.R. and Vennesland, B. (1957) J. Biol. Chem. 228, 85–96.
39 Nakamoto, T. and Vennesland, B. (1960) J. Biol. Chem. 235, 202–204.
40 Pfeil, E. (1951) Chem. Ber. 84, 229–245.
41 Woodward, R.B., Wendler, N.L. and Brutschy, F.J. (1945) J. Am. Chem. Soc. 67, 1425–1430.
42 Doering, W. Von E. and Young, R.W. (1950) J. Am. Chem. Soc. 72, 631.
43 Sarett, L.H., Feurer, M. and Folkers, K. (1951) J. Am. Chem. Soc. 73, 1777–1779.
44 Gardi, R., Vitali, R., Ercoli, A. and Klyne, W. (1965) Tetrahedron 21, 179–191.
45 Rhone, J.R. and Huffman, M.N. (1965) Tetrahedron Lett. 19, 1395–1398.
46 Jeffery, J. (1980) in J. Jeffery (Ed.), Dehydrogenases Requiring Nicotinamide Coenzymes, Birkhäuser, Basel, pp. 92–96.
47 Prelog, V. (1964) Pure Appl. Chem. 9, 119–130.
48 Dutler, H., van der Baan, J.L., Hochuli, E., Kis, Z., Taylor, K.E. and Prelog, V. (1977) Eur. J. Biochem. 75, 423–432.
49 Cronholm, T. and Fors, C. (1976) Eur. J. Biochem. 70, 83–87.
50 Helmchen, R.E. (1973) Dissertation, ETH, Zurich, No. 4991.
51 Gibb, W. and Jeffery, J. (1972) Biochim. Biophys. Acta 268, 13–20.
52 Garavito, R.M., Rossmann, M.G., Argos, P. and Eventoff, W. (1977) Biochemistry 16, 5065–5071.
53 Schulz, G.E., Schirmer, R.H., Sachsenheimer, W. and Pai, E.F. (1978) Nature (London) 273, 120–124.
54 Rossmann, M.G., Moras, D. and Olsen, K.W. (1974) Nature (London) 250, 194–199.
55 Brändén, C.-I. and Eklund, H. (1980) in J. Jeffery (Ed.), Dehydrogenases Requiring Nicotinamide Coenzymes, Birkhäuser, Basel, pp. 40–84.
56 Schulz, G. (1980) J. Mol. Biol. 138, 335–347.
57 Mannervik, B., Boggaram, V., Carlberg, I. and Larson, K. (1980) in K. Yagi and T. Yamano (Eds.), Flavins and Flavoproteins, Japan Scientific Societies Press, Tokyo, pp. 173–187.
58 Stern, B.K. and Vennesland, B. (1960) J. Biol. Chem. 235, 209–212.
59 Entsch, B., Ballou, D.P. and Massey, V. (1976) J. Biol. Chem. 251, 2550–2563.
60 Wierenga, R.K., de Jong, R.J., Kalk, K.H., Hol, W.G.J. and Drenth, J. (1979) J. Mol. Biol. 131, 55–73.
61 Ohlsson, I., Nordström, B. and Brändén, C.-I. (1974) J. Mol. Biol. 89, 339–354.
62 Hofsteenge, J., Vereijken, J.M., Weijer, W.J., Beintema, J.J., Wierenga, R.K. and Drenth, J. (1980) Eur. J. Biochem. 113, 141–150.
63 Ronchi, S., Minchiotti, L., Curti, B., Zapponi, C. and Bridgen, J. (1976) Biochim. Biophys. Acta 427, 634–643.
64 Pastore, E.J. and Friedkin, M.J. (1962) J. Biol. Chem. 237, 3802–3810.
65 Pastore, E.J., Friedkin, M. and Jardetsky, O. (1963) J. Am. Chem. Soc. 85, 3058–3059.
66 Pastore, E.J. and Williamson, K.L. (1968) Fed. Proc. 27, 764.

67 Matthews, D.A., Alden, R.A., Bolin, J.T., Freer, S.T., Hamlin, R., Xuong, N., Kraut, J., Poe, M., Williams, M. and Hoogsteen, K. (1977) Science 197, 452–455.
68 Matthews, D.A., Alden, R.A., Bolin, J.T., Filman, D.J., Freer, S.T., Hamlin, R., Hol, W.G.J., Kisliuk, R.L., Pastore, E.J., Plante, L.T., Xuong, N. and Kraut, J. (1978) J. Biol. Chem. 253, 6946–6954.
69 Matthews, D.A., Alden, R.A., Freer, S.T., Xuong, N. and Kraut, J. (1979) J. Biol. Chem. 254, 4144–4151.
70 Bitar, K.G., Blankenship, D.T., Walsh, K.A., Dunlap, R.B., Reddy, A.V. and Freisheim, J.H. (1977) FEBS Lett. 80, 119–122.
71 Adams, M.J., Archibald, I.G., Helliwell, J.R., Jenkins, S.E. and White, S.W. (1981) in G. Dodson, J.P. Glusker and D. Sayre (Eds.), Structural Studies on Molecules of Biological Interest, Oxford University Press, Oxford, pp. 328–338.
72 Brändén, C.-I. (1980) Quart. Rev. Biophys. 13, 317–338.
73 Adams, M.J., Helliwell, J.R. and Bugg, C.E. (1977) J. Mol. Biol. 112, 183–197.
74 Adams, M.J., Archibald, I.G. and Helliwell, J.R. (1977) in H. Sund (Ed.), Pyridine Nucleotide Dependent Dehydrogenases, de Gruyter, Berlin, pp. 72–83.
75 Abdallah, M.A., Adams, M.J., Archibald, I.G., Biellmann, J.-F., Helliwell, J.R. and Jenkins, S.E. (1979) Eur. J. Biochem. 98, 121–130.
76 Stern, B.K. and Vennesland, B. (1960) J. Biol. Chem. 235, 205–208.
77 Lienhard, G.E. and Rose, I.A. (1964) Biochemistry 3, 190–195.
78 White, J.L., Hackert, M.L., Buehner, M., Adams, M.J., Ford, G.C., Lentz, P.J. Jr., Smiley, I.E., Steindel, S.J. and Rossmann, M.G. (1976) J. Mol. Biol. 102, 759–779.
79 Holbrook, J.J., Liljas, A., Steindel, S.J. and Rossmann, M.G. (1975) in P.D. Boyer (Ed.), The Enzymes, 3rd Edn., Academic Press, New York, pp. 191–292.
80 Rossmann, M.G., Adams, M.J., Buehner, M., Ford, G.C., Hackert, M.L., Lentz, P.J. Jr., McPherson, A. Jr., Schevitz, R.W. and Smiley, I.E. (1971) Cold Spring Harbour Symp. Quant. Biol. 36, 179–191.
81 Taylor, S.S., Oxley, S.S., Allison, W.S. and Kaplan, N.O. (1973) Proc. Natl. Acad. Sci. U.S.A. 70, 1790–1794.
82 Taylor, S.S. (1977) J. Biol. Chem. 252, 1799–1806.
83 Hill, E.J., Tsernoglou, D., Webb, L.E. and Banaszak, L.J. (1972) J. Mol. Biol. 72, 577–591.
84 Webb, L.E., Hill, E.J. and Banaszak, L.J. (1973) Biochemistry 12, 5101–5109.
85 Rao, S.T. and Rossmann, M.G. (1973) J. Mol. Biol. 76, 241–256.
86 Lodola, A., Shore, J.D., Parker, D.M. and Holbrook, J.J. (1978) Biochem. J. 175, 987–998.
87 Graves, J.L., Vennesland, B., Utter, M.F. and Pennington, R.J. (1956) J. Biol. Chem. 223, 551–557.
88 Loewus, F.A., Levy, H.R. and Vennesland, B. (1956) J. Biol. Chem. 223, 589–597.
89 Harris, J.I. and Waters, M. (1976) in P.D. Boyer (Ed.), The Enzymes, 3rd Edn., Vol. 13, Academic Press, New York, pp. 1–49.
90 Davidson, B.E., Sajgó, M., Noller, H.F. and Harris, J.I. (1967) Nature (London) 216, 1181–1185.
91 Harris, J.I. and Perham, R.N. (1968) Nature (London) 219, 1025–1028.
92 Walker, J.E., Carne, A.F., Runswick, M.J., Bridgen, J. and Harris, J.I. (1980) Eur. J. Biochem. 108, 549–565.
93 Otto, J., Machleidt, W., Wachter, E., Rückl, G. and Machleidt, I. (1981) in preparation.
94 Otto, J., Argos, P. and Rossmann, M.G. (1980) Eur. J. Biochem. 109, 325–330.
95 Smith, E.L., Austen, B.M., Blumenthal, K.M. and Nyc, J.F. (1975) in P.D. Boyer (Ed.), The Enzymes, 3rd Edn., Vol. 11, Academic Press, New York, pp. 294–367.
96 Birktoft, J.J., Miake, F., Banaszak, L.J. and Frieden, C. (1979) J. Biol. Chem. 254, 4915–4918.
97 Birktoft, J.J., Miake, F., Frieden, C. and Banaszak, L.J. (1980) J. Mol. Biol. 138, 145–148.
98 Wooton, J.C. (1974) Nature (London) 252, 542–546.
99 Austen, B.M., Haberland, M.E., Nyc, J.F. and Smith, E.L. (1977) J. Biol. Chem. 252, 8142–8149.
100 Ohashima, T. and Soda, K. (1979) Eur. J. Biochem. 100, 29–39.
101 Cross, D.G. and Fisher, H.F. (1970) J. Biol. Chem. 245, 2612–2621.
102 Fujioka, M. and Takata, Y. (1979) Biochim. Biophys. Acta 570, 210–212.
103 Ogawa, H. and Fujioka, M. (1978) J. Biol. Chem. 253, 3666–3670.

104 Ogawa, H., Okamoto, M. and Fujioka, M. (1979) J. Biol. Chem. 254, 7030–7035.
105 Fujioka, M., Takata, Y., Ogawa, H. and Okamoto, M. (1980) J. Biol. Chem. 255, 937–942.
106 Ogawa, H. and Fujioka, M. (1980) J. Biol. Chem. 255, 7420–7425.
107 Fujioka, M. and Takata, Y. (1981) Biochemistry 20, 468–472.
108 Ogawa, H., Hase, T. and Fujioka, M. (1980) Biochim. Biophys. Acta 623, 225–228.
109 Jörnvall, H. (1980) in J. Jeffery (Ed.), Dehydrogenases Requiring Nicotinamide Coenzymes, Birkhäuser, Basel, pp. 126–148.
110 Grieshaber, M. and Gäde, G. (1976) J. Comp. Physiol. 108, 225–232.
111 Hochachka, P.W., Hartline, P.H. and Fields, J.H.A. (1977) Science 195, 72–74.
112 Gäde, G. and Grieshaber, M. (1975) J. Comp. Physiol. 102, 149–158.
113 Grieshaber, M. and Gäde, G. (1977) Comp. Biochem. Physiol. 58B, 249–252.
114 Olomucki, A., Huc, C., Lefebure, F. and Thoai, N.v. (1972) Eur. J. Biochem. 28, 261–268.
115 Biellmann, J.-F., Branlant, G. and Olomucki, A. (1973) FEBS Lett. 32, 254–256.
116 Eklund, H., Nordström, B., Zeppezauer, E., Söderlund, G., Ohlsson, I., Boiwe, T., Söderberg, B.-P., Tapia, O. and Brändén, C.-I. (1976) J. Mol. Biol. 102, 27–59.
117 Brändén, C.-I., Jörnvall, H., Eklund, H, and Furugren, B. (1975) in P.D. Boyer (Ed.), The Enzymes, 3rd Edn., Vol. 11, Academic Press, New York, pp. 103–190.
118 Cleland, W.W. (1977) Adv. Enzymol. 45, 273–387.
119 Maret, W., Andersson, I., Dietrich, H., Schneider-Bernlöhr, H., Einarsson, R. and Zeppezauer, M. (1979) Eur. J. Biochem. 98, 501–512.
120 Andersson, I., Maret, W., Zeppezauer, M., Brown, R.D. and Koenig, S.H. (1981) Biochemistry 20, 3424–3432.
121 Andersson, I., Maret, W., Zeppezauer, M., Brown, R.D. and Koenig, S.H. (1981) Biochemistry 20, 3433–3438.
122 Andersson, I., Maret, W., Zeppezauer, M., Brown, R.D. and Koenig, S.H. (1981) Biochemistry 20, 3424–3432.
123 Eklund, H. and Brändén, C.-I. (1979) J. Biol. Chem. 254, 3458–3461.
124 Brändén, C.-I. (1977) in H. Sund (Ed.), Pyridine Nucleotide-Dependent Dehydrogenases, de Gruyter, Berlin, pp. 325–334.
125 Whittle, S.R. and Turner, A.J. (1981) Biochim. Biophys. Acta 657, 94–105.
126 Boghosian, R.A. and McGuinness, E.T. (1979) Biochim. Biophys. Acta 567, 278–286.
127 Turner, A.J. and Hryszko, J. (1980) Biochim. Biophys. Acta 613, 256–265.
128 Tulsiani, D.R.P. and Touster, O. (1977) J. Biol. Chem. 252, 2545–2550.
129 Beedle, A.S., Rees, H.R. and Goodwin, T.W. (1974) Biochem. J. 139, 205–209.
130 Turner, A.J. and Hick, P.E. (1976) Biochem. J. 159, 819–822.
131 Felsted, R.L., Richter, D.R. and Bachur, N.R. (1977) Biochem. Pharmacol. 26, 1117–1124.
132 Bosron, W.F. and Prairie, R.L. (1973) Arch. Biochem. Biophys. 154, 166–172.
133 Feldman, H.B., Szczepanik, P.A., Harve, P., Corrall, R.J.M., Yu, L.C., Rodman, H.M., Rosner, B.A., Klein, P.D. and Landau, B.R. (1977) Biochim. Biophys. Acta 480, 14–20.
134 Clements, R.S. and Winegrad, A.I. (1972) Biochem. Biophys. Res. Commun. 47, 1473–1479.
135 Wermuth, B., Münch, J.D.B. and von Wartburg, J.P. (1979) Experientia 35, 1288–1289.
136 Wermuth, B. and Münch, J.D.B. (1979) Biochem. Pharmacol. 28, 1431–1433.
137 Wermuth, B., Münch, J.D.B. and von Wartburg, J.P. (1977) J. Biol. Chem. 252, 3821–3828.
138 Flynn, T.G., Shires, J. and Walton, D.J. (1975) J. Biol. Chem. 250, 2933–2940.
139 Branlant, G. and Biellmann, J.-F. (1980) Eur. J. Biochem. 105, 611–621.
140 Hoffman, P.L., Wermuth, B. and von Wartburg, J.-P. (1980) J. Neurochem. 35, 354–366.
141 Davidson, W.S., Walton, D.J. and Flynn, T.G. (1978) Comp. Biochem. Physiol. 60B, 309–315.
142 Felsted, R.L., Gee, M. and Bachur, N.R. (1974) J. Biol. Chem. 249, 3672–3679.
143 Wermuth, B. (1981) J. Biol. Chem. 256, 1206–1213.
144 Ris, M.M. and von Wartburg, J.-P. (1973) Eur. J. Biochem. 37, 69–77.
145 Ahmed, N.K., Felsted, R.L. and Bachur, N.R. (1978) Biochem. Pharmacol. 27, 2713–2719.
146 Lin, Y.M. and Jarabak, J. (1978) Biochim. Biophys. Res. Commun. 81, 1227–1234.

147 Grisebach, H. (1977) Naturwissenschaften 64, 619–625.
148 Wengenmayer, H., Ebel, J. and Grisebach, H. (1976) Eur. J. Biochem. 65, 529–536.
149 Gross, G.G. and Kreiten, W. (1975) FEBS Lett. 54, 259–262.
150 Wyrambik, D. and Grisebach, H. (1975) Eur. J. Biochem. 59, 9–15.
151 Wyrambik, D. and Grisebach, H. (1979) Eur. J. Biochem. 97, 503–509.
152 Von Wartburg, J.-P. and Wermuth, B. (1980) in W.B. Jakoby (Ed.), Enzymatic Basis of Detoxication, Vol. 1, Academic Press, New York, pp. 249–260.
153 Morpeth, F.F. and Dickinson, F.M. (1980) Biochem. J. 191, 619–626.
154 Levitt, M. and Chothia, C. (1976) Nature (London) 261, 552–557.
155 Chothia, C., Levitt, M. and Richardson, D. (1977) Proc. Natl. Acad. Sci. U.S.A. 74, 4130–4134.
156 Richardson, J.S. (1976) Proc. Natl. Acad. Sci. U.S.A. 73, 2613–2623.
157 Sternberg, M.J.E. and Thornton, J.M. (1976) J. Mol. Biol. 105, 367–382.
158 Nagano, K. (1977) J. Mol. Biol. 109, 235–250.
159 Hol, W.G.J., van Duijnen, P.T. and Berendsen, H.J.C. (1978) Nature (London) 273, 443–446.
160 Ptitsyn, O.B. and Finkelstein, A.V. (1980) Quart. Rev. Biophys. 13, 339–386.
161 Banner, D.W., Bloomer, A.C., Petsko, G.A., Phillips, D.C., Pogson, C.I., Wilson, I.A., Corran, P.H., Furth, A.J., Milman, J.D., Offord, R.E., Priddle, J.D. and Waley, S.G. (1975) Nature (London) 255, 609–614.
162 Levin, M., Muirhead, H., Stammers, D.K. and Stuart, D.I. (1981) Nature (London) 271, 626–630.
163 Lindqvist, Y. and Brändén, C.-I. (1980) J. Mol. Biol. 143, 201–211.
164 Argos, P., Garavito, R.M., Eventoff, W. and Rossmann, M.G. (1978) J. Mol. Biol. 126, 141–158.
165 Eventoff, W. and Rossmann, M.G. (1975) Crit. Rev. Biochem. 3, 111–140.
166 Moras, D., Olsen, K.W., Sabesan, M.N., Buehner, M., Ford, G.C. and Rossmann, M.G. (1975) J. Biol. Chem. 250, 9137–9162.
167 Buehner, M., Ford, G.C., Moras, D., Olsen, K.W. and Rossmann, M.G. (1973) Proc. Natl. Acad. Sci. U.S.A. 70, 3052–3054.
168 Buehner, M., Ford, G.C., Moras, D., Olsen, K.W. and Rossmann, M.G. (1974) J. Mol. Biol. 90, 25–49.
169 Eklund, H., Samama, J.-P., Wallén, L., Brändén, C.I., Åkeson, Å. and Jones, T.A. (1981) J. Mol. Biol. 146, 561–587.

Stereochemistry of pyridoxal phosphate-catalyzed reactions

HEINZ G. FLOSS and JOHN C. VEDERAS *

*Department of Medicinal Chemistry and Pharmacognosy, Purdue University, West Lafayette, IN 47907, U.S.A., and * Department of Chemistry, University of Alberta, Edmonton, Alberta, Canada T6G 2G2*

1. Introduction

In devising its synthetic strategies Nature has developed a number of molecules, enzyme substrates or cofactors, which are remarkable in the diversity of different reaction paths open to each of them. Probably the most versatile of these is pyridoxal phosphate (PLP), the essential cofactor of amino acid metabolism. This compound can initiate reactions that may lead to the cleavage of any of the four bonds at the α-carbon, to electrophilic or nucleophilic reactions at the β-carbon and even to bond cleavage and formation at the γ-carbon of α-amino acids. Thanks to this chemical versatility, PLP plays a pivotal role in connecting carbon and nitrogen metabolism, in the formation of biogenic amines, in providing an entry into the 'one-carbon pool' and in a number of other important processes. (For general reviews of various aspects of pyridoxal phosphate catalysis, see [1–5]; certain specialized aspects of pyridoxal phosphate enzymes are discussed in [6].) Underlying this multitude of different reactions is a simple, common mechanistic principle. The cofactor forms a Schiff's base with the amino group of the substrate and then acts as an electron sink transiently storing electrons which are freed in the cleavage of bonds until they are claimed again in a new bond-forming step.

The overall mechanistic features of PLP catalysis were established some time ago, primarily through the work of Braunstein and his associates in the U.S.S.R. [7] and Snell, Metzler and their coworkers in the U.S.A. [8]. The cofactor is present in the enzyme active site as a Schiff's base with the ϵ-amino group of a lysine residue. Transaldimination with the amino group of the substrate gives the coenzyme–substrate complex. Cleavage of one of the three C—C or C—H bonds at the α-carbon, aided by protonation of the pyridine nitrogen, gives rise to a resonance-stabilized carbanion at C-α (quinoid intermediate). This species can be reprotonated either at C-4′ of the cofactor, as in the transamination reaction, or at the α-carbon of the substrate (decarboxylation, α,β-cleavage, racemization) (Scheme I). Alternatively, this intermediate (1) may undergo elimination of a group from the β-carbon either as a cation or as an anion to generate the resonance-stabilized species 2 or 3,

Tamm (ed.) Stereochemistry

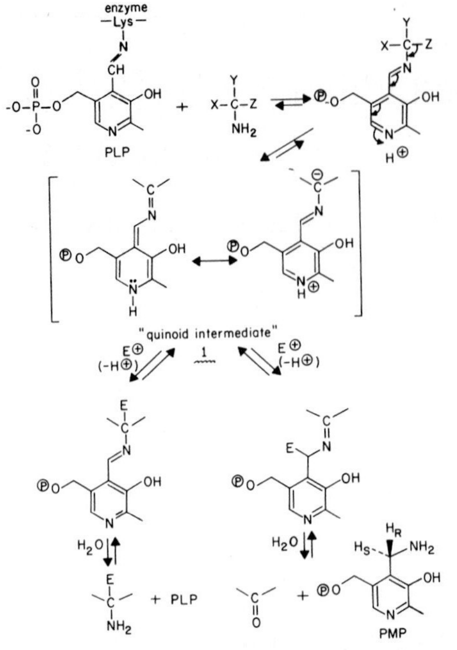

Scheme I. General mechanism of PLP-catalyzed reactions at the α-carbon atom of amino acids and amines.

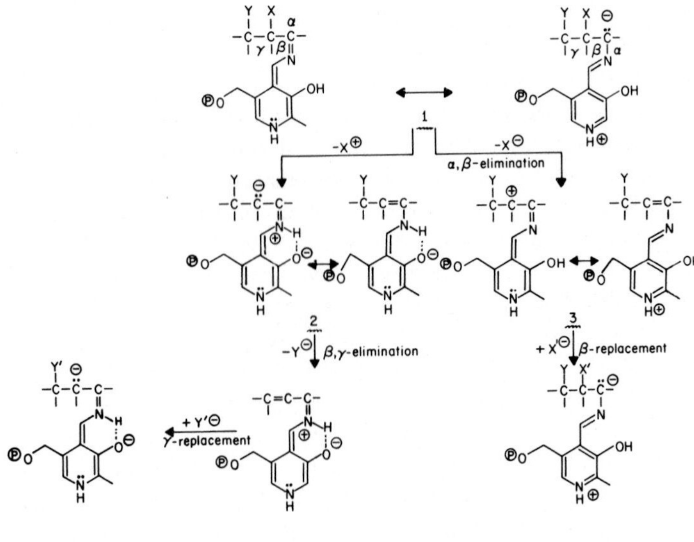

Scheme II. General mechanism of PLP-catalyzed reactions at the β- and γ-carbon atoms of amino acids.

respectively (Scheme II). Hydrolysis of species 3 gives a keto acid and ammonia, representing the α,β-elimination reaction, whereas addition of an electronegative group (X'^-) at C-β and a proton at C-α constitutes the β-replacement reaction. Species 2 can be protonated at C-β; this constitutes the essence of the aspartate-β-decarboxylase and kynureninase reactions. Alternatively, 2 may stabilize by elimination of an electronegative group at C-γ to give species 4. Protonation of the latter at C-γ followed by hydrolysis to a keto acid and ammonia represents the β,γ-elimination reaction, whereas addition of another electronegative group (Y'^-) at C-γ followed by protonation at C-β and C-α completes the γ-replacement reaction.

This, in a nutshell, summarizes the essential electronic features of PLP catalysis. Not infrequently there is crossover between different reaction paths, i.e., a given enzyme which has evolved to function in one reaction mode will, under certain conditions, also catalyze reactions of one of the other types. Such crossover is seen frequently, for example, between β-replacement and α,β-elimination or between protonation of species 1 at C-α and C-4′, in the latter case often leaving the enzyme in a catalytically inactive pyridoxamine phosphate (PMP) form.

2. *Stereochemical concepts of pyridoxal phosphate catalysis*

The stereochemistry of pyridoxal phosphate-catalyzed reactions was last summarized comprehensively in 1971 by Dunathan [2], who outlined many of the basic concepts in this field. Aspects of PLP catalysis have been discussed in other reviews on enzyme reaction stereochemistry (e.g., [9]), and a brief review, emphasizing their own work, has recently been published by the present authors [10]. Much work has been done in this field during the past ten years, most of it supporting the concepts laid out in Dunathan's review, often refining the picture and sometimes modifying the original ideas.

In light of the large number of reaction paths available to a PLP-amino acid (or amine) Schiff's base, one important function of the enzyme protein must be to impose reaction specificity upon the system. In 1966 Dunathan proposed [11] that this is achieved by control of the conformation around the C-α—N bond of the substrate–cofactor complex so as to orient the bond at C-α which is to be broken perpendicular to the plane of the conjugated π system. In this conformation the breaking σ bond achieves maximal orbital overlap with the π system, resulting in a substantial rate enhancement for the cleavage of that bond [12,13]. Implicit in this proposal is, of course, the idea that the pyridine ring, C-4′, the amino nitrogen and C-α of the coenzyme–substrate complex must lie in a plane, an essential prerequisite for resonance stabilization of species 1 (Scheme I). Hence conformers 5a, 5b and 5c (Scheme III), or the corresponding rotamers turned 180° around the C-α—N bond, must be the orientations of the complex maintained in enzymes catalyzing breakage of the C-α—H bond, the C-α—COOH bond and the C-α—C-β bond, respectively. If an enzyme binds the relatively rigid PLP cofactor at the pyridine nitrogen and at the phosphate, attachment of a single distal group on the substrate, in most cases

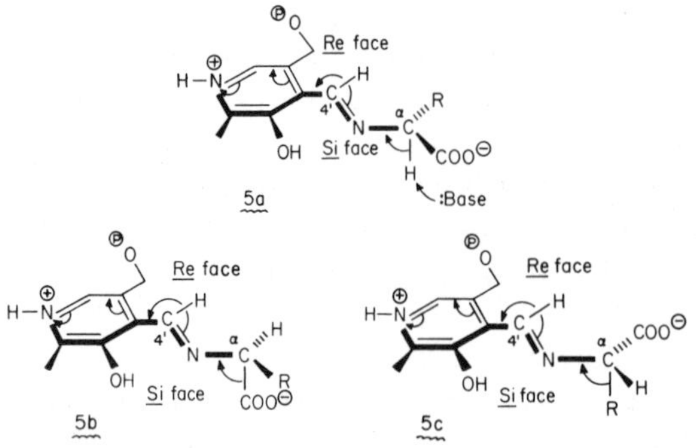

Scheme III. Optimal conformations about the C-α–N bond for cleavage of the C-α–H (5a), C-α–COOH (5b) or C-α–C-β (5c) bond of L-amino acids.

probably the carboxyl group, would result in a three-point binding of the complex fixing a particular conformation of the C-α—N bond. This concept is extremely plausible and appealing to chemical intuition. However, it must be remembered that the actual evidence supporting it is rather scant. Stereoelectronic control of reaction rates has, of course, been demonstrated in numerous systems, including amino acid/pyridoxal (PL) Schiff's bases [14] and substitution-inert metal complexes of amino acid/*o*-hydroxyaldehyde Schiff's bases [15,16]. In the first case it was shown that in the absence of enzyme the rates of racemization and α-hydrogen exchange of amino acid/PL Schiff's bases are determined by the proportion of conformer having the C-α—H-α bond orthogonal to the π plane. However, although there seem to be no contradictory data, there are only few results supporting the notion of conformational control of the C-α—N bond by the protein. Some of the best evidence comes from observed crossovers in reaction specificity, particularly the finding that L-serine transhydroxymethylase also transaminates D-alanine [17].

Another stereochemical concept which began to emerge at the time of Dunathan's review of the subject [2] is the idea that the reactions of PLP enzymes all take place on one face of the planar PLP–substrate complex (the 'exposed' or 'solvent' face), the other face being covered by the protein. This concept developed from studies on decarboxylases and transaminases, which uniformly had shown group interchanges in a retention mode and proton transfers with suprafacial geometry. It was also noted that the 'exposed' face is always the *Si* face at C-4′ of the cofactor (see Scheme III) leading to the subsequent suggestion [18] that this stereochemical constancy reflects evolution of PLP enzymes from a common progenitor, a 'grandfather enzyme'; an evolution during which an arbitrary choice between two equally likely stereochemical options has been preserved.

Finally, yet another concept, dealing with the spacial orientation of the cofactor in the protein, was developed by Ivanov and Karpeisky. These authors reasoned that the plane of the cofactor must rotate relative to the protein and the noncovalently bound substrate during the transition from the coenzyme–lysine Schiff's base to the coenzyme–substrate Schiff's base [19]. This movement of the cofactor during transaldimination, deduced from models to involve a 40° rotation around the C-2–C-5 axis and a 90° rotation around the C-4–C-4′ axis, was first proposed for the reaction catalyzed by aspartate transaminase and is supported by CD measurements [19]. It would be expected to be a feature of all PLP enzyme reactions involving transaldimination.

In the following discussion of more recent results on the stereochemistry of PLP enzyme reactions, some of the newer data will be analyzed in terms of these early concepts.

3. *Results on the stereochemistry of pyridoxal phosphate enzymes*

(a) Reactions at the α-carbon

(i) Transaminases

Aminotransferases (transaminases) catalyze the reversible interconversions of pairs of α-amino and α-keto acids or of terminal primary amines and the corresponding aldehydes by a 'shuttle mechanism' in which the enzyme alternates between its PLP form and the corresponding PMP form. In the first half-reaction the PLP form of the enzyme binds the amino acid (or amine) and forms the coenzyme–substrate Schiff's base. Cleavage of the C-α—H bond is then followed by protonation at C-4′. Hydrolysis of the resulting ketimine then gives a keto acid (or aldehyde), leaving the enzyme in the PMP form. The latter is recycled to the PLP form by condensation with an α-keto acid, deprotonation at C-4′, protonation at C-α and transaldimination to release the α-amino acid formed.

The steric course of the process involves the following five parameters (Scheme IV):

(a) Configuration at C-α of the substrate (D or L)
(b) Configuration of the C-4′—N double bond (*cis* or *trans*)
(c) Conformation of the C-α—N bond
(d) Site of proton addition at C-4′ (*Si* or *Re*)
(e) Mode of prototropic shift from C-α to C-4′ (suprafacial or antarafacial) (for definitions of stereochemical terms, see [20,21])

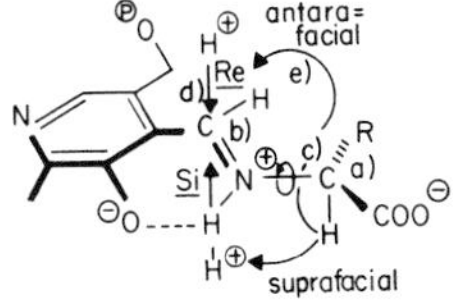

Scheme IV. Stereochemical parameters of enzymatic transamination.

Knowledge of four of these five parameters is sufficient to completely describe the system. In the case of α-amino acids the configuration at C-α is usually known; a given enzyme will transaminate either only L- or only D-amino acids. The configuration of the C-4′—N double bond must be *trans*; a *cis* double bond would be incompatible with coplanarity with the pyridine ring due to steric interference by the adjacent ring substituents. This leaves two parameters to be determined in order to fully define the stereochemistry of the process.

One of these parameters, the site of protonation at C-4′, was determined independently by the groups of Dunathan and of Arigoni for several L-amino acid transaminases. Both laboratories made use of the fact that various apo-transaminases, including apo-aspartate transaminase, can reversibly bind pyridoxal (PL) instead of PLP and convert it stoichiometrically into pyridoxamine (PM) in the presence of appropriate L-amino acids [22]. Dunathan et al. [23] showed that only one atom of deuterium is incorporated into PM when the reaction with apo-aspartate transaminase is carried out in ^2H_2O and only one atom of ^2H is removed from [4′-$^2H_2$]PM in H_2O. Experiments with synthetic samples of (4′*R*)- and (4′*S*)-[4′-$^2H_1$]PM suggested that the hydrogen mobilized by the enzyme occupies the *pro*-4′*S* position in PM. Besmer and Arigoni [24,25] generated PM tritiated stereospecifically at C-4′ by carrying out the reaction in ^3H_2O and established its configuration as 4′-*S* by a rigorous correlation with the known absolute configurations of stereospecifically labeled glycolic acid and benzylamine. The apo-aspartate transaminase system then served as a relay enzyme for the configurational analysis of PM samples generated with other transaminases. In this way it was shown that L-alanine transaminase [26], pyridoxamine-pyruvate transaminase [27] and dialkyl amino acid transaminase [28] also operate on the *pro-S* hydrogen at C-4′ of PM, the latter enzyme both in the transamination of L-alanine and in the decarboxylation/transamination of α-amino-isobutyrate. In addition, a number of 'abortive' transamination reactions, i.e., reactions involving an error in the protonation of the quinoid intermediate (1, Scheme I), catalyzed by other PLP enzymes have been shown to involve protonation at C-4′ with the same stereochemistry. These enzymes include L-glutamate decarboxylase (decarboxylation/transamination of α-methylglutamate) [29], serine hydroxymethyl transferase (transamination of D-alanine) [30], tryptophan synthase β_2 protein (dehydration/transamination of L-serine in the presence of 2-mercaptoethanol) [18] and aspartate-β-decarboxylase (decarboxylation/transamination of L-aspartate) [31]. Thus all these eight enzymes, which include 4 bona fide transaminases, protonate or deprotonate C-4′ of the coenzyme–substrate or coenzyme–intermediate complex on the *Si* face.

For two transaminases the remaining unknown stereochemical parameter was determined by demonstrating an internal transfer of tritium (dialkyl amino acid transaminase) [28] or deuterium (pyridoxamine-pyruvate transaminase) [27] from the α-position of the substrate L-alanine to C-4′ of the cofactor. Internal hydrogen transfer from the α-position of the substrate amino acid to C-4′ of PLP has also been demonstrated for two of the 'abortive' transamination reactions, those catalyzed by tryptophan synthase β_2 protein [32] and by aspartate-β-decarboxylase [31]. In addition, the same phenomenon must occur in alanine transaminase, as deduced from the observation that the enzyme catalyzes exchange of the β-hydrogens of

L-alanine at a substantially faster rate than of the α-hydrogen [33,34]. Internal proton transfer strongly suggests that the deprotonation and protonation are mediated by a single base, implying that the process must be suprafacial. Therefore the conformation of the C-α—N single bond must be such that the α-hydrogen is exposed on the side of the complex corresponding to the *Si* face at C-4′, i.e., in an L-amino acid-coenzyme complex the carboxyl group and C-4′ would be *trans* to each other (Scheme V). In the case of pyridoxamine-pyruvate transaminase the results

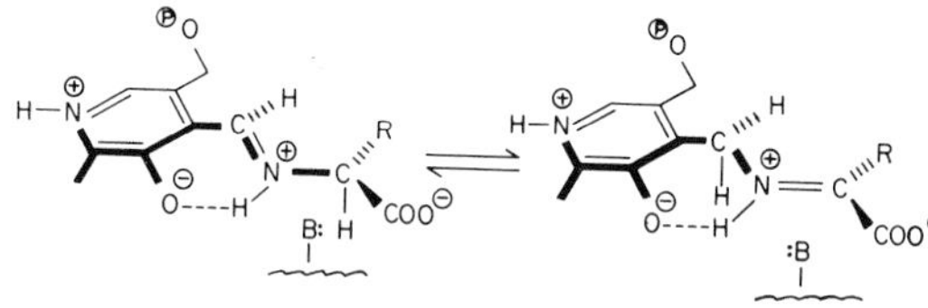

Scheme V. Stereochemistry of the aldimine–ketimine tautomerization in the transamination of an L-amino acid.

showed 2–4.5% internal transfer of deuterium when the reaction was carried out with L-[α-^{2}H]alanine in H_2O, but 50% transfer of normal hydrogen with unlabeled alanine in ^2H_2O [2]. This large difference in the transfer of ^{1}H vs. ^{2}H suggests that the transfer might be mediated by a polyprotic base, like the ϵ-amino group of a lysine residue. The hydrogen being transferred would thus be diluted by two other hydrogens, i.e., only 1/3 of the labeled hydrogen would be carried foreward. An isotope effect in the cleavage of one of the three equivalent N—H (N—^{2}H) bonds would further reduce the degree of transfer of ^{2}H from an $—NH_2{}^2H^+$ group below the statistical value of 33%, but would increase the degree of transfer of H from an $—NH^2H_2^+$ group. However, while the results are suggestive of a polyprotic base, it has to be kept in mind that they could also be merely fortuitous manifestations of conformational differences of the enzyme in H_2O and in ^2H_2O, which could in turn lead to differences in the relative rates of proton transfer and exchange with solvent.

A different approach to the determination of the conformation of the C-α—N bond was taken by Arigoni's laboratory. Working with aspartate transaminase, for which no internal hydrogen transfer could be demonstrated, these workers reduced an equilibrium mixture of enzyme, aspartate and oxalacetate with tritiated sodium borohydride [26]. From the reaction mixture phosphopyridoxylaspartic acid was isolated and hydrolyzed to pyridoxylaspartic acid. Further degradation by hydrogenolysis gave tritiated aspartic acid, whereas oxidation with hypochlorite produced tritiated pyridoxamine. The configurations at the labeled chiral centers of these two products were determined by incubation with aspartate transaminase and apo-aspartate transaminase, respectively, to be 96.5% *S* in the aspartic acid and 89% *S* at C-4′ of the PM. Analogous results were obtained upon reduction of the enzyme complex with glutamate and α-ketoglutarate. As outlined in Scheme VI these results allow a deduction of the conformation of the C-α—N single bond in the PLP–substrate aldimine, because the conformation of this bond correlates with the

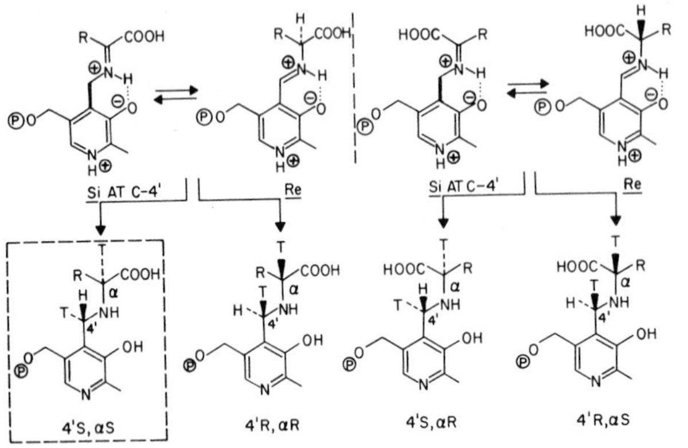

Scheme VI. Stereochemical possibilities in the reduction of an L-amino acid–transaminase equilibrium complex with tritiated $NaBH_4$.

configuration of the C-α—N double bond in the tautomeric ketimine. The finding that the tritiated chiral centers at C-4′ and C-α both have *S* configuration indicates that the C-α—H bond in the original coenzyme–substrate complex must be displayed on the *Si* face relative to C-4′. Together with the knowledge that protonation at C-4′ occurs on the *Si* face, this proves that the tautomerization step involves a suprafacial proton abstraction and addition (Scheme V). The same conclusion was reached for pig heart alanine transaminase by tritiated $NaBH_4$ reduction of the equilibrium complex of the enzyme with L-alanine and pyruvate [26].

The reduction experiments of Austermühle-Bertola also indicate that the Schiff's base is reduced primarily from one side. This is the same side on which the proton transfers take place in the enzymatic process. Hence, one face of the coenzyme–substrate complex is exposed to the water and can be approached by reagents like borohydride; the other face must be covered by the protein. Surprisingly, however, the tritiated borohydride reduction of aspartate transaminase in the absence of substrate at pH 8.1 [26] or 7.5 [35] or in the presence of the inhibitor glutarate at pH 5.05 [26] gave, after hydrolysis of the protein, pyridoxyllysine in which the tritium occupied the *pro-R* position at C-4′ to the extent of 88.6%, 90% and 70.5%, respectively. Similarly, reduction of holo-alanine transaminase gave pyridoxyllysine carrying 78% of the tritium in the *pro*-4′*R* position [26]. Hence in both these enzymes the PLP-lysine Schiff's base in the holoenzyme is reduced from the *Re* face, indicating that during the transaldimination to the PLP–substrate complex the cofactor must undergo a conformational change, relative to the protein, which exposes opposite faces of C-4′ to the solvent in the two aldimines. Analogous results have been obtained in experiments on tyrosine decarboxylase [36]. Two types of motion could account for this conformational change (Scheme VII) [36]. In the first of these [3,19,37] the imine nitrogen is always on the same side of the C-4—C-4′ bond as the phenolic OH (*cisoid* conformation), and rotation about an axis through

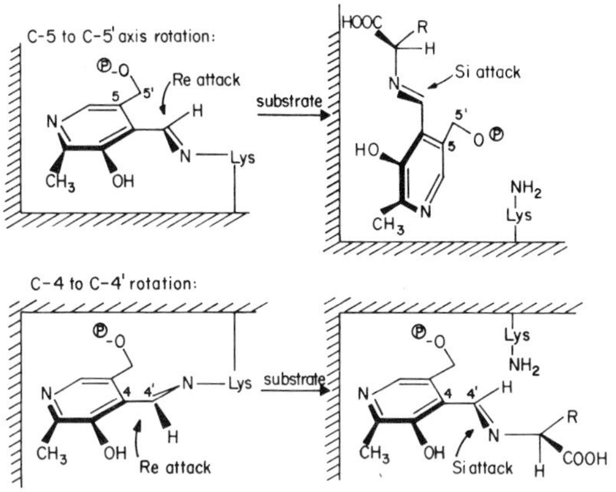

Scheme VII. Two possible modes of cofactor reorientation during transaldimination from PLP–lysine to PLP–substrate Schiff's base.

the C-5—C-5′ bond during substrate binding exposes the other face of both the imine bond and the pyridine ring. Model studies and calculations show that in the absence of enzyme, pyridoxal Schiff's bases prefer a *cisoid* conformation [37,38]. Another possibility is that the pyridine ring keeps the same side exposed while rotation about the C-4—C-4′ bond during substrate binding exposes a new face of the imine bond; this is a *transoid* to *cisoid* reorientation. Such a conformational change has been suggested for aspartate aminotransferase on the basis of absorption spectra [39,40]. Modifications of aspartate aminotransferases have also indicated conformational changes during substrate binding [41,42]. The observation of a change in the exposed face of C-4′ of the cofactor thus fits neatly into the concept [19] that a reorientation of the cofactor is an essential part of the catalytic process. Recent X-ray diffraction studies by Jansonius' group [43,44], which led to a 2.8 Å map of mitochondrial aspartate transaminase, have considerably refined the picture. In the holoenzyme the cofactor–lysine Schiff's base is in the *cisoid* conformation with the C-4′—N double bond roughly coplanar with the pyridine ring, the *Si* face against a β sheet of the protein and the *Re* face partially exposed to solvent. In the catalytic process the pyridine ring then undergoes a reorientation "in a manner similar, but not identical, to that proposed by Ivanov and Karpeisky [19]" [44]. According to the authors this reorientation is compatible with a change to exposure of the *Si* face in the PLP–substrate aldimine. The work also defines the interactions involved in substrate binding and indicates that the base mediating α-hydrogen abstraction cannot be a histidine, as had been postulated [45]. However, tyrosine-70 (and, presumably, lysine-258) are located in proximity to the C-α—H bond. On the other hand, some recent results by Zito and Martinez-Carrion [35] cast doubts on the assumption that the change in accessibility of C-4′ to $NaBH_4$ is in any simple way related to the spacial requirements for the catalytic process. These authors reduced modified aspartate transaminase, in which lysine-258, the residue binding PLP as a Schiff's base in the holoenzyme, was carbamylated, in the presence of

L-aspartate. They observed that tritium was introduced at C-4′ of the cofactor from the *Re* face to the extent of over 90%. Thus the exposed face of the coenzyme–substrate complex in the modified enzyme is the *Re* face as in the holoenzyme rather than the *Si* face as in the complex of the normal enzyme with L-aspartate. Yet this modified enzyme is still able to undergo the half-transamination reaction with conversion of active site bound PLP to PMP [45] and does so with stereospecific protonation of the cofactor from the *Si* face [46]. According to this result, a change in the exposed face of the cofactor upon transaldimination or a conformation in which the *Si* face of C-4′ is exposed in the coenzyme–substrate complex are not requirements for catalytic activity. Likewise, the face on which reactions take place in the catalytic process is not necessarily the face that is most accessible to external reagents.

In addition to the transaminases acting at the α-carbon of α-amino acids there are a number of enzymes which catalyze the transamination of other amino groups, mostly interconversions between amino groups at primary carbons and the corresponding aldehydes, for example, the terminal amino groups of diamino acids like ornithine [48] or lysine [49] or of γ-aminobutyric acid (GABA) [50]. The stereochemistry of hydrogen removal from C-γ of GABA during transamination with three enzymes, bacterial [51,52] and mammalian [52] GABA transaminase and a ω-amino acid: pyruvate transaminase from a *Pseudomonas* species [51] has been studied. Making use of the known stereochemistry of bacterial glutamate decarboxylase [53], both groups prepared (4*R*)- and (4*S*)-[4-^{3}H]GABA by decarboxylation of L-glutamate in ^3H_2O and [2-^{3}H]glutamate in H_2O, respectively. Transamination of these samples with ω-amino acid: pyruvate aminotransferase gave succinic semialdehyde with predominant loss of tritium from the 4*R* isomer and predominant tritium retention from the 4*S* form. Hence, this enzyme removes the *pro-R* hydrogen [51]. The opposite stereochemical preference, removal of the *pro*-4*S* hydrogen is shown by both GABA transaminases [51,52]. Recent unpublished work in the laboratory of Soda (K. Soda, personal communication) with 6*R*- and 6*S*-[6-^{3}H]lysine has shown that L-lysine ϵ-aminotransferase also removes the *pro-S* hydrogen whereas the tritium from the 6*R* isomer is retained in the product, Δ^1-piperideine 6-carboxylate. L-Ornithine δ-aminotransferase stereospecifically equilibrates the *pro-S* hydrogen at C-5 of L-ornithine with solvent protons and thus shows the same steric preference.

(ii) Racemases

A substantial number of PLP enzymes catalyze the racemization or epimerization of primary α-amino acids [54,55]. Of particular physiological importance are microbial alanine racemases because of their involvement in bacterial cell wall formation, which makes them a potential target for chemotherapy. An interesting substrate specificity is exhibited by diaminopimelate racemase [56] which acts only on meso- and LL-diaminopimelate, but not on the DD-isomer, i.e., the enzyme requires the L configuration at one end of the molecule in order to epimerize the chiral center at the other end. Racemization is also occasionally observed as an alternate catalytic activity of other PLP enzymes; most notable is the racemization of alanine catalyzed by tyrosine phenol-lyase from *E. intermedia* [57].

The detailed mechanism of action of these enzymes is not too well understood. α-Hydrogen exchange accompanying racemization has been demonstrated in most cases examined, indicating that the reaction probably proceeds through a resonance-stabilized carbanion at C-α (quinoid intermediate 1, Scheme I). The distinguishing feature of racemases, compared to other PLP enzymes, obviously is their ability to protonate C-α in this species on either face. This could be accomplished by two acid/base groups situated on opposite sides of the coenzyme–substrate complex, as in the case of the nonPLP-dependent proline racemase [58], or by a single base which first abstracts a proton on one face and then adds it back either on the same or the opposite face. In the latter case either the base must move relative to the substrate or the substrate relative to the base during the catalytic process. Comparative measurements of rates of α-hydrogen exchange and racemization with several enzymes have not led to any clear distinction between these possibilities, although the data obtained on alanine racemase from *Pseudomonas* have been interpreted to suggest a two-base mechanism [55] as shown in Scheme VIII. On the other hand,

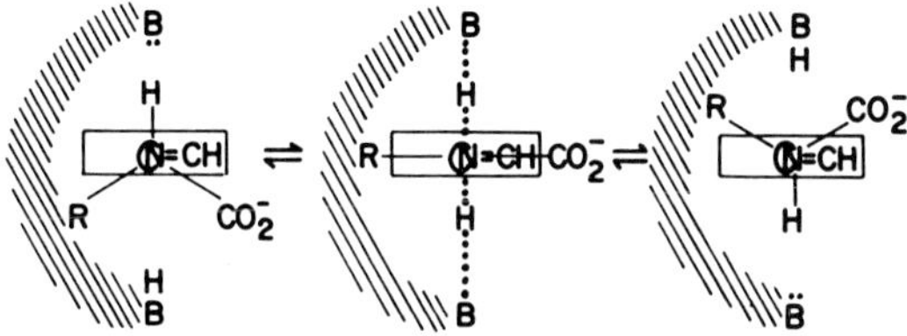

Scheme VIII. Proposed two-base mechanism for amino acid racemization by PLP-dependent enzyme (according to [55]).

Henderson and Johnston [59] based on their work with alanine racemase from *B. subtilis* proposed catalysis by a single base with motion of the coenzyme–substrate complex in a process they termed a 'swinging door' mechanism (Scheme IX). Their proposed mechanism is based on the observation of pronounced asymmetry in the interaction of many racemases with their enantiomeric substrates [55], for example, in the inactivation of the enzyme by D- and L-β-chloroalanine [59]. The position of the 'door' relative to the point of protonation/deprotonation would determine whether a D- or an L-amino acid is interacting with the enzyme, and if one position were more stable than the other, the enzyme would interact differently with a D- and an L-isomer.

There is at the moment no compelling evidence for either of these mechanisms. An important experiment which needs to be done with enzymes of this class is to probe for internal transfer of the α-hydrogen from one enantiomer to the other under single turnover conditions with trapping of the product. An experimental design to accomplish this is currently being explored with tyrosine phenol-lyase and will be discussed below. Demonstration of any internal return of the α-hydrogen

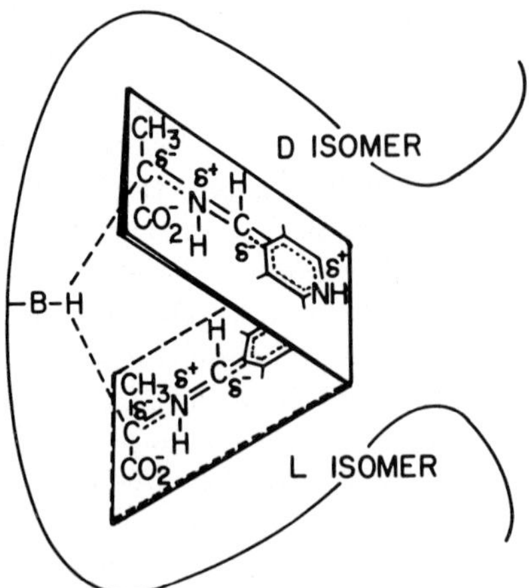

Scheme IX. 'Swinging door' mechanism for PLP-dependent enzymatic amino acid racemization (from [57]).

would strongly favor a single base mechanism. Failure to demonstrate such internal transfer, however, would be inconclusive as it would be compatible both with a two-base mechanism and with a single base mechanism in which exchange of the abstracted α-hydrogen with solvent is fast compared to foreward transfer.

(iii) Decarboxylases

PLP-dependent amino acid α-decarboxylases [60] should bind the coenzyme–substrate complex in a conformation aligning the C-α—COOH bond perpendicular to the plane of the system, favoring cleavage of that bond [11]. Protonation at C-α then completes the reaction. In accord with this concept early work on several bacterial decarboxylases has shown ([61], cf. [2]) incorporation of only one solvent proton into the decarboxylation product and lack of exchange of the other methylene hydrogen of the amine (or of the α-hydrogen of the amino acid) by the enzyme. The stereochemistry of this process, replacement of the carboxyl group by a proton in a retention or inversion mode, was first determined for the decarboxylation of L-tyrosine with tyrosine decarboxylase from *Streptococcus faecalis* by Belleau and Burba [62]. These authors determined the stereochemistry of tyramine oxidation with monoamine oxidase using synthetic *R*- and *S*-[1-^{2}H]tyramine and measurement of the isotope effect on reaction rate. They then prepared two samples of tyramine by decarboxylation of L-tyrosine in ^2H_2O and of L-[2-^{2}H]tyrosine in H_2O and showed, by measurement of their rates of oxidation with monoamine oxidase, that the first sample had *R* and the second had *S* configuration. Hence the newly introduced hydrogen occupies the *pro-R* position, replacing the carboxyl group in a retention mode. The results of this classical study have been confirmed recently by more modern methodology [63].

The stereochemistry of a number of amino acid decarboxylases acting on a variety of substrates has since been determined. These include the decarboxylation of *S*-3-(3,4-dihydroxyphenyl)-2-methylalanine (α-methyldopa) by aromatic amino acid decarboxylase in man [64], of L-tyrosine [63] and L-tryptophan [65] by the same enzyme from hog kidney, of L-lysine by L-lysine decarboxylase from *Bacillus cadaveris* [66,67] and *E. coli* [67], of L-glutamic acid [53,68] and *S*-α-methylglutamic acid [53] by glutamate decarboxylase from *E. coli* and of L-glutamic acid by the mammalian enzyme [52], of L-histidine by mammalian histidine decarboxylase [69], of L-ornithine by bacterial ornithine decarboxylase (K. Soda, personal communication), of L-tryptophan by crude tyrosine decarboxylase from *S. faecalis* [70], and, by an indirect determination, the decarboxylation of L-tyrosine by an enzyme in *Papaver somniferum* [53]. All these decarboxylases act on amino acids of *S* configuration and the steric course of the reaction was in each case found to be retention. Thus in all these enzymes cleavage of the C-α—COOH bond and proton addition occur on only one face of the coenzyme–substrate complex. Interestingly, decarboxylation in a retention mode was also observed for histidine decarboxylase from *Clostridium welchii* [71,72] and from a *Lactobacillus* [71], enzymes which do not require PLP as a cofactor but contain a pyruvyl residue in the active site.

Attempts have been made with two enzymes to define in absolute terms the face on which the reaction takes place. Glutamate decarboxylase undergoes slow abortive transamination during decarboxylation of L-glutamate or α-methylglutamate, and Sukhareva et al. [29] have shown that in this process a proton is added at C-4′ on the *Si* face. Assuming that this erroneous protonation is mediated by the same base that normally protonates C-α, one would conclude that the C-α—COOH bond must be displayed on the side corresponding to the *Si* face at C-4′. Vederas and coworkers [36] reduced several Schiff's base complexes of tyrosine decarboxylase from *Streptococcus faecalis* with tritiated $NaBH_4$. Reduction of the internal PLP–lysine Schiff's base in the absence of substrates occurred primarily (78–98%) from the *Re* face. However, after addition of either L-tyrosine or the product, tyramine, to the enzyme reduction proceeded mostly (72–77%) from the *Si* face at C-4′. Tyrosine decarboxylase thus undergoes a similar conformational change upon substrate binding and transaldimination as has been seen with the alanine and aspartate transaminases [26], in which the opposite face of the cofactor becomes accessible to external reagents. If the parallel to these two transaminases extends further to the fact that the solvent exposed face of the coenzyme–substrate complex is also the face on which protonation occurs, one would again be led to the conclusion that the carboxyl group must be aligned on the *Si* face relative to C-4′. The available evidence thus strongly suggests, although does not prove, the stereochemical mechanism shown in Scheme X for PLP-dependent L-amino acid α-decarboxylases. There is considerable evidence that the conformation of the system is controlled to a large extent by binding of a distal group in the fully extended side chain of the substrate amino acid ([73,74], cf. [75]).

Recently, the group of Soda [76] examined the stereochemistry of the reaction catalyzed by meso-α, ϵ-diaminopimelate decarboxylase from *Bacillus sphaericus*. This enzyme, which they purified to homogeneity [70], shows similarly unique substrate specificity as the metabolically related diaminopimelate racemase. It requires the L configuration at the distal center but decarboxylates only the chiral center of D configuration, i.e., only *meso*-diaminopimelate is a substrate but not the DD- or the

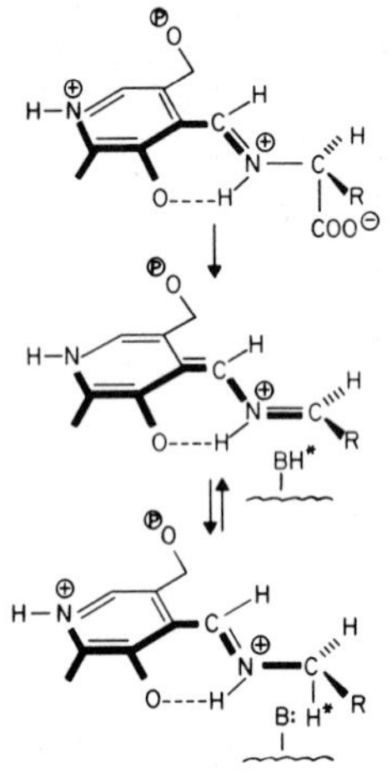

Scheme X. Stereochemical mechanism for PLP-dependent L-amino acid α-decarboxylases.

LL-isomer [77]. Decarboxylation of diaminopimelate in ^2H_2O and of [2,6-$^2H_2$]diaminopimelate in H_2O gave two samples of [6-^2H]lysine which were degraded to 5-phthalimido[5-^2H]valerate and compared by their rotations to the same compound prepared from L-glutamate with glutamate decarboxylase in ^2H_2O. The results indicated that the newly introduced hydrogen at C-6 of lysine occupies the *pro-R* position and hence, that the decarboxylation has occurred in an inversion mode [76]. This is so far the only example of a D-amino acid α-decarboxylase studied and the only case of α-decarboxylation with inversion. It is tempting to speculate that the disposition of the base protonating C-α and the binding mode of the distal group of the side chain seen in L-amino acid α-decarboxylases may be preserved in this enzyme, forcing disposition of the carboxyl group on the *Re* face relative to C-4′ and replacement of COOH by H on opposite faces of the complex (Scheme XI).

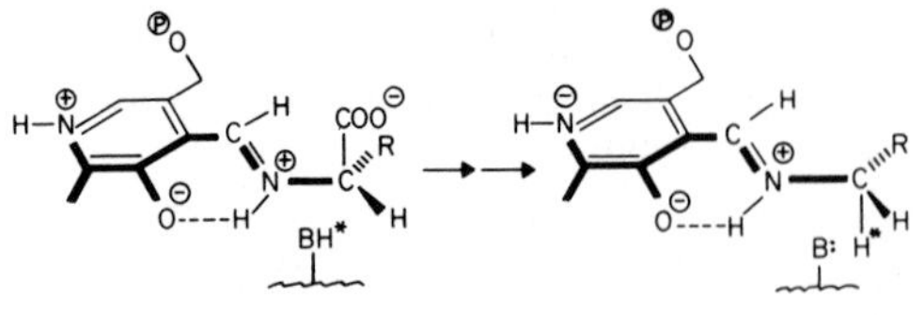

Scheme XI. Possible geometry of PLP–substrate and PLP–product complexes in meso-diaminopimelate decarboxylase.

Tritiated borohydride reduction experiments on this enzyme should be of considerable interest.

(iv) Enzymes catalyzing α,β-bond cleavage or formation

The most prominent enzyme in this class is serine transhydroxymethylase which catalyzes the reversible reaction of L-serine with tetrahydrofolate (H_4-folate) to form 5,10-methylenetetrahydrofolate (CH_2-H_4-folate) and glycine. The enzyme also cleaves several other β-hydroxy amino acids to glycine and the respective carbonyl compounds [78,79]. It was shown that during the cleavage of L-threonine to acetaldehyde the oxygen is preserved [80]. The available mechanistic evidence supports a retro-aldol cleavage of the PLP–serine Schiff's base to form a PLP–glycine anion which is then protonated at C-α ([81]; [82] and refs. therein). The rate of the overall reaction as well as that of proton abstraction from glycine in the absence of CH_2-H_4-folate is enhanced by H_4-folate ([83], and refs. therein). Protonation of the intermediate glycine–PLP anion complex occurs in a retention mode; serine transhydroxymethylase removes exclusively the *pro*-2*S* hydrogen of glycine and leaves the *pro*-2*R* hydrogen in the product L-serine [84–87]. In accord with the configurational position occupied by the hydroxymethylene group of L-serine, the enzyme catalyzes α-hydrogen exchange and transamination of D-alanine [17], with reprotonation at C-4′ of the intermediate coming from the *Si* face [30]. Surprisingly, some (but not all) L-amino acids not only exchange their α hydrogen but also react with formaldehyde to form new amino acids (presumably β-hydroxymethyl derivatives) [88]. Although L-alanine reacts in this way and D-alanine undergoes transamination, the D antipodes of all other amino acids tested exhibit no reaction or hydrogen exchange of any sort [88–90]. In contrast, the enzyme shows no specificity for the heterotopic carboxyl groups in the decarboxylation of aminomalonate [91]. If one assumes that stereoelectronic control requires orientation of the cleaving bond perpendicular to the cofactor plane, these unusual results suggest that the PLP–substrate complex may be bound to the enzyme in two different C-α—N conformations, possibly due to the presence of two binding sites for the distal group. If this idea is correct, each antipode of racemic alanine or stereospecifically carboxyl-labeled aminomalonate should react stereospecifically with retention of configuration at the α carbon.

Although the nature of the bound glycine species has been extensively studied, the nature of the aldehyde species involved in the enzymatic reaction and the role of H_4-folate are still unclear. Initial studies by Biellmann and Schuber on the serine generated in tissue slices from [^{3}H]formate (presumably via methenyl-H_4-folate and CH_2-H_4-folate) showed that 72% of the tritium at C-3 of serine was in the *pro-S* position [92]. To examine the reasons for partial stereospecificity we investigated [93] this problem using purified enzymes and L-serine samples which were stereospecifically tritiated at C-3 (>98% isomeric purity) prepared from [1-^{3}H]glucose and [1-^{3}H]mannose [94,95]. Serine transhydroxymethylase converted these substrates under single turnover conditions into CH_2-H_4-folate, which in a coupled reaction with CH_2-H_4-folate dehydrogenase was oxidized to methenyl-H_4-folate to stereospecifically remove one of the hydrogens from the 5,10-methylene bridge (Scheme XII). The results using pure serine transhydroxymethylase are in excellent agreement with those obtained with rat liver tissue, i.e., there is a 24% crossover of label from each serine isomer into the CH_2-H_4-folate isomer with the methylene unit of

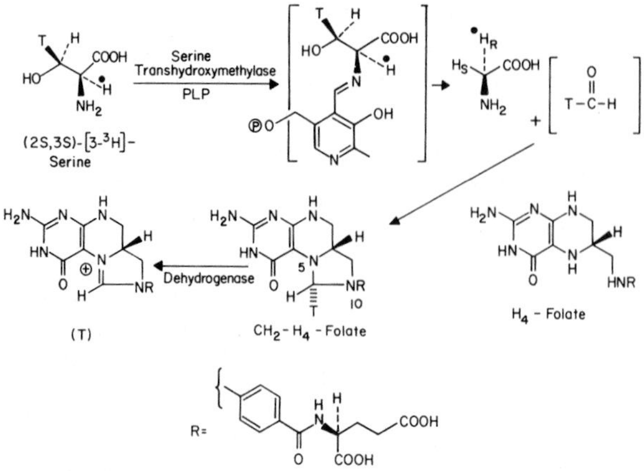

Scheme XII. Steric course of the serine transhydroxymethylase reaction.

opposite configuration. The absolute stereochemistry of the overall transformation was not apparent at the time [93], but can now be formulated as shown in Scheme XII based on the recently determined absolute configuration of H_4-folate [96,97] and NMR studies in the laboratory of Benkovic ([98]; S.J. Benkovic, personal communication). Apparently the enzyme catalyzes the cleavage of serine by two parallel pathways which are undetectable unless the methylene group is stereospecifically labeled. Other studies show that serine transhydroxymethylase cleaves both threonine and allothreonine as well as the L-isomers of *erythro-* and *threo-β*-phenylserine [88,99]. This indicates considerable latitude in the steric requirements at the β carbon and suggests that the reaction may proceed from two different conformations around the C-α—C-β bond. An alternative possibility is that cross-over occurs through the formaldehyde intermediate, which is bound on a single face of the enzyme immediately after its formation, but may be occasionally released and rebound on the opposite face.

The enzyme 5-aminolevulinate synthetase [100] catalyzes the condensation of glycine with succinyl-CoA to give 5-aminolevulinic acid, the first specific intermediate in porphyrin biosynthesis. The reaction involves both an α-decarboxylation of the amino acid and the formation of an α,β-carbon–carbon bond. These two processes might be interconnected, i.e., the thioester carbonyl group of succinyl-CoA might react with a carbanion generated by decarboxylation of the PLP–glycine Schiff's base, or sequential, i.e., a carbanion generated by deprotonation of the PLP–glycine complex would be acylated by succinyl-CoA followed by decarboxylation and protonation of the second carbanion. Work from the laboratory of Akhtar [101] supports the second pathway by showing that one hydrogen in the methylene group of glycine, the *pro-R* hydrogen, is replaced during the reaction catalyzed by the enzyme from *Rhodopseudomonas spheroides*. Further work by the same group [102] has defined the enantiotopic position at C-5 of aminolevulinate occupied by the hydrogen carried forward from glycine. 5-Aminolevulinate formed from [2-

^{3}H]glycine was converted in situ into porphobilinogen, which was acetylated at the amino group and then oxidized to *N*-acetylglycine. Hydrolysis of the latter and equilibration of the glycine with serine transhydroxymethylase in H_2O showed that the tritium occupied the *pro-S* position. Hence the hydrogen introduced in the protonation of the final PLP product carbanion must occupy the *pro-R* position at C-5 of aminolevulinate. This result indicates that of the two carbon–carbon bond forming or breaking reactions at C-α of glycine one must proceed with retention and one with inversion. As the authors point out [102], the hydrogen abstracted from glycine and the newly introduced hydrogen at C-5 of aminolevulinate occupy the same position (superimposing COO^- and COR); thus deprotonation and protonation could be mediated by a single base. If one assumes that this base is situated on the *Si* face relative to C-4′ and that the concept of stereoelectronic control by orthogonal alignment applies, the two alternative reaction pathways shown in Scheme XIII emerge. This reaction must involve a considerable amount of confor-

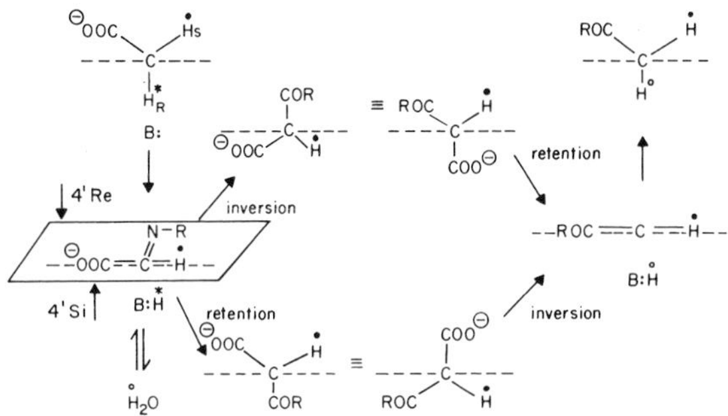

Scheme XIII. Stereochemical alternatives in the 5-aminolevulinate synthetase reaction (end-on view down the C-α–N bond; the dashed line represents the plane of the π system of the Schiff's base and pyridine ring).

mational reorientation; on either route one large group (COO^- or COR) must move 120° around the C-α—N bond, whereas the other moves a maximum of 90° and the hydrogen only 60°.

A much simpler stereochemical situation exists in the analogous condensation of serine and palmityl-CoA catalyzed by the enzyme dihydrosphingosine synthetase [103,104]. The immediate product of this reaction, 3-keto-dihydrosphingosine, is further converted to dihydrosphingosine or sphingosine and then into sphingolipids. From the known configurations of the starting material serine (L) and the product (2*S*) [105] it follows that the COOH group of the substrate has been replaced by R-CO in a retention mode. Thus the simplest stereochemical path would be direct acylation of the quinoid intermediate generated by cleavage of the C-α—COOH bond in PLP–serine on the same face from which COOH has departed. In support of this mechanism it was found [106] that L-[2-^{3}H]serine is incorporated into dihydrosphingosine with retention of tritium (Scheme XIV).

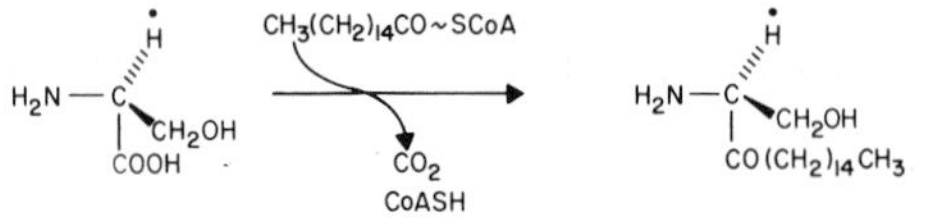

$H_2N—C(H)(CH_2OH)—COOH$ —[$CH_3(CH_2)_{14}CO{\sim}SCoA$; CO_2, CoASH]→ $H_2N—C(H)(CH_2OH)—CO(CH_2)_{14}CH_3$

Scheme XIV. Stereochemistry of formation of 3-keto-dihydrosphingosine.

(b) Reactions at the β-carbon

(i) Stereochemistry at C-β in nucleophilic β-replacements and α,β-eliminations

A number of PLP enzymes catalyze the nucleophilic displacement of a group X at C-β of an amino acid resulting in conversion into a new amino acid. The general mechanism of these β-replacement reactions [107] involves cleavage of the C-α—H bond in the PLP–amino acid complex to generate the quinoid intermediate 1, followed by elimination of X^- (or HX after protonation of X^-H to make it a better leaving group) from C-β. The resulting PLP–α-aminoacrylate intermediate 3 (Scheme II) can then undergo the reverse reaction sequence, i.e., Michael addition of X'^- and H^+ at C-β and C-α, respectively, to generate the PLP–product Schiff's base. In the closely related α,β-elimination reaction, on the other hand, the PLP–aminoacrylate undergoes hydrolysis to pyruvate and ammonia with regeneration of PLP enzyme. These two reaction types are mechanistically very similar, sharing the key intermediate PLP–α-aminoacrylate, and it is thus not surprising that a number of enzymes functioning normally in one reaction mode can under certain conditions catalyze reactions of the other type.

The concept of stereoelectronic control of bond cleavage and formation would predict that there are two conformations around the C-α—C-β bond in the quinoid intermediate in which the C-β—X bond is optimally disposed for interaction with the π system. In both conformations the C-β—X bond is aligned perpendicular to the π plane, in one case *syn* and in the other *anti* to the C-α—H bond. Furthermore, in the β-replacement reaction both the breaking bond C-β—X and the newly formed bond C-β—X′ should be aligned orthogonal to the π plane. Hence the leaving group X and the incoming group X′ can either bind on opposite faces of the π system, leading to reaction with inversion at C-β, or they can bind both on the same face resulting in retention of configuration at C-β. In the latter case the reaction must either occur by a ping-pong mechanism (X leaves the enzyme before X′ binds) or the enzyme must undergo a conformational change during the catalytic process which interchanges the positions of X and X′ relative to the PLP–aminoacrylate. This follows from the simple fact that two objects cannot be in the same place at the same time. The steric course at C-β has now been determined for a number of β-replacement reactions using five different enzymes and three different substrates; these are listed in Table 1. The stereospecifically labeled substrates of known absolute configuration at C-β were synthesized enzymatically [95], chemically [108] or by a combination of enzymatic and chemical methods [109,110]. The configurations of the products at C-β were determined by conversion into compounds of known absolute configuration and comparison by stereospecific enzymatic reactions, e.g., conversion of cysteine into serine [95] and of tryptophan and β-cyanoalanine into aspartate [110–113], or by NMR comparison with authentic stereospecifically deuterated material synthesized independently, e.g., tryptophan [109] and 2,4-dihydroxyphenylalanine [108].

The results obtained with five enzymes (Table 1) show that without exception the nucleophilic β-replacement reactions occur with retention at C-β. These enzymes include ones which normally function in this reaction type (tryptophan synthase, *O*-acetylserine sulfhydrase and β-cyanoalanine synthase) and ones which normally catalyze α,β-eliminations (tryptophanase, tyrosine phenol-lyase). The leaving group X and the incoming group X′ thus must line up on the same face of the PLP–amino acid complex. For *O*-acetylserine sulfhydrase a kinetic study [116] has given clear-cut evidence for a bi-bi ping-pong mechanism. Taken together these and the stereochemical results strongly support an elimination–addition mechanism in which the leaving group exits from the same side of the planar PLP–aminoacrylate intermediate as the nucleophile approaches. For β-cyanoalanine synthase Braunstein and coworkers [117,118] have suggested a direct displacement of the β substituent by the incoming nucleophile, based on their observation that α-hydrogen exchange of substrate in the absence of cyanide is slower than the overall reaction. A single direct displacement would result in inversion of configuration at C-β. To be compatible with the stereochemical results, this displacement would have to be a double displacement involving an anionic group of the enzyme on the ‘protein side’ of the coenzyme–substrate complex. Although both double displacement and elimination–addition mechanisms are in agreement with the stereochemical data, there is no evidence for the former process in any PLP-catalyzed β-replacement reaction.

A similar stereochemical question as in the β-replacement reactions can be asked in the α,β-eliminations where the group X is replaced by a hydrogen, i.e., is the proton added at C-β of the PLP–aminoacrylate on the same face from which X departed or on the opposite face? This question has been answered for a number of enzymes which generate either α-ketobutyrate or pyruvate as the keto acid product. Crout and coworkers [119,120] determined the steric course of proton addition in the α,β-elimination of L-threonine by biosynthetic L-threonine dehydratase and of D-threonine by an inducible D-threonine dehydratase, both in *Serratia marcescens*. Either substrate, deuterated at C-3, was converted in vivo into isoleucine, which was compared by proton NMR to a sample prepared from (3*S*)-2-amino[3-^{2}H]butyric acid. With both enzymes the hydroxyl group at C-3 was replaced by a proton in a retention mode. Although this has not been established with certainty, it is likely that both enzymes, like other bacterial threonine dehydratases [121], contain PLP as cofactor. Sheep liver L-threonine dehydratase, on the other hand, is not a PLP enzyme but contains an α-ketobutyrate moiety at the active site [122]. It replaces the hydroxyl group of L-threonine with H in a retention mode, but that of L-allothreonine in an inversion mode [123]. Snell and coworkers [124] established that the replacement of OH by H in the α,β-elimination of D-threonine catalyzed by the PLP-containing D-serine dehydratase from *E. coli* also proceeds in a retention mode. They

TABLE 1
Steric course at C-β in nucleophilic β-replacement reactions catalyzed by PLP enzymes

Enzyme	Substrate	Product	Ref.
	X–CH(T(D))–CH(NH$_2$)–COOH —Enzyme + X′H→ X′–CH(T(D))–CH(NH$_2$)–COOH + XH		
O-Acetylserine sulfhydrase	(2*S*,3*S*)-*O*-acetylserine X = OAc	(2*R*,3*S*)-cysteine X′ = SH	95
β-Cyanoalanine synthase	(2*R*,3*S*)-cysteine X = SH	(2*S*,3*S*)-cyanoalanine X′ = CN	111
Tyrosine phenol-lyase	(2*S*,3*S*)-serine X = OH	(2*S*,3*R*)-tyrosine X′ = 4′-hydroxyphenyl	109
	(2*S*,3*S*)-tyrosine X = 4′-hydroxyphenyl	(2*S*,3*R*)-dihydroxyphenylalanine X′ = 2′,4′-dihydroxyphenyl	108
Tryptophanase	(2*S*,3*S*)-serine X = OH	(2*S*,3*R*)-tryptophan X′ = 3′-indolyl	114,115
Tryptophan synthase, or β_2 subunit	(2*S*,3*S*)-serine X = OH	(2*S*,3*R*)-tryptophan X′ = 3′-indolyl	110,112,113

carried out the reaction in $^{2}H_2O$, oxidized the product to propionate and compared its rotation to that of [2-^{2}H]propionate of known absolute configuration.

In the other cases studied the keto acid produced was pyruvate. The reaction at C-β involved the conversion of a CH_2 into a CH_3 group and the stereochemical analysis required the use of all three isotopes of hydrogen to generate a chiral methyl group. (For a review of chiral methyl groups, see [125].) The configuration of the methyl group was analyzed by the methodology developed by Cornforth et al. [126] and Arigoni and coworkers [127] or a modification thereof [128]. The examples

TABLE 2

Steric course at C-β in α,β-elimination reactions catalyzed by PLP enzymes

Enzyme	Substrate	Configuration of CH_3	Ref.
	X–C(H)(T)–C(H)(NH$_2$)–COOH —ENZYME, D_2O→ D(H)–C(H)(T)–CO–COOD + ND_3 + XD(H)		
S-Alkylcysteine lyase	(2*R*,3*S*)-cystine	*S*	111
	(2*R*,3*R*)-cystine	*R*	
	X = $HO_2CCH(NH_2)CH_2SS$		
Tyrosine phenol-lyase	(2*S*,3*S*)-serine	*S*	129
	(2*S*,3*R*)-serine	*R*	
	X = OH		
Tryptophanase	(2*S*,3*S*)-serine	*S*	114,115
	(2*S*,3*R*)-serine	*R*	
	X = OH		
	(2*S*,3*R*)-tryptophan	*S*	114,115
	(2*S*,3*S*)-tryptophan	*R*	
	X = 3′-indolyl		
D-Serine dehydratase	(2*R*,3*S*)-serine	*S*	130
	(2*R*,3*R*)-serine	*R*	
	X = OH		
Tryptophan synthase	(2*S*,3*S*)-serine	achiral	113,114
	(2*S*,3*R*)-serine	achiral	
	X = OH		
	[pyruvate contains 100% (H)]		
	X–C(D)(T)–C(H)(NH$_2$)–COOH —ENZYME, H_2O→ H–C(D)(T)–CO–COOH		
Tryptophan synthase	(2*S*,3*S*)-serine	*R*	113
	(2*S*,3*R*)-serine	*S*	
	X–C(H)(D)–C(H)(NH$_2$)–COOH —ENZYME, HTO→ T–C(H)(D)–CO–COOH		
Tyrosine phenol-lyase	(2*S*,3*R*)-tyrosine	*R*	129
	(2*S*,3*S*),(2*R*,3*R*)-tyrosine	achiral	
	X = 4′-hydroxyphenyl		

investigated, the experimental setup used and the results are summarized in Table 2. In most cases a stereospecifically β-tritiated amino acid was reacted in ^2H_2O. This arrangement produced pyruvate containing a chiral methyl group in all instances except with tryptophan synthase β_2 protein, where no deuterium from the solvent was incorporated at C-β. However, a modified experimental setup, conversion of serine stereospecifically deuterated *and* tritiated at C-β in H_2O, allowed the elucidation of the steric course of this reaction as well [113]. The α,β-elimination of tyrosine catalyzed by tyrosine phenol-lyase was studied with the stereospecifically deuterated amino acid in tritiated water. As expected (2*S*,3*R*)-[3-^{2}H]tyrosine gave pyruvate containing a chiral methyl group, whereas the methyl group of pyruvate derived from (2*S*,3*R*), (2*R*,3*S*)-[3-^{2}H]tyrosine was racemic, in agreement with the known ability of the enzyme to react both D- and L-tyrosine [131]. The latter result suggests that the steric course of the replacement of X by H is the same with D- as with L-tyrosine. Comparison of the configurations of the substrates and the products reveals that in every single case the replacement of the β-substituent by a proton has occurred in a retention mode. The α,β-elimination reactions catalyzed by PLP enzymes studied so far all follow the pattern that reactions occur on only one face of the planar PLP–amino acid complex, i.e., C-β of the intermediate PLP–aminoacrylate is protonated on the same face from which the group X has departed. This is true for six enzymes that normally function in α,β-elimination reactions as well as for the reaction catalyzed by tryptophan synthase which normally carries out a β-replacement. Additional studies with selected enzymes catalyzing β-replacements and/or α,β-eliminations have further delineated the stereochemical parameters of these reactions.

(ii) Tryptophan synthase

Tryptophan synthase [132] is a tetrameric protein consisting of two α subunits which catalyze the conversion of indoleglycerol phosphate into enzyme-bound indole and glyceraldehyde 3-phosphate, and two PLP-containing β subunits which catalyze the β-replacement reaction between L-serine and indole. In the absence of α subunit the β_2 protein also catalyzes a multitude of other reactions [132] including the α,β-elimination of L-serine and an abortive transamination reaction in the presence of mercaptoethanol which leaves the enzyme in the inactive PMP form. The steric course of the β-replacement of OH by indole was found to be retention (Table 1), regardless of whether the reaction was carried out with the native enzyme or the β_2 protein and whether indoleglycerol phosphate or indole itself served as the source of the nucleophile [113]. The α,β-elimination reaction of stereospecifically β-tritiated L-serine in ^2H_2O catalyzed by the β_2 protein generated a racemic methyl group in pyruvate (Table 2). However, the initial suggestion [114] that this may reflect non-enzymatic protonation of C-β of the aminoacrylate was found to be in error when it was discovered that no deuterium from the solvent ^2H_2O was incorporated into the pyruvate [113]. Further experiments showed that the third hydrogen of the methyl group of pyruvate is derived from the α-position of the substrate, i.e., the α-hydrogen of L-serine migrates to C-β where it replaces the hydroxyl group in a retention mode. The transfer of the α-proton occurs without exchange with solvent and is strictly intramolecular. The latter point was proven by converting (2*S*,3*S*)-[3-^{2}H,3-^{3}H]serine into pyruvate in ^2H_2O in the presence of an excess of (2*R*,*S*)-[2-^{2}H]serine, conditions under which the α-hydrogen of the *same* molecule constituted the only source of ^{1}H to generate a chiral methyl group [113]. Since the transfer of the α-hydrogen is intramolecular it is presumably mediated by a single base and should be suprafacial. Together with the fact that OH is replaced by the migrating hydrogen in a retention mode, this defines the geometry of the C-α—C-β bond in the PLP–serine complex: the C-α—H bond and the C-β—OH bond must be *syn* oriented.

Information on the conformation of the C-α—N bond in the PLP–substrate complex was obtained from the abortive transamination reaction catalyzed by the β_2 protein in the presence of mercaptoethanol. Dunathan and Voet [18] showed that the newly introduced hydrogen at C-4′ of the PMP formed occupies the *pro-S* position; hence cofactor protonation occurs at the *Si* face. When the reaction was carried out with unlabeled serine in ^2H_2O, only 25% of one atom of deuterium was incorporated at C-4′ of PMP, implying that about 75% of the hydrogen must have been transferred within the PLP–substrate complex, most likely from the α position of serine [113]. Such a transfer again should be suprafacial and requires a conformation of the C-α—N bond in which the carboxyl group and C-4′ are *trans*. The enzyme thus contains a base, probably the imidazole ring of a histidine [133], which can internally transfer the α-hydrogen of the substrate either to C-β or to C-4′. The same base presumably also reprotonates C-α in the β-replacement reaction to generate the final PLP–product complex, although attempts to prove this have given inconclusive results [113]. This base must therefore be positioned about equidistant from C-α, C-β and C-4′, which form part of a quasi six-membered ring, on the *Si* face of the complex relative to C-4′. The stereochemistry of these three reactions catalyzed by tryptophan synthase is summarized in Scheme XV.

In collaboration with Dr. E.W. Miles of the National Institutes of Health we recently carried out sodium borohydride reduction experiments with tryptophan synthase in order to further define the orientation of the PLP–substrate and the internal PLP Schiff's base relative to the protein [134]. Reduction of the native holoenzyme reconstituted from apoenzyme by [4′-^{3}H]PLP in the presence of L-serine and the inhibitor indolepropanol phosphate gave, after denaturation of the protein, a radioactive phosphopyridoxyl-amino acid in addition to radioactivity bound covalently to the protein. The former was identified, surprisingly, as phosphopyridoxylalanine rather than the serine derivative, whereas the latter upon hydrolysis gave pyridoxyllysine. Both compounds were degraded to PM which was analyzed for its configuration at C-4′. The results indicate that the PLP Schiff's base in the holoenzyme alone has been reduced from the *Re* face, whereas in the presence of substrate the reduction has occurred from the *Si* face. Another reduction was carried out on unlabeled holoenzyme with tritiated sodium borohydride in the presence of indolepropanol phosphate and (3*R*)-[2,3-$^2H_2$]serine. Degradation of the resulting pyridoxylalanine, which must have arisen by reduction of both double bonds in the

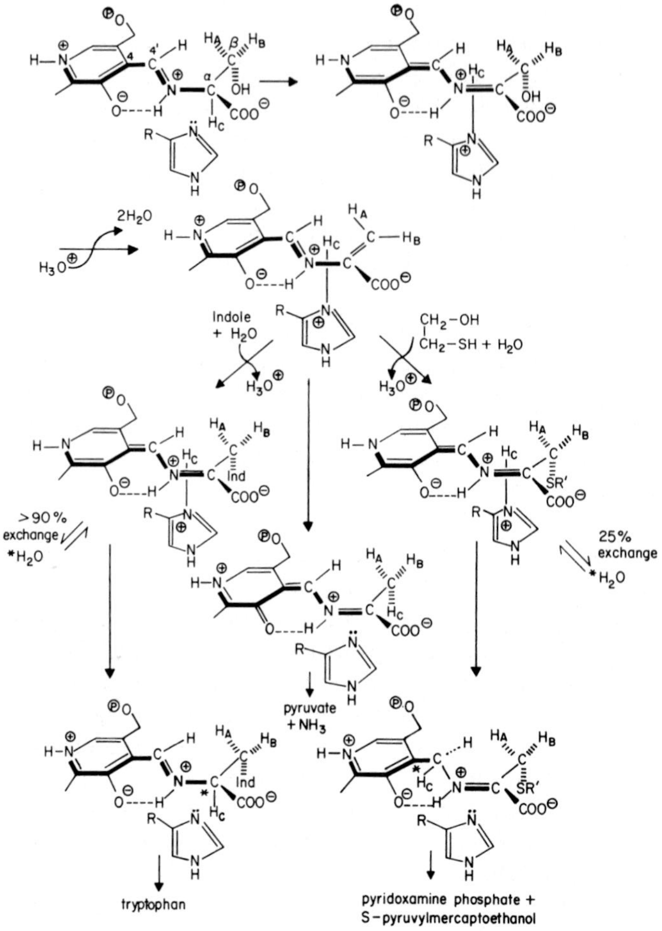

Scheme XV. Stereochemical course of reactions catalyzed by tryptophan synthase.

PLP–aminoacrylate, revealed the tritium distribution shown in Scheme XVI. The methyl group has *S* configuration, indicating hydride attack at the *Re* face of the intermediate PLP–aminoacrylate, the same side as the *Si* face at C-4′. L configuration at C-α has not yet been proven but seems extremely likely. The results provide

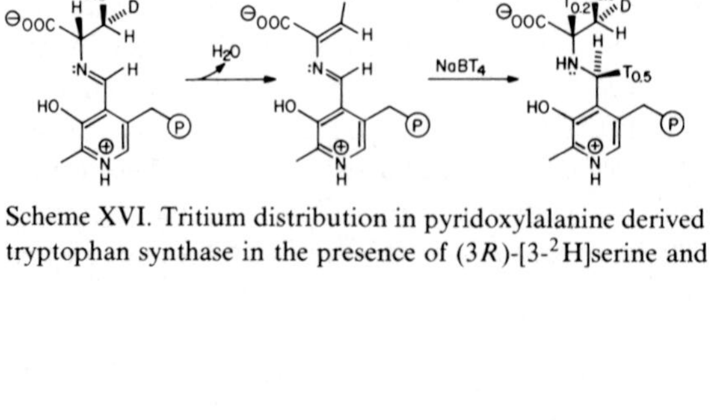

Scheme XVI. Tritium distribution in pyridoxylalanine derived from sodium [^{3}H]borohydride reduction of tryptophan synthase in the presence of (3*R*)-[3-^{2}H]serine and indolepropanol phosphate.

the first chemical evidence for the intermediacy of the PLP–aminoacrylate in β-replacement reactions. Reduction of this species occurs entirely on one face, with the initial hydride attack in 6 out of 10 times at C-β and in 4 out of 10 times at C-4′. In the latter case the remaining unconjugated enamine needs to tautomerize to the imine before the second reduction step. Finally, the data suggest that substrate binding does again result in exposure of a different face of the cofactor in tryptophan synthase.

(iii) Tryptophanase and tyrosine phenol-lyase

Tryptophanase is the protein primarily responsible for bacterial catabolism of L-tryptophan to indole, pyruvate and ammonia [135]. As discussed earlier both this reaction and the analogous α,β-elimination of serine and the β-replacement synthesis of tryptophan catalyzed by this enzyme proceed in a retention mode at C-β (cf. Tables 1 and 2). To test the possibility that the α proton of tryptophan which is removed in the initial step is involved in protonation of the aminoacrylate intermediate or the indolyl group, L-[2-^{3}H]tryptophan was incubated with tryptophanase and lactate dehydrogenase. Although less than 0.05% of the tritium could be detected in the methyl group of pyruvate, trapped as lactate, there was significant transfer of tritium from C-2 of tryptophan to C-3 of indole. Further experiments demonstrated that the hydrogen migration is at least predominantly intramolecular. These results suggest that a single base accomplishes a suprafacial 1,3 shift of the α hydrogen and that the elimination is a *syn* process [115]. In a parallel set of experiments we observed 63.5% internal transfer of α-^{1}H in a ^2H_2O medium but only 7.9% transfer of α-^{2}H in an ^2H_2O medium, suggesting that the base mediating it may be polyprotic, e.g., an amino group [114]. Since the aminoacrylate intermediate is protonated from the same side as indole leaves, the same base may also be involved in this proton transfer. If this is true, hydrogen exchange of the base with solvent must then be more rapid than decomposition of the intermediate to pyruvate. Since the β-replacement reaction (formation of tryptophan from serine and indole) also proceeds in a retention mode, all reactions appear to occur on a single side of the substrate–cofactor complex. To define further the geometry of this intermediate, we reduced the tryptophanase–PLP–L-alanine complex with sodium[^{3}H]borohydride. The L-alanine is a competitive inhibitor which exchanges its α-hydrogen and generates an aldimine–ketimine equilibrium on the enzyme surface [136] (cf. Scheme VI). Degradation of the resulting pyridoxylalanine to alanine and pyridoxamine and stereochemical analysis showed that reduction proceeds predominantly from the *Re* face at C_α and the *Si* face at C-4′. These results define the ketimine bond as *E* and identify the exposed ‘solvent side’ at C-4′ as *Si*. Reduction of the tryptophanase–PLP complex in the absence of substrate shows a greater exposure of the *Re* face at C-4′ (J.C. Vederas, I.D. Reingold and H.G. Floss, unpublished results), indicating conformational changes upon binding of substrate similar to those described previously for alanine transaminase, aspartate transaminase, tyrosine decarboxylase and tryptophan synthase. The stereochemical picture evolving from these studies for the coenzyme–substrate complex in tryp-

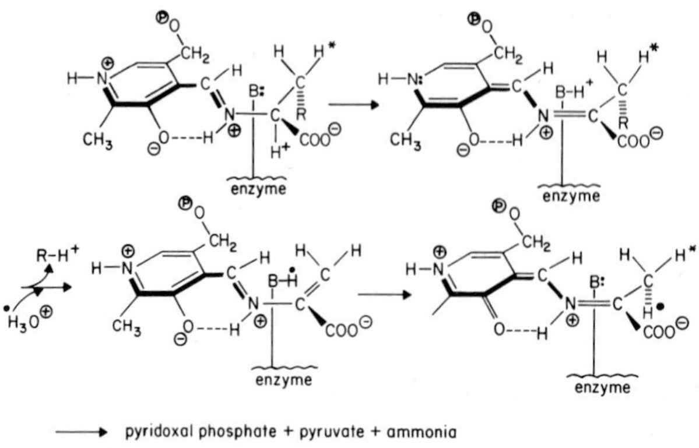

→ pyridoxal phosphate + pyruvate + ammonia

Scheme XVII. Stereochemical mechanism of the tryptophanase reaction.

tophanase (Scheme XVII) is similar to that for tryptophan synthase in terms of the conformation of the complex and the relative position of the essential base, although the nature of the base and the fate of the α-hydrogen are different [115].

The closely related bacterial enzyme tyrosine phenol-lyase [137] has an even wider substrate and reaction specificity than tryptophanase, including the remarkable ability to cleave both D- and L-tyrosine and to interconvert D- and L-alanine. As already discussed and summarized in Tables 1 and 2, the stereochemistry at C-β in all the α,β-elimination and β-replacement reactions of this enzyme studied so far is always retention [108,109,129]. This includes the α,β-elimination of L- as well as of D-tyrosine. The fate of the α-hydrogen of L-tyrosine in this reaction has been probed in preliminary experiments (H. Kumagai, E. Schleicher and H.G. Floss, unpublished results), and the results tentatively suggest transfer of deuterium from the α-position to C-4 of the resulting phenol. Attempts to demonstrate intramolecularity of this transfer have so far been inconclusive. The base abstracting H-α in this enzyme may be histidine [138].

A system has recently been developed to examine the question whether any internal return of the H-α can be demonstrated in the interconversion of L- and D-alanine catalyzed by this enzyme. L-[2-^{2}H]Alanine in H_2O or unlabeled L-alanine in ^2H_2O are converted into D-alanine with tyrosine phenol-lyase in the presence of excess D-amino acid:acetyl-coenzyme A acetyltransferase [139] and acetyl-coenzyme A to ensure single turnover conditions. The resulting *N*-acetyl-D-alanine is analyzed for its deuterium content by mass spectrometry and NMR. Initial data indicate 12% transfer of α-^{1}H from L-alanine to the D-isomer in ^2H_2O (S.-j. Shen, H. Kumagai and H.G. Floss, unpublished results), but these results need to be verified. If they are confirmed, racemization by a single base mechanism would be indicated.

(iv) Electrophilic displacement at C-β

The enzymes discussed in the previous three sections catalyze the removal of an anionic group from C-β of a substrate. PLP enzymes are also able to stabilize a negative charge at the β-carbon, and this operational mode is apparent in the enzymes aspartate-β-decarboxylase [140] and kynureninase [141], which catalyze the following two reactions, respectively:

$$\mathrm{HOOC{-}CH_2{-}\underset{\displaystyle NH_2}{\underset{|}{CH}}{-}COOH} \xrightarrow[\text{decarboxylase}]{\text{aspartate-}\beta} \mathrm{CH_3{-}\underset{\displaystyle NH_2}{\underset{|}{CH}}{-}COOH + CO_2}$$

$$\mathrm{C_6H_4(NH_2){-}\overset{\displaystyle O}{\overset{\|}{C}}{-}CH_2{-}\underset{\displaystyle NH_2}{\underset{|}{CH}}{-}COOH} \xrightarrow[\mathrm{H_2O}]{\text{kynureninase}} \mathrm{CH_3{-}\underset{\displaystyle NH_2}{\underset{|}{CH}}{-}COOH} + \mathrm{C_6H_4(NH_2)COOH}$$

Although an earlier formulation had interpreted the kynureninase reaction in terms of an α,β-elimination mechanism [142], the available evidence now points to a mechanism paralleling that of aspartate-β-decarboxylase [141] as proposed by Braunstein [143]. No stereochemical studies have been reported on kynureninase, but some work has been done on aspartate-β-decarboxylase.

Among a number of reactions catalyzed by aspartate-β-decarboxylase is the decarboxylation of aminomalonate to glycine. Meister and coworkers showed that this α-decarboxylation is stereospecific and the newly introduced hydrogen occupies the *pro-S* position at C-2 of glycine [144]. The *pro-R* carboxyl group of aminomalonate is removed in the process [145], indicating replacement of COOH by H in a retention mode. This carboxyl group occupies the position of the CH_2—COOH group in L-aspartate. In the β-decarboxylation of L-[2-^{3}H]aspartate about 1% of the tritium is transferred to the product alanine. One in every 2000 turnovers of the enzyme leads to transamination of the product to pyruvate, leaving the enzyme in the inactive PMP form. In this process 17% of the tritium from L-[2-^{3}H]aspartate is transferred to C-4′ of the cofactor, where it predominantly occupies the *pro-S* position [31]. By the arguments used before, this transfer must be suprafacial, indicating that the C-α—H bond is displayed on the 4′-*Si* face of the PLP–L-aspartate complex. Experiments with L-aspartate samples labeled stereospecifically with deuterium and/or tritium showed that the β-carboxyl group is replaced by a hydrogen in an inversion mode and no significant (< 10–15%) transfer of hydrogen from C-α to C-β was detected [31]. The chiral purity of the methyl group was low and this was traced to slow deprotonation/protonation of C-β of alanine catalyzed by the enzyme.

In attempting to interpret these results one can think in terms of a single base mechanism (Scheme XVIII A), the same base abstracts H-α and protonates C-β as well as C-α or C-4′, or the involvement of two base groups in these proton transfers. The single base mechanism is rendered very unlikely by the results, because only 1% of tritium from C-α of the substrate is transferred to the product (i.e., no more than 1% of H-α could be recycled to C-β) whereas 17% appear at C-4′, which must be protonated after C-β. Thus a two-base mechanism is indicated. An attractive version, shown in Scheme XVIII B, has two base groups situated on opposite faces of the PLP–substrate complex and the C-α—H and C-β—COOH bonds *syn* oriented. The base on the 4′-*Si* face would abstract H-α and reprotonate C-α or C-4′.

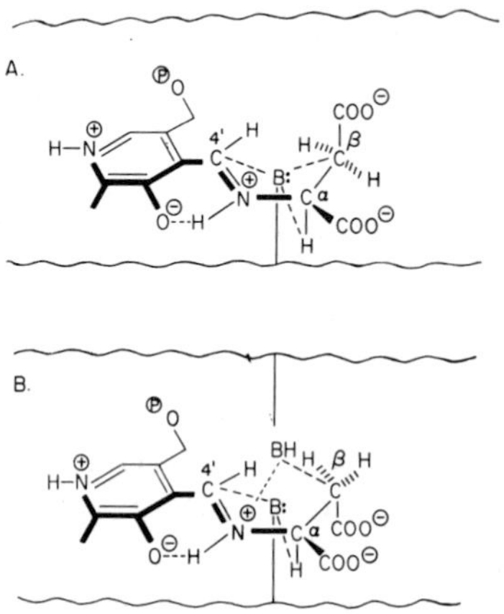

Scheme XVIII. PLP–substrate Schiff's base geometry for a single-base mechanism (A) and possible geometry for a two-base mechanism (B) for aspartate-β-decarboxylase.

The base on the opposite face must be in the conjugate acid form when it protonates C-β following or concerted with departure of the CO_2; it might acquire a proton during formation of the PLP–amino acid Schiff's base from the zwitterion form of the amino acid when protons must be removed from the $-NH_3^+$ group. This arrangement of two bases would explain the stereochemistry of aminomalonate decarboxylation – the carboxyl group removed would be displayed on the 4′-*Re* face if the other carboxyl group occupies approximately the same binding site as the α-carboxyl group of L-aspartate – and it might explain the unusual observation that aspartate-β-decarboxylase modified by reconstitution of apoenzyme with *N*-methyl-PLP can β-decarboxylate D-aspartate at a significant rate to give L-alanine [146].

(c) Reactions at the γ-carbon

As summarized in Scheme II, PLP enzymes can catalyze replacements at the γ-carbon of amino acids and eliminations of HY between C-β and C-γ. In mechanistic similarity to the aspartate-β-decarboxylase reaction, in these processes the quinoid intermediate 1 loses a proton from C-β, followed by elimination of an anionic group (Y^-) from C-γ, to generate the central intermediate PLP–vinylglycine, 4 (Scheme II). This species, the vinylogue of 1, can undergo a number of reactions. Addition of a new anionic group (Y'^-) and reversal of the reaction sequence constitutes the γ-replacement reaction, as in cystathionine-γ-synthase. On the other hand, in analogy to the protonation of 1 at C-α, 4 can be protonated at C-γ, generating a PLP–α-aminocrotonate Schiff's base, which is the equivalent of the PLP–aminoacrylate intermediate in the β-replacement and α,β-elimination reactions. As the latter, this PLP–α-aminocrotonate can add an anionic group at C-β (OH^-, threonine synthetase) to generate a β-substituted amino acid or can be protonated at C-β (γ-cystathionase) to initiate hydrolysis to α-ketobutyric acid. Only a few PLP enzymes catalyze reactions at the γ-carbon of amino acids [107]. Stereochemical studies have been reported on cystathionine-γ-synthase, which catalyzes the formation of cystathionine from cysteine and a 4-*O*-acyl-homoserine, a step in the biosynthesis of methionine:

$$\mathrm{R{-}O{-}CH_2{-}CH_2{-}\underset{\displaystyle NH_2}{\underset{|}{CH}}{-}COOH + HS{-}CH_2{-}\underset{\displaystyle NH_2}{\underset{|}{CH}}{-}COOH \rightarrow}$$

$$\mathrm{HOOC{-}\underset{\displaystyle NH_2}{\underset{|}{CH}}{-}CH_2{-}S{-}CH_2CH_2{-}\underset{\displaystyle NH_2}{\underset{|}{CH}}{-}COOH + ROH}$$

(R = succinyl, *Salmonella* enzyme; R = CH_3CO-, *B. subtilis* enzyme; R = phosphoryl or malonyl, plant enzyme),

on threonine synthetase catalyzing the conversion of phosphohomoserine to threonine:

$$\text{Ⓟ}\mathrm{O{-}CH_2{-}CH_2{-}\underset{\displaystyle NH_2}{\underset{|}{CH}}{-}COOH \rightarrow CH_3{-}\underset{\displaystyle OH}{\underset{|}{CH}}{-}\underset{\displaystyle NH_2}{\underset{|}{CH}}{-}COOH + P_i}$$

and on γ-cystathionase, which catalyzes the β,γ-elimination of cystathionine and also of homoserine to α-ketobutyrate:

$$\mathrm{Y{-}CH_2{-}CH_2{-}\underset{\displaystyle NH_2}{\underset{|}{CH}}{-}COOH + H_2O \rightarrow CH_3{-}CH_2{-}CO{-}COOH + YH + NH_2}$$

$$\mathrm{Y = OH \text{ or } {-}S{-}CH_2{-}\underset{\displaystyle NH_2}{\underset{|}{CH}}{-}COOH}$$

A β,γ-elimination of succinylhomoserine to α-ketobutyrate is also catalyzed by cystathionine-γ-synthase from *Salmonella* in the absence of L-cysteine.

The most extensive studies were carried out on cystathionine-γ-synthase from *Salmonella*. β,γ-Elimination of *O*-succinylhomoserine in ^2H_2O led to the incorporation of 1 atom of deuterium at C-β, but only 0.2 atoms of deuterium at C-γ of the α-ketobutyrate formed [147]. The deuterium at C-β was found to occupy the *pro-S* position [148]. The enzyme also catalyzed rapid hydrogen exchange between ^2H_2O and cystathionine and other amino acids. In the 4-carbon chain of cystathionine and in homoserine the α-hydrogen and one β-hydrogen are exchanged rapidly [148,149], whereas the other β-hydrogen in homoserine exchanges 100 times more slowly [148]. The rapidly exchanging β-hydrogen in homoserine, presumably the hydrogen abstracted in the reaction sequence leading to the PLP–vinylglycine complex, was characterized by its proton NMR signal [148], and subsequent assignments of the signals of the magnetically nonequivalent β-hydrogens of homoserine indicate that it is the *pro-R* hydrogen [150,151]. This hydrogen is sterically equivalent to the *pro*-3*S* hydrogen of α-ketobutyrate, i.e., proton abstraction and reprotonation at C-β seem to take place on the same face. The low deuterium incorporation at C-γ suggests that the proton added at this position originates from within the coenzyme–substrate complex. Indeed, Posner and Flavin [152] were able to demonstrate proton transfer from both C-α and C-β to C-γ; the hydrogen transferred from C-β was the same one shown to undergo faster exchange. These results indicate that deprotonation at C-α and C-β and protonation at C-γ, and presumably also reprotonation at C-β and C-α, are all catalyzed by a single, polyprotic base in the active site of the enzyme. All these proton transfers presumably are suprafacial, further defining the geometry of the PLP–substrate complex. Recent work from the laboratory of Walsh [153] has unraveled the stereochemistry of the events at the γ-carbon. Based on their finding that L-vinylglycine is an excellent alternate substrate to *O*-succinylhomoserine [154] these authors synthesized *Z*-DL-[4-^{2}H]vinylglycine and *E*-DL-[3,4-$^2H_2$]-vinylglycine and converted both samples into cystathionine with cystathionine-γ-synthase and L-cysteine. The cystathionine samples were degraded to homoserine by a stereospecific sequence of enzymatic and chemical reactions and the proton NMR spectra of the latter were compared to those of authentic (*R*)- and (*S*)-L-[4-^{2}H]homoserine prepared earlier by an independent route [155]. The results indicate that the sulfur of cysteine is added on the *Si* face at C-4 of the PLP–vinylglycine complex. Chemical succinylation of the homoserine samples so generated, enzymatic conversion to cystathionine and NMR comparison with the previous set of cystathionine samples then established that the replacement of the γ-substituent by the sulfur of cysteine in the cystathionine-γ-synthase reaction proceeds with retention of configuration. Finally, in unpublished work the same group observed that in the β,γ-elimination reaction the γ-substituent is replaced by hydrogen in a manner consistent with a net suprafacial process, i.e., in a retention mode (C. Walsh, personal communication). These results lead to a very detailed stereochemical picture for the reactions catalyzed by cystathionine-γ-synthase, as shown in Scheme XIX. The assumption is made in this scheme, not proven by experiment, that the reactions all take place on the *Si* face, rather than the *Re* face, relative to C-4′ of the cofactor in the PLP–substrate complex.

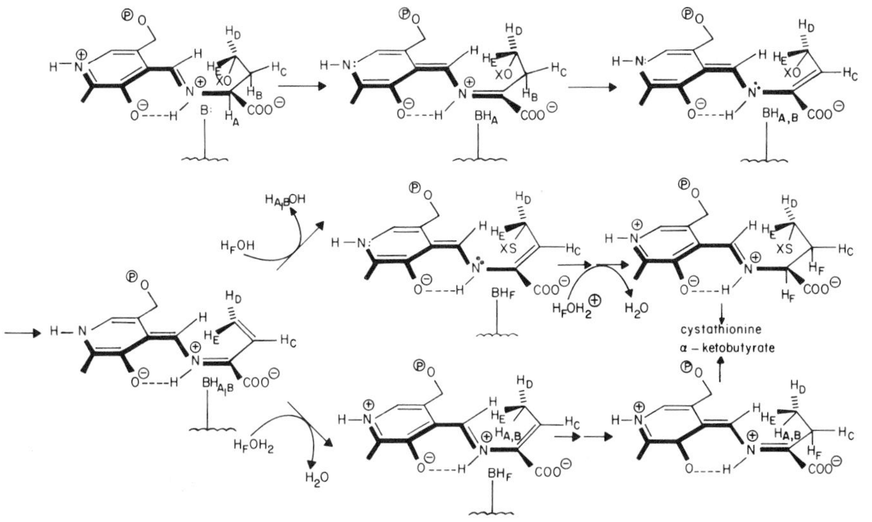

Scheme XIX. Stereochemical mechanism of reactions catalyzed by cystathionine-γ-synthase.

Studies on the β,γ-elimination reaction catalyzed by γ-cystathionase showed that in the conversion of homoserine to α-ketobutyrate one atom of deuterium from the solvent is incorporated at C-β, where it occupies the *pro-S* position [156]. The stereochemistry of protonation at C-β in this reaction is thus the same as in the β,γ-elimination catalyzed by cystathionine-γ-synthase.

Work on threonine synthetase has shown incorporation of exactly two atoms of deuterium from the solvent into the threonine formed from phosphohomoserine [157]. One of these is located at C-α and the other, according to the original work, at C-β or C-γ. Based on more recent results [158] this second hydrogen must be at C-γ, as had been assumed by Flavin and Slaughter [157]. Therefore this reaction does not involve any significant internal proton recycling. The steric course of the reaction at C-β has been established by Fuganti [158] using stereospecifically β-tritiated homoserine samples prepared by a route developed earlier in his laboratory [159]. Phosphorylation followed by conversion into threonine indicated almost complete loss of tritium from the 3*S* isomer and 95% retention of tritium from (2*S*,3*R*)-phospho-[3-^{3}H]homoserine [158]. Degradation of the threonine established that the tritium retained in the reaction was located at C-β. The replacement of a β-hydrogen by OH thus occurs in a retention mode, but the hydrogen removed is the *pro-S* hydrogen, the opposite one that is removed by cystathionine-γ-synthase. These results parallel the stereochemistry of β-protonation observed for the α,β-elimination catalyzed by bacterial L-threonine dehydratase [119,120] (see Section 3.b.i) where the proton is added on the *Re* face at C-β of the PLP–aminocrotonate intermediate. The stereochemical data on threonine synthetase can be interpreted in two ways (Scheme XX). The reactions could all take place on one face of the PLP–amino acid complex

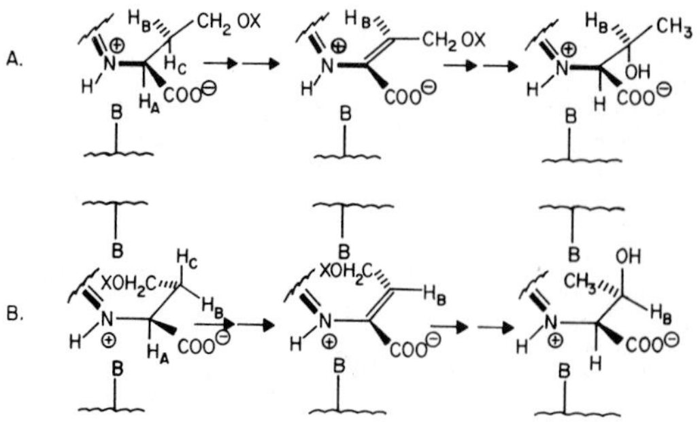

Scheme XX. Two possible steric arrangements of the PLP—amino acid Schiff's base in threonine synthetase.

(route A), resulting in an *E* configuration of the intermediate aminocrotonate. Alternatively, the aminocrotonate might have *Z* configuration, forcing proton abstraction at C-α and C-β to occur on opposiste sides of the PLP–substrate complex (route B). Experiments with stereospecifically labeled vinylglycine to establish the C-γ–oxygen bond conformation would be very desirable.

The stereochemical data discussed for the protonation/deprotonation at C-β catalyzed by the above enzymes are consistent with the results of trapping experiments with *N*-ethylmaleimide carried out by Flavin and Slaughter [160]. This reagent will react with enamines, like the PLP–aminocrotonate intermediate, with formation of an adduct, α-keto-3-[3′-(*N*′-ethyl-2′,5′-diketopyrrolidyl)]butyric acid (KEDB): This compound has two chiral centers which are generated during the reaction.

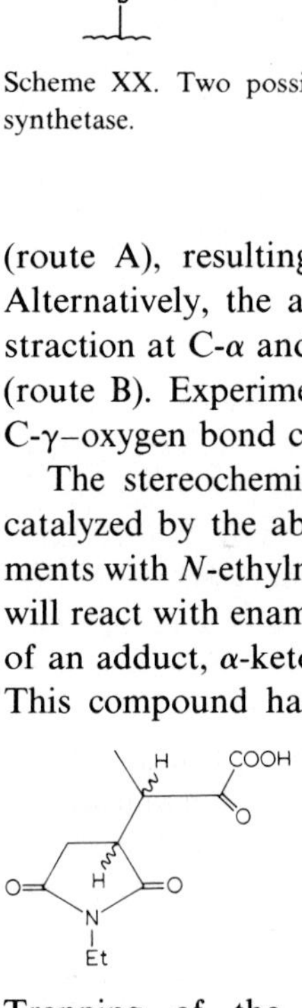

Trapping of the aminoacrylate intermediate in the reactions catalyzed by cystathionine-γ-synthase and γ-cystathionase produced the same diastereomer of KEDB which was different from the one formed with bacterial L-threonine dehydratase. Unfortunately, this experiment has apparently not been done with threonine synthetase.

Some very interesting and unusual PLP-mediated reactions at the γ-carbon of amino acids revolve around the formation and metabolism of 1-aminocyclopropanecarboxylic acid (ACC), the newly discovered intermediate in the formation of the fruit-ripening hormone ethylene from methionine [161–163]. ACC synthase catalyzes the formation of ACC from *S*-adenosylmethionine (Scheme XXI) [164,165], a process which could be formulated mechanistically either as being analogous to the PLP-catalyzed γ-replacement reactions or as involving a direct, concerted replacement of the thiomethyladenosyl group by the PLP-generated α-carbanion. ACC can fragment to ethylene and a two-carbon acid, probably glyoxylic acid or glycine, by a mechanism which is still rather obscure [161–163]. Finally, ACC can be deaminated by an inducible bacterial PLP enzyme to give

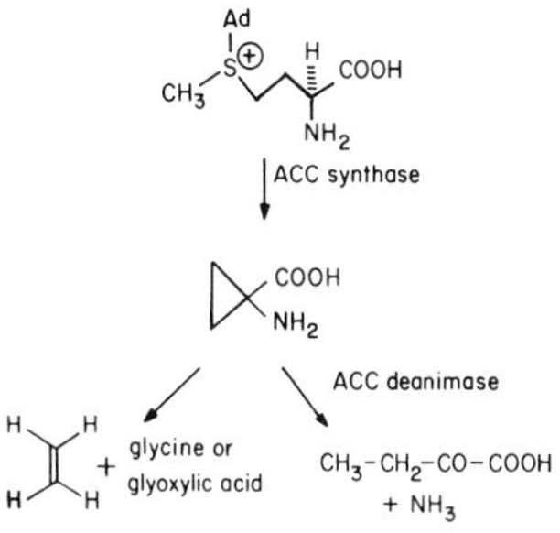

Scheme XXI. Formation and conversions of 1-aminocyclopropanecarboxylic acid (ACC).

α-ketobutyrate and ammonia [166]. Given the steric constraints built into the ACC molecule the stereochemical analysis of these reactions, now under way in several laboratories, should be extremely interesting. The only study reported so far showed that of the 4 stereoisomeric coronamic acids (1-amino-2-ethylcyclopropane-1-carboxylic acid) only the (1*S*,2*S*)-isomer is a substrate for ACC deaminase and produces 2-oxohexanoic acid as the keto acid [167]. This result would suggest that in the deaminase reaction the bond between C-α and the *pro-S* methylene group of ACC is broken and that the *pro-S* methylene group becomes the methyl group of α-ketobutyrate.

(d) Other pyridoxal phosphate-catalyzed reactions

PLP is involved catalytically in a number of other enzyme reactions of various types, for example in the degradation of glycine to CH_2-H_4-folate and CO_2 catalyzed by glycine-NAD oxidoreductase [168]. Various amine oxidases (EC 1.4.3.6) and B_{12}-dependent aminomutases [169,170] also contain PLP as a cofactor. However, no stereochemical information on the role of PLP in these reactions has been published. A particularly interesting involvement of a pyridoxal cofactor, in this case PMP, has been uncovered in the formation of deoxyhexoses for bacterial cell wall synthesis. Rubinstein and Strominger [171] described the isolation of two enzymes from *Salmonella*, E1 and E3, which jointly catalyze the conversion of cytidine diphospho(CDP)-4-keto-6-deoxyglucose into CDP-4-keto-3,6-dideoxyglucose. The evidence from their studies indicates that E1, which contains PMP as the cofactor, catalyzes the conversion of CDP-4-keto-6-deoxyglucose into a PLP–Δ3,4-glucoseen intermediate (by 1,4-elimination of H_2O from the Schiff's base between PMP and the 4-keto group of the substrate), which is then reduced by E3 at the expense of one NAD(P)H. Surprisingly, tritium from [^{3}H]NADPH is not incorporated at either C-3 of the product or C-4′ of the cofactor; rather, both positions acquire a proton from the solvent. A stereochemical comparison of this reaction with the perhaps mechanistically related α,β-elimination reactions of amino acids should be interesting.

4. Common stereochemical features of pyridoxal phosphate enzymes

Probably the most important common stereochemical feature that has evolved from all the studies on PLP enzymes is the fact that overwhelmingly these reactions take place on only one face of a relatively planar coenzyme–substrate complex. The driving force for the evolution of this feature presumably is the economy of utilizing the same acid–base group in the enzyme for multiple catalytic steps in the process [9]. As a result displacement reactions proceed generally in a retention mode and protonation/deprotonation reactions usually are suprafacial. The advantage of single base catalysis often determines the conformations of the PLP–amino acid complexes, as in the quasi six-membered ring geometry of the C-4′/N/C-α/C-β/(C-γ) array seen in tryptophan synthase, tryptophanase and cystathionine-γ-synthase. Occasionally, other factors may override these evolutionary driving forces, leading to inversion modes for group displacements and/or antarafacial proton abstractions and additions. For example, preservation of the binding mode of the side chain of L-amino acids evolved for L-amino acid α-decarboxylases in meso-2,6-diaminopimelate decarboxylase, acting on the center of D configuration, may be the reason for the inversion mode observed for the decarboxylation catalyzed by this enzyme. It is less obvious why aspartate-β-decarboxylase operates in an inversion mode.

The feature that PLP enzymes all exhibit the same steric preference between the two faces of the cofactor, i.e., they operate on the *Si* face relative to C-4′, is another well-established common trait. This has been demonstrated now for 10 enzymes and no exceptions have been found. This indicates a high degree of conservation in the mode of binding of the cofactor to the protein and supports the hypothesis of Dunathan [18] that these enzymes have evolved from a common progenitor. The notion that the face on which reactions take place is identical with the 'exposed' face, the face most accessible to external reagents like $NaBH_4$, is probably true in general, although, as the experiments with carbamylated aspartate transaminase have shown, other factors may also influence the access of $NaBH_4$ to either face of the cofactor. The cofactor rotation upon substrate binding and transaldimination, which must be postulated on theoretical grounds and has been demonstrated by the X-ray work on aspartate transaminase, results in a drastic change in accessibility of the two faces of the cofactor C-4′ to $NaBH_4$, although again a variety of other factors probably also influence the relative degree of accessibility of the two faces. The stereochemical analysis of $NaBH_4$-reduction products thus seems to be a useful tool to obtain information on the conformation of the PLP–substrate complex and on its orientation relative to the protein.

Finally, the concept that the proteins of different PLP enzymes determine reaction specificity by controlling the conformations of critical bonds of the PLP–

substrate complex, e.g., the C-α–N bond for reactions at C-α, the C-α–N and the C-α–C-β bonds for reactions at C-β, is by now widely accepted, despite the fact that there is little hard evidence at the enzyme level to prove it. It is supported by the X-ray work on aspartate transaminase and, importantly, no conflicting evidence has come to light. Hence, it is probably justified to consider this another of the common features of PLP enzymes.

Acknowledgements

Work from the authors' laboratories has been supported by the National Institutes of Health (Research Grant GM 18852 to H.G.F.) and by the Natural Sciences and Engineering Research Council of Canada (Grant NSERC A 0845 to J.C.V.).

References

1 Jencks, W.P. (1969) Catalysis in Chemistry and Enzymology, McGraw-Hill, New York, pp. 133–146.
2 Dunathan, H.C. (1971) Adv. Enzymol. 35, 79.
3 Braunstein, A.E. (1973) Enzymes 9, 379.
4 Walsh, C. (1979) Enzymatic Reaction Mechanisms, W.H. Freeman, San Francisco, pp. 777–833.
5 Metzler, D.E. (1979) Adv. Enzymol. 50, 1.
6 Various authors (1979) Methods Enzymol. 62, Part D, Section VI.
7 Braunstein, A.E. and Shemyakin, M.M. (1953) Biokhimiya 18, 393.
8 Metzler, D.E., Ikawa, M. and Snell, E.E. (1954) J. Am. Chem. Soc. 76, 648.
9 Hanson, K.R. and Rose, I.A. (1975) Acc. Chem. Res. 8, 1.
10 Vederas, J.C. and Floss, H.G. (1980) Acc. Chem. Res. 13, 455.
11 Dunathan, H.C. (1966) Proc. Natl. Acad. Sci. U.S.A. 55, 713.
12 Corey, E.J. and Sneen, R.A. (1956) J. Am. Chem. Soc. 78, 6269.
13 Fraser, R.R. and Champagne, P.J. (1978) J. Am. Chem. Soc. 100, 657.
14 Tsai, M.D., Weintraub, H.J.R., Byrn, S.R., Chang, C. and Floss, H.G. (1978) Biochemistry 17, 3183.
15 Belokon, Y.N., Belikov, V.M., Vitt, S.V., Savel'eva, T.F., Burbelo, V.M., Bakhmutov, V.I., Aleksandrov, G.G. and Struchkov, Y.T. (1977) Tetrahedron Lett. 33, 2551.
16 Fischer, J.R. and Abbott, E.H. (1979) J. Am. Chem. Soc. 101, 2781.
17 Schirch, L. and Jenkins, W.T. (1964) J. Biol. Chem. 238, 3797–3801.
18 Dunathan, H.C. and Voet, J.G. (1974) Proc Natl. Acad. Sci. U.S.A. 71, 3888.
19 Ivanov, V.I. and Karpeisky, M.Y. (1969) Adv. Enzymol. 32, p. 21.
20 Hanson, K.R. (1966) J. Am. Chem. Soc. 88, 2731.
21 Woodward, R.B. and Hoffmann, R. (1965) J. Am. Chem. Soc. 87, p. 2511.
22 Wada, H. and Snell, E.E. (1962) J. Biol. Chem. 237, 127.
23 Dunathan, H.C., Davis, L., Gilmer Kury, P. and Kaplan, M. (1968) Biochemistry 7, 4532.
24 Besmer, P. and Arigoni, D. (1969) Chimia 23, p. 190.
25 Besmer, P. (1970) Ph.D. dissertation No. 4435, ETH, Zurich.
26 Austermühle-Bertola, E. (1973) Ph.D. Dissertation No. 5009, ETH, Zurich.
27 Ayling, J.E., Dunathan, H.C. and Snell, E.E. (1968) Biochemistry 7, 4537.
28 Bailey, G.B., Kusamrarn, T. and Vuttivej, K. (1970) Fed. Proc. 29, 857.
29 Sukhareva, B.S., Dunathan, H.C. and Braunstein, A.E. (1971) FEBS Lett. 15, 241.
30 Voet, J.G., Hindenlang, D.M., Blanck, T.J.J., Ulevitch, R.J., Kallen, R.G. and Dunathan, H.C. (1973) J. Biol. Chem. 248, 841.

31 Chang, C.C., Laghai, A., O'Leary, M.H. and Floss, H.G. (1982) J. Biol. Chem., in press.
32 Tsai, M.D., Schleicher, E., Potts, R., Skye, G.E. and Floss, H.G. (1978) J. Biol. Chem. 253, 5344.
33 Cooper, A.J.L. (1976) J. Biol. Chem. 251, 1088.
34 Golichowski, A., Harruff, R.C. and Jenkins, W.T. (1977) Arch. Biochem. Biophys. 178, 459.
35 Zito, S.W. and Martinez-Carrion, M. (1980) J. Biol. Chem. 255, 8645.
36 Vederas, J.C., Reingold, I.D. and Sellers, H.W. (1979) J. Biol. Chem. 254, 5053.
37 Tumanyan, V.G., Mamaeva, D.K., Bocharov, A.C., Ivanov, V.I., Karpeisky, M. and Yakovlev, G.I. (1974) Eur. J. Biochem. 50, 119.
38 Tsai, M.D., Byrn, S.R., Chang, C., Floss, H.G. and Weintraub, J.R. (1978) Biochemistry 17, 3177.
39 Fisher, T.L. and Metzler, D.E. (1969) J. Am. Chem. Soc. 91, 5323.
40 Metzler, C.M., Metzler, D.E., Martin, D.S., Newman, R., Arnone, A. and Rogers, P. (1978) J. Biol. Chem. 253, 5251.
41 Gehring, H. and Christen, P. (1978) J. Biol. Chem. 253, 3158–3163.
42 Pfister, K., Kägi, J.H.R. and Christen, P. (1978) Proc. Natl. Acad. Sci. U.S.A. 75, 145.
43 Eichele, G., Ford, G.C., Glor, M., Jansonius, J.N., Mavrides, C. and Christen, P. (1979) J. Mol. Biol. 133, 161.
44 Ford, G.C., Eichele, G. and Jansonius, J.N. (1980) Proc. Natl. Acad. Sci. U.S.A. 77, 2559.
45 Peterson, D.L. and Martinez-Carrion, M. (1970) J. Biol. Chem. 245, 806.
46 Slebe, J.C. and Martinez-Carrion, M. (1976) J. Biol. Chem. 251, 5663.
47 Martinez-Carrion, M., Slebe, J.C. and Gonzalez, M. (1979) J. Biol. Chem. 254, 3160.
48 Strecker, H.J. (1965) J. Biol. Chem. 240, 1225.
49 Soda, K., Misono, H. and Yamamoto, T. (1968) Biochemistry 7, 4102.
50 Scott, E.M. and Jacoby, W.B. (1959) J. Biol. Chem. 234, 932.
51 Burnett, G., Walsh, C., Yonaha, K., Toyama, S. and Soda, K. (1979) J. Chem. Soc. Chem. Commun., 826.
52 Bouclier, M., Jung, M.J. and Lippert, B. (1979) Eur. J. Biochem. 98, 363.
53 Yamada, H. and O'Leary, M.H. (1978) Biochemistry 17, 669.
54 Adams, E. (1972) in P.D. Boyer (Ed.), The Enzymes, 3rd Edn., Vol. 6, Academic Press, New York, pp. 479–507.
55 Adams, E. (1976) Adv. Enzymol. 44, 69.
56 Antia, M., Hoare, D.S. and Work, E. (1957) Biochem. J. 65, 448.
57 Kumagai, H., Kashima, N. and Yamada, H. (1970) Biochem. Biophys. Res. Commun. 39, 796.
58 Cardinale, G. and Abeles, R. (1968) Biochemistry 7, 3970.
59 Henderson, L.L. and Johnston, R.B. (1976) Biochem. Biophys. Res. Commun. 68, 793.
60 Boeker, E.A. and Snell, E.E. (1972) in P.D. Boyer (Ed.), The Enzymes, 3rd Edn., Vol. 6, Academic Press, New York, pp. 217–253.
61 Mandeles, S., Koppelman, R. and Hanke, M.E. (1954) J. Biol. Chem. 209, 327.
62 Belleau, B. and Burba, J. (1960) J. Am. Chem. Soc. 82, 5751.
63 Battersby, A.R., Chrystal, E.J.T. and Staunton, J. (1980) J. Chem. Soc. Perkin Trans. 1, 31.
64 Marshall, K.S. and Castagnoli, N. Jr. (1973) J. Med. Chem. 16, 266.
65 Houck, D.R. and Floss, H.G. J. Nat. Prod., in press.
66 Leistner, E. and Spenser, I.D. (1975) J. Chem. Soc. Chem. Commun., 378.
67 Gerdes, H.J. and Leistner, E. (1979) Phytochemistry 18, 771.
68 Santaniello, E., Kienle, M.G., Manzocchi, A. and Bosisio, E. (1979) J. Chem. Soc. Perkin Trans. 1, 1677.
69 Battersby, A.R., Joyeau, R. and Staunton, J. (1979) FEBS Lett. 107, 231.
70 Tsai, M.D., Floss, H.G., Rosenfeld, H.J. and Roberts, J. (1979) J. Biol. Chem. 254, 6437.
71 Battersby, A.R., Nicoletti, M., Staunton, J. and Vleggaar, R. (1980) J. Chem. Soc. Perkin Trans. 1, 43.
72 Santaniello, E., Manzocchi, A. and Biondi, P.A. (1981) J. Chem. Soc. Perkin Trans. 1, 307.
73 Fonda, M.L. (1972) Biochemistry 11, 1304.
74 O'Leary, M.H. and Piazza, G.J. (1978) J. Am. Chem. Soc. 100, 632.
75 Bailey, G.B., Chotamangsa, D. and Vuttivej, K. (1970) Biochemistry 9, 3243.

76 Asada, Y., Tanizawa, K., Sawada, S., Suzuki, T., Misono, H. and Soda, K. (1981) Biochemistry 20, p. 6881.
77 Asada, Y., Tanizawa, K., Kawabata, Y., Misono, H. and Soda, K. (1981) Agr. Biol. Chem. 45, 1218.
78 Chen, M.S. and Schirch, L. (1973) J. Biol. Chem. 248, 3631.
79 Ulevitch, R.J. and Kallen, R.G. (1977) Biochemistry 16, 5355.
80 Jordan, P.M., El-Obeid, H.A., Corina, D.L. and Akhtar, M. (1976) J. Chem. Soc. Chem. Commun., 73.
81 Schirch, L., Tatum, C.M. and Benkovic, S.J. (1977) Biochemistry 16, 410.
82 Schirch, L., Slagel, S., Barra, D., Martini, F. and Bossa, F. (1980) J. Biol. Chem. 255, 2986.
83 Chen, M.S. and Schirch, L. (1973) J. Biol. Chem. 248, 7979.
84 Besmer, P. and Arigoni, D. (1968) Chimia 23, 190.
85 Jordan, P.M. and Akhtar, M. (1970) Biochem. J. 116, 277.
86 Wellner, D. (1970) Biochemistry 9, 2307.
87 Akhtar, M., El-Obeid, H.A. and Jordan, P.M. (1975) Biochem. J. 145, 159.
88 Ulevitch, R.J. and Kallen, R.G. (1977) Biochemistry 16, 5342.
89 Ulevitch, R.J. and Kallen, R.G. (1977) Biochemistry 16, 5350.
90 Hansen, J. and Davis, L. (1979) Biochim. Biophys. Acta 568, 321.
91 Palekar, A.G., Tate, S.S. and Meister, A. (1973) J. Biol. Chem. 248, 1158.
92 Biellmann, J.F. and Schuber, F. (1970) Bull. Soc. Chim. Biol. 52, 211.
93 Tatum, C.M., Benkovic, P.A., Benkovic, S.J., Potts, R., Schleicher, E. and Floss, H.G. (1977) Biochemistry 16, 1093.
94 Floss, H.G., Onderka, D.K. and Carroll, M. (1972) J. Biol. Chem. 247, 736.
95 Floss, H.G., Schleicher, E. and Potts, R. (1976) J. Biol. Chem. 251, 5478.
96 Fontecilla-Camps, J.C., Bugg, C.E., Temple, C., Jr., Rose, J.D., Montgomery, J.A. and Kisliuk, R.L. (1979) J. Am. Chem. Soc. 101, 6114.
97 Armarego, W.L.F., Waring, P. and Williams, J.W. (1980) J. Chem. Soc. Perkin Trans. 1, 334.
98 Poe, M., Jackman, L.M. and Benkovic, S.J. (1979) Biochemistry 18, 5527.
99 Schirch, L. and Diller, A. (1971) J. Biol. Chem. 246, 3961.
100 Jordan, P.M. and Shemin, D. (1972) in P.D. Boyer (Ed.), The Enzymes, 3rd Edn., Academic Press, New York, pp. 339–356.
101 Zaman, Z., Jordan, P.M. and Akhtar, M. (1973) Biochem. J. 135, 257.
102 Abboud, M.M., Jordan, P.M. and Akhtar, M. (1974) J. Chem. Soc. Chem. Commun., 643.
103 Braun, P.E. and Snell, E.E. (1968) J. Biol. Chem. 243, 3775.
104 Brady, R.N., DiMari, S.J. and Snell, E.E. (1969) J. Biol. Chem. 244, 491.
105 Fujino, Y. and Zabin, I. (1962) J. Biol. Chem. 237, 2069.
106 Weiss, B. (1963) J. Biol. Chem. 238, 1953.
107 Davis, L. and Metzler, D.E. (1972) in P.D. Boyer (Ed.), The Enzymes, 3rd Edn., Vol. 7, Academic Press, New York, p. 33.
108 Sawada, S., Kumagai, H., Yamada, H. and Hill, R.K. (1975) J. Am. Chem. Soc. 97, 4334.
109 Fuganti, C., Ghiringhelli, D., Giangrasso, D. and Grasselli, P. (1974) J. Chem. Soc. Chem. Commun., 726.
110 Fuganti, C., Ghiringhelli, D., Giangrasso, D., Grasselli, P. and Amisano, A.S. (1974) Chim. Ind. 56, 424.
111 Tsai, M.D., Weaver, J., Floss, H.G., Conn, E.E., Creveling, R.K. and Mazelis, M. (1978) Arch. Biochem. Biophys. 190, 553.
112 Skye, G.E., Potts, R. and Floss, H.G. (1974) J. Am. Chem. Soc. 96, 1593.
113 Tsai, M.D., Schleicher, E., Potts, R., Skye, G.E. and Floss, H.G. (1978) J. Biol. Chem. 253, 5344.
114 Schleicher, E., Mascaro, K., Potts, R., Mann, D.R. and Floss, H.G. (1976) J. Am. Chem. Soc. 98, 1043.
115 Vederas, J.C., Schleicher, E., Tsai, M.D. and Floss, H.G. (1978) J. Biol. Chem. 253, 5350.
116 Cook, P.F. and Wedding, R.T. (1976) J. Biol. Chem. 251, 2023.
117 Akopyan, T.N., Braunstein, A.E. and Gorayachenkova, E.V. (1975) Proc. Natl. Acad. Sci. U.S.A. 72, 1617.

118 Tolosa, E.A., Maslova, R.N., Gorayachenkova, E.V., Willhardt, I.H. and Braunstein, A.E. (1975) Eur. J. Biochem. 53, 429.
119 Komatsubara, S., Kisumi, M., Chibata, I., Gregorio, M.M.V., Müller, U.S. and Crout, D.H.G. (1977) J. Chem. Soc. Chem. Commun., 839.
120 Crout, D.H.G., Gregorio, M.V.M., Müller, U.S., Komatsubara, S. and Kisumi, M. (1980) Eur. J. Biochem. 106, 97.
121 Umbarger, H.E. (1973) Adv. Enzymol. 37, 349.
122 Kapke, G. and Davis, L. (1975) Biochemistry 14, 4273.
123 Kapke, G. and Davis, L. (1976) Biochemistry 15, 3745.
124 Yang, I.Y., Huang, Y.Z. and Snell, E.E. (1975) Fed. Proc. 34, 496.
125 Floss, H.G. and Tsai, M.D. (1979) Adv. Enzymol. 50, 243.
126 Cornforth, J.W., Redmond, J.W., Eggerer, H., Buckel, W. and Gutschow, C. (1970) Eur. J. Biochem. 14, 1.
127 Lüthy, J., Rétey, J. and Arigoni, D. (1969) Nature (London) 221, 1213.
128 Rose, I.A. (1970) J. Biol. Chem. 245, 6052.
129 Kumagai, H., Yamada, H., Sawada, S., Schleicher, E., Mascaro, K. and Floss, H.G. (1977) J. Chem. Soc. Chem. Commun., 85.
130 Cheung, Y.F. and Walsh, C. (1976) J. Am. Chem. Soc. 98, 3397.
131 Kumagai, H., Yamada, H., Matsui, H., Ohkrishi, H. and Ogata, K. (1970) J. Biol. Chem. 245, 1767.
132 Miles, E.W. (1979) Adv. Enzymol. 49, 127.
133 Miles, E.W. and Kumagai, H. (1974) J. Biol. Chem. 249, 2843.
134 Houck, D.R., Miles, E.W. and Floss, H.G. (1981) paper presented at Symposium on Pyridoxal Phosphate-Associated Enzymes, Knoxville, Tennessee, June 1981.
135 Snell, E.E. (1975) Adv. Enzymol. 42, 287.
136 Morino, Y. and Snell, E.E. (1967) J. Biol. Chem. 242, 2800.
137 Yamada, H. and Kumagai, H. (1975) Adv. Appl. Microbiol. 19, 249.
138 Kumagai, H., Utagawa, T. and Yamada, H. (1975) J. Biol. Chem. 250, 1661.
139 Zenk, M.H. and Schmitt, J.H. (1965) Biochem. Z. 342, 54.
140 Tate, S.S. and Meister, A. (1974) Adv. Enzymol. 35, 503.
141 Soda, K. and Tanizawa, K. (1979) Adv. Enzymol. 49, 1.
142 Longenecker, J.B. and Snell, E.E. (1955) J. Biol. Chem. 213, 229.
143 Braunstein, A.E. (1972) Enzymes: Structure and Function 29, 135.
144 Palekar, A.G., Tate, S.S. and Meister, A. (1970) Biochemistry 9, 2310.
145 Palekar, A.G., Tate, S.S. and Meister, A. (1971) Biochemistry 10, 2180.
146 Tate, S.S. and Meister, A. (1969) Biochemistry 8, 1056.
147 Guggenheim, S. and Flavin, M. (1968) Biochim. Biophys. Acta 151, 664.
148 Posner, B.I. and Flavin, M. (1972) J. Biol. Chem. 247, 6402.
149 Guggenheim, S. and Flavin, M. (1969) J. Biol. Chem. 244, 6217.
150 Hansen, P.E., Feeney, J. and Roberts, G.C.K. (1975) J. Magn. Res. 17, 249.
151 Fuganti, C. and Coggiola, D. (1977) Experientia 33, 847.
152 Posner, B.I. and Flavin, M. (1972) J. Biol. Chem. 247, 6412.
153 Chang, M.N.T. and Walsh, C. (1980) J. Am. Chem. Soc. 102, 7370.
154 Johnston, M., Marcotte, P., Donovan, J. and Walsh, C. (1979) Biochemistry 18, 1729.
155 Chang, M.N. and Walsh, C. (1980) J. Am. Chem. Soc. 102, 2499.
156 Krongelb, M., Smith, T.A. and Abeles, R.H. (1968) Biochim. Biophys. Acta 167, 473.
157 Flavin, M. and Slaughter, C. (1960) J. Biol. Chem. 235, p. 1112.
158 Fuganti, C. (1979) J. Chem. Soc. Chem. Commun., 337.
159 Coggiola, D., Fuganti, C., Ghiringhelli, D. and Grasselli, P. (1976) J. Chem. Soc. Chem. Commun., 143.
160 Flavin, M. and Slaughter, C. (1969) J. Biol. Chem. 244, 1434.
161 Adams, D.O. and Yang, S.F. (1979) Proc. Natl. Acad. Sci. U.S.A. 76, 170.
162 Lürssen, K., Naumann, K. and Schröder, R. (1979) Naturwissenschaften 66, 264.

163 Lürssen, K., Naumann, K. and Schröder, R. (1979) Z. Pflanzenphysiol. 92, 285.
164 Yu, Y.B., Adams, D.O. and Yang, S.F. (1979) Arch. Biochem. Biophys. 198, 280.
165 Boller, T., Herner, R.C. and Kende, H. (1979) Planta 145, 293.
166 Honma, M. and Shimomura, T. (1978) Agric. Biol. Chem. 42, 1825.
167 Honma, M., Shimomura, T., Shiraishi, K., Ichihara, A. and Sakamura, S. (1979) Agric. Biol. Chem. 43, 1677.
168 Robinson, J.R., Klein, S.M. and Sagers, R.D. (1973) J. Biol. Chem. 248, p. 5319.
169 Zappia, V. and Barker, H.A. (1970) Biochim. Biophys. Acta 207, 505.
170 Somack, R. and Costilow, R.N. (1973) Biochemistry 12, 2597.
171 Rubinstein, P.A. and Strominger, J.L. (1974) J. Biol. Chem. 249, 3776.

CHAPTER 5

Stereochemistry of enzymatic substitution at phosphorus

PERRY A. FREY

Department of Chemistry, Ohio State University, Columbus, OH 43210, U.S.A.

1. Introduction

(a) Enzymatic substitution at phosphorus

Among the most important reactions in biochemistry are those that involve substitution at phosphorus in phosphates. These reactions are involved in energy transduction, replication, transcription and recombination of nucleic acids, metabolic regulation, and metabolic pathways including biosyntheses of nucleotides, amino acids, proteins, complex lipids and complex carbohydrates. They are among the most intensively studied proteins in biochemistry, both because of their fundamental importance and because they present special challenges as research subjects.

Much has been learned from the kinetics of representative examples of each class of enzymes acting on phosphates, and the three-dimensional structures of several phosphotransferases have been determined. Additional information about catalytic pathways has been obtained by the use of isotopic tracers, in many cases to determine which bonds are cleaved and in others to determine whether substrate-derived phosphorylated intermediates may be involved. An important recent advance, developed for the latter purpose, is the study of positional isotope exchange or 'isotope scrambling' in ^{18}O-labeled phosphates [1–3]. This method has been used to show that γ-glutamyl phosphate is an intermediate in the glutamine synthetase reaction [2]. Magnetic resonance techniques have been used to obtain information about the nature of interactions among nucleotides, metal ions, and enzymes, with special reference to the distances between metal ions and substrate atoms complexed with enzymes and whether the metal forms a bridge between the enzyme and the substrate. ^{31}P-Nuclear magnetic resonance has recently been used to determine the partitioning of several phosphotransferases among their respective enzyme–substrate and enzyme–product complexes, showing that they exist in comparable amounts and, therefore, that binding interactions equalize the free energies of substrates and products in their enzyme-bound states, in marked contrast to their unequal solution free energies [4]. These and other aspects of phosphoryl transfer reactions have recently been reviewed by J.R. Knowles [5].

Tamm (ed.) Stereochemistry

(b) Stereochemistry and mechanisms of substitution in phosphates

Knowledge of the stereochemical course of any chemical reaction is a prerequisite for understanding the mechanism by which it occurs. Because of recent advances in the synthesis of chirally substituted phosphates and the development of methods for analyzing their configurations, such information is now becoming available for enzymatic reactions which catalyze substitution at phosphorus. This review outlines the most significant stereochemical studies appearing in the literature through 1980 and compiles the results through that date. Detailed aspects of the methods are reviewed elsewhere [6–11].

Stereochemical studies of phosphates are in some respects similar or analogous to studies of saturated carbon, insofar as both are tetrahedral centers and may be achiral, prochiral, or chiral depending upon the identities of the substituent groups. Important differences include a degree of variance in the mechanisms of substitution and their stereochemical consequences as well as many of the methods by which chiral centers are synthesized and analyzed for configuration.

Nonenzymatic mechanisms of substitution define the mechanistic possibilities for enzymatic reactions. The mechanisms that have been proposed based on nonenzymatic studies given in Fig. 1 are of two general classes, dissociative and associative [12–14]. The dissociative mechanism, the first one given in Fig. 1, is an S_N1 mechanism, in which the leaving group departs in the rate-limiting step leaving the planar monomeric metaphosphate anion as a discrete, metastable intermediate, which is immediately captured by the nucleophilic acceptor to form a stable product. This mechanism will proceed with compulsory loss of stereochemical configuration at a chiral phosphorus center, since the planar intermediate can be captured from either face if it is a truly free species. The second mechanism is a single S_N2 displacement, in which the incoming nucleophile displaces the leaving group in a single step through a trigonal-bipyramidal transition state. This associative mechanism involves compulsory inversion of configuration at a chiral phosphorus center. The third mechanism is also associative but involves two steps and a trigonal-bipyramidal intermediate. The two-step associative mechanism becomes possible when the trigonal bipyramid resulting from the addition of the nucleophilic acceptor to the reaction center is stable enough to exist as a discrete intermediate in an energy minimum rather than as the transitional energy maximum in the S_N2 process. It is known that when such an intermediate exists the incoming nucleophile enters at an apical position and the leaving group departs from an apical position [12,14]. If the intermediate is sufficiently stable, its lifetime may be long enough to allow a structural change, pseudorotation, to occur. This results in two of the equatorial groups becoming apical and the two apical groups becoming equatorial. It is then possible for a group that was equatorial in the initial adduct to enter an apical position, thereby becoming a leaving group. This mechanism is important in certain nonenzymatic reactions, especially those involving cyclic phosphates. Inversion of configuration at a chiral center is expected when the leaving group is apical in the initial intermediate, as in the case of in-line attack, and retention is expected when

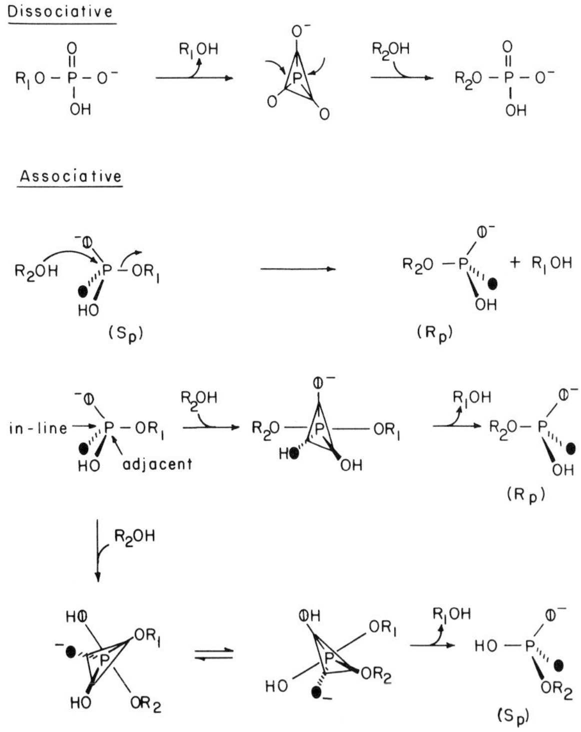

Dissociative

Associative

Fig. 1. Mechanisms of substitution in phosphate.

the leaving group is equatorial in the initial adduct, as in the case of adjacent attack, and must enter an apical position through pseudorotation prior to leaving.

It had been expected by most workers in the field that substitution at phosphorus by transferases would proceed stereospecifically with inversion or retention but not loss of configuration at chiral phosphorus centers. This was not a denial of the possible or probable involvement of metaphosphates as intermediates, but rather a recognition of the probability that such intermediates would be unlikely to be either sufficiently long-lived or sufficiently free of enzymic binding interactions to undergo the facial reorientation required for racemization to be observed. It was thought that free metaphosphates, should they appear as intermediates, would be captured by water leading to phosphohydrolase rather than phosphotransferase activity. This expectation has been borne out by the stereochemical studies completed to date. In addition, however, stereochemical studies of phosphohydrolase reactions have shown that they are stereospecific as well, with no evidence of the racemization that would be expected for a mechanism involving the rate limiting expulsion of metaphosphate anion by enzymatic catalysis followed by its nonenzymatic capture by water.

The other major mechanistic issue addressed in stereochemical work is the question of whether transferase reactions proceed by single-displacement mechanisms according to Equation 1 or by double-displacement mechanisms analogous to Equation 2,

$$E \overset{\pm A\text{-}P}{\rightleftharpoons} E \cdot A\text{-}P \overset{\pm B}{\rightleftharpoons} E \cdot A\text{-}P \cdot B \rightleftharpoons E \cdot A\ B\text{-}P \rightleftharpoons E + A + B\text{-}P \tag{1}$$

$$E \overset{\pm A\text{-}P}{\rightleftharpoons} E\ A\text{-}P \overset{\pm A}{\rightleftharpoons} E\text{-}P \overset{\pm B}{\rightleftharpoons} E\ B\text{-}P \overset{\pm B\text{-}P}{\rightleftharpoons} E \tag{2}$$

involving covalently bonded phosphoryl– or nucleotidyl–enzyme intermediates. Kinetic pathways represented by these equations can be distinguished by steady state kinetic analysis, since Equation 1 is a sequential binding pathway involving ternary complexes whereas Equation 2 is a ping-pong or double-displacement pathway. While reaction kinetics could distinguish binding pathways, it could not provide a general answer to the question of the possible involvement of covalently bonded E-*P* intermediates in kinetic pathways involving compulsory ternary or higher complexes, because it could not distinguish the number of such complexes in the pathway. Thus the interconversion of the central complexes in Equation 1 might involve a double-displacement process and an E-*P* intermediate through a third ternary complex as in Equation 3, and this could not be detected by the conventional steady-state kinetic analysis:

$$E \cdot A\text{-}P \cdot B \rightleftharpoons E\text{-}P \cdot A \cdot B \rightleftharpoons E \cdot A \cdot B\text{-}P \tag{3}$$

If it should happen that each enzymatic displacement at phosphorus occurs by an in-line mechanism with inversion of configuration, a stereochemical analysis would distinguish whether a single or double displacement had occurred, since in the single displacement net inversion of configuration would be observed while in the double displacement the result would be net retention. This is the picture that has emerged to this writing. Each enzymatic substitution at phosphorus for which there is no evidence of an E-*P* intermediate independent of stereochemistry has been found to proceed with net inversion of configuration, and each for which there is conclusive nonstereochemical evidence of an E-*P* intermediate has been found to proceed with net retention of configuration. The stereochemical test has now been accepted in the field as the best single criterion for determining whether a covalent intermediate is involved, short of isolating and characterizing it and showing that it is kinetically competent. When it is not involved all other methods provide essentially negative findings on the issue, whereas the stereochemical test provides a positive finding that cannot be questioned on the basis of inadequate experimental design or the possibility that the mechanism involves the fleeting existence of an intermediate at an undetectable steady state concentration.

(c) Stereochemistry of metal–nucleotide complexes

The true substrates for many enzymes which utilize nucleotides as substrates are metal–nucleotide complexes, often Mg(II)–nucleotides such as Mg–ATP. Numerous

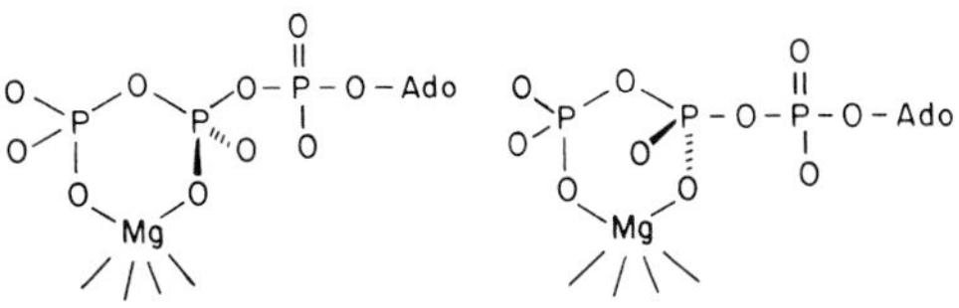

Fig. 2. Diastereomers of β,γ-bidentate Mg–ATP.

isomers of Mg–ATP are possible, many of which are stereoisomers of the several possible coordination isomers. For example, the β-phosphorus atom in ATP is a prochiral center, so that there must be two diastereoisomers of (β,γ)-coordinated Mg–ATP. Both isomers have Mg^{2+} coordinated to one of the three sterically equivalent oxygens of the γ-phosphorus, but in one of them a second coordination position of Mg^{2+} is occupied by one of the diastereotopic oxygens at P_β, i.e., the *pro-R* oxygen while in the other isomer this coordination position is occupied by the *pro-S* oxygen as illustrated in Fig. 2. It is expected that an enzyme will stereospecifically, or at least stereoselectively, bind and utilize one of these isomers as a substrate.

Concurrently with the development of research dealing with the stereochemical course of substitution at chiral phosphorus centers, the question of the stereospecificities of enzymes for metal–nucleotide complexes has also been addressed. The methodologies developed for the two types of studies differed initially, but significant overlap has appeared. The applications of the two approaches have unmasked the complete stereochemical courses of several phosphotransferase reactions.

2. *Methodologies of stereochemical investigations*

There are two basic prerequisites for completing an enzymatic investigation involving substitution at phosphorus. A substrate for the subject enzyme must first be synthesized with a chirally substituted phosphorus at the reaction center. ATP is a concrete example with three potentially chiral phosphorus centers, each of which undergoes enzymatic substitution. The enzymatic bond cleavages resulting from substitution at P_α, P_β and P_γ are illustrated in Fig. 3. Substitution at P_α is catalyzed by nucleotidyltransferases and certain ATP-dependent synthetases, and substitution at P_γ is catalyzed by phosphotransferases, other ATP-dependent synthetases and

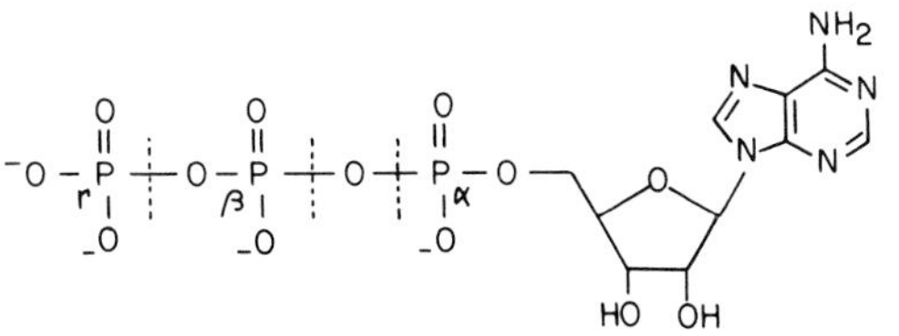

Fig. 3. Enzymatic P–O bond cleavages in ATP.

ATPases. A few pyrophosphotransferases catalyze substitution at P_β. P_α and P_β are prochiral centers and can be made chiral by stereospecific replacement of one or the other diastereotopic oxygen with sulfur or enrichment with ^{18}O or ^{17}O. P_γ is an achiral center which can be made chiral either by replacement of one of the three equivalent oxygens with sulfur and stereospecific enrichment of another with ^{18}O to give a chiral [^{18}O]phosphorothioate, or by stereospecific enrichment of one with ^{17}O and another with ^{18}O to give a chiral [^{16}O, ^{17}O, ^{18}O]phosphate.

The second prerequisite for a stereochemical investigation is that an effective method for assigning configurations at the chiral phosphorus centers in substrates and products must be available. The methods in current use include X-ray diffraction crystallography, mass spectrometry, nuclear magnetic resonance and circular dichroism. The methods by which chiral phosphorothioates can be synthesized and configurationally analyzed are discussed in this section.

(a) Chiral phosphorothioates

The first chiral phosphates to be used for stereochemical analyses were chiral phosphorothioates, which were used to determine the stereochemical courses of ribonuclease, UDP-glucose pyrophosphorylase, adenylate kinase and several other kinases and synthetases. The chiral phosphorothioates either had sulfur in place of an oxygen at an otherwise prochiral center of a phosphodiester or phosphoanhydride or stereospecifically placed sulfur and ^{18}O (or ^{17}O) in a terminal phosphoryl group. The syntheses and configurational analyses of the most important of these compounds are outlined in the following.

(i) Synthesis

Syntheses of chiral phosphorothioates are remarkable for the small number of basic chemical reactions required to prepare a large number of compounds which serve satisfactorily as substrates for enzymes. They are, briefly, (a) the thiophosphorylation of an alcohol or vicinal glycol with thiophosphoryl chloride in an organic solvent, followed by aqueous work-up to afford respectively a thiophosphate monoester or a cyclic phosphorothioate diester; (b) the coupling of a thiophosphate monoester with another phosphate to form a thiophosphoanhydride; (c) the cyclization of a vicinal hydroxy thiophosphate monoester to a cyclic phosphorothioate diester; and (d) the reaction of a phosphoroanilidate diester with a base and CS_2 to give a phosphorothioate diester. These basic reactions are illustrated in Fig. 4. Except for the last, all are conceptually straightforward, and in practice the only serious complications are those arising from the adventitious presence of water during crucial stages, a problem avoided by carefully drying solvents and reactants.

Nucleoside 5′-phosphorothioates are conveniently synthesized by thiophosphorylation of the nucleoside by $PSCl_3$ in triethylphosphate followed by aqueous work-up [15,16]. The reaction is regio-selective for the 5′-hydroxyl group, although a small percentage of 3′- or 2′-phosphorothioate is produced and removed by chromatographic purification of the major product. Product work-up in $H_2^{18}O$ or $H_2^{17}O$ yields the [^{18}O]- or [^{17}O]phosphorothioate [16].

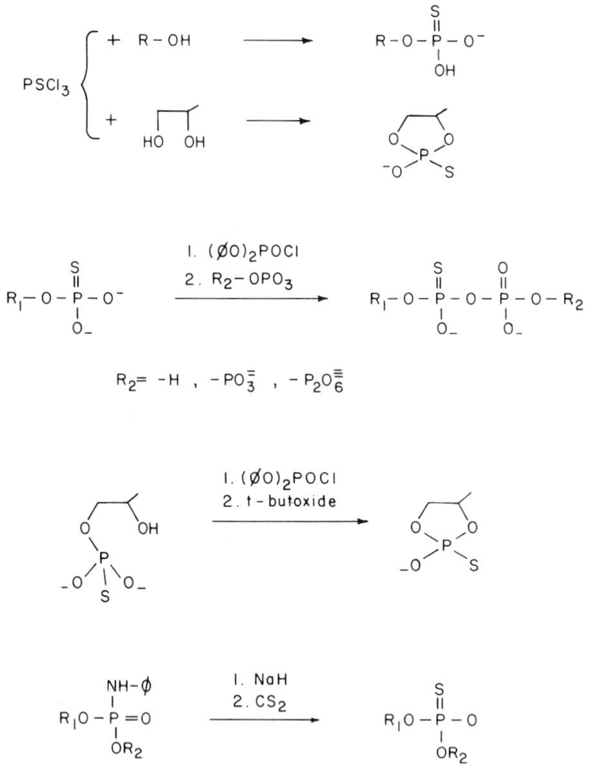

Fig. 4. Important synthetic reactions for phosphorothioates.

Nucleoside 2′,3′-cyclic phosphorothioates are prepared by reacting the 5′-acetylnucleoside with triimidazole phosphinsulfide followed by aqueous work-up and deblocking [17]. This procedure is illustrated in Fig. 5 for the *exo*- and *endo*-isomers of 2′,3′-cyclic uridine phosphorothioate, 2′,3′-cyclic UMPS, which were first synthesized by Eckstein and associates. The two could be separated because the *endo*- isomer is crystalline while the other is an oil. After repeated crystallization the pure *endo*- isomer was obtained and its crystal structure determined, giving its absolute configuration [18].

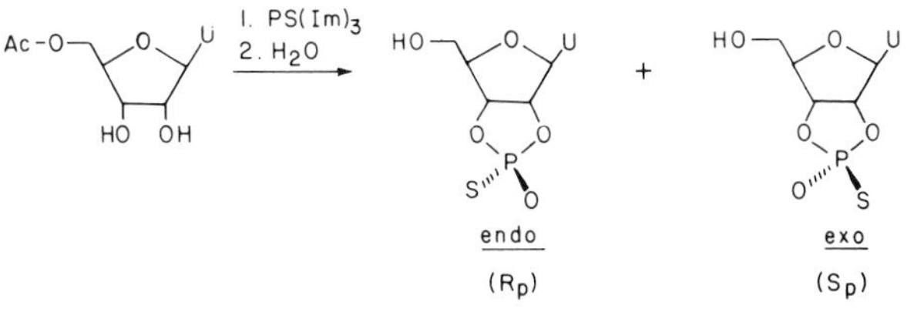

Fig. 5. Synthesis of *endo*- and *exo*-2′,3′-cyclic uridine phosphorothioate.

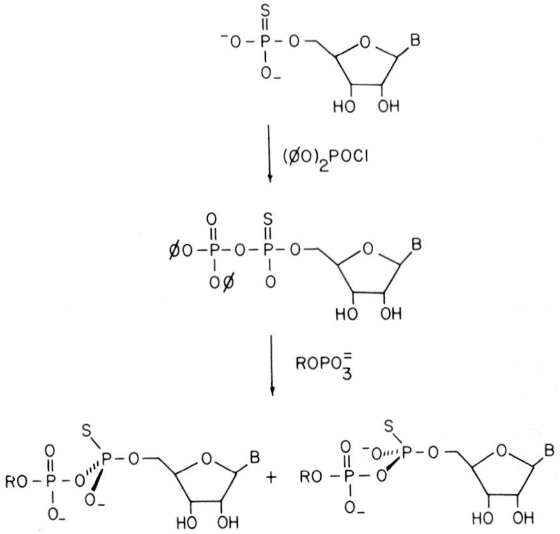

Fig. 6. Synthesis of α-thionucleotides. The compounds with B = a nucleic acid base and R = -H, -PO_3, α-D-glycosyl have been synthesized.

α-Thionucleotides are synthesized from nucleoside 5′-phosphorothioates and phosphate, pyrophosphate, or phosphate esters as outlined in Fig. 6. This is the method originally introduced by Michelson for the synthesis of the phosphoanhydride linkage [19] and is used to synthesize ADPαS, ATPαS, UDPαS, UTPαS and UDPαS-glucose [20,21]. Note that the reaction produces two diastereomers, actually epimers, differing in configuration at P_α. They are designated in the original literature as isomers A and B, isomer A being defined as that isomer of ADPαS preferentially phosphorylated by the phosphoenolpyruvate–pyruvate kinase system [20]. This isomer is now known to have the (S_P) configuration at P_α while the B isomer has the (R_P) configuration. The configurational assignment is discussed in the next section.

The (S_P) and (R_P) isomers of most of the known α-thionucleotides have been separated by enzymatic or chromatographic methods. Eckstein and Goody separated the isomers of ADPαS and ATPαS using enzymatic methods according to Equation 4 [20]:

$$(S_p)\text{-ADP}\alpha\text{S} + (R_p)\text{-ATP}\alpha\text{S} \xleftarrow[\text{kinase}]{\substack{P\text{-Arg}+\\ \text{arginine}}} (S_p + R_p)\text{-ADP}\alpha\text{S}$$

$$\xrightarrow[\text{kinase}]{\substack{\text{PEP}+\\ \text{pyruvate}}} (S_p)\text{-ATP}\alpha\text{S} + (R_p)\text{-ADP}\alpha\text{S} \qquad (4)$$

Baseline separations of ADPαS can also be achieved by reversed-phase high-performance liquid chromatography [22], which is the method of choice because enzymatic phosphorylation at the β-phosphorus is stereoselective rather than stereospecific with respect to P_α configurations, so that the separations are not complete.

The diastereomers of UTPαS and UDPαS-glucose can be separated by the action of yeast UDP-glucose pyrophosphorylase according to Equations 5 and 6 [21]:

$$(R_p + S_p)\text{-UTP}\alpha\text{S} + \text{Glc-1-}P \rightarrow (S_p)\text{-UDP}\alpha\text{S-Glc} + (S_p)\text{-UTP}\alpha\text{S} \quad (5)$$

$$(R_p + S_p)\text{-UDP}\alpha\text{S-Glc} + \text{PP}_i \rightarrow (R_p)\text{-UDP}\alpha\text{S-Glc} + (R_p)\text{-UTP}\alpha\text{S} \quad (6)$$

In this case the chiral centers are also the reaction centers and the enzyme appears to be quite stereospecific for that center.

A convenient synthesis of (S_p)-ATPαS is the enzymatic phosphorylation of AMPS, adenosine 5′-phosphorothioate by the adenylate kinase–pyruvate kinase system according to Equation 7 [23]:

$$\text{AMPS} + 2\ \text{phosphoenolpyruvate} \xrightarrow[\text{(ATP)}]{\text{adenylate and pyruvate kinases}} (S_p)\text{-ATP}\alpha\text{S} + 2\ \text{pyruvate} \quad (7)$$

This process is stereospecific and has the advantage of giving a quantitative conversion. Although AMP, the usual acceptor substrate for adenylate kinase, is achiral at phosphorus and so not subject to stereospecific phosphorylation, the phosphorus in AMPS is a prochiral center. Adenylate kinase catalyzes the phosphorylation of the *pro-R* oxygen exclusively to produce the (S_p) configuration.

β-Thionucleotides that have so far been synthesized include adenosine 5′-[2-thiodiphosphate], ADPβS, and the (R_p) and (S_p) diastereomers of adenosine 5′-[2-thiotriphosphate], ATPβS. ADPβS is synthesized by coupling AMP with *S*-carbamoylethyl phosphorothioate using the Michelson procedure of activating AMP with diphenylphosphorochloridate, coupling with *S*-carbamoylethyl phosphorothioate, and deblocking the resulting *S*-carbamoylethyl-ADPβS in base [24]. The (R_p) and (S_p) diastereomers of ATPβS can be prepared by stereoselective enzymatic phosphorylation of ADPβS [20]. With Mg^{2+} as the activating cation, (S_p)-ATPβS is produced by phosphorylation with phosphoenolpyruvate and pyruvate kinase. The (R_p) diastereomer is produced by phosphorylation with acetyl-*P*, arginine-*P*, 1,3-diphosphoglycerate or creatine phosphate in the presence of Mg^{2+} and the appropriate kinase. These transformations are illustrated schematically in Fig. 7.

The chiral γ-[^{18}O]phosphorothioate of ATP, (R_p)-ATPγS, $\gamma^{18}\text{O}(\beta,\gamma)^{18}\text{O}$, was synthesized by the procedure outlined in Fig. 8 [25]. (S_p)-ADPαS, $\alpha^{18}\text{O}(\alpha,\beta)^{18}\text{O}$ was prepared by stereospecific phosphorylation of AMPαS, $\alpha^{18}O_2$ using the adenylate and pyruvate kinase system followed by dephosphorylation with glucose and hexokinase. This was the starting material for the synthesis, with the chiral α-

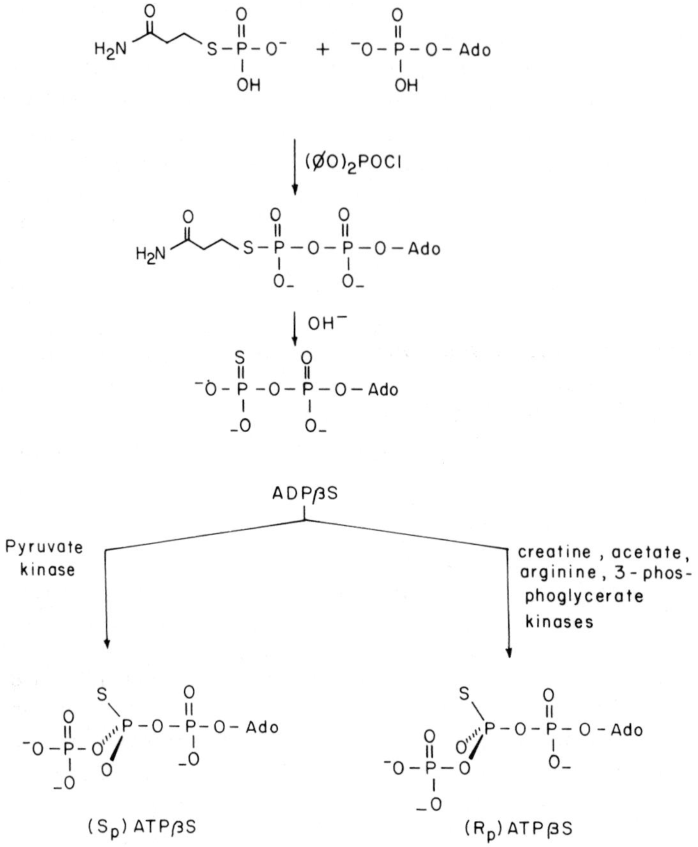

Fig. 7. Synthesis of β-thionucleotides.

phosphorus ultimately becoming the γ-[^{18}O]thiophosphoryl group in (R_p)-ATPγS, γ^{18}O(β,γ)^{18}O. It was condensed with 2′,3′-methoxymethylidene-AMP by the Michelson phosphoanhydride synthesis method to produce the half-blocked dinucleoside [^{18}O]tripolyphosphorothioate. Chemical removal of the unblocked adenosyl and the methoxymethylidene groups gave (R_p)-ATPγS, γ^{18}O(β,γ)^{18}O in very good yield, 55–58% after chromatographic purification.

(R_p)- and (S_p)-ADPβS, β^{18}O were synthesized by an analogous procedure starting with AMPS, $^{18}O_2$ and 2′,3′-methoxymethylidene-AMP as outlined in Fig. 9 [26]. After condensation by the Michelson procedure the diastereomeric mixture of half-blocked dinucleoside [^{18}O]pyrophosphorothioates was separated by chromatography. The configurations at the chiral centers were assigned by chemically degrading each of the isomers to the corresponding isomer of ADPαS, α^{18}O and determining its P_α configuration by ^{31}P-NMR spectrometry. The half-blocked dinucleoside [^{18}O]pyrophosphorothioates of known configuration were then converted to the

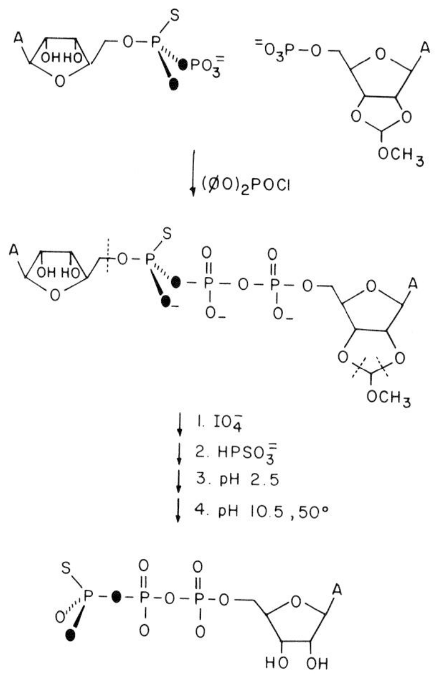

Fig. 8. Synthesis of (R_p)-ATPγS,γ^{18}O$(\beta,\gamma)^{18}$O. The symbols ○ and ● designate ^{16}O and ^{18}O, respectively.

corresponding isomers of ADPβS, β^{18}O by chemical removal of the unblocked adenosyl groups and deblocking of the methoxymethylidene nucleoside.

(R_p)- and (S_p)-AMPS, ^{18}O have also been synthesized by an analogous procedure [27]. AMP and AMPS, $^{18}O_2$ were coupled by the Michelson procedure to the diastereomeric diadenosyl [^{18}O]pyrophosphorothioates which were separated by chromatography. The separated (R_p)- and (S_p)-diadenosyl [^{18}O]pyrophosphorothioates were subjected to the hydrolytic action of *Crotalus adamanteus* nucleotide pyrophosphatase, which cleaved the compounds to AMP and (R_p)- or (S_p)-AMPS, ^{18}O with retention of configuration at the chiral phosphorus centers. (R_p)- and (S_p)-AMPS, ^{18}O have also been synthesized from the diastereomers of ADPαS, α^{18}O, which were separated by enzymatic phosphorylation of the (S_p) isomer to (S_p)-ATPαS, α^{18}O, leaving unreacted (R_p)-ADPαS, α^{18}O. After chromatographic separation the two compounds were degraded to (R_p)- and (S_p)-AMPS, ^{18}O by the hydrolytic action of alkaline phosphatase [28]. The hydrolysis was easily arrested at the AMPS stage because of the resistance of phosphorothioates to alkaline phosphatase.

D-Glycerate-2- and -3-[^{18}O]phosphorothioates have been synthesized by the route outlined in Fig. 10 [29]. Methyl D-glycerate was converted to the diastereomeric mixture of *exo*- and *endo*-D-glycerate cyclic phosphorothioates, which were separated by chromatography. Each isomer was then subjected to hydrolysis with Li^{18}OH, which was known to occur largely by an in-line mechanism with inversion of

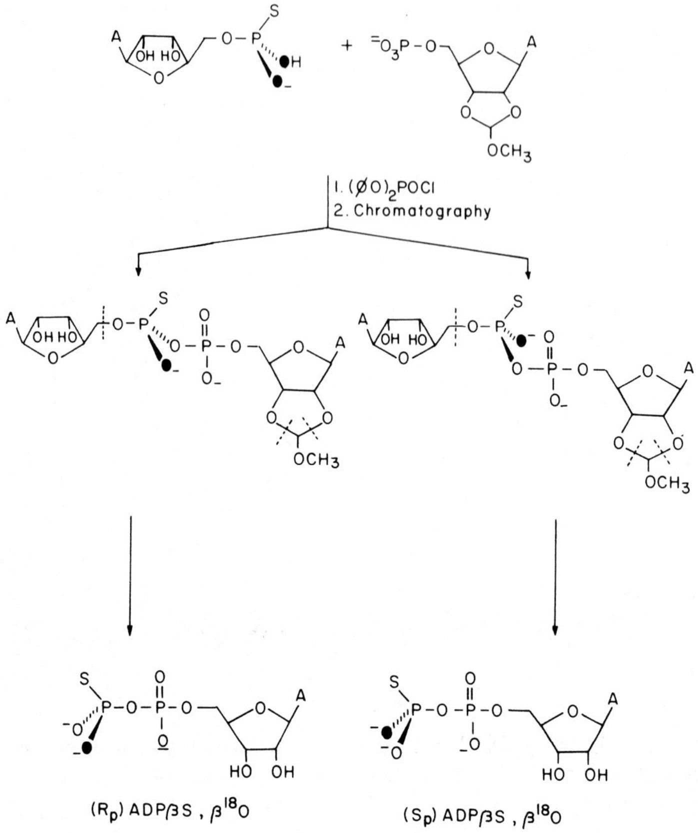

Fig. 9. Synthesis of (R_p)- and (S_p)-ADPβS,β^{18}O. Symbols are as in Fig. 8.

configuration, and which actually occurred with 80–90% stereoselectivity [28,29]. In each case a mixture of D-glycerate-2- and 3-[^{18}O]phosphorothioates was obtained which could be separated by chromatography. Once the absolute configuration of the major isomer of D-glycerate cyclic phosphorothioate was shown to be *endo*-, the configurations of both products could be assigned [30]. This procedure was similar to that employed earlier by Usher and Eckstein for analyzing the stereochemical course of hydrolysis by ribonuclease [31]. They synthesized *endo*-2′,3′-cyclic UMPS and showed that hydrolysis by base or ribonuclease action in $H_2^{18}O$ produced uridine 3′-[^{18}O]phosphorothioate having the same, inverted, configuration.

(R_p)- and (S_p)-3′,5′-cyclic AMPS have been synthesized from 3′,5′-cyclic AMP by the procedure in Fig. 11 [32]. 3′,5′-Cyclic AMP was first reacted with benzoyl chloride in pyridine to block the C-2 hydroxyl group and the adenine ring. It was then converted to the diastereomeric mixture of phosphoroanilidates which were separated by thin layer chromatography. The phosphoroanilidates were converted to

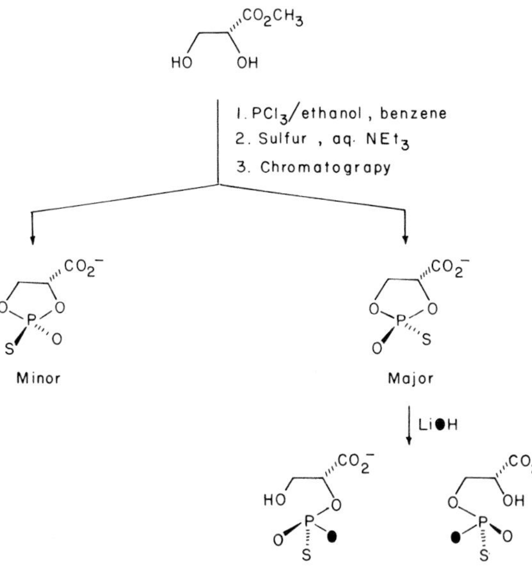

Fig. 10. Synthesis of chiral D-glycerate-2- and -3-[^{18}O]phosphorothioates.

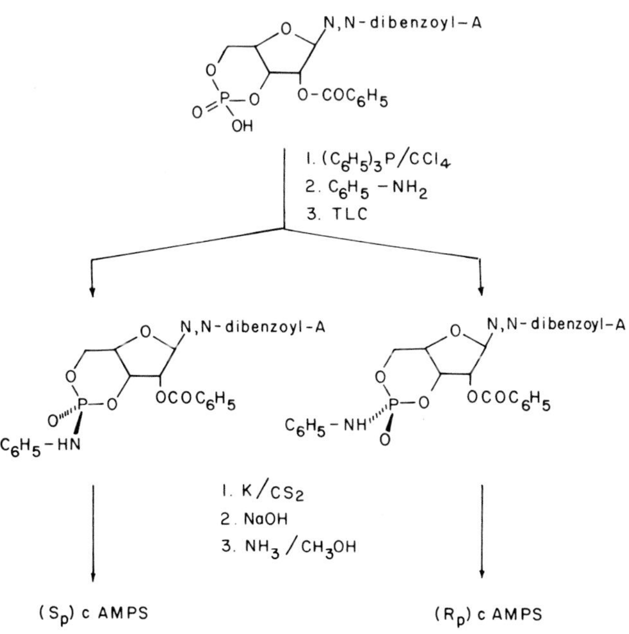

Fig. 11. Synthesis of (R_p)- and (S_p)-3′,5′-cyclic adenosine phosphorothioate.

the corresponding phosphorothioates by reaction with potassium metal and CS_2 and then deblocked to (R_p)- and (S_p)-3′,5′-cyclic AMPS. The treatment of phosphoroanilidates with a strong base followed by CS_2, CO_2, benzaldehyde or CSe_2 has been developed as a general procedure to replace the anilidate group with S, O or Se with retention of configuration [32–34]. This procedure has also been used with [^{18}O]benzaldehyde to prepare chirally enriched 3′,5′-cyclic AMP, ^{18}O and with $C^{18}O_2$ to prepare 3′,5′-cyclic deoxyAMP, ^{18}O [35,36]. The cyclization of adenosine 5′-*O*,*O*-bis(*p*-nitrophenyl)phosphorothioate and deblocking to 3′,5′-cyclic AMPS has also been described [37].

(ii) Configuration assignments

All assignments of absolute configuration are based ultimately upon correlations with the crystal structures of a few phosphorothioates whose structures have been determined. Relative configurational assignments are made by the application of enzymatic and spectroscopic techniques. Examples of the enzymatic technique are the specificities of kinases for the diastereomers of ATPαS and ADPαS as well as ATPβS. Pyruvate kinase preferentially accepts one of the two isomers of ADPαS with Mg^{2+} as the activating metal. This isomer, initially designated diastereomer A and now known to have the (S_p) configuration, can be identified and distinguished from diastereomer B, the (R_p) isomer, by its substrate activity [20]. Similarly, pyruvate kinase is known to produce and/or cleave a single isomer of ATPβS with Mg^{2+} as the activator. This isomer, initially designated diastereomer A and now known to have the (S_p) configuration, has the opposite configuration from that of (S_p)-ATPαS, although their configurational symbols are the same. Adenylate and 3-phosphoglycerate kinases exhibit the same specificities for diastereomers A of ATPαS and ATPβS as pyruvate kinase with Mg^{2+} as the activator, while arginine and creatine kinases exhibit the opposite specificities, preferentially accepting diastereomers B of ATPαS and ATPβS [20,39].

These isomers and all α-thionucleotide diastereomers can be conveniently distinguished by ^{31}P-NMR spectrometry, as shown by Sheu et al. [23] and also by Jaffe and Cohn [38]. The P_α chemical shifts for the diastereomers of α-thionucleotides differ by 0.25–0.4 ppm, (R_p) upfield, and similar differences exist between the diastereomers of ATPβS. ^{31}P-NMR is probably the most positive way of assigning configuration to these nucleotides. This is because the enzymes often do not exhibit absolute specificity for one or the other isomer, and a false positive result can be obtained if too much test enzyme is used with an unknown. A typical set of spectra for a mixture of α-thionucleotides is given in Fig. 12.

Fig. 13 shows how the absolute configurations of ATPαS isomers A and B were assigned by correlation with the configuration of *endo*-2′,3′-cyclic UMPS [40]. One diastereomer of the dinucleoside phosphorothioate U-*P* (S)-A had been shown to be produced from *endo*-2′,3′-cyclic UMPS by the reaction of ribonuclease and to be cyclized by base exclusively to *endo*-2′,3′-cyclic UMPS. Since both reactions were known to proceed by in-line, inversion mechanisms, this isomer was assigned the (R_p) configuration, and it was also shown to be a far better substrate for snake

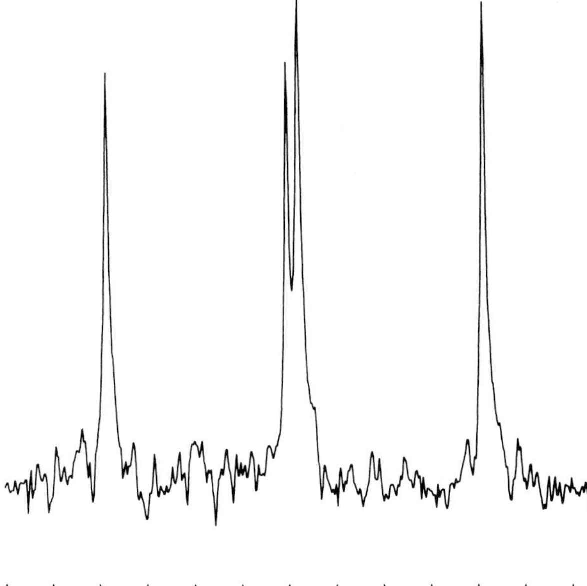

Fig. 12. Use of ^{31}P-nuclear magnetic resonance to distinguish α-thionucleotide diastereomers. Shown is the P_α region for the mixture of ADPαS diastereomers. The upfield doublet is the (S_p) diastereomer.

venom phosphodiesterase than the (S_p) diastereomer. Upon testing the diastereomers of ATPαS as substrates snake venom phosphodiesterase was found to cleave diastereomer B to AMPS and PP_i much faster than it acted on diastereomer A. Diastereomer B was, therefore, also assigned the (R_p) configuration and diastereomer A the (S_p) configuration. This assignment was confirmed independently by alternative methods [41,42].

There are two basic methods by which configurations are assigned to chiral [^{18}O]thiophosphoryl groups. The first one is the one introduced by Usher, Richardson and Eckstein [31], which is applicable to vicinal hydroxy [^{18}O]phosphorothioates when the absolute configurations of the corresponding diastereomeric cyclic phosphorothioates are known. The principle is illustrated in Equation 8 for *sn*-glycerol-3-[^{18}O]phosphorothioate:

1. $(C_2H_5O)_2POCl$
2. *t* - Butoxide

endo – + *exo* – (8)

When the (R_p) isomer is cyclized and the diastereoisomeric cyclic phosphates are separated the *endo*-isomer will contain ^{18}O while the *exo*-isomer will not. When the

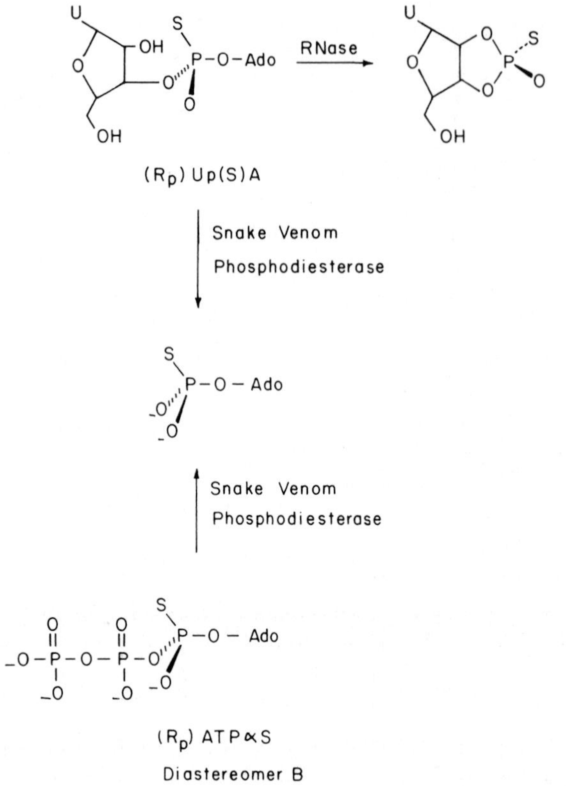

RNase

(R_p) Up(S)A

Snake Venom Phosphodiesterase

Snake Venom Phosphodiesterase

(R_p) ATPαS

Diastereomer B

Fig. 13. Determination of the absolute configuration of ATPαS by correlation with *endo*-2′,3′-cyclic UMPS.

(S_p) isomer is cyclized the *exo*-cyclic phosphorothioate will contain the ^{18}O. The labeling pattern follows from the mechanism of this cyclization, specifically by the fact that the base catalyzed internal displacement of the ^{16}O- or ^{18}O-activated intermediate proceeds by an in-line mechanism [12,14,31].

The other method for assigning configurations of [^{18}O]thiophosphoryl groups is by stereospecific enzymatic phosphorylation of the chiral [^{18}O]thiophosphoryl group. Many stereochemical studies can be designed in such a way that the chiral center whose configuration must be determined is a nucleoside 5′-[^{18}O]phosphorothioate such as AMPS, ^{18}O or a nucleoside 5′-(2-thio[2-^{18}O]diphosphate) such as ADPβS, $\beta^{18}O$. Stereospecific phosphorylation of such centers by enzymes such as adenylate kinase, pyruvate kinase and acetate kinase, as shown in Fig. 14, reduces the problem of configurational assignment to the determination of whether ^{18}O associated with the chiral center is bridging or nonbridging in the phosphorylated product. Adenylate kinase catalyzes the specific phosphorylation of the *pro-R* oxygen in AMPS [23], so that phosphorylation of (R_p)-AMPS, ^{18}O leads to (S_p)-ADPαS, $(\alpha, \beta)^{18}O$, that is

ATP → O

PEP → O

Acetyl − P → O

Fig. 14. Orientations of enzymatic phosphorylations of chiral [^{18}O]phosphorothioates. Shown are the orientations of phosphorylation of AMPS by adenylate kinase and ATP and the phosphorylation of ADPβS by pyruvate kinase and phosphoenolpyruvate (PEP) and by acetate kinase and acetyl-*P*.

with α, β bridging ^{18}O. Similar phosphorylation of (S_p)-ADPαS, α^{18}O leads to (S_p)-ADPαS, α^{18}O with nonbridging ^{18}O. The configuration of an unknown sample of AMPS, ^{18}O can, therefore, be determined by adenylate and pyruvate kinase-catalyzed phosphorylation to (S_p)-ATPαS, α^{18}O and analysis for bridging and nonbridging ^{18}O.

Several techniques are available to analyze for bridging and nonbridging ^{18}O in nucleoside phosphorothioates. One developed in this laboratory involves the chemical degradation of the nucleoside di- or triphosphate α- or β-S to trimethylphosphate and trimethylphosphorothioate by a procedure in which bridging ^{18}O is partitioned between trimethylphosphate and trimethylphosphorothioate, while nonbridging ^{18}O bonded to the phosphorothioate center remains associated exclusively with trimethylphosphorothioate. The degradation products are analyzed for ^{18}O content by gas chromatographic-mass spectroscopy [6,26]. The method was first used in the assignment of absolute configurations to ATPβS diastereomers A and B [26]. The (R_p) and (S_p) isomers of ADPβS, β^{18}O were synthesized as in Fig. 9 and subjected to stereoselective phosphorylation by the phosphoenolpyruvate/pyruvate kinase and acetylphosphate–acetate kinase systems. The resulting samples of ATPβS were analyzed for bridging and nonbridging ^{18}O, which established their absolute configurations. The assignments were made as shown in Fig. 15, which depicts the results

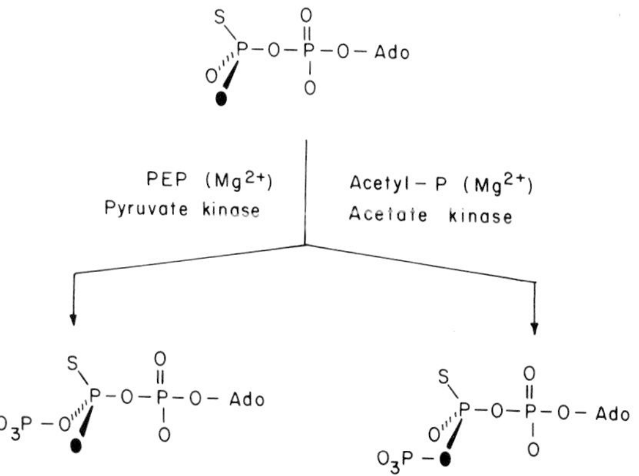

Fig. 15. Configurational analysis of chiral [^{18}O]phosphorothioates by stereospecific phosphorylation.

obtained with (R_p)-ADPβS, β^{18}O. Phosphorylation with phosphoenolpyruvate and pyruvate kinase, using Mg^{2+} as the activating ion, produced diastereomer A in which the ^{18}O was found to be nonbridging. The (S_p) configuration, therefore, was assigned to diastereomer A. Phosphorylation with acetyl-*P* and acetate kinase, again with Mg^{2+} activation, produced diastereomer B, within which ^{18}O was bridging. Diastereomer B was therefore assigned the (R_p) configuration. The same configurational assignments were made simultaneously by an entirely different approach, in which the configurations of ATPβS diastereomers A and B were correlated with the absolute configurations of the Λ and Δ isomers of $Co(NH_3)_4$ATP [43].

Bridging and nonbridging ^{18}O bonded to P_α in ATPαS or dATPαS, or the diphosphates, can also be distinguished by enzymatically cleaving the α, β bond using snake venom phosphodiesterase, certain ATP-dependent synthetases, or a nucleotidyltransferase [16,44,45]. The bridging oxygen departs with the pyrophosphate or phosphate while nonbridging ^{18}O remains associated with AMPS. The products are chemically degraded to phosphate and derivatized to trimethyl- or triethylphosphate for gas chromatographic-mass spectral analysis of ^{18}O content.

There are also two important ^{31}P-NMR techniques for determining bridging and nonbridging ^{18}O or ^{17}O. The isotope shift effect of ^{18}O on the chemical shift of phosphorus atoms to which it is bonded can be used for this purpose [46]. The ^{18}O-induced isotope shift for phosphorus in phosphates is about 0.020 ppm per ^{18}O atom for nonbridging ^{18}O and about 0.012 ppm per ^{18}O for bridging ^{18}O. When ^{18}O is nonbridging the isotope shift appears only for the phosphorus to which it is bonded, whereas when it is bridging it is smaller and observed for both phosphorus atoms to which it is bonded. The ^{18}O-shift is exemplified in Fig. 16 by the ^{31}P-NMR spectrum of an ATP derivative with nonbridging ^{18}O at all three phosphorus atoms.

The other ^{31}P-NMR technique involves the use of ^{17}O rather than ^{18}O. It is based upon the observation that interaction between ^{17}O, which has a large electric quadrupole moment, and ^{31}P to which it is bonded results in a greatly broadened NMR signal for ^{31}P, so broadened that it is difficult to detect in the ^{31}P-NMR spectrum [10,47]. This phenomenon is very useful for determining whether ^{17}O is bridging or nonbridging, since bridging ^{17}O will attenuate the NMR signal of only the one phosphorus to which it is bonded. The technique can be used with lower resolution NMR spectrometers, whereas the ^{18}O-induced isotope shifts are best detected with high field instruments.

The ^{18}O-induced isotope shift and the ^{17}O-induced attenuation of the ^{31}P signal due to broadening have been combined to provide a method for distinguishing (R_p)- and (S_p)-[^{16}O, ^{17}O, ^{18}O]phosphorothioate [48]. The (R_p) and (S_p) enantiomers of [^{16}O, ^{17}O, ^{18}O]phosphorothioate were synthesized as in Fig. 17 and then converted to (S_p)-ATPβS, β^{17}O, β^{18}O by a series of enzymatic transformations whose stereochemical courses were known or could be determined. The product mixtures shown in Fig. 18 were obtained and distinguished by their ^{31}P-NMR spectra. The distinguishing factor was the effect of ^{17}O in rendering certain species silent in the measured spectra by its line broadening effect. In each mixture two of the three species were silent because of the presence of ^{17}O. The third species contained only

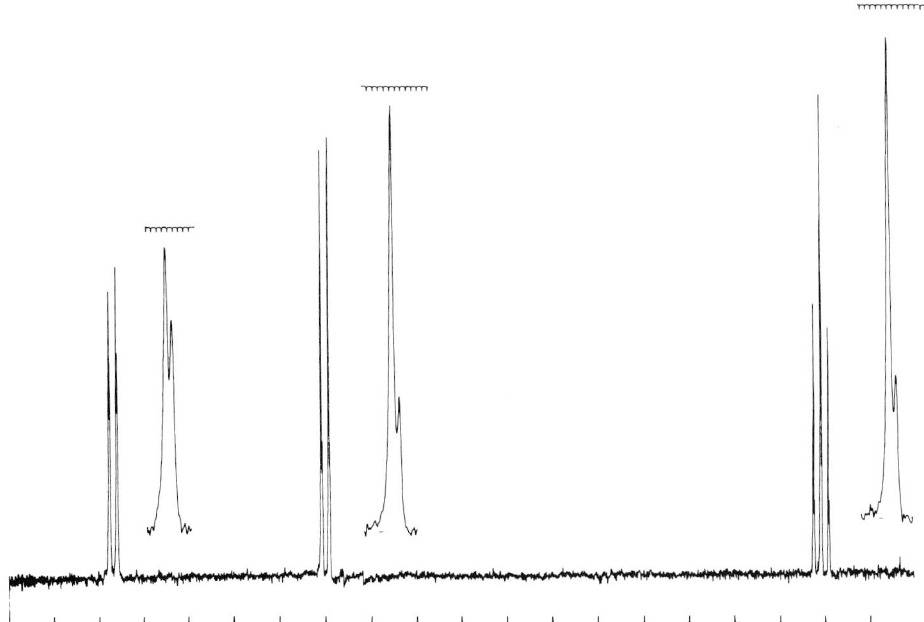

Fig. 16. ^{18}O-isotope shift effects on the α, β and γ phosphorus atoms of ATP. Each phosphorus position is enriched with nonbridging ^{18}O. One signal from each pattern is expanded.

^{18}O and ^{16}O and exhibited the ^{18}O isotope shifts. However, the predominant one originating with the S_p enantiomer contained largely *nonbridging* ^{18}O while the predominant one originating with the R_p enantiomer contained largely *bridging* ^{18}O, resulting in detectable differences in the spectra.

The actual compositions were more complex than those given in Fig. 18 because of the presence of substantial amounts of ^{18}O in all samples of $[^{17}O]H_2O$ used in the syntheses of $[^{17}O]$phosphate. This additional ^{18}O tended to reduce the differences in the spectra, but the isomers could nevertheless be distinguished.

(iii) Phosphorothioates as substrates

Phosphorothioates such as those described above do not arise in nature and so are not naturally occurring substrates for enzymes. Questions arise concerning the advisability of using them to study the stereochemical courses of enzymatic reactions. The two most common fears are, first, whether a given target phosphorothioate will serve as a substrate for the enzyme of interest and, second, if it does whether the stereochemical results will reflect the stereochemical course followed by all substrates for the enzyme, including the naturally occurring substrates. The two questions are related but different, and the first one is clearly the one that has caused the more concern among researchers contemplating or undertaking to trace the stereochemical course of an enzymatic reaction. This is because the failure

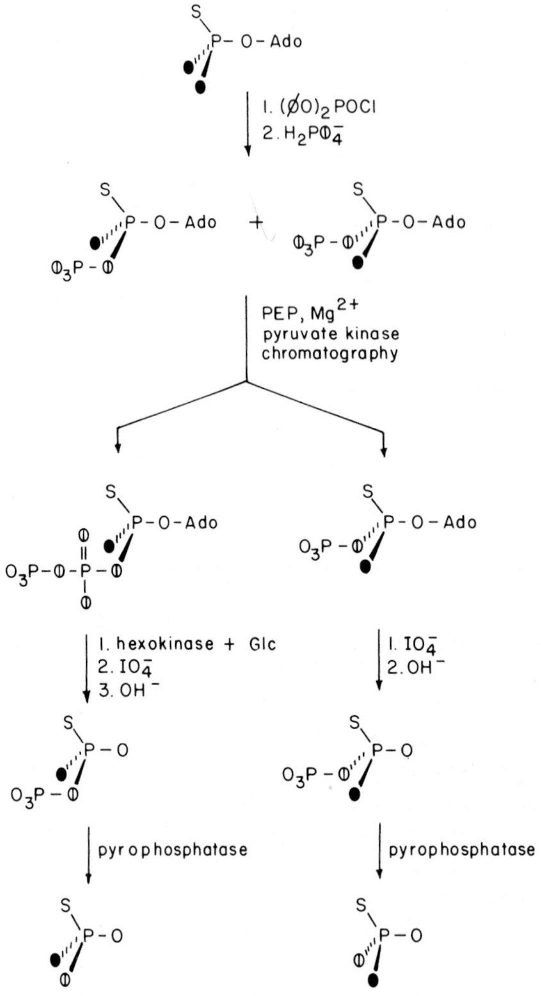

Fig. 17. Synthesis of (R_p)- and (S_p)-[$^{16}O,^{17}O,^{18}O$]phosphorothioate.

of a phosphorothioate as a substrate would undermine the experiment itself, whereas the second question concerns the interpretation of a basically successful experiment.

To date phosphorothioates have with very few exceptions, been found to be acceptable substrates for stereochemical experiments. They generally react more slowly than naturally occurring substrates, from 1 to 10% of the rates for phosphates in many cases of phosphotransferases and 10 to 100% of the normal rates in the cases of nucleotidyltransferases acting on α-thionucleotides. In a very few cases, notably alkaline phosphatase and phosphoglycerate mutase, phosphorothioates are unacceptable as substrates, necessitating the development of methods for synthesiz-

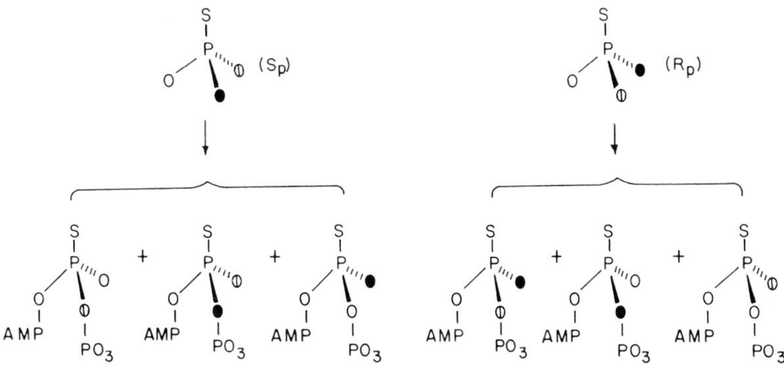

Fig. 18. Configurational analysis of chiral [^{16}O,^{17}O,^{18}O]phosphorothioates.

ing and configurationally characterizing chiral [^{16}O, ^{17}O, ^{18}O]phosphates in order to determine the stereochemical course of phosphoryl group transfer by these enzymes.

There is little reason to expect the substitution of sulfur for oxygen to change the stereochemical course of an enzymatic phosphate substitution reaction. Such a change would reflect a major difference in reaction mechanisms. The 2- to 100-fold rate differences often observed, while substantial in terms of their experimental manifestations, do not represent major fractions of the 10^8- to 10^{12}-fold rate acceleration factors characteristic of enzymes. The differences are comparable, for example, to the difference in activities of horse liver and yeast alcohol dehydrogenases toward ethanol as a substrate, both of which are thought to follow similar basic mechanisms of bond cleavage, though differing in rate-limiting steps. The substitution of sulfur for oxygen will certainly have its effects on the detailed aspects of the reaction mechanism, resulting in different steady-state levels of intermediates and perhaps a different rate-limiting step or a transition from coupled steady-state steps to a single, localized rate-limiting step. However, such major differences as double versus single displacements or in-line versus adjacent substitution mechanisms are not expected to result from sulfur substitution in substrates. This expectation is borne out by the results of three comparative stereochemical studies in which the stereochemical courses for chiral phosphorothioate substrates were shown to be the same as those for chiral phosphate substrates with the enzymes glycerokinase, adenylyl cyclase, and 3′,5′-cyclic AMP phosphodiesterase [30,49,50].

(b) Chiral phosphates

Despite the broad utility of chiral phosphorothioates, certain enzymes, such as alkaline phosphatase and phosphoglycerate mutase, do not accept phosphorothioates as substrates. Stereochemical studies of these enzymes awaited the development of methods to synthesize and assign configurations to chiral phosphates. Chiral [^{16}O, ^{17}O, ^{18}O]phosphomonoesters and [^{18}O]phosphodiesters have been elegantly synthesized and successfully used to determine the stereochemical courses of a number of enzymes.

(i) Synthesis

Chiral [^{16}O, ^{17}O, ^{18}O]phosphomonoesters and ATPγ[^{16}O, ^{17}O, ^{18}O] have been synthesized by Knowles and associates, who devised the procedure outlined in Fig. 19 [51–55]. The procedure has been used to synthesize phenyl[^{16}O, ^{17}O, ^{18}O]phosphate and 2-[^{16}O, ^{17}O, ^{18}O]phospho-D-glycerate as well as the propylene glycol ester shown. The starting cyclic adduct was prepared by reaction of (−)-ephedrine with $P^{17}OCl_3$, giving a separable mixture of 2-chloro-1,3,2-oxazaphospholidin-2-ones whose chemistry had been described [56]. The major isomer was converted to (R_p)-1-[^{16}O, ^{17}O, ^{18}O]phospho-1,2-propanediol and (S_p)-ATPγ[^{16}O, ^{17}O, ^{18}O] by the reactions shown. The stereochemistry at each step of the synthesis was well precedented in the literature; nevertheless, the configurations were verified by independent methods described in the next section.

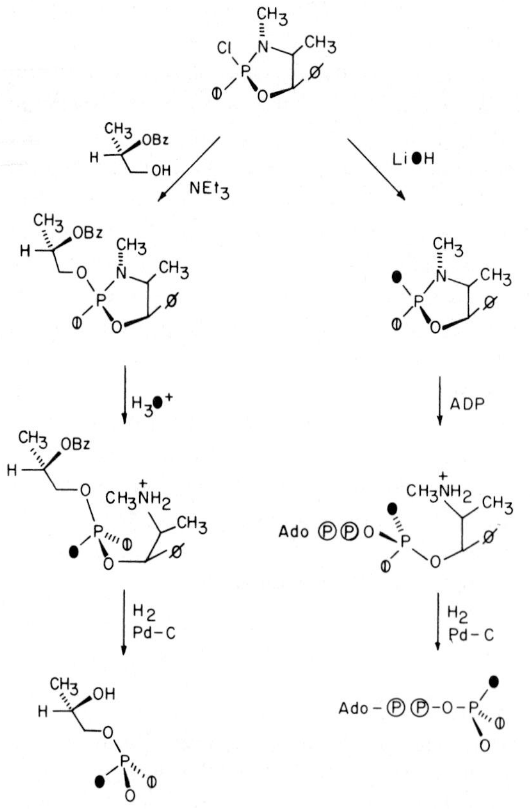

Fig. 19. Synthesis of chiral [^{16}O,^{17}O,^{18}O]phosphates by the (−)-ephedrine method.

An alternative route to (R_p)-methyl-[^{16}O, ^{17}O, ^{18}O]phosphates, presented in Fig. 20, was devised by Lowe and coworkers, who began with (S)-mandelic acid and converted it by asymmetric synthesis to (1R, 2S)-[1-^{18}O]-1,2-dihydroxy-diphenylethane. This was treated with $P^{17}OCl_3$ followed by methanolysis to give the crystalline 5-member cyclic methyl triester of known configuration which was cleaved by hydrogenolysis to (R_p)-methyl-[^{16}O, ^{17}O, ^{18}O]phosphate. The configuration and enantiomeric purity were not established independently of the synthesis, but the configurational assignment was secured by the known configuration of the cyclic methyl hydrobenzoin triester and the fact that no bonds to phosphorus were cleaved in its hydrogenolysis [57].

(R_p)- and (S_p)-3′,5′-cyclic AMP^{18}O have been synthesized by Stec and associates as described in Fig. 11 for 3′,5′-cyclic AMPS, substituting [^{18}O]benzaldehyde for CS_2 [35]. The corresponding isomers of 3′,5′-cyclic deoxy AMP,^{18}O have also been synthesized by this procedure substituting $C^{18}O_2$ for CS_2 [36].

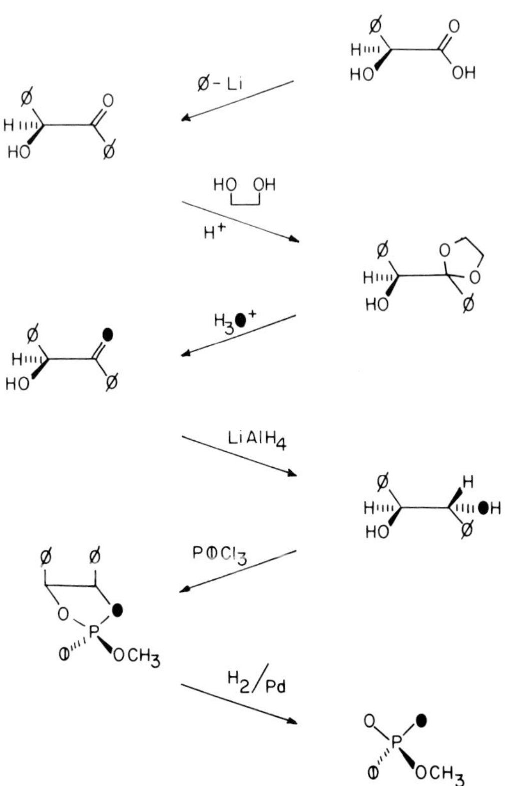

Fig. 20. Synthesis of chiral [^{16}O,^{17}O,^{18}O]phosphates by the hydrobenzoin method.

(ii) Configuration assignments

J.R. Knowles and associates have devised two methods for analyzing the configurations of the chirally enriched phosphate group in (*S*)-1,2-propanediol-1-[^{16}O, ^{17}O, ^{18}O]phosphate [58,59]. They have described their methods in detail elsewhere [5], so the present discussion has been abridged, with the intention of conveying the principles upon which the analyses are based, while avoiding an extended discussion of factors which complicate them.

Both methods involve the cyclization of (*S*)-1,2-propanediol-1-[^{16}O, ^{17}O, ^{18}O]phosphate by random activation of the peripheral phosphorus oxygens and internal displacement of activated oxygen by the vicinal C-2 hydroxyl group. Cyclization of a given isomer leads to a mixture of three cyclic phosphates which differ in their isotopic compositions. The cyclic products from (*S*)-1,2-propanediol-1-(R_p)-[^{16}O, ^{17}O, ^{18}O]phosphate are presented in Fig. 21. Those derived from the (S_p) epimer would have the opposite configurations at phosphorus; for example, in the species containing both ^{17}O and ^{18}O the ^{17}O would be *syn*- to the methyl group rather than *anti*- as in Fig. 21. Since the *syn*- and *anti*- oxygens are chemically equivalent they are not easily distinguished, however, methylation makes them

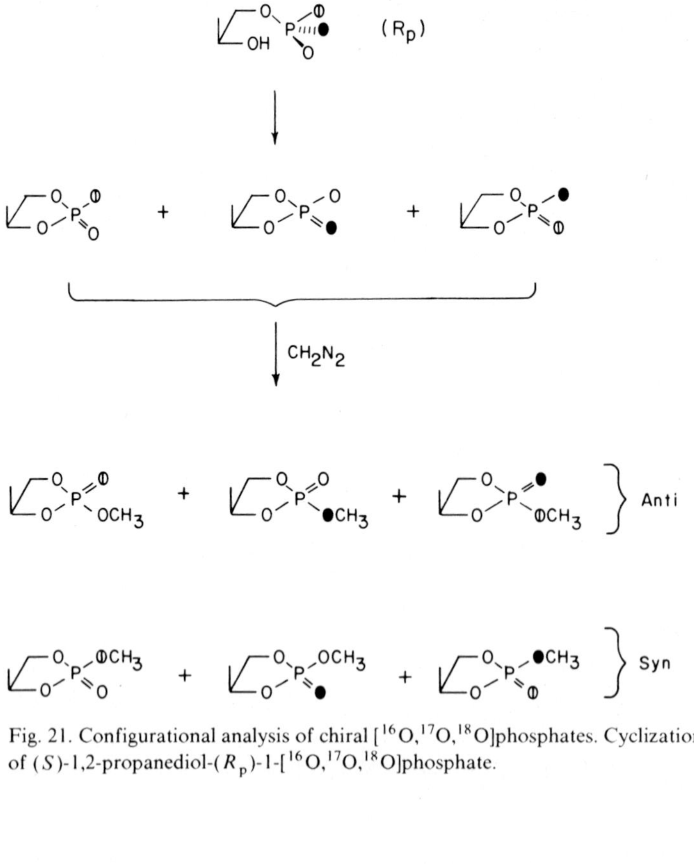

Fig. 21. Configurational analysis of chiral [^{16}O,^{17}O,^{18}O]phosphates. Cyclization and methylation products of (*S*)-1,2-propanediol-(R_p)-1-[^{16}O,^{17}O,^{18}O]phosphate.

chemically inequivalent and thereby more easily distinguishable. This leads to six methylation products, three *syn*- and three *anti*- methyl esters. The *syn*- and *anti*-isomers are diastereomeric and, therefore, separable, in this case by chromatography. The separation and analysis of *syn*- and *anti*-isomers is fundamental to the analysis. For the present purposes the *syn*-isomers will be used to exemplify the configurational assignment, recognizing that the *anti*- isomers contain the same information.

The *syn*- isomers from the (R_p) and (S_p) epimers of *S*-1,2-propanediol-1-[^{16}O, ^{17}O, ^{18}O]phosphate are compared in Fig. 22, and can be seen to differ. For example, the $m+2$ species differ in that the one derived from the (R_p) epimer has an [^{16}O]methoxyl group while that from (S_p) epimer has an [^{18}O]methoxyl group. This is not alone sufficient to distinguish the (R_p) and (S_p) epimers, however, since the $m+3$ species from the (S_p) epimer also has an [^{18}O]methoxyl group. It is necessary in identifying the S_p epimer to show that it is the $m+2$ species that contains an [^{18}O]methoxyl group. Similarly, the $m+1$ species contains an [^{16}O]methoxyl and the $m+3$ contains an [^{17}O]methoxyl.

It happens that a common fragmentation mode for methyl phosphates is the loss of formaldehyde from the methoxyl group. If it can be shown that for one set of *syn*-isomers the $m+1$ parent ion loses [^{16}O]formaldehyde to form a $m'+1$ daughter while the $m+2$ ion loses [^{18}O]formaldehyde to form the m' daughter and the $m+3$ parent loses [^{17}O]formaldehyde to form the $m'+2$ daughter, the (S_p) configuration

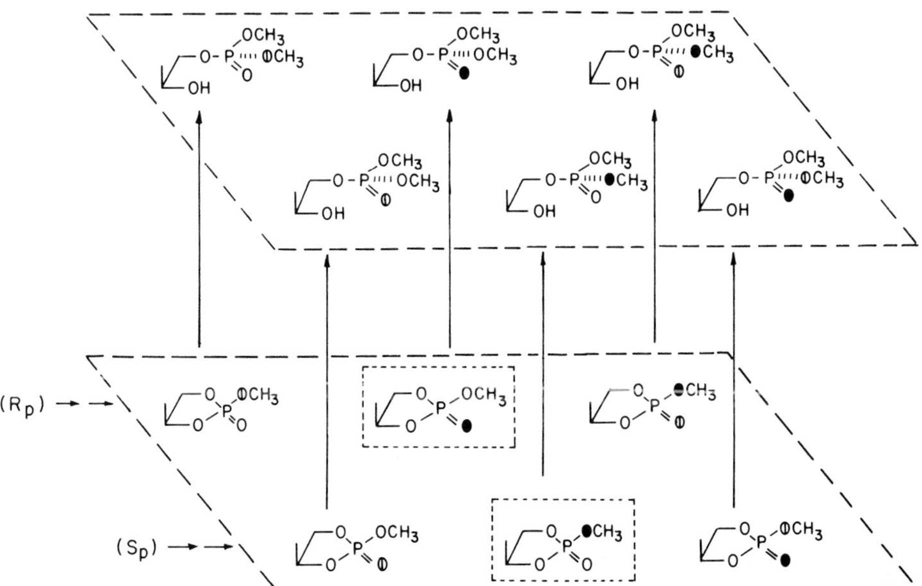

Fig. 22. Configurational analysis of chiral [^{16}O,^{17}O,^{18}O]phosphates. Configurations of *syn*- isomers of the methyl cyclic phosphates (lower plane) and their methanolysis products (upper plane) derived from the (R_p) and (S_p) isomers of *S*-1,2-propanediol-1-[^{16}O,^{17}O,^{18}O]phosphate.

can be assigned to the precursor of that set. Such parent–daughter relationships can be established by metastable ion mass spectrometry. Unfortunately, the cyclic methyl esters do not exhibit the necessary metastable transitions; however, their methanolysis products compared in the upper plane of Fig. 22 contain exactly the same information and do exhibit the necessary transitions. Metastable-ion mass spectrometry of these dimethyl-(*S*)-1,2-propanediol-1-[^{16}O, ^{17}O, ^{18}O]phosphates leads to an unambiguous assignment of configuration.

The other method of configurational assignment is experimentally much simpler. It is accomplished simply by obtaining a high resolution ^{31}P-NMR spectrum of the *syn-* (or *anti-*) methyl cyclic phosphate esters in Fig. 22. Each set of *syn-* isomers consists of two species which contain ^{17}O and are, therefore, silent in the ^{31}P-NMR. In each set the one remaining ester, set off in Fig. 22 by enclosure in boxes, contains only ^{16}O and ^{18}O, and only this species of each set is seen in the NMR spectrum. But they give different ^{31}P-NMR spectra, since the magnitude of the ^{18}O isomer shift is larger for double-bonded than for single-bonded ^{18}O, and the species from the (R_p) epimer contains double-bonded ^{18}O while that from the (S_p) epimer contains only single-bonded isotope. It is not necessary in this analysis to separate the *syn-* and *anti-* isomers, since their ^{31}P-chemical shifts are different and easily resolved.

The foregoing discussion refers to the idealized situations depicted in Figs. 21 and 22. Other species actually appear in the analyses because of the fact that all samples of ^{17}O also contain ^{18}O, and several other experimental complications introduce additional elements of nonideality. All can be quantitatively taken into account in analyzing the data, however, so that the assignments are secure.

Configurational assignment to cyclic [^{18}O]phosphodiesters is made by alkylating the phosphate group to a mixture of axially and equatorially alkylated phosphate diesters. The triesters in Equation 9 exemplify the nature of the product mixture:

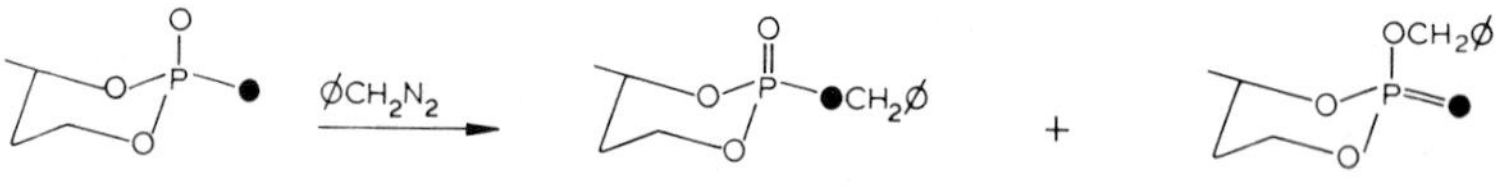

$$\xrightarrow{\phi CH_2N_2} \quad \bullet CH_2\phi \quad + \quad OCH_2\phi \qquad (9)$$

The ^{31}P-NMR chemical shifts of the equatorially and axially alkylated isomers are widely separated so they are easily distinguished in the spectrum of the mixture. Moreover, the single-bonded ^{18}O in the equatorially alkylated isomer in Equation 9 induces an isotope shift in the ^{31}P signal that is less than one half the magnitude of that induced by the double-bonded ^{18}O in the axially alkylated isomer. Were the ^{18}O to be in the axial position the reverse would be found; that is, the axially alkylated isomer would exhibit the smaller and the equatorially alkylated isomer the larger isotope shift. The triesters in Equation 9 can also be distinguished by electron-impact mass spectrometry. Gerlt and Coderre used the former method and Stec and coworkers the latter to verify that configuration is retained in the conversion of chiral phosphoanilidates to chiral [^{18}O]phosphates [35,36].

(c) Chiral metal–nucleotides

The diastereomers of Mg–ATP in Fig. 2 exemplify in a limited way the stereochemical problem in metal nucleotides. The two isomers shown are in rapid exchange equilibrium, however, so it is not possible to separate them and study their individual interactions with enzymes. The problem is further complicated by the fact that these diastereomers represent the stereochemical possibilities in only one of the coordination isomers. Others are possible, including the α-, β- and γ-monodentate isomers and α,β,γ-tridentate isomers, most of which exist as two or more diastereoisomers. All of the coordination isomers and their diastereoisomers of Mg–ATP are in rapid exchange equilibrium.

A central question in phosphotransferases and nucleotidyltransferases is the structure of the metal–nucleotide complex which is the true substrate for the enzyme. It is unlikely that all of the possible Mg–ATP complexes could serve as substrates for a given enzyme, but until recently there has been no way to determine which isomer is active. The difficulty is the coordination exchange equilibrium, which is rapidly set up and dynamically maintained in solutions of Mg–ATP. To avoid this problem, metal–nucleotide complexes have been synthesized using coordination exchange-inert metals such as Cr(III) and Co(III) in place of Mg(II) [7,60]. The resulting complexes are structurally stable and can be separated by chromatographic methods into their coordination isomers and stereoisomers. The isomers can then be investigated as substrates or inhibitors of specific enzymes.

Before discussing their synthesis, the stereochemical notation for these complexes should be mentioned. Referring again to Fig. 2 it is evident that coordination of Mg^{2+} with one or the other diastereotopic oxygen at P_α creates a chiral center. The *R*,*S* stereochemical designations could be used, but a complication arises in that when this is done a given configuration can be either (*R*)- or (*S*)-, depending upon whether the metal is Mg(II) or Cr(III). This is because their priorities are reversed relative to phosphorus. To facilitate the description of these compounds a new configurational designation was introduced by Cleland, in which two isomers are designated left- or right-handed, Λ or Δ, depending upon the screw sense relating the position of the nucleoside and the chelated metal [61]. The rule for defining handedness is the following. Consider the generalized metal–phosphate chelate ring in Fig. 23. Viewing the chelate ring from the face at which the nucleoside or nucleoside monophosphoryl group is attached, if the shortest path around the ring

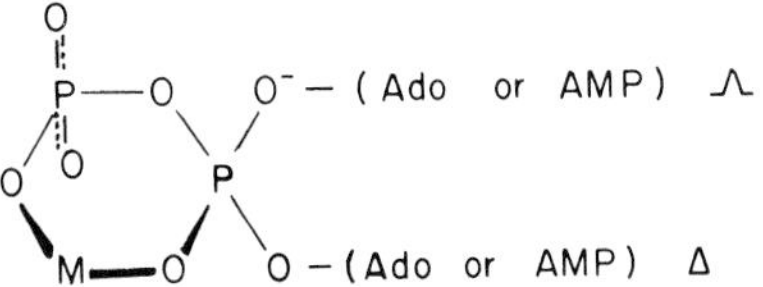

Fig. 23. Definitions of the Λ and Δ stereochemical designations for diastereomeric metal–nucleotide complexes.

from the metal to this group is clockwise the isomer is right-handed, Δ. If the shortest path is counter-clockwise the isomer is left-handed, Λ. These designations consistently correlate the configurations of metal–nucleotide complexes regardless of which metal is chelated.

(i) Synthesis and separation

A variety of Cr(III)– and Co(III)–nucleotides have been synthesized and their diastereomers separated and characterized. Since these metals form octahedral hexacoordinate complexes, some of the coordination positions must be occupied by ligands other than the nucleotide in monodentate, bidentate and tridentate metal–nucleotide complexes. The most widely used Cr(III) complexes contain coordinated water at these positions, although those with coordinated ammonia are also known. The known, stable Co(III) complexes contain ammonia at the additional coordination positions. Although other nitrogen ligands can also be used, complexes with coordinated water are too unstable to be isolated and characterized under the usual laboratory conditions.

The γ-monodentate and bidentate Cr–ATP complexes can be prepared by heating $Cr(H_2O)_6^{3+}$ salts with ATP, brief heating producing mainly γ-monodentate and longer heating the β,γ-bidentate complexes [62]. The γ-monodentate Cr–ATP can be essentially quantitatively converted to the diastereomeric mixture of β,γ-bidentate complexes by heating briefly at pH 3. The diastereomeric mixture of α,β,γ-tridentate Cr–ATP can be prepared by acid treatment of the β,γ-bidentate complexes. The diastereomeric mixture of Cr–ADP complexes have also been prepared by heating ADP with $Cr(H_2O)_6^{3+}$; however, the β- and α-monodentate complexes have not been described.

The corresponding Co(III)–ammine–nucleotide complexes have been synthesized by similar heating of nucleotide with an ammine complex of cobalt [61,64]. The tetrammine complexes such as the bidentate $Co(NH_3)_4$ADP have been prepared as well as the tridentate, triammine $Co(NH_3)_3$ATP complexes. Since Co(III) is diamagnetic in contrast to Cr(III) which is paramagnetic, the ^{31}P-NMR spectra of these complexes are narrow line spectra which reveal differences in chemical shifts among diastereomers [64].

A variety of chromatographic and enzymatic procedures have been devised for separating the coordination isomers and diastereoisomers of Cr(III)– and Co(III)–nucleotide complexes. Dowex 50-H^+ and cross-linked cycloheptamylose are the most widely used chromatographic media [7,60]. The diastereoisomers of such coordination isomers as β,γ-bidentate Cr–ATP are sterically different and can in principle be separated by physical methods such as chromatography, especially on optically active chromatographic media such as cross-linked cycloheptamylose. Another means of separating them is by the use of enzymes, since only one component of a diastereomeric mixture normally serves as a substrate for a given enzyme.

(ii) Configurations of metal–nucleotides

The Cr(III)– and Co(III)–nucleotides are colored complexes, with weak absorption bands in the blue to green region for Cr(III)–nucleotides and the violet to red region for Co(III)–nucleotides. The Cr(III)–nucleotides exhibit two visible absorption maxima, one between 415 and 445 nm and the second between 595 and 630 nm. The corresponding regions in the case of Co(III)–nucleotides are 380–355 nm and 520–560 nm. The ultraviolet absorption spectra are characteristic of the heterocyclic bases.

Inasmuch as these complexes contain chiral centers in the chelate rings of diastereomers, circular dichroic bands are expected to be observed at wavelengths corresponding to the visible bands of the chelated metals. The CD bands are prominent spectral properties that reliably reflect the different configurations of the Λ and Δ isomers of these nucleotides. For example, two Λ and Δ diastereomers of bidentate Cr–ATP exhibit CD bands with molar ellipticities of -1000 for one and $+1000$ $\mathrm{deg \cdot cm^2 \cdot dmol^{-1}}$ for the other at 575 nm. The bands are nearly mirror images and serve as a reliable spectral method for distinguishing the two isomers [7,60].

Although the visible CD spectra can distinguish Λ and Δ isomers they do not by themselves enable one to decide which isomers are Λ and which are Δ. The absolute configurations of the β,γ-bidentate Λ and Δ isomers of $Co(NH_3)_4ATP$ have been determined by crystallographic analysis of the chelate ring and polyphosphate backbone of the isomer of $Co(NH_3)_4ATP$ that is *not* active as a substrate for hexokinase [65]. The mixture of diastereomers was subjected to hexokinase action in the presence of glucose. One isomer reacted and the unreacted isomer was repurified and degraded according to Equation 10:

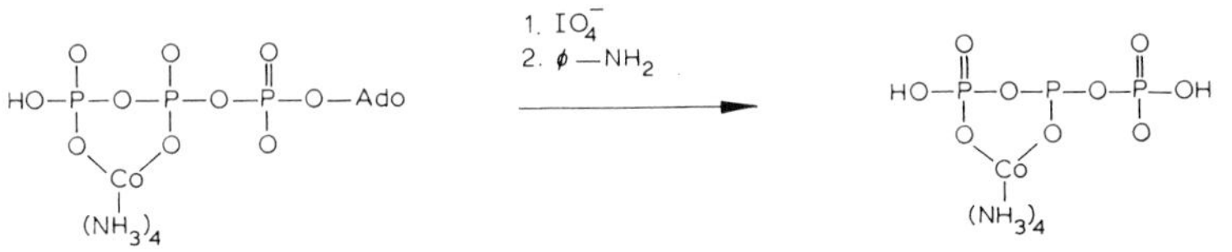

$$\text{HO–P(O)(O)–O–P(O)(O)–O–P(=O)(O)–O–Ado} \;[\beta,\gamma\text{-O chelated to } Co(NH_3)_4] \xrightarrow[2.\ \phi\text{–}NH_2]{1.\ IO_4^-} \text{HO–P(=O)(O)–O–P(O)(O)–O–P(=O)(O)–OH} \;[\text{chelated to } Co(NH_3)_4] \qquad (10)$$

Periodate cleavage of the ribose ring followed by aniline-catalyzed β-elimination of the polyphosphate backbone gave (dihydrogen tripolyphosphate) tetraamminecobalt(III), which was crystallized and structurally analyzed by X-ray diffraction. This analysis established its configuration as Δ, which meant that the other isomer, the substrate for hexokinase, had the Λ configuration. A number of correlations to this structure have been made which have led to the empirical rule that a positive ellipticity for the longer wavelength CD band corresponds to a Λ isomer, while a negative band corresponds to a Δ isomer. This rule is presently based on a limited body of data but appears to hold for adenine nucleotides and Co(III) and Cr(III) complexes of polyphosphates.

3. Selected stereochemical investigations

The stereochemical course of enzymatic substitution at phosphorus has been determined for over thirty enzymes. While most of these studies have been completed within the past three years, and so would be timely subjects for review, the available space does not permit complete discussions of all of them. Therefore, the earliest and most representative studies for each of four classes of enzymes, phosphohydrolases, phosphotransferases, nucleotidyltransferases and ATP-dependent synthetases are described here. These studies exemplify the principles and methods employed in the field to date.

(a) Phosphohydrolases

The first determination of the stereochemical course of enzymatic substitution in a phosphate was the study of pancreatic ribonuclease completed by Usher, Richardson and Eckstein and their associates in 1970 [31]. The enzyme had been shown to catalyze the endonucleolytic cleavage of RNA by the two-step catalytic pathway of Equation 11:

(11)

The first step is an internal displacement of the leaving 5′-OH end of the cleaved RNA (designated R_2OH) by the 2′-hydroxyl group to produce the 2′,3′-cyclic nucleotide as an intermediate. Hydrolysis then opens the cyclic structure to form the 3′-phosphate end of the cleaved RNA. Usher, Eckstein and associates showed that each of these steps proceeds by a mechanism involving stereochemical inversion at phosphorus. The method by which the second step was shown to proceed with inversion is outlined in Fig. 24 [31]. The diastereomers of 2′,3′-cyclic uridine phosphorothioate had been synthesized as a mixture and separated by repeated crystallization of one isomer, while the other remained as an oil. The crystalline isomer was subjected to X-ray diffraction analysis and shown to have the *endo*-configuration [18]. It was also shown to serve as a substrate for ribonuclease [66]. The stereochemical analysis proceeded by carrying out both the enzymatic and OH^- catalyzed hydrolysis of *endo*-2′,3′-cyclic UMPS in $H_2^{18}O$, producing two samples of uridine 3′-[^{18}O]phosphorothioate. These were recyclized to the mixtures of *exo*- and *endo*-2′,3′-cyclic UMPS by activation with diethylphosphorochloridate followed by *t*-butoxide-catalyzed ring closure. The isomers were separated and subjected to analysis for ^{18}O. In both the ribonuclease and OH^- catalyzed hydrolyses the recyclized *endo*- isomer contained very little ^{18}O, whereas the *exo*- isomer was highly enriched. The ^{18}O-enrichments showed that the stereochemical course of the introduction of ^{18}O into uridine 3′-[^{18}O]phosphorothioate by RNase-catalyzed hydrolysis of *endo*-2′,3′-cyclic UMPS must have been the same as that by which it was removed in its recyclization to *endo*-2′,3′-cyclic UMPS. More importantly both ribonuclease and OH^- catalyzed hydrolyses followed the same stereochemical course. According

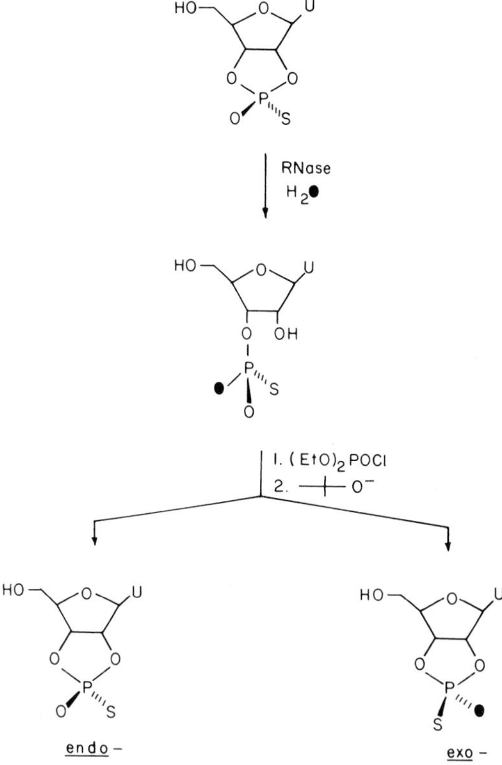

Fig. 24. Stereochemical inversion in the ribonuclease-catalyzed hydrolysis of *endo*-2′,3′-cyclic UMPS in $H_2^{18}O$.

to the rules for pseudorotation the OH^- catalyzed reaction should have proceeded with inversion, so that ribonuclease must also have catalyzed hydrolysis with inversion.

Although the absolute configuration of *endo*-2′,3′-cyclic UMPS had been determined prior to the foregoing study, knowledge of this configuration was not essential to the determination of the stereochemistry of ring opening by $H_2^{18}O$. The minimal prerequisites were that a means for separating the diastereomers be available and that the stereochemistry of the OH^- catalyzed hydrolysis be known. If it had not been known which was the *endo*- isomer, then whichever isomer was found to be a substrate could have been carried through the transformations in Fig. 24, or either isomer if both were substrates, and the stereochemical course could have been determined in the same way by separating the recyclized diastereomers and determining their ^{18}O contents. The absolute configuration of *endo*-2′,3′-cyclic UMPS was required, however, for another study in which RNase was used to catalyze its methanolysis in methanolic water to uridine methylphosphorothioate. The absolute configuration of the product was determined by crystallographic analysis and shown to be inverted relative to that of the substrates [67].

J.R. Knowles and coworkers used chiral [^{16}O, ^{17}O, ^{18}O]phosphate esters to show that the transphosphorylation catalyzed by *E. coli* alkaline phosphatase proceeds with overall retention of configuration [52]. They synthesized phenyl-[^{16}O, ^{17}O, ^{18}O]phosphate by the procedure in Fig. 19 and used it as the phosphoryl-donor substrate with 1,2-propanediol as acceptor according to Equation 12. They determined the configuration of the 1,2-propanediol-1-[^{16}O, ^{17}O, ^{18}O]phosphate by the procedure described in Figs. 21 and 22 and found it to be the same as that of the phenyl-[^{16}O, ^{17}O, ^{18}O]phosphate they used as the phosphate donor.

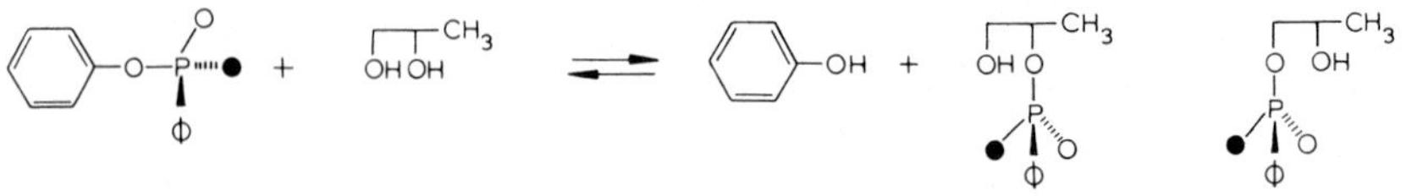

(12)

Alkaline phosphatase was known to catalyze phosphohydrolase and transphosphorylation reactions by a double-displacement mechanism, which was consistent with the above finding of net retention of configuration. The foregoing was a key study for other reasons as well, however, because the broad substrate specificity of this enzyme and its known stereochemistry have been used to good advantage by Knowles and associates to determine the configurations of several [^{16}O, ^{17}O,^{18}O]phosphate esters produced in a variety of phosphotransferase reactions. The chiral phosphate ester of unknown configuration is used in place of chiral phenylphosphate as the transphosphorylation donor in Equation 12. The configuration of the 1,2-propanediol-1-[^{16}O, ^{17}O, ^{18}O]phosphate produced is then determined as above and taken as the configuration of the phosphate donor.

Tsai and Chang showed that 5′-nucleotidase-catalyzed hydrolysis of AMPS to adenosine and phosphorothioate proceeds with overall inversion of configuration [28]. They synthesized the (R_p) and (S_p) isomers of AMPS, ^{18}O used them as substrates for 5′-nucleotidase in $H_2^{17}O$, and determined the configurations of the two samples of [^{16}O, ^{17}O, ^{18}O]phosphorothioate by the procedure of Webb and Trentham outlined in Fig. 18 [48]. The configurations of chiral phosphorothioate samples obtained from the (R_p) and (S_p) isomers of AMPS, ^{18}O were (R_p) and (S_p), respectively, the configurations corresponding to inversion. The stereochemistry was thereby shown to be that of Equation 13:

H₂Φ + O P(S)(●)—O—Ado ⟶ HΦ—P(S)(●)O + adenosine (13)

R_p R_p

The stereochemical course of 3′,5′-cyclic AMP phosphodiesterase was shown to involve inversion of configuration. This was first concluded by Eckstein and Stec and their associates, who synthesized (S_p)-3′,5′-cyclic AMPS and carried out its cyclic AMP phosphodiesterase-catalyzed hydrolysis in $H_2^{18}O$. The resulting AMPS, ^{18}O was subjected to stereospecific phosphorylation to (S_p)-ATPαS, α^{18}O by the adenylate and pyruvate kinase system, and this was analyzed for bridging and nonbridging ^{18}O, with the result shown in Fig. 25. Since the ^{18}O was nonbridging the configuration of AMPS, ^{18}O must have been (S_p), which corresponds to inversion of configuration [68]. A similar approach was used to show that snake venom phosphodiesterase catalyzed hydrolysis proceeds with retention of configuration [41,45].

Inversion of cAMP phosphodiesterase has recently been confirmed by Coderre and Gerlt, who used chiral cyclic [α-^{17}O,^{18}O]-AMP as the substrate in H_2O [50]. The configurational analysis of the resulting chiral [^{16}O, ^{17}O, ^{18}O]AMP is depicted in Fig. 26. After phosphorylating the product to the triphosphate level, the mixture of three isomers of [^{16}O, ^{17}O, ^{18}O]ATP was cyclized by adenylate cyclase to 3′,5′-cyclic [^{16}O, ^{17}O, ^{18}O]AMP as a mixture of isomers, a process known to involve inversion of configuration [49,69]. This mixture was alkylated with diazomethane to the mixture of axial and equatorial alkylated species, analogous to those in Equation 9, and subjected to ^{31}P-NMR analysis. All ^{17}O-containing species were silent and the ^{18}O-induced perturbations of the ^{31}P chemical shift showed that the major ^{16}O,^{18}O-containing species had the same configuration as the starting 3′,5′-cyclic AMP, ^{18}O. Therefore, since the configuration at P_α was known to have been inverted by adenylyl cyclase, it must have been inverted by cyclic AMP phosphodiesterase.

A claim that the stereochemical course of cyclic AMP phosphodiesterase action on 3′,5′-cyclic AMPS and 3′,5′-cyclic AMP, ^{18}O differed, inversion for the chiral cyclic phosphorothioate and retention for the chiral cyclic phosphate, resulted from

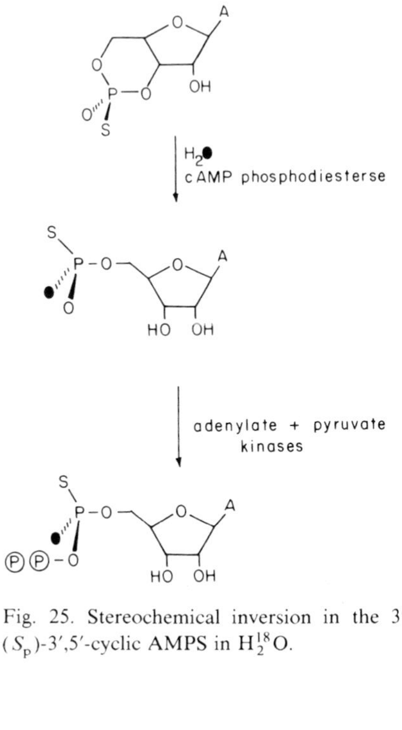

Fig. 25. Stereochemical inversion in the 3′,3′-cyclic AMP phosphodiesterase-catalyzed hydrolysis of (S_p)-3′,5′-cyclic AMPS in $H_2^{18}O$.

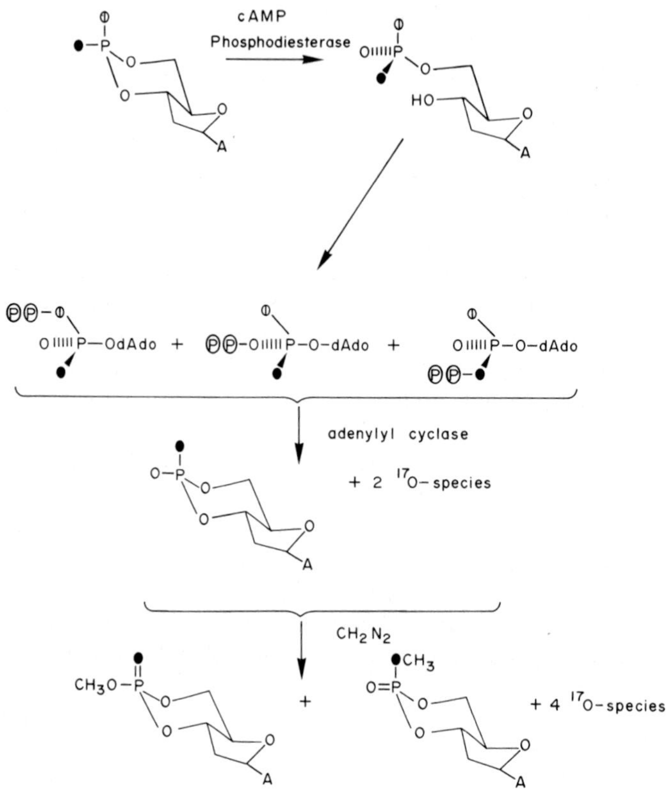

Fig. 26. Confirmation of cAMP phosphodiesterase stereochemistry using (R_p)-3′,5′-cyclic AMP, ^{18}O.

a mistaken configurational assignment to [^{16}O, ^{17}O, ^{18}O]AMP [70,71]. The configurational assignment was later corrected [72].

(b) Phosphotransferases

The stereochemical consequence of [^{18}O]thiophosphoryl and [^{16}O, ^{17}O, ^{18}O]phosphoryl group transfer catalyzed by seven phosphotransferases were simultaneously determined in the author's laboratory and in the laboratory of J.R. Knowles. The first to be completed was the demonstration of inversion by adenylate kinase; however, prior to that glycerokinase, hexokinase and pyruvate kinase had been shown to catalyze [^{18}O]thiophosphoryl group transfer with the same stereochemical consequences, either all with inversion or all with retention. Glycerokinase was later shown to catalyze both [^{16}O, ^{17}O, ^{18}O]phosphoryl and [^{18}O]thiophosphoryl group transfer with inversion of configuration, which meant that hexokinase and pyruvate kinase also transfer with inversion.

The strategy outlined in Fig. 27 was designed to determine the stereochemical course of [^{18}O]thiophosphoryl group transfer by adenylate kinase [25]. This plan required that ATPγS, γ^{18}O and ADPβS, β^{18}O be synthesized with the same configurations at P_γ and P_β, or known relative configurations, and it was for this purpose that the synthetic procedures in Figures 8 and 9 were devised. ATPγS, γ^{18}O was used as the [^{18}O]thiophosphoryl donor substrate for adenylate kinase with AMP as acceptor to produce a second sample of ADPβS, β^{18}O of unknown configuration at P_β. The configurations of the synthetic and enzymatic samples of ADPβS, β^{18}O were compared by stereospecific enzymatic phosphorylation of both samples using pyruvate kinase and acetate kinase, which were known to have opposite P_β-specificities for ATPβS with Mg^{2+} as the activator, and the resulting samples of ATPβS, ^{18}O were subjected to analysis for bridging and nonbridging ^{18}O. The configurations were found to be opposite, showing that [^{18}O]thiophosphoryl transfer occurred with inversion.

The foregoing procedure was designed to produce the desired information without knowing the absolute configurations of any of the [^{18}O]thiophosphoryl groups. It had to rely upon comparing relative configurations because the precursor of the chiral [^{18}O]thiophosphoryl group in ATPγS, γ^{18}O was ADPαS, $\alpha^{18}O_2$ diastereomer A, whose absolute configuration had not yet been determined. By the time of the completion of the analysis the configuration of diastereomer A had been shown to be (S_p), so that the absolute configurations of ATPγS, γ^{18}O and synthetic ADPβS, β^{18}O were known as well. The stereochemical courses of nucleoside diphosphate kinase [21], nucleoside phosphotransferase [27], and adenosine kinase [73], were subsequently determined by analogous procedures.

The stereochemical courses of the glycerokinase, pyruvate kinase, and hexokinase reactions were elucidated simultaneously in a two-stage study. The enzymes first

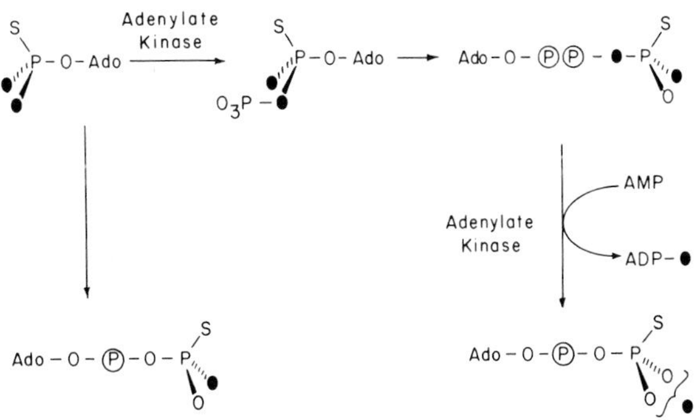

Fig. 27. Strategy for determining the stereochemical course of [^{18}O]thiophosphoryl transfer by adenylate kinase.

were shown by the transformations depicted in Fig. 28 to catalyze [^{18}O]thiophosphoryl group transfer stereospecifically and with the same stereochemical consequences, either all inversion or all retention [29]. D-Glycerate-2-[^{18}O]phosphorothioate and D-glycerate-3-[^{18}O]phosphorothioate having opposite relative configurations at phosphorus but unknown absolute configurations were converted to *sn*-glycerol-3-[^{18}O]phosphorothioate by three different routes, one involving pyruvate kinase and hexokinase, a second involving pyruvate kinase and glycerokinase, and a third involving no kinase action, i.e., no bond cleavage to phosphorus. The configurations of the three samples of *sn*-glycerol-3-[^{18}O]phosphorothioate were compared by the procedure of Equation 8. Samples *x* and *y* were found to have identical phosphorus configurations which were opposite that of *z*. This meant that all three pathways resulted in net retention of configuration, since the pathway to *z* did not involve a phosphotransferase. Inasmuch as the other two pathways each involved two phosphotransferases, one of which was common to both pathways, it had to be concluded that all three enzymes catalyzed the transfers of the [^{18}O]thiophosphoryl groups with the same stereochemical consequences.

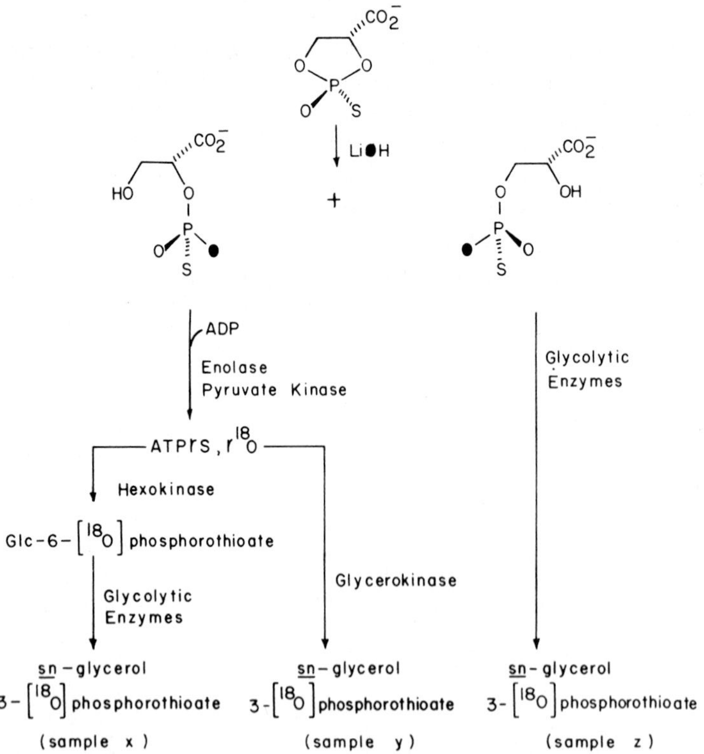

Fig. 28. Strategy for comparing the stereochemical courses of [^{18}O]thiophosphoryl transfer by glycerokinase, hexokinase, and pyruvate kinase.

In the second stage of the study glycerokinase was shown to catalyze both phosphoryl and thiophosphoryl transfer with inversion of configuration. In the $[^{16}O,^{17}O,^{18}O]$phosphoryl study ATP, $\gamma[^{16}O,^{17}O,^{18}O]$ was synthesized and used with glycerokinase to produce a sample of *sn*-glycerol-3-$[^{16}O,^{17}O,^{18}O]$phosphate. The $[^{16}O,^{17}O,^{18}O]$phosphoryl group was transferred to 1,2-propanediol, using alkaline phosphatase to catalyze transphosphorylation and substituting glycerophosphate for phenyl phosphate in Equation 12, producing the mixture of 1- and 2-$[^{16}O,^{17}O,^{18}O]$phosphates of 1,2-propanediol. The configuration of 1,2-propanediol-1-$[^{16}O,^{17}O,^{18}O]$phosphate was determined by the procedures in Figs. 21 and 22. This analysis showed the configuration to be opposite that in ATP, $\gamma[^{16}O,^{17}O,^{18}O]$ used as the donor substrate for glycerokinase [53]. The $[^{18}O]$thiophosphoryl study of glycerokinase proceeded by using synthetic (R_p)-ATPγS, $\gamma^{18}O$ to phosphorylate glycerol to *sn*-glycerol-3-$[^{18}O]$phosphorothioate, which was cyclized to the *syn*- and *anti*- isomers of glycerol-2,3-cyclic phosphorothioate. After separating the isomers, analysis for ^{18}O showed that one contained ^{18}O, and this was shown to be the *anti*-isomer by correlation with the crystal structure of *sn*-D-glycerate-2,3-cyclic phosphate. This meant that the thiophosphoryl configuration in *sn*-glycerol-3-^{18}O phosphorothioate must have been (S_p), the opposite of the configuration in (R_p)-ATPγS, $\gamma^{18}O$ [30].

Methods similar to those employed for glycerokinase utilizing chiral $[^{16}O,^{17}O,^{18}O]$phosphates were also used to elucidate the stereochemical courses of acetate kinase (inversion) [54] and phosphoglycerate mutase (retention) [55].

(c) Nucleotidyltransferases

The stereochemical courses of nucleotidyltransferase action have been pursued concurrently in this laboratory and that of F. Eckstein. My students and I became interested in uridylyltransferase stereochemistry because of our interest in the mechanism of action of galactose-1-*P* uridylyltransferase, the galactosemia defect enzyme, and in its mechanistic relationship with UDP-glucose pyrophosphorylase. The first nucleotidyltransferase to have its stereochemical course elucidated was UDP-glucose pyrophosphorylase from yeast, which was subjected to stereochemical analysis by the procedure outlined in Fig. 29 [74]. The enzyme, which catalyzes the interconversion of UTP and glucose-1-*P* with UDP-glucose and pyrophosphate, was shown to catalyze uridylyl-transfer with inversion of configuration at P_α. The diastereomeric mixture of UTPαS was submitted to the action of UDP-glucose pyrophosphorylase in the presence of glucose-1-*P* and pyrophosphatase. Approximately half of the UTPαS was converted to UDPαS-glucose. The unreacted UTPαS was shown by ^{31}P-NMR spectral analysis to be one of the two diastereomers, the same one produced by the action of pyruvate kinase on UDPαS and phosphoenolpyruvate, which had been designated diastereomer A and is now known to have the (S_p) configuration at P_α. It was concluded that the other isomer, diastereomer B with the (R_p) configuration had reacted with glucose-1-*P* to produce UDPαS-glucose. The UDPαS-glucose was degraded by mild acid hydrolysis to UDPαS, and shown by

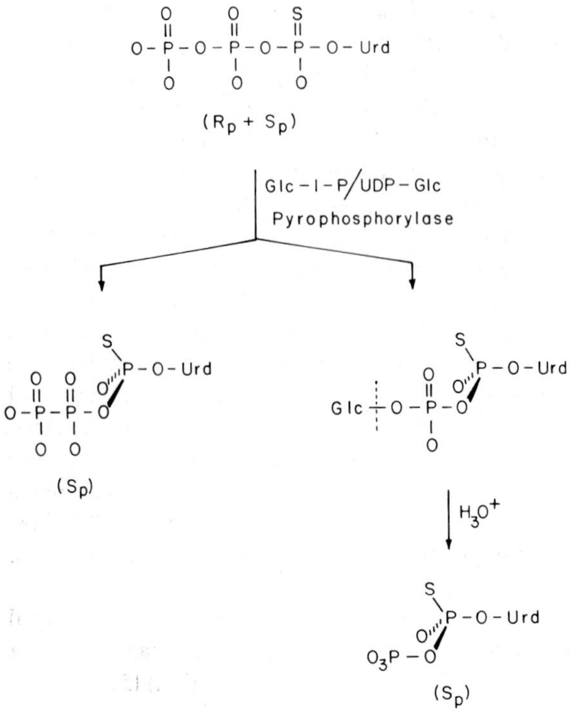

Fig. 29. Stereochemical course of uridylyl transfer by UDP-glucose pyrophosphorylase.

^{31}P-NMR spectral analysis to have the (S_p) configuration. Inasmuch as the mild acid hydrolytic degradation of UDPαS-glucose to UDPαS was known to proceed with C–O bond cleavage between glycosyl C-1 and P_β and, therefore, not to involve bond cleavage at P_α, it was concluded that the (S_p) isomer of UDPαS-glucose had been produced in the reaction of the (R_p) isomer of UTPαS with glucose-1-*P*; that is, the configuration at P_α had been inverted.

An analogous approach was used to show that galactose-1-*P* uridylyltransferase catalyzes the interconversion of UDP-glucose and galactose-1-*P* with UDP-galactose and glucose-1-*P* with retention of configuration at P_α [21]. The stereochemical transformations used in the analysis are shown in Equation 14.

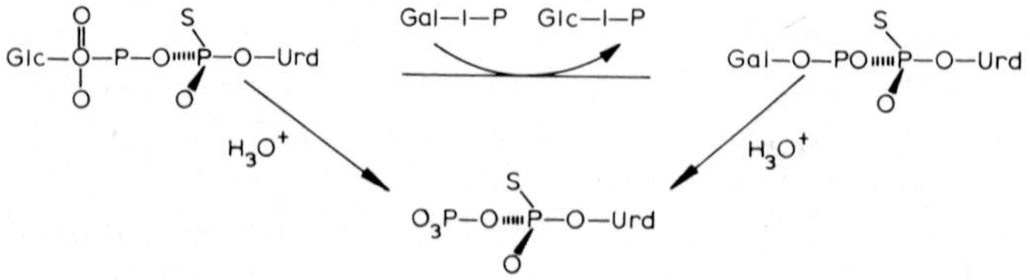

(14)

The R_p isomer of UDPαS-glucose was prepared from synthetic UDPαS-glucose by reacting the diastereomeric mixture with pyrophosphate and UDP-glucose pyrophosphorylase, which had been shown to be specific for (S_p)-UDPαS-glucose. Galactose-1-*P* uridylyltransferase catalyzed the conversion of (R_p)-UDPαS-glucose with galactose-1-*P* to produce UDPαS-galactose. Samples of the substrate, (R_p) and the product of unknown configuration were hydrolyzed to UDPαS, which were shown by ^{31}P-NMR spectral analysis to be identical; that is, they both had the (R_p) configuration.

The procedures used in Fig. 29 and Equation 14 were designed to determine the stereochemical courses of these reactions knowing only the relative configurations of the diastereomers of uridine α-thionucleotides, and the studies were carried out before the absolute configurations were known. The configurations shown in Fig. 29 and Equation 14 are those subsequently assigned to diastereomers A and B.

The stereochemistries of tRNA nucleotidyltransferase, which adds the terminal adenylyl group at the -CCA end of tRNA, and DNA-dependent RNA polymerase were elucidated by Eckstein and associates in a two stage study. The *E. coli* DNA-dependent RNA polymerase study exemplifies their approach. Diastereomer A of ATPαS was copolymerized with UTP using poly(dA-dT)–poly(dA-dT) as the template. The resulting alternating polymer (see Equation 15) was hydrolyzed by ribonuclease A to a cyclic phosphorothioate:

$$\text{ATP}\alpha\text{S} + \text{UTP} \xrightarrow{\text{poly(dA-dT)}} \text{—Up(S)ApUp(S)A—} \xrightarrow{\text{RNase A}} \text{ApU} > \text{pS} \qquad (15)$$

diastereomer A

The known stereochemistry of ribonuclease suggested that this dinucleotide had an *endo*-2′,3′-cyclic phosphorothioate end, corresponding to cleavage of the (R_p) configuration in the backbone phosphorothioate diester linkage. That this was indeed the case was shown by subjecting the dinucleotide to spleen phosphodiesterase to remove the adenylyl group and produce *endo*-2′,3′-cyclic UMPS. This was crystallized and characterized, confirming the (R_p) configuration in the polymer [75]. In the second stage the absolute configuration of ATPαS diastereomer A was shown to be (S_p), opposite that of the polymer, which meant that polymerization had proceeded with inversion [40]. Similar studies showed that tRNA nucleotidyltransferase and nucleotide phosphorylase also transfer the adenylyl group with inversion at P_α [76]. *E. coli* DNA polymerase I was shown by two different methods to catalyze polymerization of (S_p)-dATPαS with inversion of configuration at P_α [16,77].

Adenylyl cyclase of *Brevibacterium liquefaciens* catalyzes the conversion of (S_p)-ATPαS to (R_p)-3′,5′-cyclic AMPS, i.e., with inversion at P_α, and this result was confirmed using chiral 3′,5′-cyclic AMP, ^{18}O [49,69]. A novel stereochemical analysis was developed in the confirmatory study. (R_p)-3′,5′-Cyclic AMP, ^{18}O was converted with pyrophosphate and adenylyl cyclase to ATP, α^{18}O of unknown configuration. This was dephosphorylated to ADP, α^{18}O and converted to the diastereomers of $Co(NH_3)_4$ADP, α^{18}O shown in Fig. 30. ^{31}P-NMR analysis of these compounds,

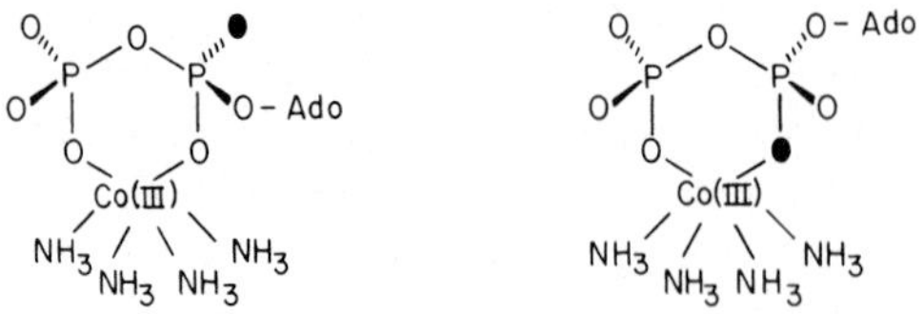

Fig. 30. Determination of the configurations of (R_p)- and (S_p)-ADP, ^{18}O by conversion to their Λ and Δ Co(III) tetrammine complexes.

whose absolute configurations were known, distinguished bridging and nonbridging ^{18}O by virtue of the differential isotope shifts induced by double versus single bonded ^{18}O [49].

(d) ATP-dependent synthetases

The first ATP-dependent synthetase to be subjected to analysis for substitution stereochemistry was phosphoribosylpyrophosphate synthetase [78]. This analysis was novel in that it utilized a coordination exchange-inert Co–ATP complex for this purpose and circular dichroic analysis for relative configurations of substrate and product. The reaction of Equation 16 was catalyzed by this enzyme.

$$\Delta\text{-}(\beta,\gamma)\text{Co(NH}_3)_4\text{–ATP} + \text{ribose-}P \rightarrow \Lambda\text{-1-Co(NH}_3)_4\text{PP-ribose-5-}P + \text{ADP} \quad (16)$$

The circular dichroic bands of $(\beta,\gamma)Co(NH_3)_4$–ATP and the Co complex of phosphoribosyl pyrophosphate were nearly mirror images, showing that they were of opposite configuration.

Yeast acetyl-CoA synthetase is known to catalyze the conversion of ATP, acetate, and coenzyme A to acetyl-CoA, AMP and pyrophosphate by a mechanism involving the intermediate formation of acetyl-adenylate according to Equations 17 and 18:

$$CH_3CO_2^- + ATP \rightleftharpoons CH_3-\overset{\overset{\displaystyle O}{\|}}{C}-OAMP + PP_i \quad (17)$$

$$CH_3-\overset{\overset{\displaystyle O}{\|}}{C}-OAMP + CoASH \rightleftharpoons CH_3-\overset{\overset{\displaystyle O}{\|}}{C}-SCoA + AMP \quad (18)$$

The first step involves the displacement of pyrophosphate from P_α of ATP by acetate. The stereochemical course of this step was shown to involve inversion by carrying out the reaction of (R_p)-ATPαS with $CH_3C^{18}O_2^-$ and coenzyme A according to Equation 19 [44].

$$CH_3C^{18}O_2^- + (R_p)\text{-ATP}\alpha\text{S} + \text{CoASH}$$

$$\rightarrow CH_3^-COSCoA + PP_i + (R_p)\text{-AMPS}, {}^{18}O \quad (19)$$

The configuration of AMPS, ^{18}O was shown to be (R_p), corresponding to inversion of configuration, by stereospecific phosphorylation, using the adenylate kinase–pyruvate kinase system, and analysis for bridging and nonbridging ^{18}O. Similar approaches have been applied to amino acyl-tRNA synthetases, which also catalyze inversion of P_α of ATP [79].

(e) Structure of enzyme-bound metal nucleotides

The synthesis and configurational assignments for Cr(III)– and Co(III)–nucleotides have made it possible to determine which of the many coordination and stereoisomers serve as the true substrates for enzymes which utilize metal–nucleotides as substrates. The procedure is relatively straightforward once the relevant Cr(III)– or Co(III)–nucleotide stereoisomers have been synthesized. One determines which of the isomers serves as a substrate for the enzyme in question. It usually happens that one is accepted at a slow rate. Of course it often happens that the enzyme cleaves a P–O bond—this is the catalytic role for enzymes acting on nucleotides—but does not cleave the Cr–O or Co–O bonds, which do not require catalysis to cleave when Mg(II) is the metal. In such cases unusual products arise, as in hexokinase, where the product derived from $Co(NH_3)_4ATP$ and glucose is Co(III)-bridged glucose-6-*P* and ADP, glucose-6-PO_3–$Co(III)(NH_3)_4ADP$ [61].

An alternative procedure is also available for determining which of two diastereotopic oxygens at P_α or P_β of ATP is coordinated to the metal. Jaffe and Cohn introduced the use of ATPβS for this purpose [43,80]. They found that the specificities of enzymes for the (R_p) and (S_p) diastereomers are sometimes dependent upon which metal ion is used as the activator. Using hexokinase as a case in point, Jaffe and Cohn showed that (R_p)-ATPβS is the preferred substrate when Mg(II) is the activator, whereas (S_p)-ATPβS is preferred when Cd(II) is the activator. Since Cd(II) preferentially coordinates sulfur and Mg(II) preferentially coordinates oxygen, Jaffe and Cohn reasoned that the explanation for metal-ion dependence was that analogous screw-sense isomers resulted from Cd(II) coordinating sulfur in the (S_p) isomer and Mg(II) coordinating oxygen in the (R_p) oxygen. The situation is illustrated in Fig. 31. Both structures are Λ isomers but one is derived from (R_p)- and the other from (S_p)-ATPβS. It was on this basis that Jaffe and Cohn were able to assign absolute configurations to the diastereomers of ATPβS based on the known config-

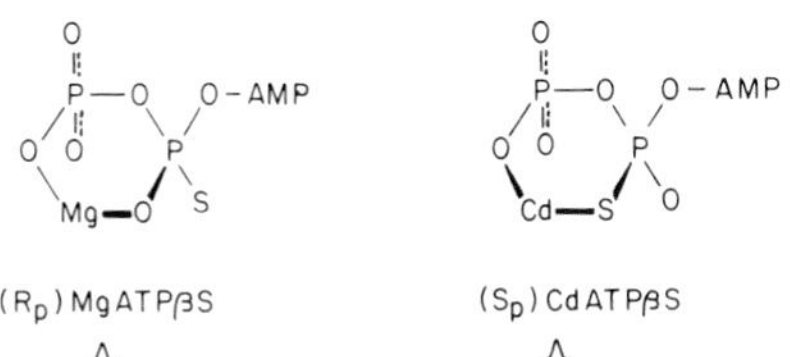

Fig. 31. Configurational correlation of the Mg(III) and Cd(II) complexes of ATPβS which exhibit substrate activity with hexokinase.

TABLE 1
Stereospecificities of enzymes for metal–nucleotides

Enzyme	P_α specificity [a]	P_β specificity [a]	Metal–nucleotide specificity		
			ATP	ADP	Ref.
Hexokinase	S_p(Mg–Cd)	R_p(Mg)–S_p(Cd)	Λ	β-mono	43, 61, 63, 79
Pyruvate kinase	S_p(Mg)–R_p(Cd)	S_p(Mg)–R_p(Cd)	Δ	Λ	43, 63, 96
Creatine kinase	R_p(Mg)–S_p(Cd)	R_p(Mg)–S_p(Cd)	Λ	Δ	63, 39
Adenylate kinase	S_p(Mg)	S_p, R_p(Mg)	Δ	—	20, 63
Arginine kinase	R_p(Mg)	R_p(Mg)	Λ	—	20, 63
Phosphofructokinase	—	—	Δ	—	63
Glycerokinase	—	—	Λ	—	63
Carbamate kinase	S_p, R_p(Mg)	R_p(Mg–Cd)	—	—	97
Protein kinase	—	S_p(Mg)–R_p(Cd)	Δ	—	98
Myosin	S_p, R_p(Mg–Co)	S_p(Mg–Co)	—	β-mono	99
Glutamine synthetase	R_p(Mg)–S_p, R_p(Co)	R_p(Mg)–S_p, R_p(Co)	Λ	Δ	97
Carbamoyl-P synthetase	R_p(Mg)–S_p(Cd)	S_p(Mg)–R_p(Cd)	—	—	97, 100
Phosphoribosyl pyrophosphate synthetase	S_p(Mg)–R_p(Cd)	—	Δ	—	78, 101
Phe-tRNA synthetase	S_p(Mg–Co)	R_p, S_p(Mg)–S_p, R_p(Co)	—	—	102
E. coli DNA polymerase I	S_p(Mg–Co)	S_p(Mg)–R_p, S_p(Co)	Δ	—	77
E. coli RNA polymerase	S_p(Mg–Co)	S_p, R_p(Mg)–R_p(Co)	Δ	—	103

[a] Stereospecificity for thionucleotides with reference to metals indicated parenthetically.

uration of isomers (see Section 2.a.ii for the alternative procedure). Conversely, the changeover from R_p to S_p specificity, or from S_p to R_p, with a change from Mg(II) to Cd(II) activation can be used to determine whether the enzyme binds the Λ or Δ isomers of metal–nucleotides. Table 1 compiles data on the Λ, Δ specificities of enzymes.

4. Conclusions

Compiled in Table 2 are the stereochemical courses of substitution at phosphorus for over thirty enzymes. These data constitute the experimental basis for two related mechanistic conclusions growing out of these investigations. One is that a single enzymatic substitution at phosphorus appears always to occur with stereochemical inversion. This conclusion follows from the fact that in every case in which no evidence can be obtained for the occurrence of a double-displacement catalytic pathway, the reaction has been found to proceed with *inversion* at the locus of substitution. Conversely, in every case for which conclusive nonstereochemical evidence for a double-displacement pathway is available, including isolation and characterization of a phosphoryl- or nucleotidyl-E, the reaction has been found to proceed with *retention* of configuration. No evidence has been obtained for pseudorotatory steps in enzymatic reactions, and all substitutions appear to be in-line. This does not mean that adjacent substitution accompanied by pseudorotation is not allowed, only that if such cases are discovered in future experiments they will prove to be exceptions reflecting some special aspects of mechanism in those cases. Thus, the stereochemical test is a strong indicator of whether phosphoryl group transfer proceeds by a single-displacement or a double-displacement pathway. The test is not decisive, however, because it can only indicate whether an odd or even number of displacements has occurred. Net inversion is consistent with an odd number and retention with an even number of displacements.

Two specific examples from Table 2 are worthy of further mention. Nucleoside phosphotransferase, which catalyzes Equation 20, had been proposed not to involve a double-displacement mechanism or an intermediate phosphoryl-enzyme.

$$N_1MP + N_2 \rightleftharpoons N_1 + N_2MP \tag{20}$$

This proposal was based on kinetic data and failure to detect an intermediate. The stereochemical test showed the reaction to proceed with retention of configuration and led to the proposal that such an intermediate is involved [27]. The phosphoryl-enzyme was subsequently isolated in good yield. This was the first study in which stereochemistry cast doubt on an earlier mechanistic proposal and provided the first evidence of a double-displacement pathway, later verified by the isolation of the intermediate.

E. coli acetate kinase has been the subject of controversy regarding whether a double- or single-displacement pathway is followed. On the one hand a phosphoryl-

TABLE 2
Stereochemistry of enzymatic substitution at phosphorus

Enzyme	Stereochemistry	Ref.
Phosphohydrolases		
Ribonuclease A, T_1, T_2	Inv–Inv (2 steps)	31, 104, 107, 108
Alkaline phosphatase	Ret.	52
E. aerogenes nonspecific phosphohydrolase	Inv.	110
Snake venom phosphodiesterase	Ret.	41, 45
3′,5′-cAMP phosphodiesterase	Inv.	68
5′-Nucleotidase	Inv.	28
Myosin ATPase	Inv.	105
Mitochondrial ATPase	Inv.	106
Phosphotransferases		
Adenylate kinase	Inv.	25
Glycerokinase	Inv.	29, 53, 30
Pyruvate kinase	Inv.	29, 53, 30
Hexokinase	Inv.	29, 53, 30
Nucleoside diphosphate kinase	Ret.	21
Nucleoside phosphotransferase	Ret.	27
Adenosine kinase	Inv.	73
Acetate kinase	Inv.	54
Phosphoglycerate mutase	Ret.	55
Polynucleotide kinase	Inv.	22
Phosphoglycerate kinase	Inv.	48
Nucleotidyltransferases		
UDP-Glc pyrophosphorylase	Inv.	74
Gal-1-*P* uridylyltransferase	Ret.	21
RNA polymerase	Inv.	40, 75
DNA polymerase	Inv.	16, 77
tRNA nucleotidyltransferase	Inv.	40, 76
Adenylyl cyclase	Inv.	49, 69
Polynucleotide phosphorylase	Inv.	109
T_4 RNA ligase	Inv.	111
ATP-dependent synthetases		
PRPP synthetase	Inv.	78
Acetyl-CoA synthetase	Inv.	44
Aminoacyl tRNA synthetases	Inv.	80

enzyme has been isolated and partially characterized as an acyl-phosphate, and the kinetics in one study appeared to be consistent with a double displacement [81–85]. On the other hand a more complete kinetic study cast doubt upon this conclusion by showing that the kinetics could not be accounted for by Equation 2 [86]. The kinetic pathway had to involve ternary complexes as in Equation 1, and so, if double displacement is compulsory, it would have to occur within the steps interconverting ternary complexes. Stereochemical analysis revealed that the transfer is accompanied by inversion of configuration, consistent with a single displacement on the main catalytic pathway [54]. The phosphoryl-enzyme appears not to be on this main pathway. Spector has countered these arguments with the proposal that acetate kinase catalyzes phosphoryl transfer by a triple-displacement pathway [87].

Inasmuch as all but a few bisubstrate phospho- and nucleotidyl-group transfer reactions proceed with net inversion of configuration, it appears that most of them follow single-displacement pathways. If these are single and not triple or pentuple displacements, then it must be concluded that any chemical advantages in catalysis that may be offered by the double-displacement pathway have not been decisive factors in the evolution of the catalytic mechanisms by which these enzymes act. The appearance of a few double-displacement pathways must be accounted for on the basis of factors other than any intrinsic chemical advantage of nucleophilic catalysis. Since all the bisubstrate enzymes that catalyze double displacements follow ping-pong kinetic pathways, in which the group donor and group acceptor are thought to share the same binding subsite in the active center, it is most likely that this sharing of sites, as between A and B in Equation 2, is the most important factor leading to the evolution of double displacements in simple bisubstrate reactions. Since both acceptors A and B could not occupy their common subsite simultaneously, there must be a provision for preserving the covalent bond energy of the group being transferred during the changeover between acceptors. The preservation of this energy is accomplished by the evolutionary appearance of an enzymic nucleophile so positioned as to enable it to form a covalent bond to the group being transferred, thus stabilizing it in the interval between the departure of A and the arrival of B.

Two pairs of chemically matched reactions exemplify the importance of shared binding sites in the evolution of double-displacement pathways. They are the reactions catalyzed by UDP-glucose pyrophosphorylase (inversion) [74] and galactose-1-*P* uridylyltransferase (retention) [21], which catalyze Reactions 21 and 22, respectively,

$$\text{UDP-Glc} + \text{PP}_i \rightleftharpoons \text{UTP} + \text{Glc-1-}P \qquad (21)$$

$$\text{UDP-Glc} + \text{Gal-1-}P \rightleftharpoons \text{UDP-Gal} + \text{Glc-1-}P \qquad (22)$$

and adenylate kinase (inversion) [25] and nucleoside diphosphate kinase (retention) [21], which catalyze Reactions 23 and 24:

$$\text{ATP} + \text{AMP} \rightleftharpoons \text{ADP} + \text{ADP} \qquad (23)$$

$$\text{ATP} + \text{GDP} \rightleftharpoons \text{ADP} + \text{GTP} \qquad (24)$$

Although each pair of reactions is chemically matched their reaction mechanisms are quite different. The first member of each pair, 21 and 23, follow sequential kinetic pathways involving ternary complexes and inversion of configuration [88,89]. They seem to proceed by single displacement mechanisms. Their chemically matched partners, Equations 22 and 24, follow ping-pong kinetic pathways involving covalent intermediates and overall retention of configuration [88–95]. In the latter two cases the group acceptors are sterically and electrostatically similar and share a single binding subsite.

The foregoing analysis should not be extended to more complex cases such as tersubstrate or polymerization reactions. Covalent intermediates may play other important roles in those reactions. To date only a few tersubstrate and polymerizing reactions involving phosphates have been subjected to stereochemical analysis. Such investigations should in the near future provide improved insights into the mechanisms by which those enzymes catalyze complex reactions.

Acknowledgements

The research described herein as emanating from this laboratory was performed in collaboration with K.F.R. Sheu, J.P. Richard, R.S. Brody, H.T. Ho and R.D. Sammons. The author is grateful to the National Institute of General Medical Sciences and the National Foundation of the United States, as well as to the American Cancer Society for their financial support.

NMR spectra appearing in the figures were obtained with the assistance of R.D. Sammons on a Bruker WP-200 spectrometer supported by the National Institute of General Medical Sciences Grant GM-27431.

References

1 Wimmer, M.J. and Rose, I.A. (1978) Annu. Rev. Biochem. 47, 1031–1078.
2 Midelfort, C.F. and Rose, I.A. (1976) J. Biol. Chem. 251, 5881–5887.
3 Kokesh, F.C. and Kakuda, Y. (1977) Biochemistry 16, 2467–2473.
4 Nageswara Rao, B.D., Kayne, F.J. and Cohn, M. (1979) J. Biol. Chem. 254, 2689–2696.
5 Knowles, J.R. (1980) Annu. Rev. Biochem. 49, 877–919.
6 Frey, P.A. (1981) Methods Enzymol., in press.
7 Cleland, W.W. (1981) Methods Enzymol., in press.
8 Buchwald, S.L., Hansen, D.E., Hassett, A. and Knowles, J.R. (1981) Methods Enzymol., in press.
9 Webb, M.R. (1981) Methods Enzymol., in press.
10 Tsai, M.D. (1981) Methods Enzymol., in press.
11 Eckstein, F., Romaniuk, P.J. and Connolly, B.A. (1981) Methods Enzymol., in press.
12 Westheimer, F.H. (1968) Accts. Chem. Res. 1, 70–78.
13 Benkovic, S.J. and Schray, K.J. (1971) The Enzymes 8, 201–238.
14 Westheimer, F.H. (1980) in P. DeMayo (Ed.), Rearrangements in Ground and Excited States, Vol. 2, Academic Press, New York, pp. 229–271.
15 Murray, A.W. and Atkinson, M.R. (1968) Biochemistry 7, 4023–4029.
16 Brody, R.S. and Frey, P.A. (1981) Biochemistry 20, 1245–1252.
17 Eckstein, F. and Gindl, H. (1968) Chem. Ber. 101, 1670–1673.
18 Saenger, W. and Eckstein, F. (1970) J. Am. Chem. Soc. 92, 4712–4718.
19 Michelson, A.M. (1963) Biochim. Biophys. Acta 91, 1–13.
20 Eckstein, F. and Goody, R.S. (1976) Biochemistry 15, 1685–1691.

21 Sheu, K.F.R., Richard, J.P. and Frey, P.A. (1979) Biochemistry 18, 5548–5556.
22 Bryant, R., Sammons, R.D., Frey, P.A. and Benkovic, S.J. (1981) J. Biol. Chem. 256, in press.
23 Sheu, K.F.R. and Frey, P.A. (1977) J. Biol. Chem. 252, 4445–4448.
24 Goody, R.S. and Eckstein, F. (1971) J. Am. Chem. Soc. 93, 6252–6257.
25 Richard, J.P. and Frey, P.A. (1978) J. Am. Chem. Soc. 100, 7757–7758.
26 Richard, J.P., Ho, H.T. and Frey, P.A. (1978) J. Am. Chem. Soc. 100, 7756–7757.
27 Richard, J.P., Prasher, D.C., Ives, D.H. and Frey, P.A. (1979) J. Biol. Chem. 254, 4339–4341.
28 Tsai, M.D. and Chang, T.T. (1980) J. Am. Chem. Soc. 102, 5416–5419.
29 Orr, G.A., Simon, J., Jones, S.R., Chin, G. and Knowles, J.R. (1978) Proc. Natl. Acad. Sci. U.S.A. 75, 2230–2233.
30 Pliura, D.H., Schomburg, D., Richard, J.P., Frey, P.A. and Knowles, J.R. (1980) Biochemistry 19, 325–329.
31 Usher, D.A., Richardson, D.I. and Eckstein, F. (1970) Nature (London) 228, 663–665.
32 Baraniak, J., Kinas, R.W., Lesiak, K. and Stec, W.J. (1979) J. Chem. Soc. Chem. Commun. 940–941.
33 Stec, W.J., Okruszek, A., Lesiak, K., Usnanski, B. and Michalski, J. (1976) J. Org. Chem. 41, 227–233.
34 Lesiak, K. and Stec, W.J. (1978) Z. Naturforsch. B 33, 782–785.
35 Baraniak, J., Lesiak, K., Sochacki, M. and Stec, W.J. (1980) J. Am. Chem. Soc. 102, 4533–4534.
36 Gerlt, J.A. and Coderre, J.A. (1980) J. Am. Chem. Soc. 102, 4531–4533.
37 Eckstein, F., Simonson, L.P. and Bar, H.P. (1974) Biochemistry 13, 3806–3810.
38 Jaffe, E.K. and Cohn, M. (1978) Biochemistry 17, 652–657.
39 Burgers, P.M.J. and Eckstein, F. (1980) J. Biol. Chem. 255, 8229–8233.
40 Burgers, P.M.J. and Eckstein, F. (1978) Proc. Natl. Acad. Sci. U.S.A. 75, 4798–4800.
41 Bryant, F.R. and Benkovic, S.J. (1979) Biochemistry 18, 2825–2828.
42 Jarvest, R.L. and Lowe, G. (1979) J. Chem. Soc. Chem. Commun. 364–366.
43 Jaffe, E.K. and Cohn, M. (1978) J. Biol. Chem. 253, 4823–4825.
44 Midelfort, C.F. and Sarton-Miller, I. (1978) J. Biol. Chem. 254, 6889–6893.
45 Burgers, P.M.J., Eckstein, F. and Hunneman, D.H. (1979) J. Biol. Chem. 254, 7476–7478.
46 Cohn, M. and Hu, A. (1978) Proc. Natl. Acad. Sci. U.S.A. 75, 200–203.
47 Tsai, M.D. (1979) Biochemistry 18, 1468–1472.
48 Webb, M.R. and Trentham, D.R. (1980) J. Biol. Chem. 255, 1775–1779.
49 Coderre, J.A. and Gerlt, J.A. (1980) J. Am. Chem. Soc. 102, 6594–6597.
50 Coderre, J.A., Mehdi, S. and Gerlt, J.A. (1981) J. Am. Chem. Soc., in press.
51 Abbott, S.J., Jones, S.R., Weinman, S.A. and Knowles, J.R. (1978) J. Am. Chem. Soc. 100, 2558–2560.
52 Jones, S.R., Kindman, L.A. and Knowles, J.R. (1978) Nature (London) 275, 564–565.
53 Blättler, W.A. and Knowles, J.R. (1979) J. Am. Chem. Soc. 101, 510–511.
54 Blättler, W.A. and Knowles, J.R. (1979) Biochemistry 18, 3927–3932.
55 Blättler, W.A. and Knowles, J.R. (1980) Biochemistry 19, 738–743.
56 Cooper, D.B., Hall, C.R., Harrison, J.M. and Inch, T.D. (1977) J. Chem. Soc. Perkin Trans. 1, 1969–1980.
57 Cullis, P.M. and Lowe, G. (1978) J. Chem. Soc. Chem. Commun. 512–514.
58 Abbott, S.J., Jones, S.R., Weinman, S.A., Bockhoff, F.M., McLafferty, F.W. and Knowles, J.R. (1979) J. Am. Chem. Soc. 101, 4323–4332.
59 Buchwald, S.L. and Knowles, J.R. (1980) J. Am. Chem. Soc. 102, 6601–6602.
60 Cleland, W.W. and Mildvan, A.S. (1979) in G.L. Eickhorn and L.G. Marzilli (Eds.), Advances in Inorganic Biochemistry, Vol. 1, Elsevier, Amsterdam, pp. 163–191.
61 Cornelius, R.D. and Cleland, W.W. (1978) Biochemistry 17, 3279–3286.
62 DePamphilis, M.L. and Cleland, W.W. (1973) Biochemistry 12, 3714–3723.
63 Dunaway-Mariano, D. and Cleland, W.W. (1980) Biochemistry 19, 1496–1505.
64 Cornelius, R.D., Hart, P.A. and Cleland, W.W. (1977) Inorg. Chem. 16, 2799–2805.
65 Merritt, E.A., Sundaralingam, M., Cornelius, R.D. and Cleland, W.W. (1978) Biochemistry 17, 3274–3278.

66 Eckstein, F. (1970) J. Am. Chem. Soc. 92, 4718–4723.
67 Saenger, W., Suck, D. and Eckstein, F. (1974) Eur. J. Biochem. 46, 559–567.
68 Burgers, P.M.J., Eckstein, F., Hunneman, D.H., Baraniak, J., Kinas, R.W., Lesiak, K. and Stec, W.J. (1979) J. Biol. Chem. 254, 9959–9961.
69 Gerlt, J.A., Coderre, J.A. and Wolin, M.S. (1980) J. Biol. Chem. 255, 331–334.
70 Jarvest, R.L., Lowe, G. and Potter, B.V.L. (1980) J. Chem. Soc. Chem. Commun. 1142–1145.
71 Jarvest, R.L. and Lowe, G. (1980) J. Chem. Soc. Chem. Commun. 1145–1147.
72 Cullis, P.M., Jarvest, R.L., Lowe, G. and Potter, B.V.L. (1981) J. Chem. Soc. Chem. Commun. 245–246.
73 Richard, J.P., Carr, M.C., Ives, D.H. and Frey, P.A. (1980) Biochem. Biophys. Res. Commun. 94, 1052–1056.
74 Sheu, K.F.R. and Frey, P.A. (1978) J. Biol. Chem. 253, 3378–3380.
75 Eckstein, F., Armstrong, V.W. and Sternbach, H. (1976) Proc. Natl. Acad. Sci. U.S.A. 73, 2987–2990.
76 Eckstein, F., Sternbach, H. and Von der Haar, F. (1977) Biochemistry 16, 3429–3432.
77 Burgers, P.M.J. and Eckstein, F. (1979) J. Biol. Chem. 254, 6889–6893.
78 Li, T.M., Mildvan, A.S. and Switzer, R.L. (1978) J. Biol. Chem. 253, 3918–3923.
79 Langdon, S.P. and Lowe, G. (1979) Nature (London) 281, 320–321.
80 Jaffe, E.K. and Cohn, M. (1979) J. Biol. Chem. 254, 10839–10845.
81 Anthony, R.S. and Spector, L.B. (1970) J. Biol. Chem. 245, 6739–6741.
82 Webb, B.C., Todhunter, J.A. and Purich, D.L. (1976) Arch. Biochem. Biophys. 173, 282–292.
83 Anthony, R.S. and Spector, L.B. (1972) J. Biol. Chem. 247, 2120–2125.
84 Todhunter, J.A. and Purich, D.L. (1974) Biochem. Biophys. Res. Commun. 60, 273–280.
85 Purich, D.L. and Fromm, H.J. (1972) Arch. Biochem. Biophys. 149, 307–315.
86 Janson, C.A. and Cleland, W.W. (1974) J. Biol. Chem. 249, 2567–2571.
87 Spector, L.B. (1980) Proc. Natl. Acad. Sci. U.S.A. 77, 2626–2630.
88 Tsuboi, K.K., Fukunaga, K. and Petricciani, J.C. (1969) J. Biol. Chem. 244, 1008–1015.
89 Rhoads, D.G. and Lowenstein, J.M. (1968) J. Biol. Chem. 243, 3963–3972.
90 Mourad, N. and Parks, R.E. Jr. (1966) J. Biol. Chem. 241, 271–278.
91 Garces, E. and Cleland, W.W. (1969) Biochemistry 8, 633–640.
92 Wong, L.J. and Frey, P.A. (1974) J. Biol. Chem. 249, 2322–2324.
93 Wong, L.J. and Frey, P.A. (1974) Biochemistry 13, 3889–3894.
94 Wong, L.J., Sheu, K.F.R., Lee, S.L. and Frey, P.A. (1977) Biochemistry 16, 1010–1016.
95 Yang, S.L. and Frey, P.A. (1979) Biochemistry 18, 2980–2984.
96 Dunaway-Mariano, D., Benovic, J.L., Cleland, W.W., Gupta, R.K. and Mildvan, A.S. (1979) Biochemistry 18, 4347–4354.
97 Pillai, R.P., Rauschel, F.M. and Villafranca, J.J. (1980) Arch. Biochem. Biophys. 199, 7–15.
98 Granot, J., Mildvan, A.S., Brown, E.M., Korido, H., Bramson, H.N. and Kaiser, E.T. (1979) FEBS Lett. 103, 265–268.
99 Yee, D. and Eckstein, F. (1980) Hoppe-Seyler's Z. Physiol. Chem. 361, 353–354.
100 Rauschel, F.M., Anderson, P.M. and Villafranca, J.J. (1978) J. Biol. Chem. 253, 6627–6629.
101 Gibson, K.J. and Switzer, R.L. (1980) J. Biol. Chem. 255, 694–696.
102 Connolly, B.A., Von der Haar, F. and Eckstein, F. (1980) J. Biol. Chem. 255, 11301–11307.
103 Armstrong, V.W., Yee, D. and Eckstein, F. (1979) Biochemistry 18, 4120–4123.
104 Usher, D.A., Erenrich, E.S. and Eckstein, F. (1972) Proc. Natl. Acad. Sci. U.S.A. 69, 115–119.
105 Webb, M.R. and Trentham, D.R. (1980) J. Biol. Chem. 255, 8629–8632.
106 Webb, M.R., Grubmeyer, C., Penefsky, H.S. and Trentham, D.R. (1980) J. Biol. Chem. 255, 11637–11639.
107 Eckstein, F., Schulz, H.H., Rüterjans, H., Haar, W. and Maurer, W. (1972) Biochemistry 11, 3507–3512.
108 Burgers, P.M.J. and Eckstein, F. (1979) Biochemistry 18, 592–596.
109 Burgers, P.M.J. and Eckstein, F. (1979) Biochemistry 18, 450–455.
110 Gerlt, J.A. and Wan, W.H.Y. (1979) Biochemistry 18, 4630–4638.
111 Bryant, F.R. and Benkovic, S.J. (1981) J. Am. Chem. Soc. 103, 697–699.

Vitamin B_{12}: Stereochemical aspects of its biological functions and of its biosynthesis

J. RÉTEY

Department of Biochemistry, Institute of Organic Chemistry, University of Karlsruhe, D-7500 Karlsruhe, Federal Republic of Germany

1. The stereochemical course of the coenzyme B_{12}-catalysed rearrangement

The rearrangements catalysed by the coenzyme B_{12} (adenosyl cobalamin, AdoCbl) dependent enzymes are unique in several ways. The coenzyme involved is more complex than any of its congeners, and it is the only known natural product containing a covalent metal–carbon bond * (Fig. 1). But quite apart from its structural complexity, the type of reaction promoted by AdoCbl in conjunction with a specific enzyme was unprecedented in organic chemistry until only a few years ago when similar rearrangements were discovered, for the first time in non-enzymic processes [1–3]. Moreover, the determination of the stereospecificity of several AdoCbl-dependent enzymes has not only helped to place these fascinating transformations on a firm mechanistic foundation, but has also recently led to a new and deeper insight into the interdependence of stereochemical course and reaction mechanism in enzyme chemistry. One of the major goals of this chapter is to describe the crucial experiments that have made these advances possible.

Vitamin B_{12} (1) was first isolated in 1948. But it was not until 1958 that the biologically active form of this coenzyme was identified by H.A. Barker and his colleagues [4], who at that time were studying the metabolism of *Clostridium tetanomorphum* growing on glutamate.

In fact, their endeavours were rewarded by a double discovery; that of AdoCbl, and also the first AdoCbl-dependent enzymic rearrangement – the reversible interconversion of L-glutamate and (2*S*,3*S*)-3-methylaspartate (Fig. 2).

With appropriate isotopic labelling experiments it was quickly demonstrated that this rearrangement involves the transposition of a glycine moiety between two adjacent carbon atoms rather than that of a carboxyl group. Furthermore, since no exchange of protons occurs between the substrate and the medium during the

* Apart from methyl cobalamin.

Tamm (ed.) Stereochemistry

reaction, the migration of the organic group must be accompanied by the migration of a hydrogen atom in the opposite direction (Fig. 3). In fact, these two key features are now known to be characteristic of all the AdoCbl-dependent enzymic rearrange-

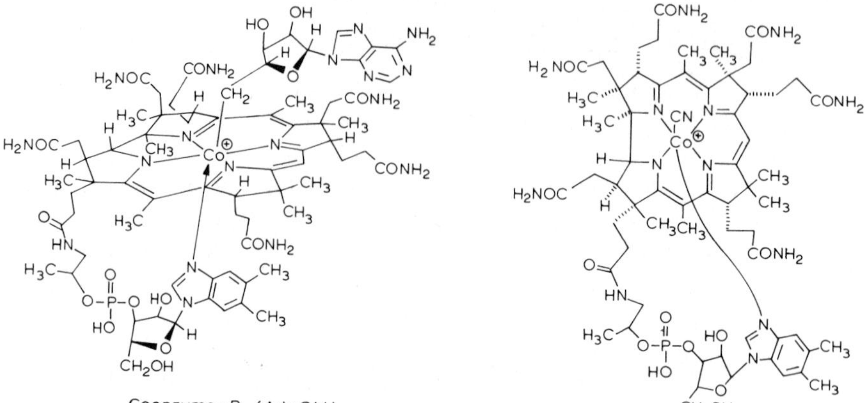

Coenzyme-B_{12}(AdoCbl)

(1)

Fig. 1. The structure of coenzyme B_{12}.

ments and the term vicinal interchange has been coined to describe such an exchange of a group R and a hydrogen atom between two adjacent carbon centres, each of which is in effect undergoing a substitution reaction.

Following the discovery of this new coenzyme and of one of the novel reactions in which it is involved, attention turned to an elucidation of its mechanism of action. Naturally, it was expected that, as had been the case with substitution reactions in free solution, stereochemical studies would provide valuable mechanistic information.

In the present case an experimental approach to the elucidation of the steric course was facilitated by the observation that a cell-free extract from *C. tetanomorphum* contained not only the AdoCbl-dependent glutamate mutase, but also a specific ammonia lyase that catalysed the reversible interconversion of mesaconate (2) and (2*S*,3*S*)-3-methylaspartate (3) (Fig. 4). In the presence of this

Fig. 2. The glutamate mutase reaction.

Fig. 3. The vicinal interchange reaction of coenzyme B_{12}-dependent rearrangements.

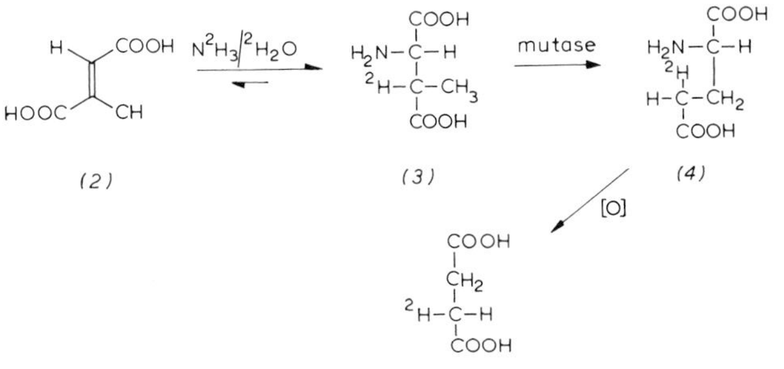

H COOH / HOOC CH (2) $\xrightarrow{N^2H_3/^2H_2O}$ COOH, H_2N-C-H, $^2H-C-CH_3$, COOH (3) $\xrightarrow{\text{mutase}}$ COOH, H_2N-C-H, ^2H, $H-C-CH_2$, COOH (4) $\xrightarrow{[O]}$ COOH, CH_2, $^2H-C-H$, COOH

(−)-(*R*)-succinate

Fig. 4. The sequence of reactions that elucidated the steric course of the glutamate mutase reaction.

lyase, a solvent proton is incorporated into (2*S*,3*S*)-3-methylaspartate, clearly evident when the reaction is performed in deuterium oxide [5]. Under such conditions a coupled enzymic reaction employing the ammonia lyase and glutamate mutase afforded the stereospecifically deuterated glutamic acid (4), which was subsequently degraded into [$^2H_1$]succinic acid by treatment with chloramine-T (Fig. 4). The negative ORD curve of this mono-deutero succinic acid confirmed the (*R*) configuration at the deuterated centre. The enzymically active 3-methylaspartate, on the other hand, was converted by reductive elimination of the amino group into (−)-2-methylsuccinic acid [6]. Since this enantiomer was known to have the (*S*) absolute configuration it follows that the substitution at C-3 in (2*S*,3*S*)-3-methylaspartate proceeds with inversion of configuration. No attempt was made to elucidate the steric course of the substitution at the other migration centre; at the methyl group in (2*S*,3*S*)-3-methylaspartate.

(a) Dioldehydratase

Dioldehydratase was discovered in certain strains of *Klebsiella pneumoniae* [7,8]. It catalyses the irreversible conversion of vicinal glycols into the corresponding 2-deoxyaldehydes (Fig. 5). The best substrates are (*R*)- and (*S*)-1,2-propanediol [9], although a number of other vicinal glycols are also accepted by the enzyme [10,11].

First of all, the reaction of the enantiomeric 1,2-propanediols will be discussed because, quite apart from their unique stereochemical interaction with the enzyme, two important discoveries were made during a detailed investigation of the way they are transformed.

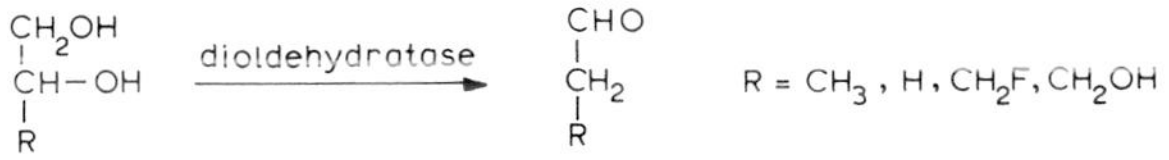

CH_2OH, $CH-OH$, R $\xrightarrow{\text{dioldehydratase}}$ CHO, CH_2, R R = CH_3, H, CH_2F, CH_2OH

Fig. 5. Reactions catalysed by dioldehydratase.

Using racemic 1,2-[1-^{3}H]propanediol as a substrate, Frey and Abeles discovered that during the reaction AdoCbl becomes tritiated in the cobalt-bound methylene group. They also showed [12] that the kinetics of this tritium transfer are compatible with a mechanism in which the AdoCbl is an obligatory acceptor site for the migrating hydrogen atom. Indeed this role of the coenzyme is now recognized to be another general feature of all the AdoCbl-dependent enzymic rearrangements. The stereochemical aspects of this phenomenon will be further elucidated later on in this chapter.

The second finding of general importance was made using 1,2-propanediols specifically labelled with ^{18}O [16]. Racemic 1,2-[2-^{18}O]propanediol (6) was prepared by equilibrating acetoxyacetone (5) in [^{18}O]water followed by reduction with $LiAlH_4$ (Fig. 6). And starting from (*R*)- and (*S*)-lactaldehydes (7 and 9), similar procedures led to (*R*)- and (*S*)-1,2-[1-^{18}O]propanediols (8 and 10). The dioldehydratase-promoted conversion of these labelled 1,2-propanediols (6, 8 and 10) into propionaldehydes was coupled with an in-situ enzymic reduction using NADH and yeast alcohol dehydrogenase, to yield the corresponding 1-propanol. This latter step was necessary to avoid an undesired and deleterious exchange of the aldehyde oxygen atom in the propionaldehydes with oxygen atoms from the medium. The extent of ^{18}O-retention in the products could then be determined by mass spectroscopy on a suitable derivative.

In the first experiment using racemic 1,2-[2-^{18}O]propanediol a 43% retention of ^{18}O was observed in the propanol. This important result demonstrates that a 1,2-migration of the secondary hydroxyl group to C-1 does indeed occur during the rearrangement, and an explanation for the loss of almost 50% of the label, when using a racemic substrate, was also forthcoming from two subsequent experiments.

Starting from (*R*)-1,2-[1-^{18}O]propanediol (8), the ^{18}O label was almost completely lost in the final product, whereas in the complementary experiment using (*S*)-1,2-[1-^{18}O]propanediol (10) most of the ^{18}O label was retained. These two results indicate that the changes which occur at C-1 take place stereospecifically and they can be

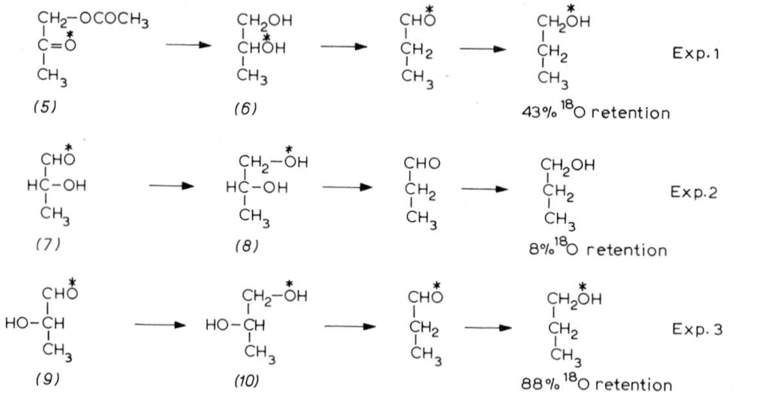

$$\underset{(5)}{CH_2{-}OCOCH_3{-}\overset{*}{C}{=}O{-}CH_3} \longrightarrow \underset{(6)}{CH_2OH{-}CH\overset{*}{O}H{-}CH_3} \longrightarrow CH\overset{*}{O}{-}CH_2{-}CH_3 \longrightarrow CH_2\overset{*}{O}H{-}CH_2{-}CH_3 \quad \text{Exp. 1}$$

43% ^{18}O retention

$$\underset{(7)}{CH\overset{*}{O}{-}HC{-}OH{-}CH_3} \longrightarrow \underset{(8)}{CH_2{-}\overset{*}{O}H{-}HC{-}OH{-}CH_3} \longrightarrow CHO{-}CH_2{-}CH_3 \longrightarrow CH_2OH{-}CH_2{-}CH_3 \quad \text{Exp. 2}$$

8% ^{18}O retention

$$\underset{(9)}{CH\overset{*}{O}{-}HO{-}CH{-}CH_3} \longrightarrow \underset{(10)}{CH_2{-}\overset{*}{O}H{-}HO{-}CH{-}CH_3} \longrightarrow CH\overset{*}{O}{-}CH_2{-}CH_3 \longrightarrow CH_2\overset{*}{O}H{-}CH_2{-}CH_3 \quad \text{Exp. 3}$$

88% ^{18}O retention

Fig. 6. Transformation on dioldehydratase of 1,2-propanediols specifically labelled with ^{18}O.

nicely integrated with the result obtained in the first experiment, in the following way.

Bearing in mind the rearrangement pattern discussed earlier for the glutamate mutase reaction (Fig. 2), it is tempting to postulate that stereospecific migration of the secondary OH group takes place in both enantiomers. The two ^{18}O-labelled geminal diols that are transiently produced (Fig. 7) will, however, be enantiomeric. In a subsequent dehydration step chiral recognition by the enzyme leads to a stereospecific loss of one of the enantiotopic hydroxyl groups from C-1. In other words, the AdoCbl-dependent stereospecific rearrangement to give the geminal diol 11 would be followed by a stereospecific dehydration in which the catalytic group responsible for the reaction recognizes only the topographic positions of the hydroxyl groups and not their origin.

One should also note that, although each step in the enzymic rearrangement can be inferred to occur stereospecifically, the absolute sense of the substitution at C-1, i.e., retention or inversion, and the dehydration of 11, i.e., loss of OH_{Re} or OH_{Si}, must remain unknown.

On the other hand, the absolute stereochemical course of the substitution at C-2 has been amenable to study [17,18]. In fact both (*R*)- and (*S*)-1,2-[1-$^{2}H_{2}$]-propanediols and also (*S*)-1,2-[2-$^{2}H_{1}$]propanediol were treated with dioldehydratase. The resulting labelled propionaldehydes were then converted into the corresponding chiral monodeuterated propionic acid (Fig. 8). The absolute configuration of these acids could be deduced by comparing their ORD curves with that of a reference compound, prepared by a method completely analogous to that used for the synthesis of the known (*S*)-[2-^{3}H]propionic acid [19] (Fig. 8). The experimental results clearly showed that the migrating hydrogen atom displaces the secondary hydroxyl group in both enantiomers with inversion of configuration, as depicted in Fig. 8.

The last stereochemically cryptic feature of this transformation concerns the specificity of the enzyme for the diastereotopic hydrogen atoms at C-1 of 1,2-propanediol. To resolve this point Zagalak et al. [18] prepared (1*R*,2*R*)- and (1*R*,2*S*)-1,2-[1-$^{2}H_{1}$]propanediols (12 and 13) by reducing (*R*) and (*S*)-lactaldehydes with (4*R*)-[4-$^{2}H_{1}$]NADH and liver alcohol dehydrogenase (Fig. 9). The cyclic acetals of 12 and 13, formed from *p*-nitrobenzaldehyde, gave different ^{1}H-NMR spectra, and their configurations were determined by spectral comparison [20] with racemic reference compounds of known (relative) configuration.

Enzymic dehydration of the labelled 1,2-propanediols 12 and 13 afforded [2-$^{2}H_{1}$]-

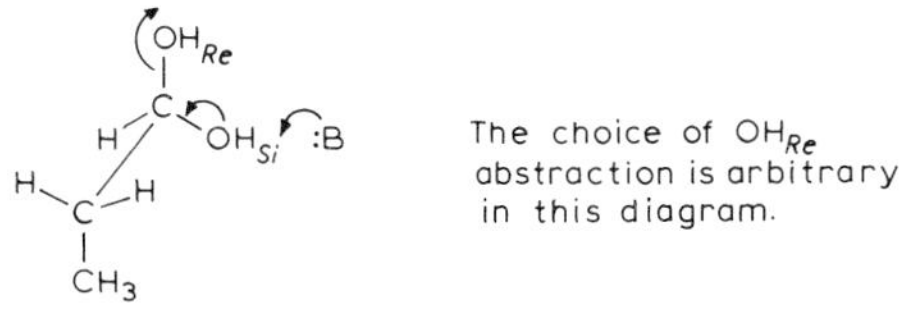

OH_{Re}, C, H, OH_{Si}, :B, H, C, H, CH_3

(11)

The choice of OH_{Re} abstraction is arbitrary in this diagram.

Fig. 7. Stereospecific dehydration of the geminal diol formed at the enzymes' active site.

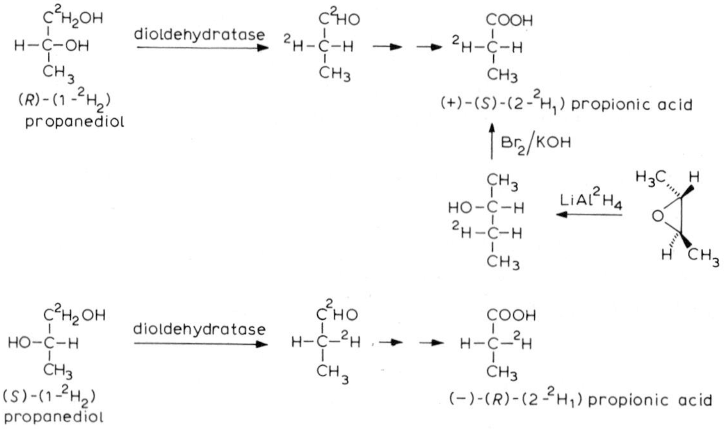

Fig. 8. Steric course at C-2 of the dioldehydratase promoted dehydration of (*R*)- and (*S*)-[1-$^2H_2$]propanediol.

and [1-$^2H_1$]propionaldehydes (14 and 15), respectively. Therefore, it follows that deuterium migrates in 12 and protium in 13. This is an interesting result because, in contrast to alcohol dehydrogenase, dioldehydratase exhibits a diastereospecificity towards the hydrogen atoms at C-1 that is dependent on the absolute configuration of the adjacent chiral centre; H_{ReR}- and the H_{SiS}-atoms are abstracted preferentially. In other words, the migrating groups (H and OH) are in an identical geometric relationship with respect to each other in both enantiomers of 1,2-propanediol.

A very interesting stereochemical picture now begins to emerge from these results. In order to account for the inversion at C-2, an *anti*-periplanar arrangement of the migrating groups was postulated (Fig. 10) [17,18]. This has the advantage of allowing the enzyme-bound AdoCbl to abstract and return the migrating hydrogen atom from

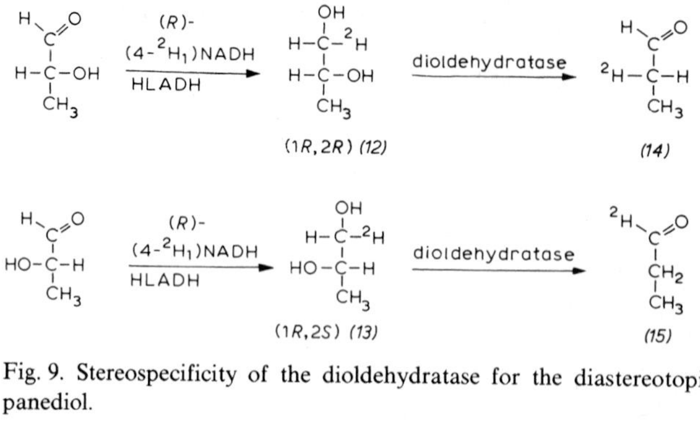

Fig. 9. Stereospecificity of the dioldehydratase for the diastereotopic hydrogen atoms at C-1 in propanediol.

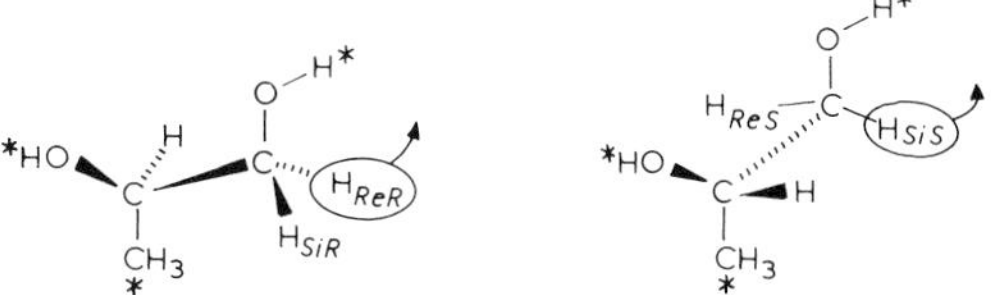

Fig. 10. Two binding modes for 1,2-propanediol at the active site of dioldehydratase. The asterisks symbolise geometrically fixed binding sites of the enzyme.

the same side of the molecule. Furthermore, the change in absolute stereochemical course observed with the two enantiomeric 1,2-propanediols can be rendered consistent with this idea by assuming that two binding modes are possible, as represented in Fig. 10, one for each enantiomer. The relative dispositions of three possible binding sites in the active site (marked with asterisks) for the primary and the secondary hydroxyl, and the methyl group, do not significantly change in these two binding modes. A crucial change in relative disposition between the catalytically active group responsible for hydrogen atom abstraction and the two hydrogen atoms at C-1 does occur. Further stereochemical evidence for the correctness of this picture was forthcoming from investigations into the transformation of other diols on dioldehydratase.

The achiral substrate glycerol could be arranged at the active site of dioldehydratase in the binding mode of either (*R*)- or (*S*)-1,2-propanediol (imagine a hydroxymethyl group in Fig. 10 instead of the methyl group).

To test whether both or only one of these binding modes is realized, specifically dideuterated glycerols were prepared by reducing (*R*)- and (*S*)-glyceric acid methyl esters with $LiAl^2H_4$ [21]. (*R*)-[1-$^2H_2$]glycerol was also prepared by an indirect route [10]. Dioldehydratase (also referred to as glycerol dehydratase) from *Klebsiella pneumoniae* strain ATCC 25955 [8] and from *Klebsiella pneumoniae* strain ATCC 8724 converted (*R*)-[1-$^2H_2$]glycerol mainly, if not exclusively, into 3-[1-H_1,2-$^2H_1$]hydroxypropionaldehyde, thereby revealing that glycerol reacts at the active site in the (*R*)-1,2-propanediol binding mode in spite of a large kinetic deuterium isotope effect ($k_H/k_{^2H} \approx 7$) associated with deuterium migration. On the other hand, (*S*)-[1-$^2H_2$]glycerol was dehydrated exclusively to 3-[3-$^2H_2$]hydroxypropionaldehyde [21,22] (Fig. 11).

This also clearly shows that the enzyme distinguishes between the enantiotopic hydroxymethyl groups in glycerol transforming only the one residing in the *Si* half-space (see Chapter 1, Fig. 18).

One research group [21,22] oxidized these 3-hydroxypropionaldehydes to the corresponding acids (Fig. 11) and examined their ORD spectra. Whereas the acid originating from (*S*)-[1-$^2H_2$]glycerol was optically inactive (as expected), the product from (*R*)-[1-$^2H_2$]glycerol showed a positive ORD curve. Starting from (2*S*,3*R*)-[3-$^2H_1$]aspartate (available via the aspartase reaction), a reference sample of (*S*)-3-[2-$^2H_1$]hydroxypropionic acid was prepared by a straightforward sequence of reactions (Fig. 12).

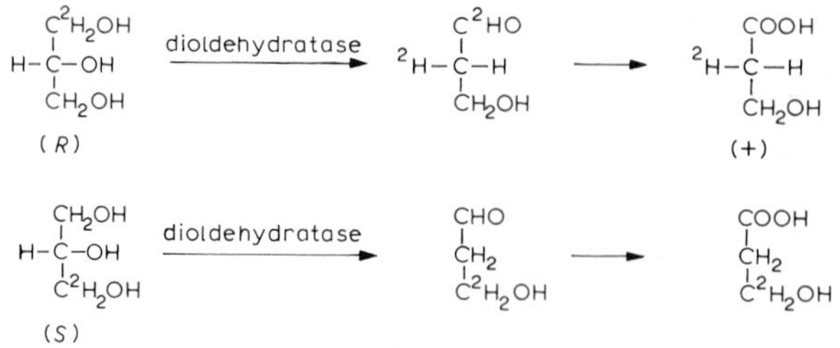

C^2H_2OH / H–C–OH / CH_2OH (*R*) —dioldehydratase→ C^2HO / ^2H–C–H / CH_2OH → COOH / ^2H–C–H / CH_2OH (+)

CH_2OH / H–C–OH / C^2H_2OH (*S*) —dioldehydratase→ CHO / CH_2 / C^2H_2OH → COOH / CH_2 / C^2H_2OH

Fig. 11. Transformation on dioldehydratase of (*R*)- and (*S*)-[1-$^2H_2$]glycerol.

The positive ORD curve of this reference compound established the (*S*) configuration of the enzymically derived sample and also demonstrated that stereochemical inversion at the C-2 of glycerol takes place during the dioldehydratase reaction, exactly as was observed during the rearrangement of 1,2-propanediol to propionaldehyde.

The specificity of the enzyme for the diastereotopic hydrogen atoms at the C-1 of glycerol was also determined using (1*R*,2*R*)-[1-$^2H_1$]glycerol as a substrate (see Chapter 3 for the method of preparation). Again, in spite of an isotope effect deuterium migration was observed exclusively to furnish, after chemical oxidation, (+)-(*S*)-3-[2-$^2H_1$]hydroxypropionic acid (Fig. 13).

As in the case of 1,2-propanediols, the migration of the secondary hydroxyl was also proven using glycerol samples specifically labelled with ^{18}O. Assuming that glycerol reacts on dioldehydratase in the (*R*)-1,2-propanediol binding mode, it was expected that the secondary hydroxyl group would be completely retained in the aldehyde product. This expectation was largely fulfilled in practice, although the ^{18}O retention was not so high (probably due to less efficient trapping of the aldehyde product with yeast alcohol dehydrogenase). But support for this result was obtained in a complementary experiment where complete loss of the heavy oxygen isotope was observed when (*R*)-[1-^{18}O]glycerol was used as substrate.

Bachovin et al. [10] have also investigated the stereospecificity of an inactivation process that is associated with the transformation of most of the substrates on dioldehydratase, particularly with glycerol. Whereas [$^2H_5$]glycerol exhibited a large kinetic isotope effect in the inactivation process ($k_H/k_{^2H} = 14$), (*R*)-[1-$^2H_2$]- and (*R*,*S*)-[1-$^2H_2$]glycerol exhibited only a small one ($k_H/k_{^2H} = 1.8$). This was taken as evidence for the involvement of the (*S*)-1,2-propanediol binding mode during the inactivation process *.

Ethylene glycol has a much higher symmetry than either 1,2-propanediol or glycerol, and a study of its stereochemical interaction with dioldehydratase required the most sophisticated labelling techniques. The two hydroxyl groups are homotopic and cannot be differentiated by an enzyme, whereas although the hydrogen atoms in

* The chemical nature of this inactivation is unknown.

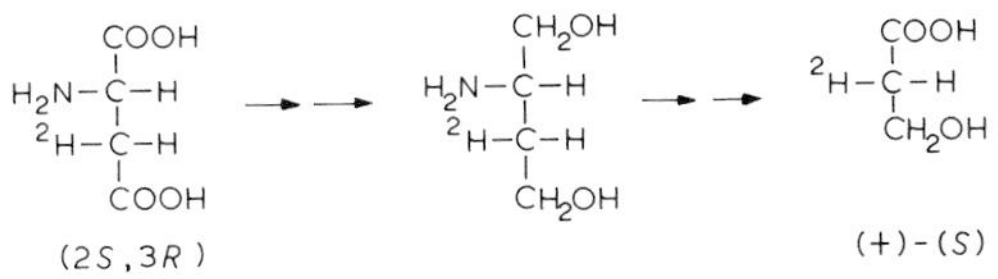

Fig. 12. Preparation of (*S*)-3-hydroxy[2-$^2H_1$]propionic acid from (2*S*,3*R*)-[3-$^2H_1$]aspartate.

one geminal pair are enantiotopic each is homotopic with respect to one in the other geminal pair.

With [$^2H_4$]ethylene glycol as a substrate one observes a moderate isotope effect ($k_H/k_{^2H} = 2$) on the V_{max} of the dioldehydratase reaction. This is much less than the value of ca. 10 observed with 1,2-propanediol. The application of partially deuterated substrates allows one to measure an isotope effect arising through intramolecular competition between otherwise homotopic groups. Such intramolecular isotope effects can be substantiated by appropriate product analysis for deuterium.

For this purpose, the specifically deuterated ethylene glycols 16–18 were synthesized (Fig. 14). It is easy to see that protium and deuterium migration will result in different products in each case and that the determination of the product ratio will give an apparent intramolecular deuterium isotope effect. The procedure is illustrated for [1-$^2H_2$]ethylene glycol (16) (Fig. 15). A straightforward analysis of the product ratio was achieved by running a mass spectrum of the dimedone derivatives, and comparing the intensities of the M^+ and M^+-15 peaks. The apparent isotope effect for 16 was $72/28 = 2.6$, whereas (*S*,*S*)- and (*R*,*R*)-[1,2-$^2H_2$]ethylene glycol (17 and 18) both had a $k_H/k_{^2H}$ value equal to 10 ± 1. The explanation for the difference in the apparent isotope effects must lie in the fact that substrate 16 can be bound at the enzyme's active site in two distinct ways, whereas only one way is possible with substrates 17 and 18 due to rotational symmetry of these molecules.

If substrate-binding were essentially irreversible the product ratio achieved with 16 as the substrate should not reflect any kinetic isotope effect, since regardless of which half of the molecule was fixed in the reactive position the substrate would be converted into product. If, on the other hand, substrate-binding were reversible and faster than hydrogen abstraction, then the product ratio should reflect the maximum isotope effect. The observed isotope effects indicate that neither of these extreme situations is realized, i.e., the substrate binding and H-transfer must have comparable activation energies as illustrated in Fig. 16.

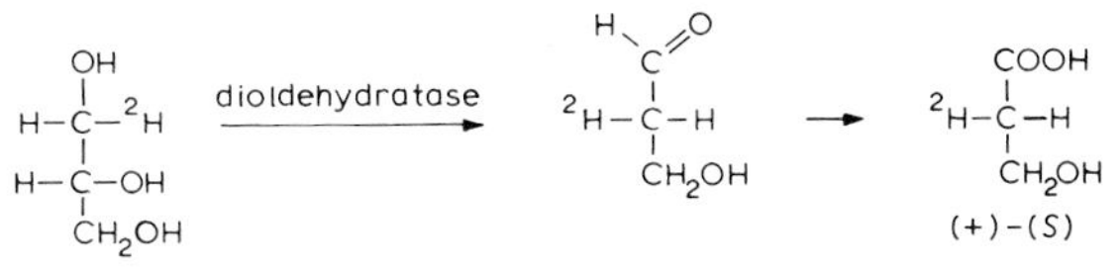

Fig. 13. Stereospecificity of dioldehydratase for the diastereotopic hydrogen atoms at C-1 of glycerol.

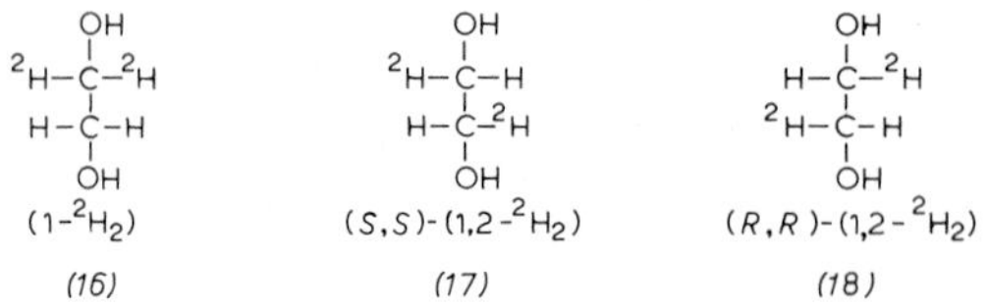

(1-$^2H_2$) (16)

(S,S)-(1,2-$^2H_2$) (17)

(R,R)-(1,2-$^2H_2$) (18)

Fig. 14. Stereospecifically deuterated ethylene glycols which were tested on dioldehydratase.

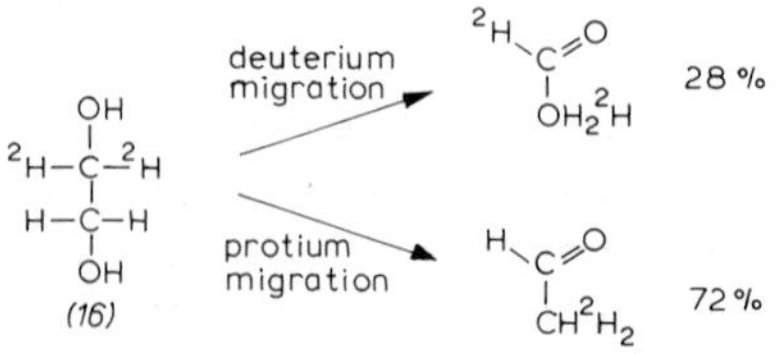

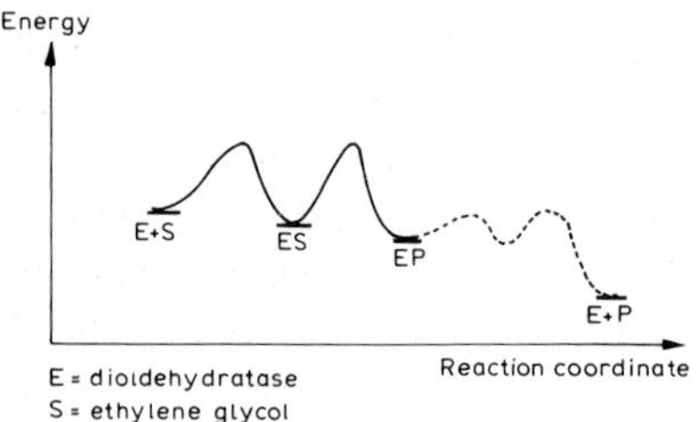

Fig. 15. Transformation on dioldehydratase of [1-$^2H_2$]ethylene glycol.

Energy

E+S

ES

EP

E+P

Reaction coordinate

E = dioldehydratase

S = ethylene glycol

Fig. 16. Reaction profile of the transformation of ethylene glycol on dioldehydratase.

A further inference from the identical intramolecular isotope effects ($k_H/k_{^2H} = 10 \pm 1$) observed with (*S,S*)- and (*R,R*)-[1,2-$^2H_2$]ethylene glycols (17 and 18) is that dioldehydratase does not exhibit stereospecificity for the enantiotopic methylene hydrogen atoms of the substrate. A similar conclusion could also be drawn from the product analysis of the reactions with (*S*)- and (*R*)-[1-$^2H_1$]ethylene glycols, albeit the evaluation of the results was much more complicated [22].

All these results can be rationalized by the following postulates: (1) ethylene glycol can easily oscillate between the energetically similar binding modes of (*R*)- and (*S*)-1,2-propanediol (Fig. 10) while residing at the active site; (2) on the other hand, no interchange of the two homotopic halves of this substrate is possible after it has been fixed on the enzyme; (3) in the case of unlabelled substrate 80% of the molecules in the ES complex undergo hydrogen abstraction before dissociation occurs, (4) since the deuterium isotope effect on V_{max} is 2, the hydrogen transfer from substrate to AdoCbl with its isotope effect of $k_H/k_{^2H} = 10 \pm 1$ cannot be rate limiting in the overall reaction (Fig. 16).

Dioldehydratase is not only stereospecific for the diastereotopic methylene hydrogens of the enantiomeric 1,2-propanediols, but also for the enantiotopic hydroxyl groups of the postulated geminal diol intermediate. The contrasting lack of stereospecificity for the enantiotopic methylene hydrogens of the ethylene glycol prompted investigations on the stereospecificity of hydroxyl migration in this substrate. Along lines similar to those discussed in the previous pages, the ^{18}O-labelled ethylene glycol substrates shown in Fig. 17 were prepared. The same technique was used for the trapping and the analysis of the products as described for the experiments with ^{18}O-labelled 1,2-propanediols (vide supra).

In the achiral species 19, 20 and 21 the (*R*)- and (*S*)-1,2-propanediol binding modes are equivalent and no preferred retention or loss of the heavy oxygen isotope should be expected, even upon stereospecific migration. This has been confirmed experimentally, i.e., about 50% of the ^{18}O was retained in all three cases. In the chiral specimens 22, 23, 24 and 25, however, stereospecific migration of the *unlabelled* hydroxyl group should lead to chirally labelled geminal diols. Such a process should be preferred by the substrates 23 and 25 (Fig. 18).

Assuming the same steric course for substitution and dehydration observed with [1-^{18}O]-labelled 1,2-propanediols, and using the known intramolecular discrimination factors for protium and deuterium, one can calculate that 21% and 79% of the ^{18}O should be retained in the products derived from 23 and 25, respectively. Similarly, ^{18}O-retentions of 35% and 65% can be calculated for 22 and 24. The experimental results were not in accord with these expectations since roughly 50% retention was observed for all multiply labelled samples. Such a lack of stereospecificity can be explained either by a non-stereospecific dehydration of the geminal diol intermediate or, more likely, by equilibration of the two rotameric intermediates A and B arising from the (*R*)- and (*S*)-1,2-propanediol-like binding modes of ethylene glycol, respectively (Fig. 19). In other words, not only the enzyme-bound substrate, but also the trigonal intermediate could easily flip between the two binding modes.

The last problem that was tackled concerned the steric course of the substitution of one of the hydroxyls by the migrating hydrogen atom. The required substrates were the stereospecifically doubly labelled specimens 26 and 27 (Fig. 20).

Although both hydroxyl groups can in principle migrate in these specimens, preferred migration of the hydroxyl attached to the isotopically labelled methylene group should be expected. In [1,1-$^{2}H_2$]ethylene glycol, protium migrated in 72% of the time. This was, of course, coupled with a 1,2-shift of the oxygen atom attached to

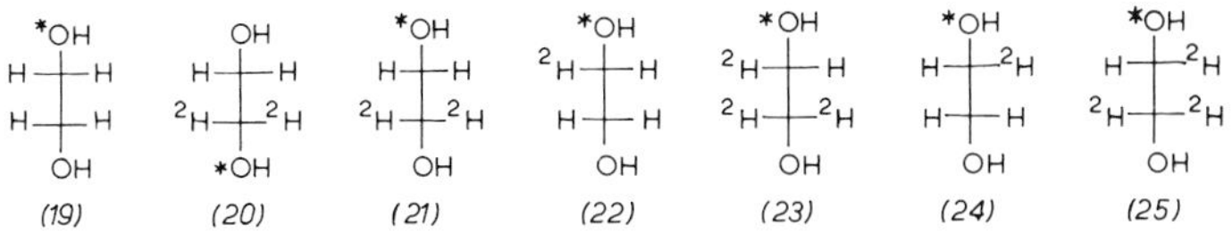

Fig. 17. Doubly labelled ethylene glycol substrates used to investigate the stereospecificity of hydroxyl migration.

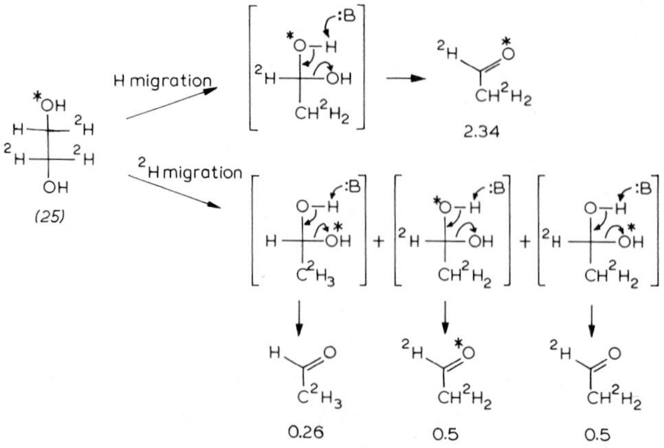

Fig. 18. The expected percentages of various labelled products of the dioldehydratase reaction using 25 as substrate. The calculation was based on the following facts and assumptions: (1) The enzyme does not differentiate between the enantiotopic hydrogen positions (conclusion from experiments with species 17 and 18 shown in Fig. 14); (2) in the competition between vicinal hydrogen atoms there is an intramolecular kinetic deuterium isotope effect of 2.6 (Fig. 15); (3) this effect is 10 for geminal hydrogen atoms; (4) the migrating hydroxyl group substitutes one of the hydrogen atoms in the vicinal position stereospecifically (i.e., with inversion).

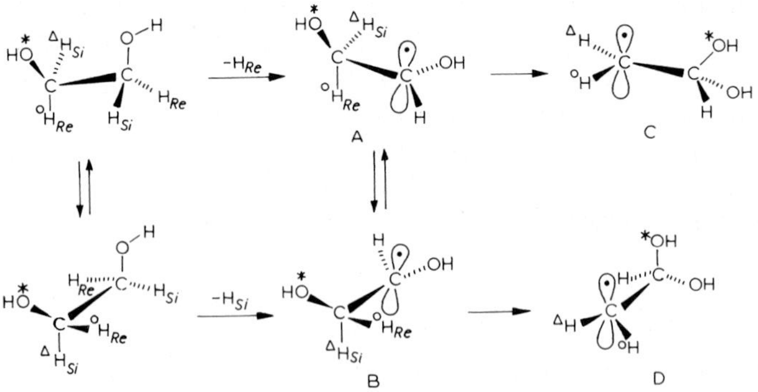

Fig. 19. Equilibration of substrate and intermediate between the two binding modes in the active site of diol dehydratase.

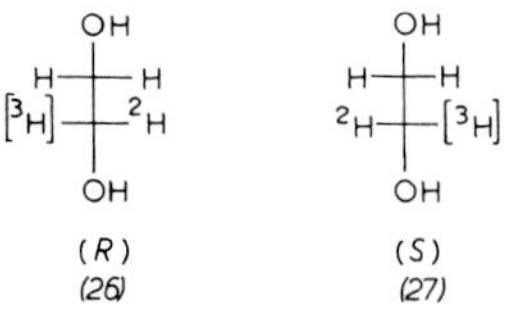

Fig. 20. Doubly labelled samples of ethylene glycol required to establish the steric course of the substitution of the hydroxyl group by the migrating hydrogen atom.

the C^2H_2 group (see Fig. 15). However, even deuterium migration should not be deleterious, since it would afford [1-^3H,2-^2H]acetaldehyde, which could be oxidized to tritium-free acetate indistinguishable from the carrier material [23,24]. Only tritium migration would lead to a singly-labelled [^3H]methyl group, but this should happen in not more than 10% of the cases.

Hence, it is to be expected that specimens 26 and 27 will afford enantiomerically different and almost pure chiral [^2H,^3H]acetaldehydes if the dioldehydratase reaction involves stereospecific substitution. Unfortunately, the experimental results did the malate synthase–fumarase system [23,24] determined that the dioldehydratase products were practically racemic [21,22]. Such a result can be rationalized by involving the intermediates C and D (Fig. 19) in which torsion around the C—C single bond should be faster than hydrogen radical transfer from the C-5′ of the modified coenzyme. The faces of the C^2H^3H group are theoretically stereoheterotopic but for all practical purposes homotopic, since no reagent can effectively differentiate between deuterium and tritium on the basis of their properties in the ground state. It is of interest that the corresponding radical intermediate starting from (*R*)- or (*S*)-1,2-propanediol must have considerably higher rotational barriers for this C—C bond, since inversion of configuration is observed. An interaction between the methyl group and the active site must be responsible for this higher torsional barrier.

(b) Methylmalonyl-CoA mutase

Methylmalonyl-CoA mutase is of special importance because it is the only known AdoCbl-dependent enzyme that is found in animals and man, as well as in bacteria. It catalyses the reversible interconversion of (*R*)-methylmalonyl-CoA and succinyl-CoA (Fig. 21).

Through appropriate labelling experiments it was shown [25] that the CoSCoA group in the substrate migrates to the methyl carbon atom while a hydrogen atom moves in the opposite direction; a process consistent with the general scheme of AdoCbl-dependent rearrangements (Fig. 3). The intramolecular nature of the CoSCoA group migration became evident from experiments in which a portion of the available substrate was doubly labelled with ^{13}C. The negative ion mass spectrum of the product [26,27] indicated no scrambling of the heavy isotopes between the doubly labelled and unlabelled portions. On the other hand, the accompanying hydrogen atom migration is intermolecular and is mediated by the C-5′ atom of the coenzyme [28,29]. Further details of this hydrogen transfer will be discussed later.

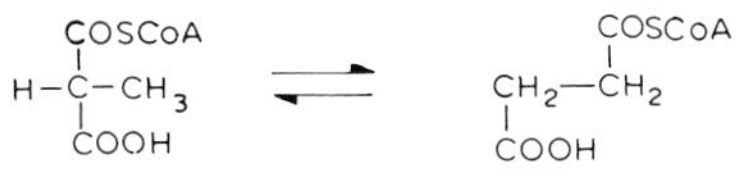

$$H\text{-}C(COSCoA)(COOH)\text{-}CH_3 \rightleftharpoons HOOC\text{-}CH_2\text{-}CH_2\text{-}COSCoA$$

Fig. 21. The methylmalonyl-CoA mutase reaction.

The steric course of the methylmalonyl-CoA mutase reaction, as it affects the C-2 of methylmalonyl-CoA, can be elucidated by determining the absolute configuration of the substrate and of a suitably labelled product. The former problem was solved by two groups [30,31]. Briefly, the (2*S*) configuration of the epimeric product obtained from enzymic carboxylation of propionyl-CoA was established and the (2*R*) configuration of the mutase-active methylmalonyl-CoA followed per exclusionem.

Sprinson and coworkers [30] conducted the methylmalonyl-CoA mutase reaction in deuterium oxide using a crude mitochondrial preparation. The presence of methylmalonyl-CoA epimerase insured that (1) all substrate molecules incorporated one atom of deuterium into position 2, and (2) in the course of the reaction the (2*R*)-epimer of methylmalonyl-CoA was continuously supplied by epimerization of the (2*S*)-epimer, which was in turn generated by the enzymic carboxylation of propionyl-CoA. Alkaline hydrolysis of the product and subsequent purification furnished succinic acid which was mainly monodeuterated (70% $^{2}H_{1}$-, 15% $^{2}H_{2}$-labelled and 13% unlabelled species). A positive ORD curve revealed its (*S*) configuration indicating stereochemical retention for the AdoCbl-dependent rearrangement (Fig. 22). No plausible explanation could be offered for the formation of doubly deuterated and unlabelled species. Essentially the same results were later obtained with a highly purified mutase preparation from *Propionibacterium shermanii* (J. Rétey, unpublished).

The second migration terminus in methylmalonyl-CoA is the methyl group. The steric course of the substitution at this centre could, in principle, be determined, but additional difficulties would arise from the reversibility of the reaction and from the intermolecular nature of the hydrogen migration. It was therefore a welcome discovery to find that ethylmalonyl-CoA is also a substrate for methylmalonyl-CoA mutase [32]. With this substrate analogue the steric course could be investigated at both migration terminii by stereospecific labelling with only one hydrogen isotope (e.g., deuterium). The synthesis of stereospecifically deuterated substrates is outlined in Fig. 23.

The chiral monodeuterated ethanols 28 and 33 were obtained by Simon's method [33] and their tosylates, 29 and 34, reacted with malonic ester anion to afford 30 and 35. The expected inversion of configuration in the malonic ester synthesis was confirmed by decarboxylating the derived acids 31 and 36 to (3*S*)- and (3*R*)-[3-$^{2}H_{1}$]butanoic acids, respectively, the chiroptical properties of which were already known [34]. The chirally deuterated CoA esters 32 and 37, prepared from 31 and 36, were rearranged on methylmalonyl-CoA mutase from *P. shermanii* and, after hydrolysis, the methylsuccinate products were isolated. In a parallel experiment the

COSCoA
^{2}H–C–CH$_3$
COOH
⟶
H COSCoA
^{2}H–C–CH$_2$
COOH

Fig. 22. The steric course of the methylmalonyl-CoA mutase reaction.

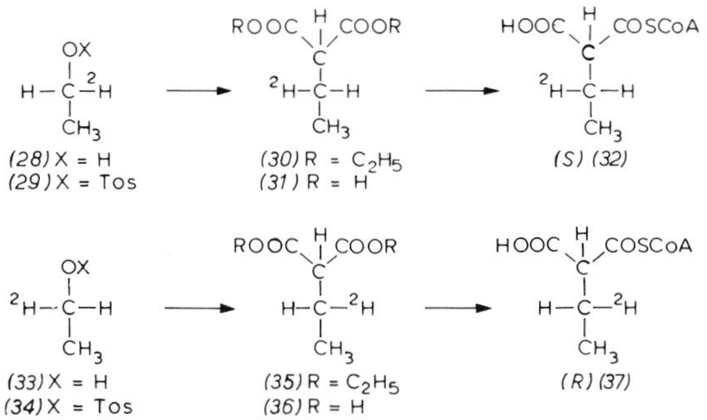

Fig. 23. Synthesis of stereospecifically deuterated ethylmalonyl-CoA.

easily accessible [2-$^2H_1$]ethylmalonyl-CoA was processed in the same manner. All product samples were submitted to mass and ^{1}H-NMR spectroscopy and to ORD/CD measurements. The unambiguous assignment of their ^{1}H-NMR spectra, however, required some additional experiments (see Fig. 24).

In the ^{1}H-NMR spectrum (270 MHz) the methine and the two methylene protons of methylsuccinic acid gave rise to a well-resolved ABC system. In the two enantiomers 38 and 39 the H atoms H_{ReR}/H_{SiS} and H_{SiR}/H_{ReS} are pair-wise reflection equivalent, i.e., identical in the NMR spectrum, whereas the diastereotopic geminal protons can be distinguished by their different chemical shifts. For the assignment of the signals to the diastereotopic protons, reference compounds of known configuration were synthesized by treating mesaconic and citraconic acids (40 and 42) with deuterated diimide. The known *syn*-addition of deuterium [35,36] afforded the racemic but stereospecifically dideuterated methylsuccinic acids (41 and 43) (Fig. 25).

From the ^{1}H-NMR spectra it then followed that the proton at the C-3 in the enantiomeric pair 41 ($\delta = 1.89$) corresponded to the H_{ReR}/H_{SiS} pair in 38 and 39, and the proton at the C-3 in the enantiomeric pair 43 ($\delta = 2.3$) coincided with the H_{SiR}/H_{ReS} pair. Furthermore, the deuterium content of each diastereotopic position can be quantitatively estimated by integration of the corresponding ^{1}H-NMR signals, and the absolute configuration and optical purity of the products can be deduced from the ORD/CD curves.

Table 1 summarizes the results obtained from the transformation of each of the specifically deuterated ethylmalonyl-CoA samples on the mutase. Surprisingly, the

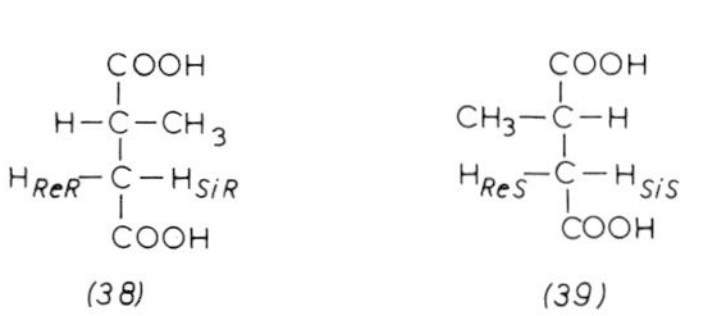

Fig. 24. Specification of the diastereotopic hydrogen atoms in (*R*)- and (*S*)methylsuccinic acid.

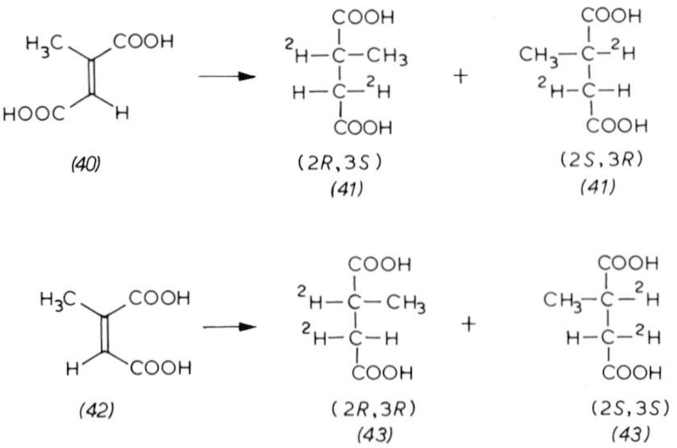

(40) → (2R,3S) (41) + (2S,3R) (41)

(42) → (2R,3R) (43) + (2S,3S) (43)

Fig. 25. Synthesis of stereospecifically dideuterated methylsuccinic acids.

isolated methylsuccinic acids were not optically pure, although an excess of the (2*R*)-enantiomer was observed in each case. (2*RS*,3*S*)-[3-$^2H_1$]Ethylmalonyl-CoA (32) was converted into methylsuccinyl-CoA with no detectable deuterium migration, and with 77% retention of configuration at C-3, as inferred from the chiroptical measurements. The almost exclusive migration of protium from the *pro*-3*R* position may be due to the inherent stereospecificity of the enzyme, or possibly to isotopic discrimination against deuterium. Indeed, the influence of such an isotope effect was demonstrated in the experiment in which (3*R*)-[3-$^2H_1$]ethylmalonyl-CoA (37) was the substrate. In 25% of all transformations the deuterium did not migrate, but

TABLE I

Products obtained from stereospecifically deuterated ethylmalonyl-CoA species

Substrate	Products	
H_{Re}, ^2H, COSCoA; H_{Si}, CH_3, COOH	COSCoA, ^2H, H; H, CH_3, COOH ~75%	COSCoA, H, ^2H; H_3C, H, COOH ~25%
H, H, COSCoA; ^2H, (S), CH_3, COOH	COSCoA, H_{Si}, H_{Re}; ^2H, CH_3, COOH ~77%	COSCoA, H_{Si}, H_{Re}; H_3C, ^2H, COOH ~23%
^2H, H, COSCoA; H, (R), CH_3, COOH	COSCoA, H, ^2H; H, CH_3, COOH ~39%	COSCoA, H_{Si}, H_{Re}; H_3C, ^2H, COOH ~25%

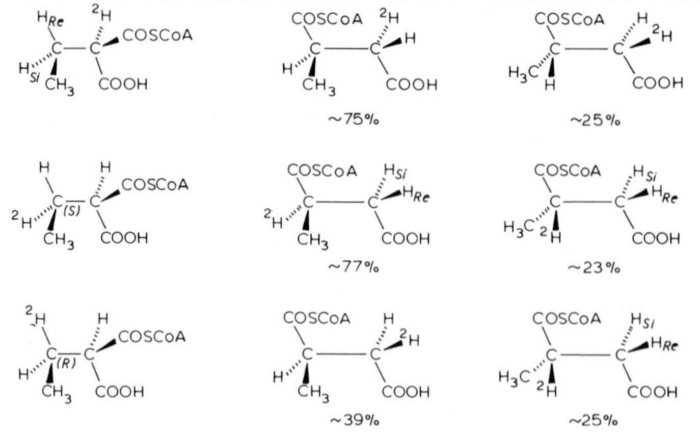

The values are corrected for 100% deuterium content of the substrate. From (3*R*)-[3-$^2H_1$]ethylmalonyl-CoA 36%, unlabelled product was also obtained. (Reproduced from [48]).

instead remained bonded to the original carbon atom. Somewhat more unexpected, however, was the fact that only 39% of the deuterium migrated to the H_{ReR}/H_{SiS} position (as seen by NMR) which means that about 36% was lost during migration. Also, the optical purity of this methylsuccinic acid sample was particularly low (R:S = 60:40). A further conclusion from the results of this experiment is that the substitution of the CoSCoA group by the migrating deuterium took place with retention if the reactive diastereomer of ethylmalonyl-CoA had the (2R) configuration. Of course, such an assumption is based only on analogy to the natural substrate, (2R)-methylmalonyl-CoA. The stereospecificity of the hydrogen migration was confirmed in a third experiment in which [2-$^2H_1$]ethylmalonyl-CoA was the substrate. This experiment was conducted in deuterium oxide. The resulting methylsuccinate was completely monodeuterated. (Recall that in the analogous experiment with methylmalonyl-CoA, unlabelled and doubly deuterated product was also formed by a mechanism not yet understood.) From the ^{1}H-NMR spectrum of the product, the deuterium distribution between the H_{SiR}/H_{ReS} and H_{ReR}/H_{SiS} positions (see 38 and 39) was 94% and 6%, respectively [37]. The absolute configuration of the 6% portion is unknown and for the sake of simplicity this species is not recorded in Table 1. The results of these experiments with specifically deuterated ethylmalonyl-CoA specimens are consistent with the existence of a trigonal intermediate (44), the conformation of which is not completely controlled by the enzyme.

The finding that about 36% of the migrating deuterium was lost during the rearrangement was confirmed by using [*ethyl*-$^2H_5$]ethylmalonyl-CoA as a substrate. Moreover, the fate of this lost deuterium was clarified by running the analogous experiment with a tritiated substrate and showing that 80% of the migrating tritium was lost to the water.

It is surprising that on reversing the labelling pattern, i.e., using water containing a heavy isotope (^{2}H or ^{3}H), practically no exchange with the migrating protium can be detected. A plausible explanation for this is the operation of different isotope effects for the washing-out and the washing-in processes.

H, H, H_3C, C, C, COSCoA, COOH

(44)

(c) β-Lysine mutase

β-Lysine mutase is the first of a group of AdoCbl-dependent enzymes that catalyses the 1,2-migration of an amino group (Fig. 26). It has been isolated from *Clostridium sticklandii* [39] and consists of a cobalamin-binding 'orange' protein and a smaller 'yellow' protein. Apart from AdoCbl, several other essential cofactors have been identified, such as pyridoxal phosphate, ATP, FAD, thiols, Mg^{2+} and K^+ [38]. The function of the 'yellow' protein and some of these cofactors is to renew continuously the cobalt–carbon bond of AdoCbl, which becomes irreversibly cleaved after about twenty catalytic cycles [39].

β-Lysine can be synthesized by an Arndt–Eistert elongation of L-ornithine, which

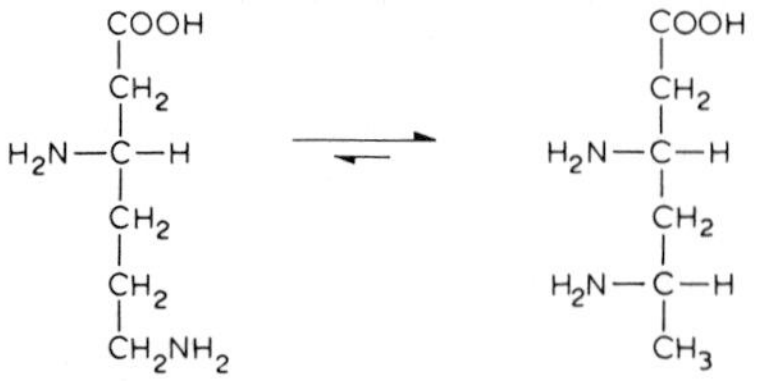

Fig. 26. The β-lysine mutase reaction.

in addition establishes its absolute configuration [40]. The configuration of 3,5-diaminohexanoic acid has been determined in several ways. Microbiological degradation [41] furnished L-3-aminobutanoic acid, indicating the presence of a (5*S*) configuration in the starting material. In an independent investigation [42] a synthetic mixture of diastereomeric 3,5-diaminohexanoic acids was separated as the dihydrochloride by fractional recrystallization. Subsequent cyclization afforded the corresponding *cis*- and *trans*-lactams (45 and 46 in Fig. 27), the latter being spectroscopically identical with the product prepared from enzymatically formed 3,5-diaminohexanoate. Originally, however, the configurational assignment of 45 and 46 was based only on the interpretation of their ^{1}H-NMR spectra, so unequivocal evidence was needed. 46 was therefore reacted with optically active phenylethylisocyanate to form the ureas 47 and 48, which could then be separated by column chromatography.

Pyrolysis followed by catalytic hydrogenation then furnished the optically active

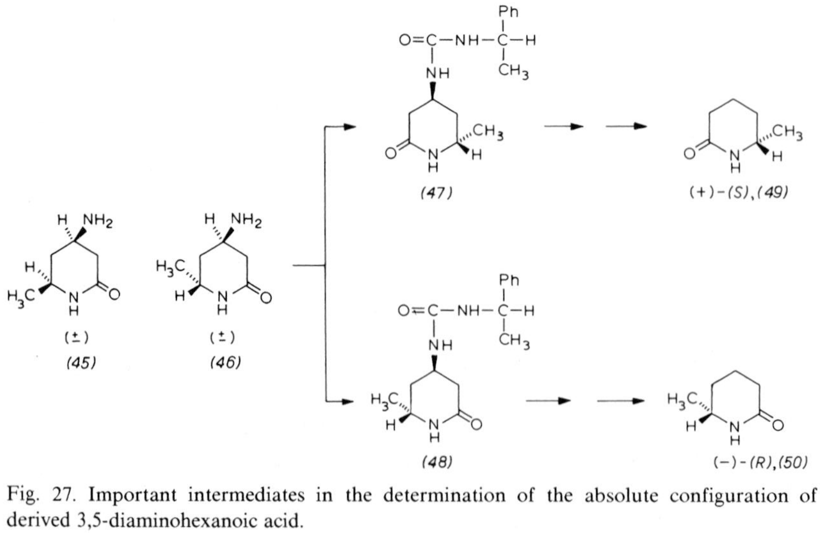

Fig. 27. Important intermediates in the determination of the absolute configuration of enzymically derived 3,5-diaminohexanoic acid.

6-methylpiperidones (49 and 50 in Fig. 27) whose absolute configurations were already known [43].

The final link in this cycle of circumstantial evidence was found by showing that a radioactive sample of enzymic origin, when processed to the urea derivative, co-chromatographed with 47 but not with 48.

After establishing the configurations of the enzymically formed 3,5-diaminohexanoic acid to be (3*S*,5*S*), the remaining stereochemical ambiguity surrounding this transformation concerned the identity of the hydrogen atom at the C-5 of β-lysine, the one which is replaced during the rearrangement. Since stereospecific labelling of this methylene group by deuterium or tritium is a difficult task, it was a pleasant surprise to find that β-lysine mutase itself could be employed for this purpose. With [5′-^{3}H]AdoCbl as coenzyme, the mutase catalysed the transfer of tritium both to 3,5-diaminohexanoate and to β-lysine [44] (Fig. 28). The tritium was located in the methyl group of 3,5-diaminohexanoate by Kuhn–Roth oxidation to acetic acid. On the other hand, the enzymically tritiated β-lysine, when isolated and again reacted on the mutase in the presence of unlabelled AdoCbl, afforded 3,5-diaminohexanoate exclusively labelled in the methyl group. This established that β-lysine mutase is stereospecific and that the same diastereotopic hydrogen atom at the C-5 of β-lysine is involved in the exchange and the transfer reaction. Then, the enzymically tritiated β-lysine was carefully oxidized with $KMnO_2$ to succinic acid without any loss of tritium.

A procedure for determining the absolute configuration of the tritiated succinic acid had already been developed, which involved partial oxidation on succinate dehydrogenase [45]. Applying this method it was then possible to show that the tritiated succinic acid of interest had the (*S*) configuration, and this in turn established the (5*S*) configuration of the tritiated β-lysine from which it had been derived. Inspection of Fig. 28 reveals that the β-lysine mutase reaction takes place with inversion at the C-5 of the substrate.

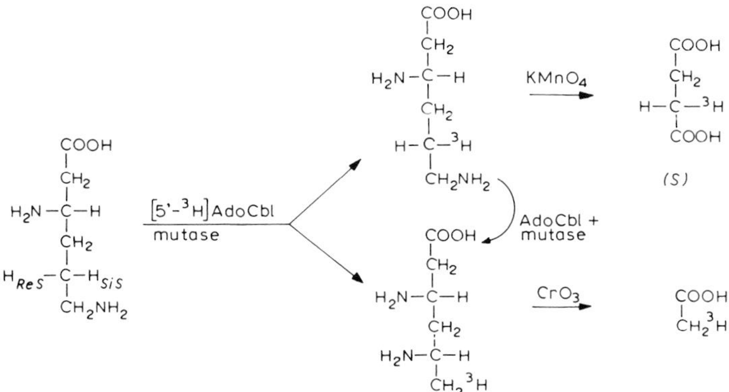

Fig. 28. Steric course of the β-lysine mutase reaction as it affects the C-5 of β-lysine.

(d) Ethanolamine ammonia lyase

Ethanolamine ammonia lyase is also found in a *Clostridium* species and catalyses the conversion of aminoethanol into acetaldehyde and ammonia (Fig. 29). Through isotopic labelling it had been demonstrated that the hydroxymethyl group of the substrate became the aldehyde group of the product [46,47]. Although there is no direct evidence for the 1,2-migration of the amino group, the likelihood of such a process is supported not only by analogy with the β-lysine mutase and dioldehydratase reactions, but also by the fact that ammonium ions are necessary before tritium transfer from [5′-^{3}H]AdoCbl to the product can be observed in the reverse reaction. The geminal aminoalcohol 51 has therefore been invoked as a reactive intermediate. [^{3}H]Acetaldehyde is also formed when aminoethanol is deaminated on the lyase in the presence of [5′-^{3}H]AdoCbl.

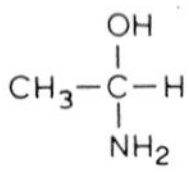

$$CH_3-C(OH)(NH_2)-H \quad (51)$$

Of stereochemical interest here is the substitution that occurs at the C-2, and a study of this required the use of substrates stereospecifically labelled with both deuterium and tritium [49]. To achieve this the aid of another enzyme was called upon, namely alanine aminotransferase (Fig. 30). [$^2H_2$]Glycine (52) was incubated with the transaminase in tritiated water, whereupon a stereospecific exchange reaction occurred to afford (R)-[$^2H_1$,^{3}H]glycine (53) (see chapter by Floss and Vederas). Esterification of 53 followed by the reduction with $LiAlH_4$ gave the desired (R)-2-[2-$^2H_1$,2-^{3}H]aminoethanol (54). By reversing the order in which the isotopic labels were introduced, the same method was used to prepare the enantiomer 55 (Fig. 30). Enzymic deamination of these doubly labelled aminoethanols then furnished the corresponding labelled acetaldehydes, which were reduced in situ by NADH on YADH to prevent the loss of the labels through spontaneous enolization. The isolated ethanol samples were oxidized to doubly labelled acetates which then were submitted to the usual stereochemical analysis with the malate synthase–fumarase system [23,24]. Both acetates turned out to be completely racemic; an unusual result indeed for an enzyme-catalysed substitution reaction.

And yet a rational explanation can be found that is also consistent with and complementary to the stereochemical and mechanistic data on other AdoCbl-dependent rearrangements already discussed in this chapter. Once again a loss of stereospecificity in the enzymic reaction points to the transient existence of a trigonal intermediate the faces of which are quasi-homotopic because rotation about the C—C bond is faster than the subsequent hydrogen-atom transfer step (Fig. 31).

The correctness of this hypothesis could be tested since both enantiomers of

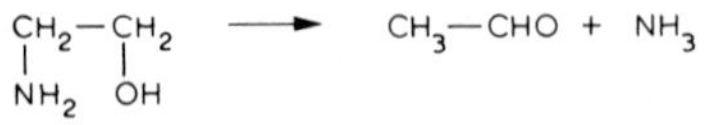

$$CH_2(NH_2)-CH_2(OH) \longrightarrow CH_3-CHO + NH_3$$

Fig. 29. The ethanolamine ammonia lyase reaction.

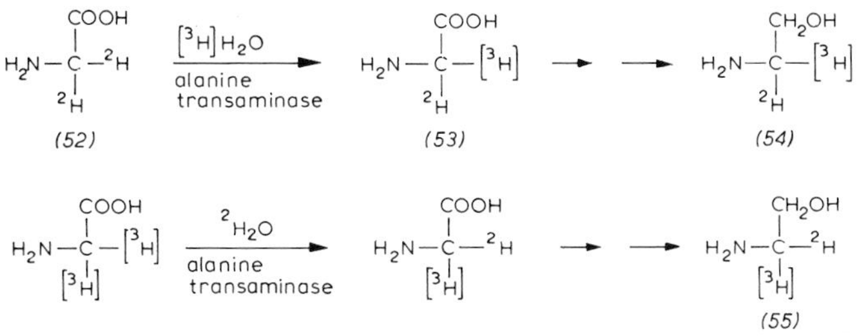

Fig. 30. Synthesis of stereospecifically deuterated and tritiated ethanolamine.

2-aminopropanol are also substrates for ethanolamine ammonia lyase [50]. These substrate analogues, if processed by the same mechanism, should give rise to trigonal intermediates with truly stereoheterotopic faces. The presence of the additional methyl group at the active site can then be expected to raise the torsional barrier in the corresponding intermediates (due to non-bonding interactions with the protein backbone). Furthermore, if this torsional barrier becomes higher than the energy barrier for the subsequent hydrogen atom transfer, the stereochemical outcome will follow only from the configuration of the original substrate and not from any chiral recognition by the enzyme of the stereoheterotopic faces of the intermediate. But if, on the other hand, the torsional barrier is still low enough, the observed stereospecificity will depend on such chiral recognition by the enzyme, and not on the configuration of the substrate.

When (*S*)- and (*R*)-2-aminopropanols were enzymically deaminated in the presence of [5′-^{3}H]AdoCbl, both the products and the remaining starting material became radioactively labelled. A work-up procedure analogous to the one applied for the 2-aminoethanol substrates led to [2-^{3}H]propionic acids. Taking advantage of the fact that propionyl-CoA carboxylase specifically removes the $2H_{Re}$ atom of propionyl-CoA [19], it was shown that these tritiated propionic acids had the (*S*) configuration irrespective of whether one started from (*S*)- or from (*R*)-2-aminopropanol. This remarkable result means that substitution of the amino group by the migrating hydrogen atom took place with retention in the case of (*S*)-2-aminopropanol, but with inversion when (*R*)-2-aminopropanol was the substrate.

Before one tries to construct a steric picture of the events that occur at the active site of the lyase, it is necessary to discuss the history of the re-isolated substrate, which also became labelled with tritium during the above experiments. In order to locate the tritium in these 2-aminopropanols, they were dissolved in deuterium oxide and incubated with HLADH and diaphorase [33]. Although NMR monitoring

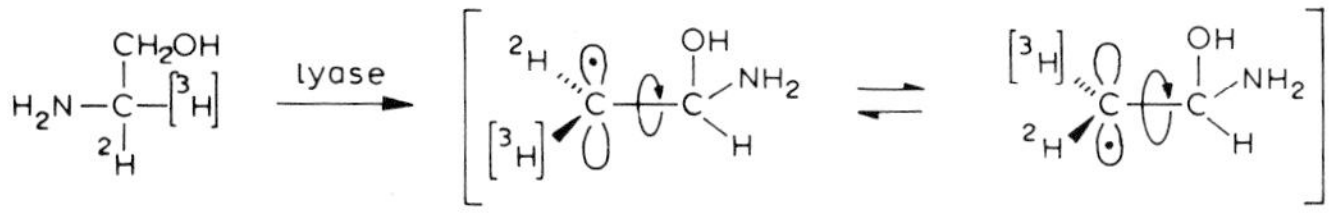

Fig. 31. A metastable trigonal intermediate in the lyase reaction results in the loss of stereospecificity.

confirmed the full exchange of the 1-H_{Re} atom, most of the tritium was retained in both the (*S*)- and the (*R*)-2-aminopropanol samples. This implied that the tritium was introduced into the 1-*pro-S* position in both cases. (Introduction of the tritium into positions 2 or 3 is a priori unlikely.) An attempt, however, to determine the exact tritium content of these samples brought an interesting detail to light. After dilution with carrier 2-aminopropanol of the appropriate configuration these samples were transformed into the 2-methyloxazolidone derivatives by reaction with phosgene and recrystallized several times. Whereas the radioactivity remained constant in the sample of the (*S*)-enantiomer, the (*R*)-enantiomer lost almost all of the radioactivity to the mother liquor. Subsequently it was shown that the tritiated molecules in this latter sample also had the (*S*) configuration. In other words, the enzymic tritium incorporation into the (*R*) substrate was concomitant with a stereochemical inversion at the C-2 to afford (*S*)-2-[1-^{3}H]aminopropanol.

These results can now be integrated into a unifying stereochemical picture of the processes that occur at the enzymes' active site, and this is depicted in Fig. 32. The 1-H_{Si} atom of both enantiomeric substrates is transferred to the C-5′ of AdoCbl, followed by the migration of the amino group. Since the hydrogen atom transfer from the modified cofactor takes place exclusively to the *Si*-face of this intermediate, irrespective of the original substrate configuration, two conclusions follow: (1) the torsional barrier around the C—C bond is low enough to allow equilibration between rotamers A and B; and (2) the topology of the enzyme's active site favours rotamer A over B. Thus both the product, and in the back reaction (*S*)-2-aminopropanol, will be formed exclusively from intermediate A. The stereochemical inversion of (*R*)-2-aminopropanol by the enzyme can then be explained, since this will also take place via intermediates B and A.

It is worth noting here that in contrast to the enantiomeric propanediols used in the dioldehydratase reaction, the two migrating groups here (1-H_{Si} and -NH_2) should have a *syn*-clinal arrangement if the coenzyme is to abstract and deliver the hydrogen atom on the same side of the molecule.

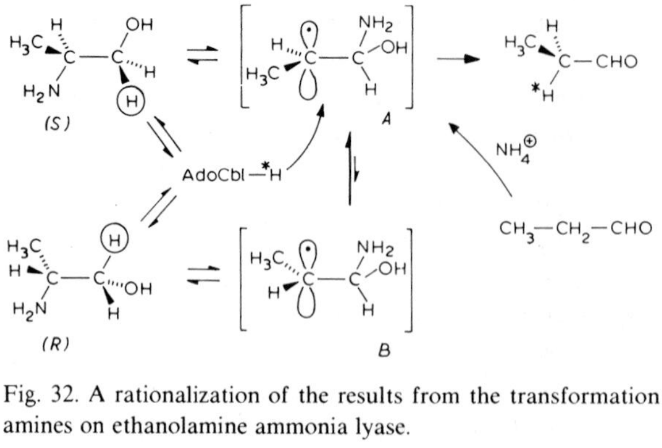

Fig. 32. A rationalization of the results from the transformation of stereospecifically labelled propanolamines on ethanolamine ammonia lyase.

(e) Conclusions

Stereochemical investigations on the AdoCbl-dependent rearrangements led to unexpected results, the mechanistic significance of which was initially obscured. However, the final demonstration that the same enzyme will catalyse reactions with different steric courses (depending on the substrate) forces one to revise one's former view of the relationship between the steric course and the mechanism of enzymic reactions. Whereas the steric course of organic reactions is believed to be a direct reflection of the mechanism, such a simple correlation is certainly not applicable for enzyme chemistry. The stereospecificity of most enzymes is based on the ability to differentiate between enantiotopic (or diastereotopic) faces or groups in metastable intermediates. Concerted reactions, the steric courses of which are orbital controlled and often predictable, are rather rare in enzymology. The AdoCbl-dependent rearrangements greatly helped recognize this state of affairs.

2. *Stereospecificity of some enzymes in the biosynthesis of the corrin nucleus*

(a) General outline of corrin biosynthesis

Tetrapyrrolic macrocycles, coordinated to certain metal ions, are involved in many essential processes of life. Chlorophylls, complexes of magnesium, are the most important pigments of photosynthesis in plants and green algae. The prosthetic groups of the cytochromes, of haemoglobin and of myoglobin are iron complexes which are involved in oxidative energy-generating processes as well as in the transport and storage of oxygen in animals. It seems that the most ancient — in evolutionary terms — among the pigments of life are the corrins the best known of which is vitamin B_{12} (1). The coenzyme function and the stereospecificity of this cobalt-containing vitamin have been discussed in detail in the first part of this chapter; the stereochemical aspects of its biosynthesis will be highlighted in the following.

Although the corrin nucleus of vitamin B_{12} (1) contains nine asymmetric carbon atoms, it is constructed by nature from achiral building blocks, and the first macrocyclic intermediate in the biosynthetic pathway, uroporphyrinogen III (60), is also achiral. Since uroporphyrinogen III is a common precursor of all tetrapyrrolic macrocycles, it is convenient to discuss its biosynthesis first and then summarize the known facts of the transformation of uroporphyrinogen III to the corrin nucleus.

In most organisms the biosynthesis of tetrapyrroles begins with the condensation of succinyl-CoA and glycine to δ-aminolaevulinic acid (56). δ-Aminolaevulinate synthase [51–53], the enzyme which catalyses this process, contains pyridoxal phosphate as a prosthetic group. Two molecules of δ-aminolaevulinic acid (ALA) are then condensed to porphobilinogen (PBG) (57) by ALA dehydratase [54–56]. Starting from glycine labelled in the methylene group (e.g., with ^{13}C or ^{14}C) PBG will carry the label as indicated in 57 [57]. PBG is the building block of the unrearranged linear tetrapyrrole, hydroxymethylbilane (58), which is formed on the enzyme PBG ammonia-lyase by successive addition of four PBG units. The dimer comprising rings A and B of the later product, urogen III, and the trimer comprising rings A, B and C, are presumptive enzyme-bound intermediates. The last and most intriguing step, the cyclization of the hydroxymethylbilane (58) with the concomitant rearrangement of ring D, is catalysed by the enzyme urogen III co-synthase. A large number of mechanistic schemes has been proposed for this reaction, but, since it has been established [58,59] that hydroxymethylbilane (58) is the substrate, the mechanism indicated in Fig. 33 appears to be the most plausible. It should be noted that hydroxymethylbilane (58) is not stable and cyclizes spontaneously to urogen I (61) under physiological conditions. This chemical cyclization occurs without rearrangement of ring D and much more slowly than the enzyme-catalyzed reaction leading to urogen III [60,61]. It seems that the co-synthase forces hydroxymethylbilane (58) to adopt a conformation (58a) in which the substituted α-position of pyrrole ring D is close to the exocyclic methylene group of ring A. The addition would lead to the still hypothetical spiro compound (59) [62] which could be first fragmented and then recyclized to urogen-III.

Urogen III is at a fork in the biosynthetic pathway leading to corrins and other macrocyclic tetrapyrroles. The breakthrough in elucidating the nature of intermediates beyond urogen-III came from the detection of isobacteriochlorins as products from organisms which construct vitamin B_{12}. The first detection of such systems was made by Bykhovsky [63–65] and pure materials were isolated in several centres [66–70] from *Propionibacterium shermanii* and *Clostridium tetanomorphum*. The three key pigments are now known as Factor I (62), sirohydrochlorin (Factor II, 63) and 20-methylsirohydrochlorin (Factor III, 64) and their structures were fully elucidated [68–74]. It is fascinating that sirohydrochlorin had previously been found to be the prosthetic group of certain bacterial enzymes [75], and the proof of this identity [66,67,70] also settled the structure of this enzymic cofactor. The newly introduced methyl groups in sirohydrochlorin originate from *S*-adenosylmethionine, as do seven methyl groups of the corrin nucleus. The further methylation of sirohydrochlorin to the trimethylated stage occurs in a surprising manner: the bridge between pyrrole rings A and D, which is later eliminated, carries the third methyl group [72–74]. The reduced forms of Factor I (62) and the two isobacteriochlorins (63 and 64) are effective precursors of cobyrinic acid. It is likely that their tetrahydro- or dihydro-derivatives are the actual intermediates of the biosynthetic pathway. Recently it has been found [76,77] that the methylated methine bridge between pyrrole rings A and D (C-20) is extruded as acetic acid. That such an extrusion is chemically feasible has been elegantly demonstrated by Eschenmoser's group [78] with a synthetic model hydroporphinoid (65) which ring-contracted to a corrinoid complex (66) on melting and lost acetate after alkaline treatment. In Factors II and III (63 and 64) four of the asymmetric centres of vitamin B_{12} (1) have already been created. Apart from the ring-contraction the following steps are required to complete the biosynthesis of cobyrinic acid: introduction of five methyl

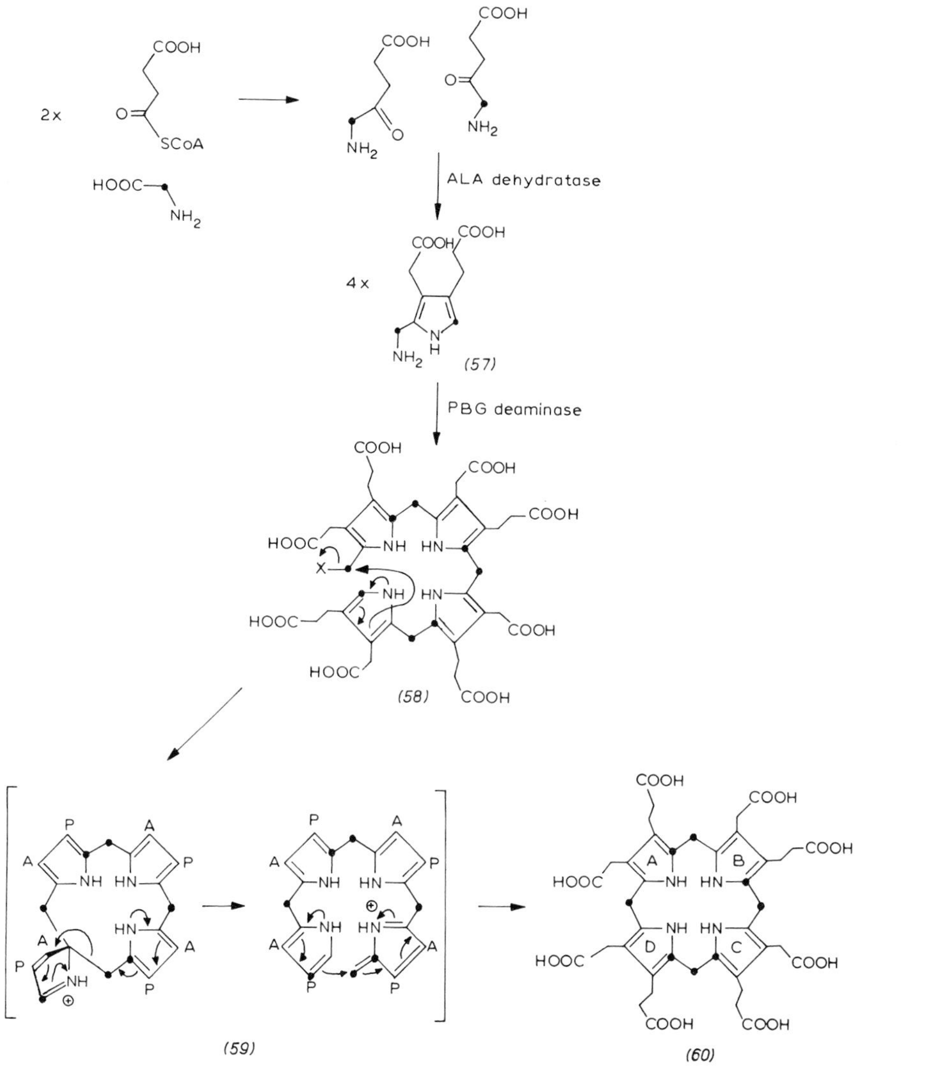

Fig. 33. Outline of the biosynthesis of uroporphyrinogen III (abbreviated urogen III). Starting with glycine labelled at position 2. The label is distributed as indicated with filled circles. The carboxyl group of glycine is lost as carbon dioxide. A = acetic acid side chain; P = propionic acid side chain.

groups into positions 1, 5, 12, 15 and 17, and decarboxylation of the acetic acid side chain at position 12. The chronological order and the mechanism of these steps are unknown and will certainly be the next target of future research.

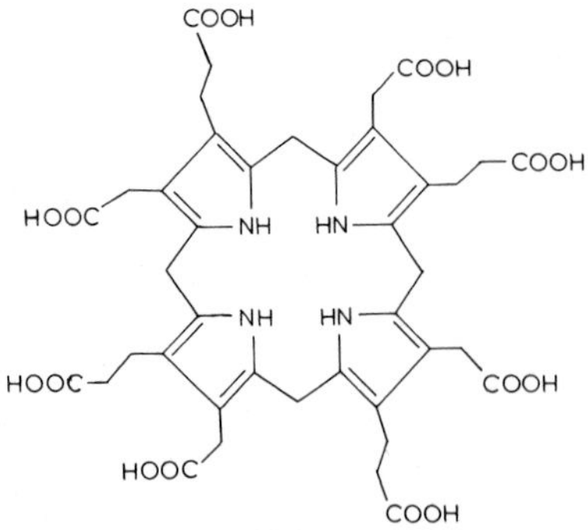

(61)

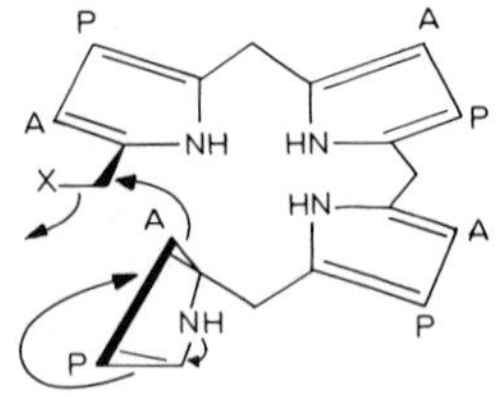

(58a)

Fig. 34. The structure of uroporphyrinogen I (urogen I) and the conformation of hydroxymethylbilane which reacts to the spiro-intermediate 59. A=acetic acid side chain; P=propionic acid side chain.

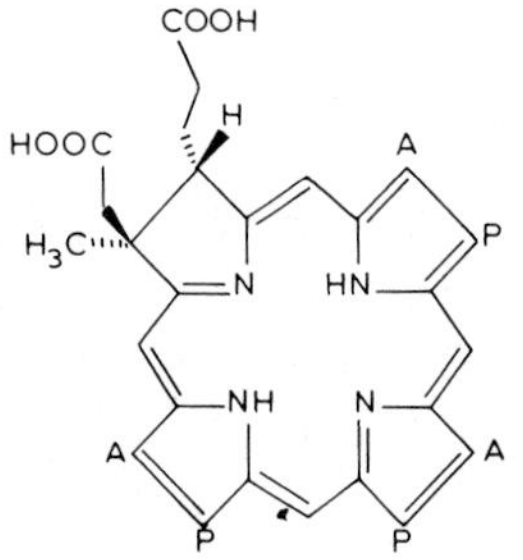

(62)

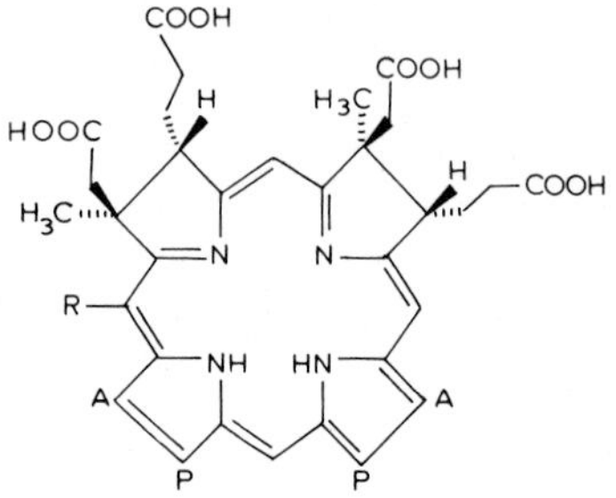

(63) R = H
(64) R = CH_3

Fig. 35. The structures of factors I (62), II (63, sirohydrochlorin) and III (64). A=acetic acid side chain; P=propionic acid side chain.

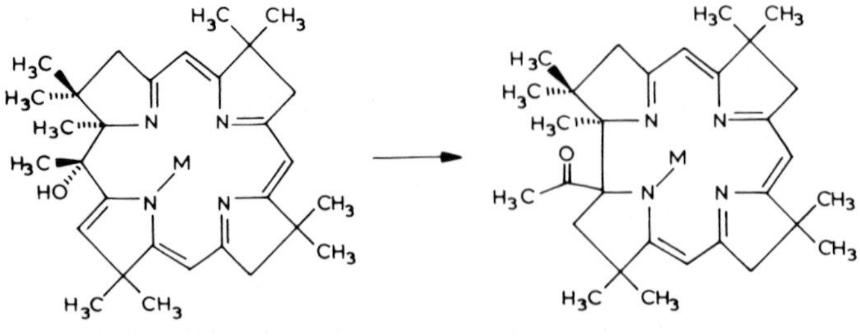

(65)

(66)

Fig. 36. Ring-contraction of a model hydroporphinoid to a corrinoid on melting.

(b) The use of stereospecifically labelled precursors

As mentioned in the general outline both urogen III (60) and its building blocks are achiral. This means that any possible stereospecificity in the biosynthesis of urogen III must be cryptic and its elucidation will require stereospecific isotope labelling. Both building blocks of urogen III, i.e. glycine and succinic acid, are available in stereospecifically labelled forms and have been used in biosynthetic studies. Moreover, *S*-adenosylmethionine with a chiral methyl group has been applied to the second stage of biosynthesis in which urogen III is transformed by subsequent methylations, inter alia, into vitamin B_{12} (1).

(i) Labelled glycine
ALA synthase is a pyridoxal phosphate-dependent enzyme and promotes Schiff-base formation between its coenzyme and glycine (67 in Fig. 37). Nucleophilicity at C-2 of the glycine could be generated either by decarboxylation or by abstraction of a proton. In the first case δ-aminolaevulinic acid would retain both methylene protons of glycine, in the second, one of the protons would be lost to the medium (Fig. 37). Acylation of the pyridoxal-bound intermediate (68 or 69) by succinyl-CoA would constitute the next step and this could be followed either by direct hydrolysis of the Schiff-base or by decarboxylation with subsequent hydrolysis depending on which course was chosen in the first stage of the reaction.

To elucidate the mechanistic details (*R*)- and (*S*)-[2-^{3}H]glycines were prepared [79] (see also chapter by Floss and Vederas) and reacted with succinyl-CoA on ALA synthase. It was shown that tritium was retained in ALA when (*S*)-[2-^{3}H]glycine was used as substrate but it was lost when (*R*)-[2-^{3}H]glycine was the starting material.

This result ruled out the conceptually simpler reaction pathway involving initial decarboxylation of the pyridoxal-glycine Schiff-base intermediate (67a in Fig. 37), but was not sufficient to elucidate the overall steric course of the reaction. For this purpose the absolute configuration of [5-^{3}H]ALA formed from (*S*)-[2-^{3}H]glycine had to be determined. This was achieved [80] by trapping the enzymically obtained [5-^{3}H]ALA with a second enzyme, ALA dehydratase, to yield the tritiated PBG 57a (Fig. 38). The aminomethylene group in PBG, which is configurationally more stable than that in ALA, was then extruded by an oxidative pathway to yield [2-^{3}H]glycine. Enzymic analysis of this tritiated glycine indicated that its absolute configuration, and, hence also that of the parent PBG (57a), was (*S*).

Since a configurational change at C-5 of ALA during conversion into the aminomethyl group of PBG is unlikely, it is established that (*S*)-[2-^{3}H]glycine was transformed into (5*S*)-[5^{3}H]ALA by ALA synthase. Of the two sequential reactions, (i) substitution of the H_{Re} atom with the succinyl group and (ii) decarboxylation, one had to take place with retention, the other with inversion. Until the absolute configuration of the intermediate pyridoxal-bound β-keto acid (70) can be determined, a definitive assignment of the steric course to the two processes is not possible. Several enzymic decarboxylations, some dependent on pyridoxal phos-

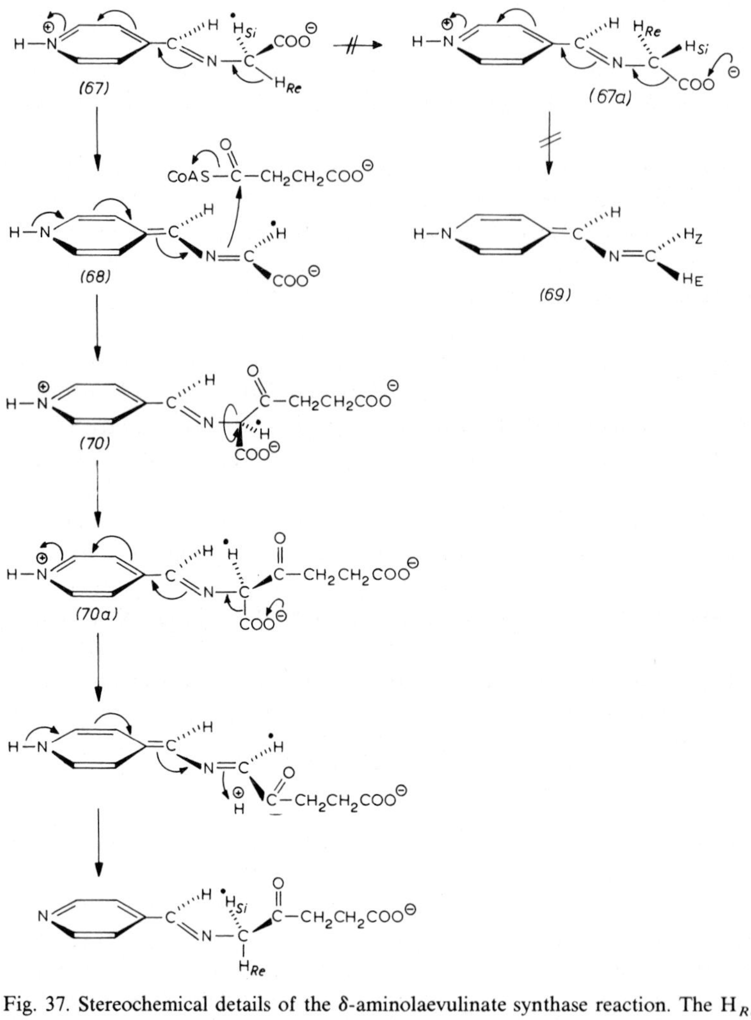

Fig. 37. Stereochemical details of the δ-aminolaevulinate synthase reaction. The H_{Re}-atom of glycine is lost to the solvent; the H_{Si}-atom becomes the 5-H_{Si}-atom of δ-aminolaevulinate.

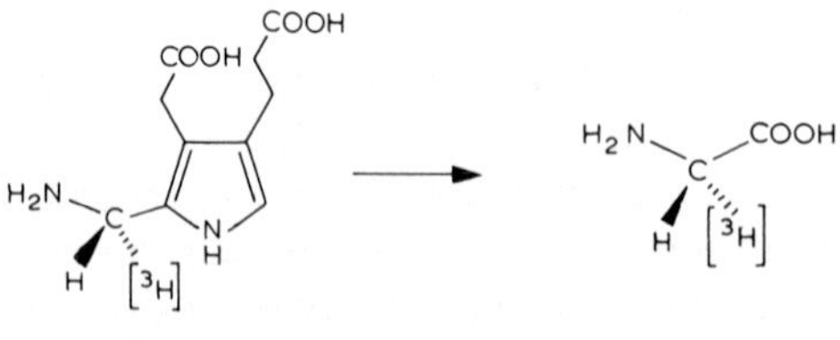

Fig. 38. The degradation of tritiated porphobilinogen to glycine.

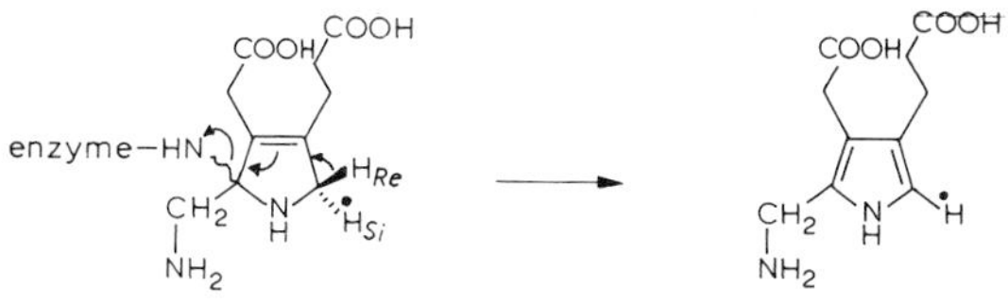

Fig. 39. The last step of the δ-aminolaevulinate dehydratase reaction takes place stereospecifically.

phate, some not, have been stereochemically analysed, however, and all of them were found to take place with retention of configuration. If this holds also for the β-keto acid intermediate 70 then the stereochemical events in the ALA synthase reaction can be specified (Fig. 37). It is noteworthy that after substitution with the succinyl residue the β-keto acid intermediate (70) must rotate about the C_α—N σ-bond by at least 60° in order to establish a favourable stereoelectronic set-up for the subsequent decarboxylation (70a).

A further question concerns the fate of the tritium in the second molecule of (5*S*)-[5-^{3}H]ALA during the dehydratase reaction. The C-5 atom of the ALA molecules becomes the C-2 atom of PBG and loses one of its enantiotopic protons, while the ring is aromatized. When (5*S*)-[5-^{3}H]ALA was reacted on the dehydratase and the product was isolated by careful chromatography on cellulose, the resulting PBG retained the tritium also at position 2 [81]. This means that the aromatization of the presumptive intermediate 71 takes place in an enzyme-bound state and involves specific abstraction of the 2-H_{Re} atom (Fig. 39).

(ii) Doubly labelled succinate

Of the eight succinate molecules which are incorporated into the corrin nucleus, four are transformed into propionate side chains. In these the eight original methylene groups remain intact. The three acetate side chains also originate from succinate. Each of these side chains retains one intact succinate methylene group, while the second methylene group of each original succinate molecule becomes a quarternary carbon atom (C-2, C-7 and C-18). The fate of the eighth succinate building block is different, however. One of its two methylene groups becomes the *pro-S* methyl group attached to the C-12 atom of ring C of the corrin [82,83], a process which involves decarboxylation. The steric course of this decarboxylation could be determined if succinate stereospecifically labelled with deuterium and tritium in the same methylene group were to be used as a precursor.

Though the pertinent experiment in the corrin series has not yet been reported, a record of the analogous experiment in the porphyrin series has appeared [84]. In contrast to corrins, in haems all four acetate side chains of the parent urogen III have been decarboxylated to methyl groups. In principle, these could be carved out as acetic acid and analysed for chirality if the starting succinate had been stereospecifically labelled with deuterium as well as with tritium. The synthesis of (2*R*)-[2-$^2H_1$,2-^{3}H]succinic acid was achieved by incubating (3*R*,*S*)-[3-^{3}H]-

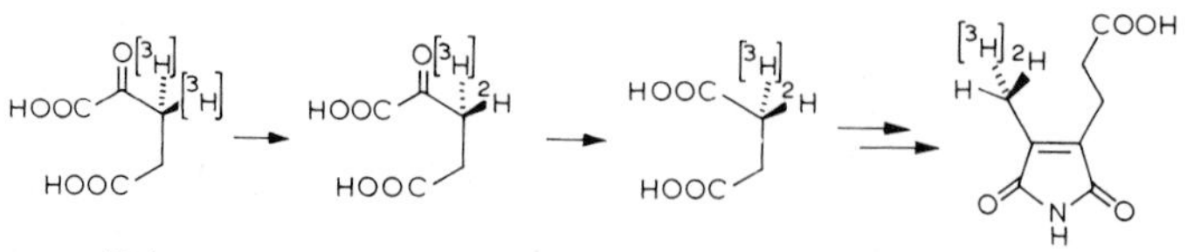

Fig. 40. Preparation and use of (2R)-[2-$^2H_1$,2-^{3}H]succinate in haem biosynthesis.

oxoglutarate (72) in deuterium oxide with NADPH and isocitrate dehydrogenase until half of the tritium was washed out into the solvent. Since isocitrate dehydrogenase is known [85] to promote specifically the exchange of the 3-H_{Si}-atom, the doubly labelled product was (3R)-[3-$^2H_1$,3-^{3}H]-2-oxoglutarate (72a). Treatment of this sample with hydrogen peroxide yielded (2R)-[2-$^2H_1$,2-^{3}H]succinic acid (73) which was used as a substrate in the biosynthesis of haem by a haemolysed erythrocyte preparation. In a known reaction sequence [86] the labelled haem was degraded to haematic acid (74) which, in turn, yielded [$^2H_1$,^{3}H]acetate upon ozonolysis in the presence of H_2O_2. Stereochemical analysis by the malate synthase–fumarase system [23,24] showed that the doubly labelled acetate was predominantly (S), specifying the steric course of the decarboxylation as retention.

Barnard and Akhtar [84] speculate that the same group (X) of the enzyme is involved in the deprotonation of the carboxyl group and in the stereospecific protonation of the vinylogous enamine (75) (Fig. 41).

*(iii) Chiral [*methyl*-$^2H_1$,^{3}H]methionine*

Seven of the eight methyl groups attached to the corrin nucleus originate from *S*-adenosylmethionine. Moreover, an eighth molecule of *S*-adenosylmethionine is required to methylate the bridging C-20 atom of sirohydrochlorin (63) which is extruded as acetate in the course of the ring-contraction to corrin. All four pyrrole rings of urogen-III are methylated in a reductive manner, i.e., a hydrogen atom is added vicinal to the methyl group. The additions take place in an *anti* fashion and this is true also for ring C, where the *pro-R* methyl group is the one derived from

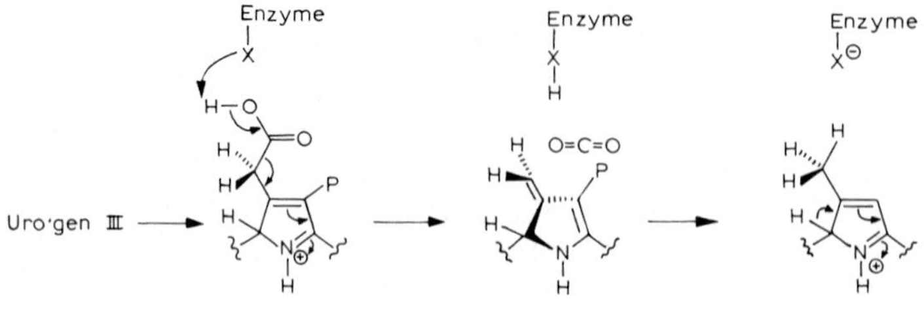

Fig. 41. Proposed mechanism for the retentive decarboxylation of acetic acid side chain during haem biosynthesis.

S-adenosylmethionine [82,83]. By these peripheric methylations seven centres of chirality are created in a strict stereospecific manner. Two further chirality centres arise at the direct junction of rings A and D.

The final stereochemical problem to be discussed is the steric course at the methyl C-atoms during the methyl transfers. To solve this problem a methionine sample with a chiral methyl group was required. The synthesis was achieved by two groups but in a very similar manner ([87]; D. Arigoni as cited in [88]) (Fig. 42). Chiral [H,^{3}H]acetate was submitted to a Schmidt degradation to yield [^{2}H,^{3}H]methylamine with retention of configuration. The ditosylate of this chiral methylamine was substituted in an S_N2 reaction by the S^- anion of homocysteine generated in situ either from *S*-benzyl homocysteine in the presence of a sodium–potassium alloy or from homocysteine and sodium hydride. Arigoni's group (cited in [88]) used the chiral [^{2}H,^{3}H]methionine as a precursor in the biosynthesis of vitamin B_{12} and some of the methyl groups of the vitamin were subsequently carved out as acetate by an oxidative pathway. It has been found that the resulting chiral acetic acid had the same absolute configuration as the one that served as the starting material for the synthesis of [$^2H_1$,^{3}H]methionine. Since the substitution of the ditosylamino group by the S^- anion of homocysteine occurred with inversion, the biological methyl transfer from *S*-adenosylmethionine to the vitamin B_{12} precursors must also have taken place with inversion at the methyl group.

(c) Conclusions

During the last few years many interesting details of the biosynthesis of corrins and porphyrins have been discovered. The methods by which nature builds up such complex molecules are sometimes conventional, even monotonous, and sometimes surprising. Monotonous are, for example, the repeated reductive methylations of the four pyrrole rings with the same regio- and stereospecificity, surprising the introduction into position C-20 of a methyl group which is extruded later as acetate.

It is not impossible that this fascinating biosynthetic pathway still hides a few surprises, but its main outline stands on a solid basis. Future research will be devoted to the identification and characterization of the individual enzymes that catalyse the conversion of urogen-III to the corrin nucleus.

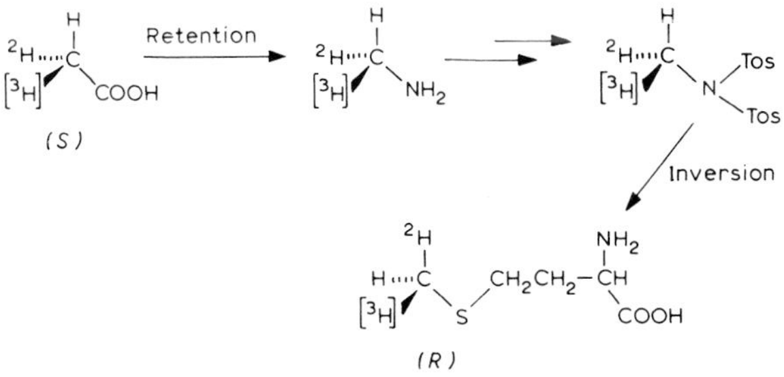

Fig. 42. Preparation of (*R*)-[*methyl*-$^2H_1$,*methyl*-^{3}H]methionine.

Acknowledgements

I am grateful to Professor I.D. Spenser and Dr. J.A. Robinson for reading parts of the manuscript, and to Verlag Chemie, Weinheim, F.R.G., for giving permission to reprint the first part of this chapter from the monograph entitled *Stereospecificity in Chemistry and Enzymology* by J. Rétey and J.A. Robinson.

The cited work from the authors' laboratory was generously supported by the Deutsche Forschungsgemeinschaft and by the Fonds der Chemischen Industrie.

References

1 Dowd, P., Shapiro, M. and Kang, K. (1975) J. Am. Chem. Soc. 97, 4754.
2 Bidlingmaier, G., Flohr, H., Kempe, U.M., Krebs, T. and Rétey, J. (1975) Angew. Chem. 87, 877 (Angew. Chem. Int. Ed. Engl. 14, 822).
3 Golding, B.T., Kemp, T.J., Nocchi, E. and Watson, W.P. (1975) Angew. Chem. 87, 841 (Angew. Chem. Int. Ed. Engl. 14, 813).
4 Barker, H.A., Weissbach, H. and Smyth, R.D. (1958) Proc. Natl. Acad. Sci. U.S.A. 44, 1093.
5 Sprecher, M., Switzer, R.L. and Sprinson, D.B. (1966) J. Biol. Chem. 241, 864.
6 Sprecher, M. and Sprinson, D.B. (1966) J. Biol. Chem. 241, 868.
7 Abeles, R.H. and Lee, H.A. (1961) J. Biol. Chem. 236, 2347.
8 Pawelkiewicz, J. and Zagalak, B. (1965) Acta Biochim. Polon. 12, 207.
9 Eagar, R.G. Jr., Bachovchin, W.W. and Richards, J.H. (1975) Biochemistry 14, 5523.
10 Bachovchin, W.W., Eagar, R.G. Jr., Moore, K.W. and Richards, J.H. (1977) Biochemistry 16, 1082.
11 Poznanskaya, A.A. and Korsova, T.L. (1979) in B. Zagalak and W. Friedrich (Eds.), Vitamin B_{12}, Proceedings of the Third European Symposium on Vitamin B_{12} and Intrinsic Factor, Walter de Gruyter, Berlin, p. 431.
12 Frey, P.A. and Abeles, R.H. (1966) J. Biol. Chem. 241, 2732.
13 Abeles, R.H. and Dolphin, D. (1976) Acc. Chem. Res. 9, 114.
14 Barker, H.A. (1972) Ann. Rev. Biochemistry 41, 55.
15 Friedrich, W. (1975) in R. Ammon and W. Dirscherl (Eds.), Vitamin B_{12} und Verwandte Corrinoide, 3rd Edn., Vol. III/2, Georg Thieme Verlag, Stuttgart.
16 Rétey, J., Umani-Ronchi, A., Seibl, J. and Arigoni, D. (1966) Experientia 22, 502.
17 Rétey, J., Umani-Ronchi, A. and Arigoni, D. (1966) Experientia 22, 72.
18 Zagalak, B., Frey, P.A., Karabatsos, G.L. and Abeles, R.H. (1966) J. Biol. Chem. 241, 3028.
19 Rétey, J. and Lynen, F. (1965) Biochem. Z. 342, 256.
20 Karabatsos, G.L., Fleming, J.S., Hsi, N. and Abeles, R.H. (1966) J. Am. Chem. Soc. 88, 849.
21 Bonetti, V. (1974) Dissertation, ETH, Zurich, No. 5366.
22 Arigoni, D. (1979) in B. Zagalak and W. Friedrich (Eds.), Vitamin B_{12}, Proceedings of the Third European Symposium on Vitamin B_{12} and Intrinsic Factor, Walter de Gruyter, Berlin, p. 389.
23 Cornforth, J.W., Redmond, J.W., Eggerer, H., Buckel, W. and Gutschow, C. (1969) Nature (London) 221, 1212.
24 Lüthy, J., Rétey, J. and Arigoni, D. (1969) Nature (London) 221, 1213.
25 Eggerer, H., Overath, P., Lynen, F. and Stadtman, E.R. (1960) J. Am. Chem. Soc. 82, 2643.
26 Kellermeyer, R.W. and Wood, H.G. (1962) Biochemistry 1, 1124.
27 Phares, E.F., Long, M.V. and Carson, S.F. (1962) Biochem. Biophys. Res. Commun. 8, 142.
28 Rétey, J. and Arigoni, D. (1966) Experientia 22, 783.
29 Cardinale, G.J. and Abeles, R.H. (1967) Biochim. Biophys. Acta 132, 517.
30 Sprecher, M., Clark, M.J. and Sprinson, D.B. (1964) Biochem. Biophys. Res. Commun. 15, 581; ibid (1966) J. Biol. Chem. 241, 872.

31 Rétey, J. and Lynen, F. (1964) Biochem. Biophys. Res. Commun. 16, 358; ibid (1965) Biochem. Z. 342, 256.
32 Rétey, J., Smith, E.H. and Zagalak, B. (1978) Eur. J. Biochem. 83, 437.
33 Günther, H., Alizade, M.A., Kellner, M., Biller, F. and Simon, H. (1973) Z. Naturforsch. 28, 241.
34 Bücklers, L., Umani-Ronchi, A., Rétey, J. and Arigoni, D. (1970) Experientia 26, 931.
35 Corey, E.J., Mock, W.L. and Pasto, D.J. (1961) J. Am. Chem. Soc. 83, 2957 (1961).
36 Hünig, S., Müller, H.R. and Thier, W. (1961) Tetrahedron Lett., 353.
37 König, A. (1979) Dissertation, University of Karlsruhe.
38 Stadtman, T.C. and Renz, P. (1968) Arch. Biochem. Biophys. 125, 226.
39 Baker, J.J., van der Drift, C. and Stadtman, T.C. (1973) Biochemistry 12, 1054.
40 Van Tamelen, E.E. and Smissman, E.E. (1953) J. Am. Chem. Soc. 75, 2031.
41 Hong, S.L. and Barker, H.A. (1973) J. Biol. Chem. 248, 41.
42 Kunz, F., Rétey, J., Arigoni, D., Tsai, L. and Stadtman, T.C. (1978) Helv. Chim. Acta 61, 1139.
43 Cervinka, O., Fábryová, A. and Novák, U. (1965) Coll. Czech. Chem. Commun. 30, 1742.
44 Rétey, J., Kunz, F., Arigoni, D. and Stadtman, T.C. (1978) Helv. Chim. Acta 61, 2989.
45 Zeylemaker, W.P., Veeger, C., Kunz, F., Rétey, J. and Arigoni, D. (1970) Chimia 24, 33.
46 Kaplan, B.H. and Stadtman, E.R. (1968) J. Biol. Chem. 243, 1974.
47 Babior, B.M. (1969) J. Biol. Chem. 244, 449.
48 Rétey, J. (1979) in T. Swain and G. Waller (Eds.), Recent Advances in Phytochemistry, Vol. 13, Plenum Publishing Corp., New York, p. 1.
49 Rétey, J., Suckling, C.J., Arigoni, D. and Babior, B.M. (1974) J. Biol. Chem. 249, 6359.
50 Diziol, P., Haas, H., Rétey, J., Graves, S.W. and Babior, B.M. (1980) Eur. J. Biochem. 106, 211.
51 Lascelles, J. (1959) Biochem. J. 72, 508.
52 Granick, S. (1966) J. Biol. Chem. 241, 1359.
53 Irving, E.A. and Elliott, W.H. (1969) J. Biol. Chem. 244, 60.
54 Shemin, D. and Russell, C.S. (1953) J. Am. Chem. Soc. 75, 4873.
55 Neuberger, A. and Scott, K.K. (1953) Nature (London) 170, 1093.
56 Del C. Battle, A.M., Ferramola, A.M. and Grinstein, M. (1967) Biochem. J. 104, 244.
57 Battersby, A.R., Hunt, E. and McDonald, E. (1973) J. Chem. Soc. Chem. Commun., 442.
58 Battersby, A.R., Fookes, C.J.R., Gustafson-Potter, K.E., Matcham, G.W.J. and McDonald, E. (1979) J. Chem. Soc. Chem. Commun., 1155.
59 Battersby, A.R., Brereton, R.G., Fookes, C.J.R., McDonald, E. and Matcham, G.W.J. (1980) J. Chem. Soc. Chem. Commun., 1124.
60 Battersby, A.R., Fookes, C.J.R., Matcham, G.W.J., McDonald, E. and Gustafson-Potter, K.E. (1979) J. Chem. Soc. Chem. Commun., 316.
61 Battersby, A.R., Fookes, C.J.R., Matcham, G.W.J. and McDonald, E. (1980) Nature (London) 285, 17.
62 Mathewson, J.H. and Corwin, A.H. (1961) J. Am. Chem. Soc. 83, 135.
63 Bykhovsky, V.Ya., Zaitseva, N.I. and Bukin, V.N. (1975) Dokl. Acad. Sci. U.S.S.R. 224, 1431.
64 Bykhovsky, V.Ya., Zaitseva, N.I., Umrikhina, A.V. and Yavorskaya, A.N. (1976) Prikl. Biokhim. Mikrobiol. 12, 825.
65 Bykhovsky, V.Ya. and Zaitseva, N.I. (1976) Prikl. Biokhim. Mikrobiol. 12, 365.
66 Deeg, R., Kriemler, H.-P., Bergmann, K.-H. and Müller, G. (1977) Hoppe-Seyler's Z. Physiol. Chem. 358, 339.
67 Bergmann, R.-H., Deeg, R., Gneuss, K.D., Kriemler, H.-P. and Müller, G. (1977) Hoppe-Seyler's Z. Physiol. Chem. 358, 1315.
68 Battersby, A.R., McDonald, E., Morris, H.R., Thompson, M., Williams, D.C., Bykhovsky, V.Ya., Zaitseva, N.I. and Bukin, V.N. (1977) Tetrahedron Lett., 2217.
69 Battersby, A.R., Jones, K., McDonald, E., Robinson, J.A. and Morris, H.R. (1977) Tetrahedron Lett., 2213.
70 Scott, A.I., Irwin, A.J., Siegel, L.M. and Shoolery, J.N. (1978) J. Am. Chem. Soc. 100, 316 and 7987.
71 Imfeld, M., Arigoni, D., Deeg, R. and Müller, G. (1979) in B. Zagalak and W. Friedrich (Eds.), Vitamin B_{12}, Proceedings of the Third European Symposium on Vitamin B_{12} and Intrinsic Factor, Walter de Gruyter, Berlin, p. 315.

72 Battersby, A.R., Matcham, G.W.J., McDonald, E., Neier, R., Thompson, M., Woggon, W.-D., Bykhovsky, V.Ya. and Morris, H.R. (1979) J. Chem. Soc. Chem. Commun., 185.
73 Lewis, N.G., Neier, R., Matcham, G.W.J., McDonald, E. and Battersby, A.R. (1979) J. Chem. Soc. Chem. Commun. 541.
74 Müller, G., Gneuss, K.D., Kriemler, H.-P., Scott, A.I. and Irwin, A.J. (1979) J. Am. Chem. Soc. 101, 3655.
75 Siegel, L.M., Murphy, M.J. and Kamin, H. (1973) J. Biol. Chem. 248, 251.
76 Mombelli, L., Nussbaumer, C., Weber, H., Müller, G. and Arigoni, D. (1981) Proc. Natl. Acad. Sci. U.S.A. 78, 11.
77 Battersby, A.R., Bushell, M.J., Jones, C., Lewis, N.G. and Pfenninger, A. (1981) Proc. Natl. Acad. Sci. U.S.A. 78, 13.
78 Rasetti, V., Pfaltz, A., Kratky, C. and Eschenmoser, A. (1981) Proc. Natl. Acad. Sci. U.S.A. 78, 16.
79 Zaman, Z., Jordan, P.M. and Akhtar, M. (1973) Biochem. J. 135, 257.
80 Abboud, M.M., Jordan, P.M. and Akhtar, M. (1974) J. Chem. Soc. Chem. Commun., 643.
81 Abboud, M.M. and Akhtar, M. (1976) J. Chem. Soc. Chem. Commun., 1007.
82 Battersby, A.R., Ihara, M., McDonald, E., Stephenson, J.R. and Golding, B.T. (1973) J. Chem. Soc. Chem. Commun., 404.
83 Scott, A.I., Townsend, C.A. and Cushley, R.J. (1973) J. Am. Chem. Soc. 95, 5759.
84 Barnard, G.F. and Akhtar, M. (1975) J. Chem. Soc. Chem. Commun., 494.
85 Lienhard, G.E. and Rose, I.A. (1964) Biochem. J. 3, 185.
86 Shemin, D. (1954) 'The Harvey Lectures', Academic Press, New York, p. 258.
87 Mascaro, L. Jr., Hörhammer, R., Eisenstein, S., Sellers, L.K., Mascaro, K. and Floss, H.G. (1977) J. Am. Chem. Soc. 99, 273.
88 Floss, H.G. and Tsai, M.D. (1979) Adv. Enzymol. 50, 243.

The stereochemistry of vision

VALERIA BALOGH-NAIR and KOJI NAKANISHI

Department of Chemistry, Columbia University, New York, NY 10027, U.S.A.

1. Introduction

The impinging of light on a visual pigment initiates the process of vision. This initial absorption of light is followed by a train of events culminating in the integration, by the brain, of the pattern of perception. The visual pigments are located in the light-sensitive cells, the photoreceptors. There are two basic types, the ciliary cell of the vertebrates and the rhabdomeric cell of the invertebrates, both possessing an inner and an outer segment. It is the outer segment membranes which contain the photoreceptors. In the case of the vertebrate eye (Fig. 1), the light entering the eye must travel through the cornea, iris, lens, vitreous humor and various layers of nerve cells in the retina before it reaches the photoreceptors. These receptor cells are of two types, rods which are responsible for scotopic vision (in dim light) and cones which work under bright illumination and are concentrated in the region of the fovea. The human retina has three types of cone cells, with differing absorption maxima at 450, 535 and 560 nm, which provide the basis for color discrimination.

Visual transduction is the process by which rods and cones convert light to a neural signal, which is transmitted to the brain via the optic nerve. This process is initiated by the absorption of light by the visual pigments, and causes a change in the electrical properties of the photoreceptor cell membranes by changing their permeability. In the case of rods, the bulk of the pigment (96.5% in rat to 99% in frog) is contained in the disc membranes. In the dark, Na^+ ions flow into the outer segments (ROS), diffuse into the inner segment (RIS) where they are pumped out by an ATP-dependent pump. Upon illumination of the receptors this dark current is shut off.

Two hypotheses have been forwarded to account for this blocking of the Na^+ channels ([1]; for a recent review, see [2]). The first is that Ca^{2+} ions, which are released from the inside of the discs into the cytoplasmic space upon illumination, are responsible for the shut off of the dark current. The second hypothesis casts cyclic GMP into the role of the transmitter; upon absorption of light by a visual pigment, activation of a phosphodiesterase occurs which reduces the levels of cyclic GMP (5×10^4 molecules of cyclic GMP disappear upon bleaching of one molecule of pigment). This in turn results in a reduced activity of a protein kinase which is

Tamm (ed.) Stereochemistry

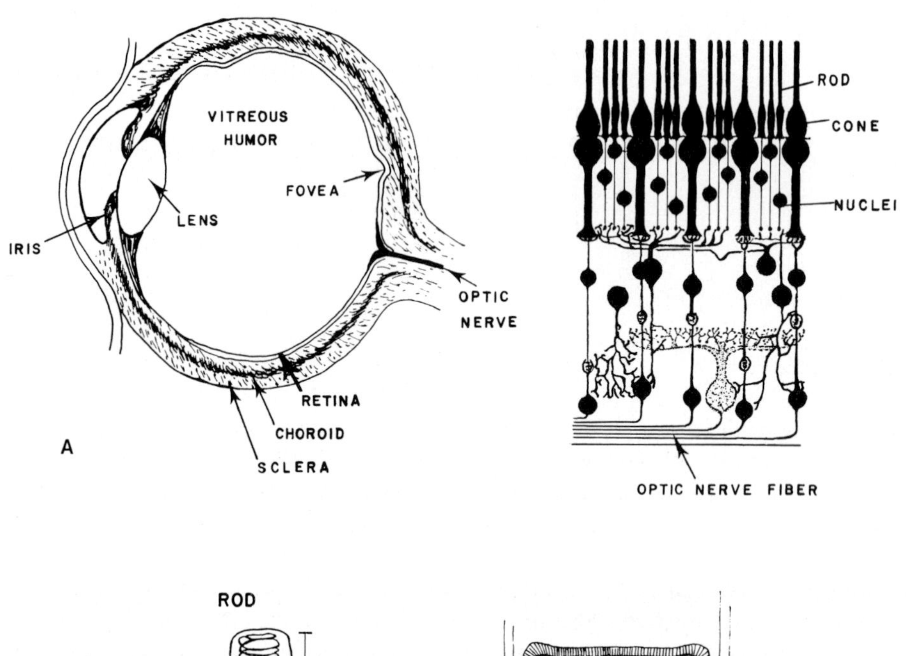

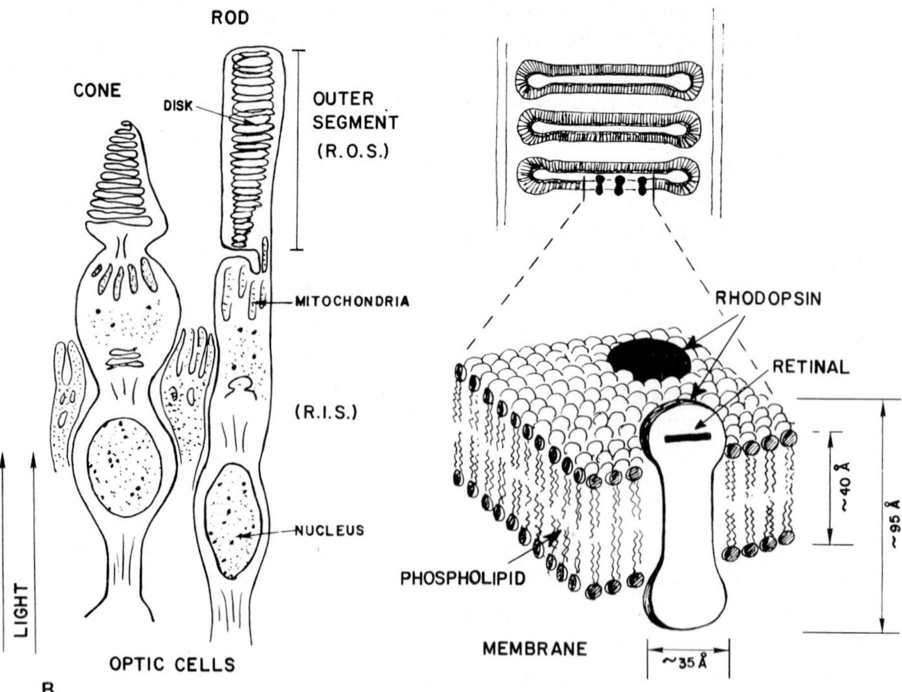

Fig. 1. Schematic diagram of a horizontal cross-section of the eye and retina (A) and of the photoreceptor cells (B).

responsible for the dephosphorylation of two small proteins (12000–13000 daltons) assumed to be present in the plasma membrane, and it is proposed that this dephosphorylation is responsible for the reduced permeability of the plasma membrane. While the mechanism by which cyclic GMP maintains the high permeability to Na^+ of ROS membrane has gained considerable support recently, the nature and function of the transmitter, which should also explain the observed signal amplification in a vertebrate rod, is complicated by the observation that Ca^{2+} levels and cyclic GMP levels are interacting, i.e., removal of Ca^{2+} increases the cyclic GMP levels.

(a) The properties of visual pigments

In 1933 G. Wald discovered the presence of vitamin A in the retina [3]. Subsequently, it was also found that it is the 11-*cis* isomer of vitamin A aldehyde, 11-*cis*-retinal, which is the chromophoric group in rhodopsins [4]. Now it is established that all species contain the same 11-*cis*-retinal and/or its 3-dehydro- derivative as the chromophore bound to various lipoproteins to form light-absorbing systems. All visual pigments studied contain one retinal per molecule of opsin [5]. Pigments containing 11-*cis*-retinal are called rhodopsins, and pigments based on 3-dehydroretinal are called porphyropsins. *Cis* isomers of retinal other than 11-*cis* are also capable of forming pigments with visual proteins; these are the visual pigment analogs (see Section 5) but only the 11-*cis* isomer has been found in nature. The rod cells are more numerous and larger (2–3 times by weight) than cone cells in the same retina, and the rod pigments are more stable; therefore, the great majority of studies have been carried out on these readily available rhodopsins, such as bovine and frog rhodopsins.

The opsin consists of protein (ca. 80–85% of which is rhodopsin), phospholipids and carbohydrates and contains very little cholesterol (1–3%) (for a review, see [6]). While the molecular weight (e.g., 40000 for bovine rhodopsin) [7], carbohydrate [8,9], lipid and amino acid [10–12] composition have been established for some rhodopsins, there is as yet no example of a visual pigment for which the full amino acid sequence is known. Only a quarter of about the 300 residues of rhodopsin have been sequenced [13,14], 39 residues at the N-terminus and 40 residues at the C-terminus. The structure of the moiety containing retinal, i.e., retinal–lysine–alanine, which is located in the carboxy-terminal region has, however, been elucidated ([15]; see also [78] and references therein). The N-terminal residue was identified as acetylmethionine [16].

Visual pigments display characteristic absorption spectra which result from the very specific interactions between protein and chromophore in the binding site, i.e., the absorbance spectrum of retinal at ca. 380 nm is red-shifted to ca. 500 nm in bovine rhodopsin. However, depending on species, rhodopsins absorb from 440 to ca. 600 nm. Porphyropsins show a similar spread in their absorption maxima, absorb at longer wavelengths than the corresponding rhodopsins, and have lower extinction coefficients (ca. 75%) than rhodopsins (e.g., bovine rhodopsin ϵ, ca. 40500) [17] as shown in Fig. 2.

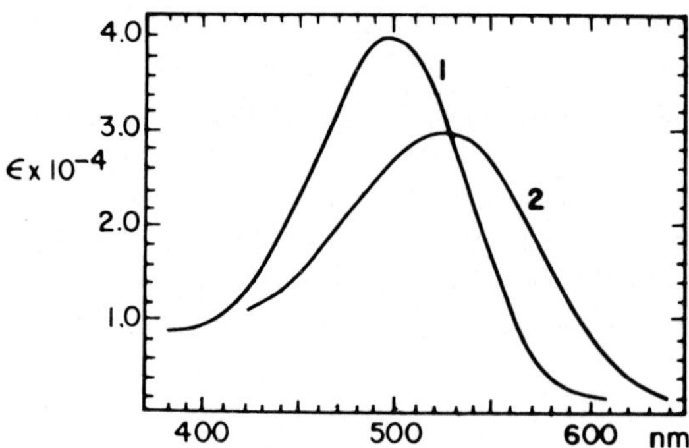

Fig. 2. Absorbance spectra of bovine rhodopsin [17] (curve 1) and fish porphyropsin [18] (curve 2).

The shape of the long wavelength absorption band is yet another characteristic property of visual pigments. Thus, Dartnall constructed nomograms [19], which showed that almost all rhodopsins (although differing in absorption maxima) have almost identical bandshapes when their spectra are plotted on a frequency basis instead of the usual wavelength scale. More convenient nomograms can also be constructed where A_x/A_{max} is directly related to λ_{max} (where A_{max} and A_x are the absorbances at the λ_{max} and at any wavelength x, respectively, and A_x/A_{max} is the relative absorbance at wavelength x). Such a nomogram is shown in Fig. 3 [20].

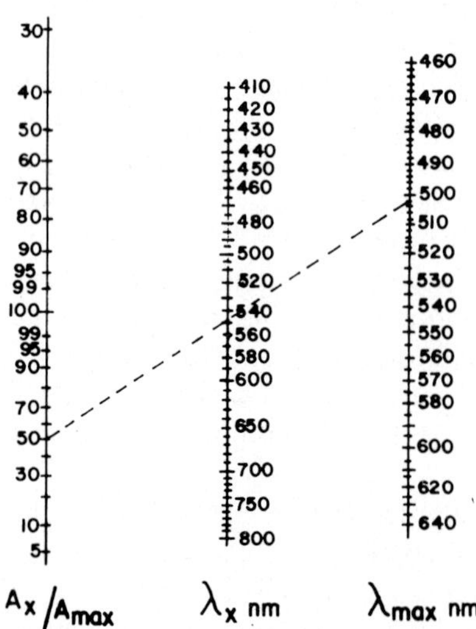

Fig. 3. Nomogram for the calculation of the spectra of rhodopsins. The dashed line shows an example: a rhodopsin with an absorbance of 50% of maximum at 550 nm on the long-wave side of the peak has a λ_{max} of 503 nm. From Knowles and Dartnall [20].

Although porphyropsins show a somewhat broader band shape than the rhodopsins absorbing at about the same wavelength, nomograms could be constructed for porphyropsins as well [18,21]. Except for pigments which absorb at either very short or very long wavelengths, all visual pigments have spectra which show a close fit with the nomograms. Thus the nomograms can be used to estimate the absorption spectrum for an unknown pigment in the receptors or in the extract.

In order to observe the individual spectrum of a pigment in a mixture of pigments with different maxima or different photosensitivities, the technique of partial bleaching can be used with advantage [22]. In addition to the long wavelength absorption band, visual pigments also show a so-called '*cis*-band' (or B-band or β-band) at around 350 nm. This *cis* peak is considered to be associated with the chromophore and has an intensity of only 0.267 of the main peak as measured for the case of pure bovine rhodopsin [12]. Visual pigments also show absorbance in the UV region due to the absorption of amino acids of opsin.

Extensive studies by various methods, such as chemical labeling [23,24], anti-rhodopsin antibody labeling [25,26], proteolytic digestion [27–29], freeze fracture [30,31], X-ray diffraction [32–34] and neutron diffraction [35,36] were carried out to determine the spatial arrangement of rhodopsin in the disc membrane. On the basis of some of these studies, it was proposed that rhodopsin was spherical. Overwhelming evidence, especially with enzymatic labelling studies [37], however, indicate that rhodopsin is a transmembrane protein; it spans the disc membrane with its carbohydrate moiety on the intradiscal surface and its phosphorylation sites on the external surface. In vertebrates, the pigments have a fixed orientation in the ROS membrane with their absorption vectors lying in the plane of the discs. The pigments in the microvilli of the compound eyes are also oriented. This dichroism of the receptor is lost upon bleaching of the visual pigments.

To probe the secondary structure of the protein, ORD and CD measurements have been carried out in the 185–250 nm region of the spectra. The α-helical content of bovine rhodopsin was thus estimated to be ca. 30% by some workers [38] and ca. 60% by others [12]. A predominantly perpendicular orientation of rhodopsin α-helices relative to the membrane plane was indicated by the disappearance of 190–210 alpha-helix Moffitt bands when irradiation was carried out with a light beam parallel to the alpha-helix axis [39]. The value of helical content in pigments was found to be also dependent on the nature of the detergent used to solubilize the membrane. A drop in helicity was seen when extracts were bleached, suggesting that the presence of the retinal in the pigment forces it to assume a more orderly structure. However, no significant change in helicity was seen when ROS suspension was irradiated. Therefore, it was suggested that the loss of helicity seen in extracts merely reflects the denaturing effect of the detergent [12,40,41]. As a corollary, these studies suggest that experiments carried out in detergents may give results unfit to describe the physiological function of the intact membrane system. Based on results derived from CD measurements on ROS membranes, conformational changes localized in the binding site, in which only a few amino acid residues participate [42,43], seem more probable than overall conformational distortions which once

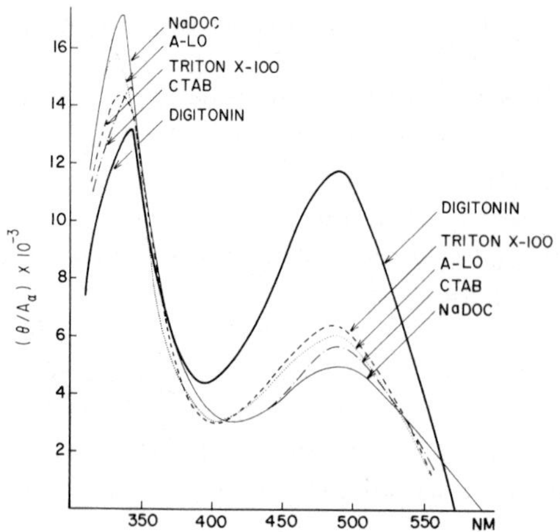

Fig. 4. Circular dichroism spectra of bovine ROS in various detergents at room temperature. $\theta/A_\alpha \times 10^{-3}$ is the ellipticity in millidegrees of α (or β) band/absorbance of α band. From Waddel et al. [44].

were thought to be an essential feature in the mechanism of light-induction of the neural response.

Bovine ROS membranes show a CD band at ca. 280 nm attributed to π–$\pi^\star$ transitions of aromatic residues and n–$\pi^\star$ transitions of cysteine, as well as two maxima in the visible region at ca. 340 nm and 490 nm (the β and α bands) corresponding to the *cis* and main peaks of the absorption spectrum. The intensity of the α band in the CD is species-dependent, but is always somewhat blue-shifted relative to the λ_{max} in all species, and this shift is promoted by detergent solubilization. Strong micellar effects have been observed in the intensity of both α and β CD bands [44] (Fig. 4).

Further, a suspension of bovine ROS displayed an α CD peak 2.3 times more intense than the corresponding extract [40,41]. It is the conformational stability of rhodopsin which is affected by detergent solubilization, the smallest divergence from ROS being observed in digitonin.

(b) Bleaching and bleaching intermediates

In intact eyes, excised retina, ROS membranes or in detergent extracts, the photolysis of visual pigments results in the formation of an initial photoproduct which then decays thermally, in the dark, through several spectrally distinct intermediates, until final separation of the isomerized retinal from the apoprotein occurs. In the process, called bleaching, the geometry of the retinal chromophore is changed from the 11-*cis* to the all-*trans* form with a quantum efficiency of 0.67 [45]. At exactly what stage during the bleaching process this isomerization occurs will be discussed later in the section of the so-called primary event. The visual pigments are then subsequently

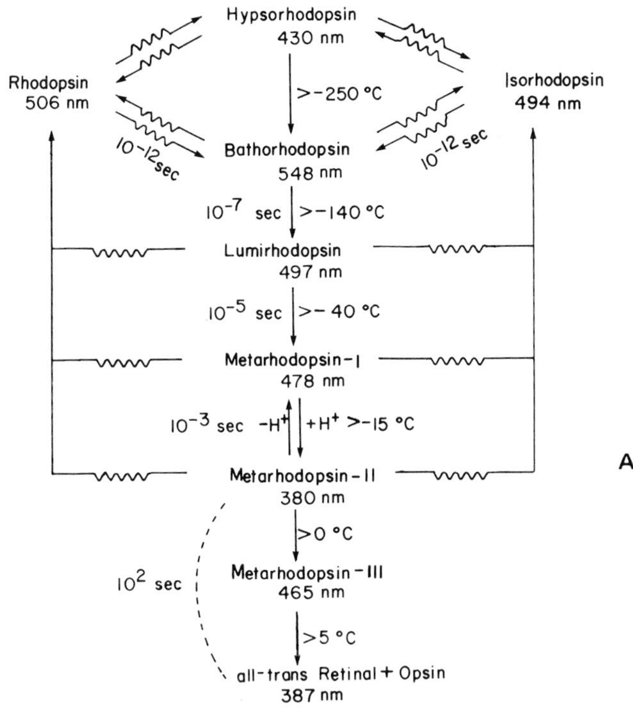

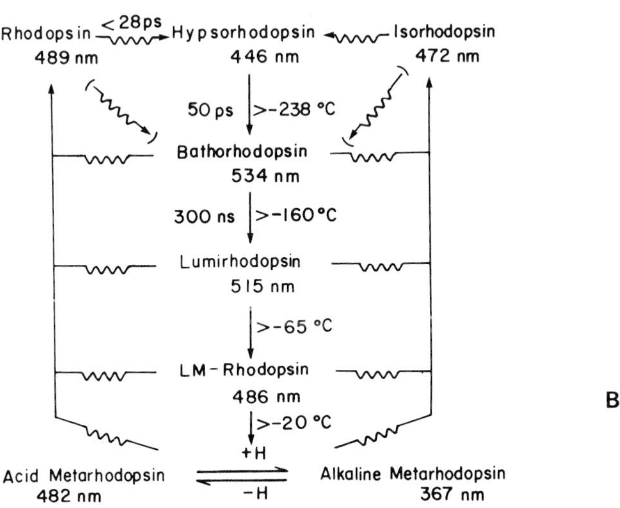

Fig. 5. (A) Bleaching intermediates in the photolysis of bovine rhodopsin extracts. Adapted from Yoshizawa and Horiuchi [61]. (B) Bleaching intermediates in the photolysis of squid rhodopsin extracts. Adapted from Shichida et al. [105]. Photoreactions are symbolized by wavy lines, thermal reactions by straight lines.

regenerated in vivo from opsin and 11-*cis*-retinal as part of the process called dark adaptation.

In spite of the differences in the behavior of the pigments, in situ, as compared to extracts, studies on extracts provide useful information, not only on the structure of the bleaching intermediates, but also on the possible roles they may be playing in the photoreceptors. The bleaching intermediates from several species have been investigated extensively by using flash photolysis techniques and both low temperature and ultrafast kinetic spectroscopy. As an example, Fig. 5 shows the sequence of the intermediates in the photolysis of bovine and squid rhodopsin extracts.

Neural excitation in the retina can be detected within a few milliseconds. Therefore, the decay of the meta-II intermediate is too slow to be involved in neural transduction. It is likely that this and subsequent steps in the bleaching process are part of the regeneration process in the receptors. The meta-I to meta-II stage could correspond to the stage when visual transduction is initiated. However, the kinetics of decay of the intermediates were found to be different in detergent extracts, in ROS and in the retina [46–48]. Further, not all the intermediates found in the extracts could be detected in the retina [49]. In extracts it is believed that in the meta-I to meta-II transition, a major conformation change occurs concurrently with the protonation. Thus, it was argued [50–53] that the entropy change is far larger than expected for a simple protonation reaction. Further in the meta-II intermediate the aldimine bond becomes reducible with sodium borohydride [54], indicating some unfolding of the opsin occurred.

The intermediates of bleaching can photochemically, but not thermally, be reconverted into rhodopsin with the exception of chicken batho-iodopsin, which is the sole intermediate which can also thermally (above −180°C) revert to chicken iodopsin [55]. It is assumed that in this case, because of the particular structural features of chicken opsin, the twisted all-*trans* form of the batho product is strained towards the 11-*cis* form, and raising the temperature can relieve this strain by conversion to the 11-*cis* chromophore.

Doubts have been raised concerning the homogeneity of the intermediates of bleaching shown in Fig. 5A. On the basis of kinetic data [56] (the meta-I to meta-II conversion can be expressed by two exponentials), it was proposed that both meta-I and meta-II are composed of two molecular species differing for example in their capability to regenerate rhodopsin. This led to the suggestion that multi-forms of rhodopsin may exist as well [57]. However, while two spectroscopically distinct forms (λ_{max} = 538 nm and λ_{max} = 555 nm) of the batho intermediate, attributed to a different twist of the chromophore, were detected recently [58,59], no multi-forms in the rhodopsin were observed. In detergent extracts the multi-forms of the thermal intermediates such as encountered for meta-I and meta-II were attributed to different amounts of associated phospholipids [60]. The exact nature of all bleaching intermediates is still but a matter of speculation.

The intermediates of bleaching also display characteristic CD spectra. The α band in the CD spectra of bathorhodopsins investigated up till now is opposite in sign (negative) of rhodopsins [61,62]. This reversal of sign was interpreted as an indication of a geometry change of the chromophore in the binding site, i.e., from 11-*cis* to a twisted all-*trans* configuration. This, however, is difficult to reconcile with the observation that intermediates containing a trans or transoid chromophore (e.g., meta-I, meta-II, etc.) also display positive α peaks. The same intermediate in the bleaching sequence can have α peak with opposite sign depending on species, e.g., bovine lumirhodopsin has a positive α band but squid lumirhodopsin has a small negative α band and an intense β band. Bovine meta-III has no CD while in frog rods a positive CD is observed for this intermediate [63].

In order to understand the circular dichroism spectra of the bleaching intermediates and pigments, first a rational explanation of the origin of the optical activity in pigments would be necessary (retinals and opsins lack the CD which is displayed by the pigments in the visible region and which disappears upon bleaching). Two mechanisms have been forwarded to account for the optical activity for the CD Cotton effects of visual pigments:

(1) Optical activity is induced by the protein, since upon binding of the retinal to opsin, the retinylidene chromophore could become inherently chiral [64–67].

(2) The electronic transitions of the bound chromophore are coupled to the transitions of the protein, e.g., aromatic amino acid side chains and/or peptide bonds by the coupled oscillator mechanism [68–70].

The 11-*cis*- and 9-*cis*-retinal chromophores cannot be planar around the 6,7 single bond due to steric interactions; in addition, 11-*cis*-retinal is twisted around the 12,13 single bond by steric hindrance. The fact that the magnitudes and signs of Cotton effects of pigments derived from 5,6-dihydroretinal [70] (conjugation between the ring and side chain is absent) and 5,6-epoxy-3-hydrorhodopsin [71] (ring and side chain conjugation is weak) are about the same as that of rhodopsin shows that a twist around the 6,7 bond cannot account for the CD. The CD of isorhodopsin, the chromophore of which is 9-*cis*-retinal, is similar to that of rhodopsin (which has an 11-cis chromophore that cannot be planar around the 12,13 bond); it follows that a chirality around the 12,13 bond also cannot account for the CD in a straightforward manner.

The coupled oscillator mechanism was criticized mainly because of the lack of reciprocal relationship between the visible and the far or near UV bands upon bleaching, i.e., according to the Kuhn sum rule, the rotatory strengths of all positive and negative Cotton effects of an optically active molecule should add up to be zero. However, due to the high intensity of the far-ultraviolet CD bands (ca. 200 times more intense than the bands in the visible region), intensity changes upon bleaching, if any, could be too minute to detect by the presently available instrumentation. The intensity of the band at 280 nm does not change upon bleaching, at least when ROS suspensions are bleached instead of detergent extracts; therefore, the coupled oscillator model involving retinal and an aromatic amino acid residue was argued to be highly unlikely [43,62]. Moreover, the 280 nm band may well involve transitions to higher excited states of retinal [72] in addition to a contribution from the protein moiety. It is thus not clear at this moment whether a twisted chromophore, a coupled oscillator mechanism or a combination of both are responsible for the CD extrema, although it is probably the third mechanism which is the most likely.

Since the CD of oriented molecules are extremely sensitive to the interaction of chromophores, such data could be very useful in the investigation of exciton phenomena. The technique also allows one to determine the degree of order and orientation of chromophores in a macromolecule. Thus, CD studies on oriented visual pigments, and possibly on some selected visual pigment analogs, could be very informative in clarifying the role played by induced optical activity of the chromophore and/or exciton coupling.

(c) The binding of retinal to opsin

It is now generally accepted that 11-*cis*-retinal is covalently bound to a lysine residue on opsin [15,73–76], and it is thought that in the dark this linkage is protected from hydrolysis and other chemical attack by a specific conformation of the opsin. The binding is through an aldimine bond, which can be reduced by sodium borohydride [54] or sodium cyanoborohydride [77], thus fixing the chromophoric group in the binding site through a stable retinyl-opsin linkage. The retinal recovered after alkaline hydrolysis, which will not cleave the carbon–nitrogen bond but cleave the polypeptide chain, was found to be bound to a lysine residue ([78] and refs. therein).

11 12 $\overset{+}{N}$H–Lys-OPSIN $\xrightarrow{h\nu / NaBH_4}$ 11 12 NH–Lys-OPSIN

N-retinylopsin

There is still some controversy regarding the state of protonation of the aldimine linkage. Application of the resonance Raman technique, which allows the investigation of a colored species in a complex biological environment by taking advantage of the resonance enhanced Raman effect, provided evidence for a protonated Schiff base linkage in rhodopsin ([79] and refs. therein). It was argued, however, that the resonance Raman experiments may not be suitable to establish that rhodopsin is protonated in the ground state, i.e., increase of basicity of the Schiff base upon electronic excitation can be very important, and protonation may result from perturbations caused by electronic excitations inherent with the resonance Raman technique ([80] and refs. therein). Recent ^{13}C-NMR results [81] also suggested that the observed NMR shifts are incompatible with a protonated Schiff base linkage; however, according to the recently proposed external point-charge model [82], it has been shown that these same NMR results can be interpreted to support the protonated Schiff base linkage (see p. 328).

2. *In vitro regeneration of visual pigments*

The regeneration in vivo of visual pigments is a complex and yet not well understood part of the process of dark adaptation. In vitro regeneration using isolated retinas, receptor cell suspensions or visual pigment extracts, however, can give useful information regarding processes occurring in vivo.

Rhodopsin can be regenerated in vitro by adding 11-*cis*-retinal to bleached rod outer segment membranes (ROS) in suspension or in a suitable detergent. The requirements for efficient regeneration are the following:

(i) Whenever possible, freshly bleached preparations of ROS should be used because opsin deteriorates upon storage in suspension or in detergent, and this leads to a considerable decrease in the regeneration yield [83]. For short periods of time, not exceeding a week, opsin can be stored at 0°C under inert atmosphere. However, in case of regenerations with modified retinals, it is important to use freshly prepared opsin to establish the conditions of binding.

(ii) The amount of alcohol [84] used to solubilize retinals should not exceed 3% of the incubation volume; moreover, only a single geometrical isomer of retinal should be used because otherwise it is not clear which isomer yielded the pigment. It is

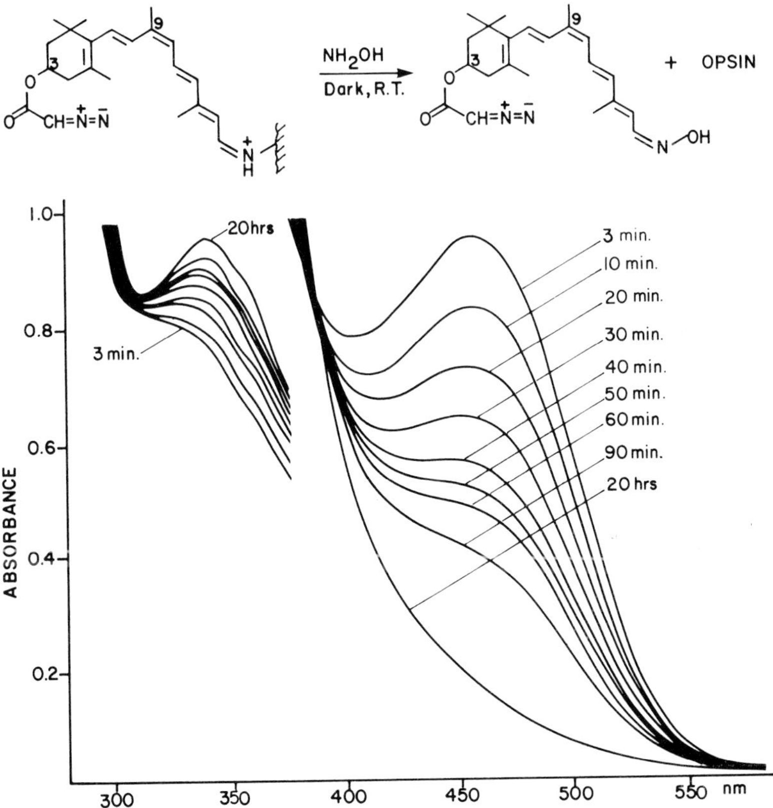

Fig. 6. Bleaching of 9-*cis*-3-diazoacetoxyrhodopsin (R. Sen, unpublished results) by hydroxylamine (0.05 M) in 2% digitonin/67 mM phosphate buffer, pH 7.0, in dark, at room temperature.

recommended to test the isomeric purity of the retinal used just prior to incubation with opsin. This is best carried out by high-performance liquid chromatography. Since retinal isomers and retinal analogs are prone to decomposition during storage even at low temperature, a purity test of retinals each time before a binding experiment is essential. When 11-*cis*-retinal is used to reconstitute rhodopsin, at least 100% excess has to be employed. In the reconstitution of visual pigment analogs, larger molar excesses may be necessary. The amount to be used has to be established by adding increasing amounts of the retinal to opsin until maximum pigment formation is observed. Further, contrary to rhodopsin, most of the visual pigment analogs are not stable in presence of hydroxylamine. Fig. 6 shows the bleaching of a rhodopsin analog, in the dark, at room temperature by 0.05 M NH_2OH.

(iii) When reconstitution is carried out in extracts, the nature of the detergent is crucial. Digitonin [84], sodium cholate at low concentration [85], or alkyl glucosides [86] are suitable detergents. Others, such as Emulphogene, CTAB, Ammonyx LO, Triton X-100 should not be used, as they inhibit regeneration. Even when pigments are formed in receptor cell suspensions, it is important to carry out solubilization of the pigment in a detergent to which the pigment is stable. For example, the formation of a visual pigment analog from adamantyl allenic retinal [87] takes place in bleached rod outer segment suspensions, and the absorption maximum of the pigment can be measured after solubilization in digitonin. However, if the pigment is solubilized in Triton X-100, thermal decomposition occurs at room temperature, as evidenced by the disappearance of the absorption maximum of the pigment within

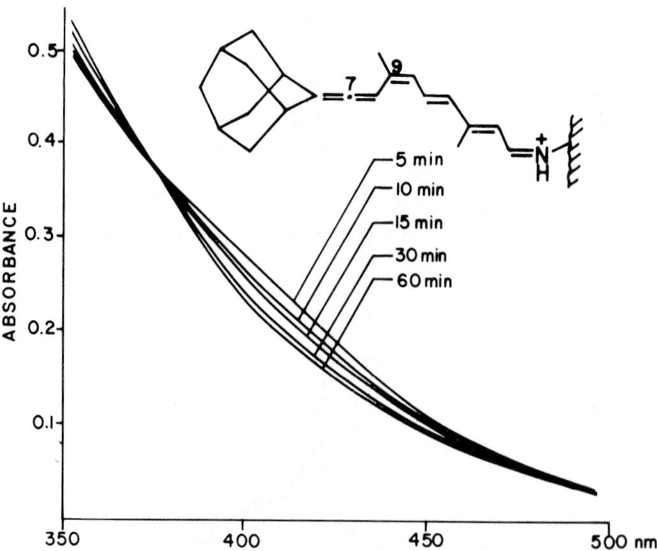

Fig. 7. Decomposition of adamantyl allenic rhodopsin [87] in Triton X-100/67 mM phosphate buffer, pH 7.0, in dark, at room temperature.

CH_2Cl_2 method

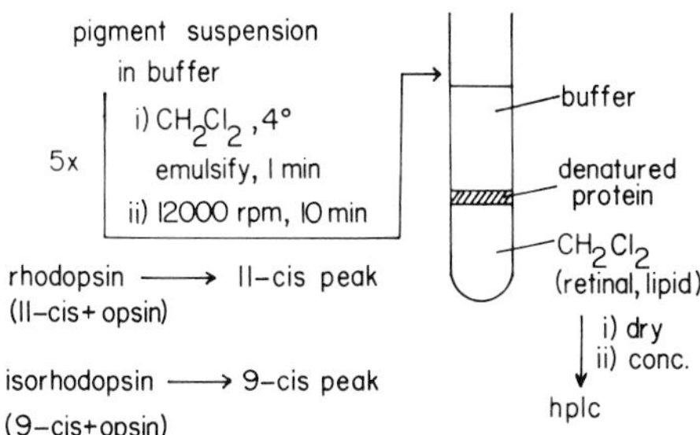

Fig. 8. A nonisomerizing procedure for the extraction of retinals from their binding sites in visual pigments [90].

an hour (Fig. 7). In Ammonyx LO, decomposition is immediate and no absorption maximum can be observed.

(iv) The lipid content of opsin is an important factor for regeneration. While it was shown that an 80% delipidated opsin preparation allows regeneration, more efficient delipidation by phospholipase A totally inhibits regeneration [88]. Partial removal of lipid by organic solvents such as toluene, hexane or petroleum ether allows regeneration. It was also shown that some if not all sulfhydryl groups of rhodopsin are involved in the regenerability of opsin [89].

In order to ascertain the nature of the bound chromophore, the pigments can be hexane-washed at low temperature to eliminate excess unbound retinal, and the bound retinal can then be extracted by shaking with methylene chloride procedure [90] (Fig. 8), which denatures the protein but does not isomerize the chromophore. Another procedure in which isomerization of the chromophore is unlikely to occur used ethanol denaturation at 0°C to identify 11-*cis*-retinal as the chromophore in rhodopsin [91].

Inhibition experiments and competitive binding studies can be used to show that a particular model retinal occupies the same binding site on opsin as does the chromophore in rhodopsin [92]. Namely, after pigment formation is complete and the excess unbound chromophore is washed out, another chromophore is added to the pigment preparation and the changes in absorption maxima are monitored. In many pigment analogs, addition of the natural chromophore to the preparation of the pigment analog does not result in the formation of rhodopsin, thus indicating that the same binding site is occupied by the retinal analog (Fig. 9). Conversely, natural pigment can first be reconstituted and then treated with modified retinal. In some cases, however, displacement of one chromophore by the other chromophore occurs. It is also possible to carry out competitive binding site studies by adding both natural and modified retinals to the same incubation, and then follow the competitive formation of pigments from the two retinals by monitoring their respective absorption maxima.

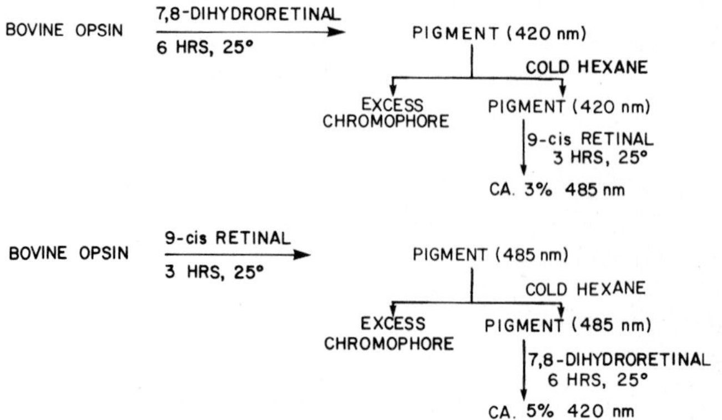

Fig. 9. Procedure for ascertaining that a retinal analog occupies the natural binding site [92].

3. The primary event

It is of crucial importance to understand the molecular basis of the very first occurrence which triggers vision, the so-called primary event. It is a photochemical event that distinguishes it from subsequent steps in the bleaching process, which are all thermal. Although the onset of neural signals, registered as transduction, are observed on a longer time scale than the primary event, this event begins the sequence of events leading to transduction.

The observation of an isomerized chromophore after bleaching of rhodopsin led to the general statement that vision is initiated by the photoisomerization of the retinal. On the basis of low temperature spectroscopic studies, which allow the observation of spectral intermediates that are extremely short-lived at room temperature, it was proposed that *cis* to *trans* isomerization of the chromophore occurs during the primary event, in which rhodopsin is transformed into a red-shifted species, bathorhodopsin. The development of mode-locked laser permitted the study of molecular dynamics in the time range of 8×10^{-9}–1×10^{-13} s; thus ultrafast kinetic spectroscopy could also be used to probe the nature of the fast processes involved in the primary event of vision. These studies raised the question of whether the primary event is indeed a *cis* to *trans* isomerization or whether other processes, such as proton translocation via tunneling, could also be implicated. The study of visual pigment analogs was also used to advantage in the investigation of the nature of the primary event.

(a) Low temperature studies of the primary event

In 1958, a new intermediate in the bleaching sequence of rhodopsin was discovered, the "liquid air illuminated rhodopin" [93]. This red-shifted transient ($\lambda_{max} = 543$ nm) could be observed when cattle rhodopsin was irradiated in aqueous-glycerol glasses at 77 K, and was later called prelumirhodopsin and finally bathorhodopsin. Several workers [61,94–100] confirmed the existence and investigated the properties of this species, and until the discovery of a still controversial new intermediate, hypsorhodopsin [61,99], the batho species was the undisputed first intermediate in the bleaching sequence.

Irradiation of rhodopsin with blue light (437 nm) at 77 K gave a photostationary mixture containing in addition to rhodopsin (33%) and isorhodopsin (16%) about 51% of batho intermediate. This process was found to be thermally irreversible but photochemically reversible. Indeed, irradiation of bathorhodopsin with red light (>650 nm) gave mainly rhodopsin and isorhodopsin. Irradiation of isorhodopsin, which contains a 9-*cis* chromophore, afforded the same batho product as the one obtained by irradiation of rhodopsin:

$$\underset{506\text{ nm}}{\text{Rhodopsin}} \underset{650\text{ nm}}{\overset{440\text{ nm}}{\leftrightarrows}} \underset{548\text{ nm}}{\text{Bathorhodopsin}} \underset{650\text{ nm}}{\overset{440\text{ nm}}{\leftrightarrows}} \underset{494\text{ nm}}{\text{Isorhodopsin}} \qquad \text{at 77 K}$$

The observation of this common batho product was used as the strongest argument in favor of the involvement of *cis* to *trans* isomerization in the primary event. The hypothesis that bathorhodopsin is a protonated, ground-state species was based on the observations that all bathorhodopsins studied [101–108] from various species (e.g., chicken, squid, frog) had red-shifted absorption maxima as compared to the parent rhodopsin, and all were stable indefinitely at liquid nitrogen temperature.

Multi-form intermediates in the bleaching sequence of rhodopsin were postulated as a result of kinetic studies carried out on later intermediates such as lumi, meta-I and meta-II at 37°C using flash photolysis techniques [56]. This led to the hypothesis that two conformeric states of rhodopsin itself exist which are in thermal equilibrium [57]. Further, three isomeric forms of rhodopsin, differing in their lipid content, were isolated by ECTEOLA cellulose chromatography [60], two conformational forms of rhodopsin in the disc membrane were implicated by the observation of two rotational relaxation times [109] and two isochromic forms of meta-I have also been postulated recently [110]. Owing to the inherent difficulties encountered in low temperature spectroscopy (such as the broadness of the absorption bands, problems of overlap and the changes of the extinction coefficients upon cooling), it was more difficult to establish whether the earlier intermediates of bleaching are homogeneous species. Careful study of the kinetics of conversion of rhodopsin and isorhodopsin to bathorhodopsin concluded that these pigments were single species. However, the kinetic analysis of photoconversion of bathorhodopsin to a mixture of rhodopsin and isorhodopsin (Fig. 10) gave two rate constants, indicating that a fast and a slow component are present in the bathorhodopsin which absorbed at 548 nm [58,59]. Their relative proportion was found to be detergent dependent. The fast component, bathorhodopsin$_1$ ($\lambda = 555$ nm) is predominant in ROS preparations, while bathorhodopsin$_2$ ($\lambda_{max} = 538$ nm) was ca. 75% of the mixture in experiments carried out in digitonin solutions. Yoshizawa et al. [58,59] suggested that these two spectrally distinct molecular species possibly differ from each other by the extent of the

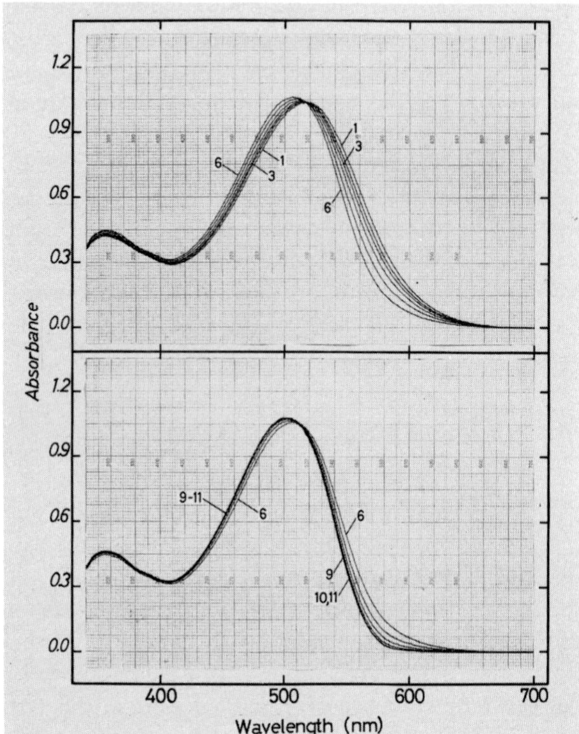

Fig. 10. Curve 1: A mixture of rhodopsin, isorhodopsin and bathorhodopsin produced by irradiation of rhodopsin with blue light (437 nm) at 77 K. Curves 2–6: Conversion of bathorhodopsin$_1$ to rhodopsin by irradiation with red light ($\lambda > 650$ nm) for 10, 10, 20, 40 and 80 s. Curves 6–11: Conversion of bathorhodopsin$_2$ to mainly rhodopsin by irradiation with red light ($\lambda > 650$ nm) for 80, 160, 320, 640, 1280 and 2560 s. From Sasaki et al. [59].

twist around the 11,12 bond as well as other bonds. Work in progress [59] indicates that each of these batho species decays to a corresponding lumi intermediate, then each lumi to the corresponding meta-I, etc. This corroborates the results of Williams and co-workers [56], who first proposed the existence of multi-form intermediates. A pigment analog, 7-*cis*-rhodopsin, despite its greatly blue-shifted absorption maximum (450 nm) as compared to rhodopsin, also gave the same batho product as the one obtained from rhodopsin [111], thus further supporting the assumption that a *cis* to *trans* isomerization can be involved in the rhodopsin to bathorhodopsin conversion. A set of two spectrally differing batho intermediates was found not only in cattle rhodopsin preparations but also in frog retina [58]. The existence of multi-form intermediates is therefore probably not a particular feature of cattle rhodopsin.

In an attempt to ascertain that bathorhodopsin is the earliest intermediate, the bleaching of cattle rhodopsin was investigated at liquid helium temperature (4 K). Irradiation at wavelengths longer than 530 nm gave a photostationary state containing in addition to rhodopsin and isorhodopsin, a species with a much blue-shifted absorption maximum (430 nm). This transient was called hypsorhodopsin [61,99]:

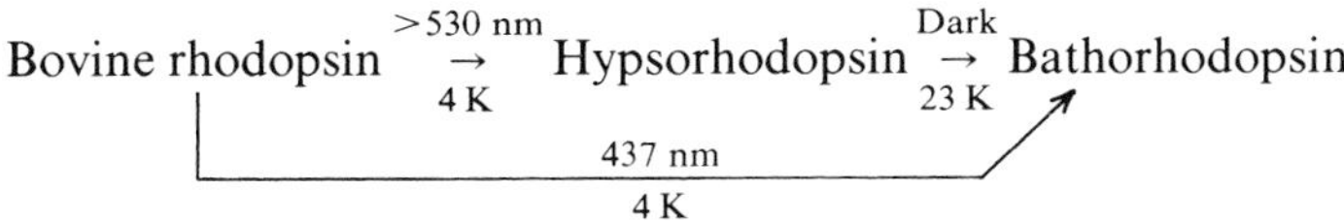

Whether hypsorhodopsin precedes bathorhodopsin in the bleaching sequence is still an open question. While hypsorhodopsin could be obtained by irradiating cattle rhodopsin with orange light (530 nm, at 4 K) irradiation with blue light (437 nm, at 4 K) gave a photosteady mixture containing mainly bathorhodopsin [105,106]. Thus it was not clear whether at 4 K bathorhodopsin is produced from rhodopsin solely through hypsorhodopsin or directly from rhodopsin.

Low temperature experiments have shown the formation of hypso intermediates from several species [99,103,105–107]. The study of early photoconversion processes in squid [108], which also involved the evaluation of the relative quantum yields among the four pigments (squid rhodopsin, squid batho-, hypso- and isorhodopsin) showed that hypsorhodopsin is a common intermediate of rhodopsin and isorhodopsin; there is no direct conversion between rhodopsin and isorhodopsin; bathorhodopsin is not converted directly to hypsorhodopsin; and both rhodopsin and isorhodopsin convert more efficiently to bathorhodopsin than to hypsorhodopsin. While a temperature dependence of the relaxation processes from the excited state of rhodopsin, and an assumption that batho could be formed from one of the high vibrational levels of the ground state hypso have been invoked to explain these findings [108], the final clarification of this matter awaits results from subpicosecond laser photolysis experiments at liquid helium temperature.

It was assumed that hypsorhodopsin contains a deprotonated form of the Schiff base because of its blue-shifted absorption maximum. As for the geometry of the chromophore in the hypso intermediate, it was speculated that it may be *trans*-like as in bathorhodopsin [112]. However, until conclusive evidence (such as could be secured by resonance Raman experiments) becomes available, it is unclear whether hypso is protonated or not and whether isomerization of the retinal moiety is involved at the hypso stage.

(b) Ultrafast kinetic spectroscopy for the study of the bleaching intermediates at room temperature

The room temperature kinetics of formation and decay of the early bleaching intermediates was investigated by pulsed laser excitation methods with nanosecond [113–116] and picosecond resolution. In these experiments, the sample is first excited by an intense, picosecond pulse and subsequently is probed by a weaker pulse; this allows study of the events occurring as a result of the excitation, including the relaxation of the excited state.

The first picosecond experiments investigating bathorhodopsin were carried out by Bush et al. [117] who found that this species formed within 6×10^{-12} s when rhodopsin was excited by a 530 nm pulse, and decayed with a lifetime of 3×10^{-8} s at room temperature. Therefore, the batho intermediate was acknowledged as a physiological intermediate. Its presence was also confirmed in the bleaching of the frog retina close to room temperature [113]. Subsequently several groups carried out picosecond measurements on visual pigments under various experimental conditions [118–126] and confirmed that bathorhodopsin is formed within a few picoseconds at room temperature and within 36 ps at liquid helium temperature.

The extremely fast formation of the batho intermediate raised the still ongoing controversy concerning the nature of the primary event. Several investigators argued that the formation of bathorhodopsin is much too fast to be associated with simple *cis–trans* isomerization as was proposed originally [94]. Thus, alternative mechanisms were put forward to explain the primary event, such as proton translocation, [80,117–119,125–127,228] concerted multibond isomerization [128] or light-induced charge reorganization to give a stabilized protein chromophore complex [129].

The nature of the primary event is further complicated by the unsolved question as to whether bathorhodopsin or the hypso intermediate should be considered to be the primary photoproduct. Picosecond studies carried out on squid rhodopsin (where the formation of batho product is an order of magnitude slower than in the case of cattle rhodopsin) showed hypsorhodopsin to be formed earlier than bathorhodopsin [120,121], since the decay time constant of hypsorhodopsin matched the formation time constant of bathorhodopsin. However, no hypsorhodopsin formation could be detected by picosecond studies on cattle rhodopsin [118,119]. This divergence of results was partly explained by the different nature of the opsins involved, the difference in the excitation wavelengths used (347 nm pulse for squid versus 530 nm for cattle), and the nature of detergents used to extract the rhodopsins (digitonin for squid versus LDAO for cattle). A reinvestigation of the picosecond data, in which a 6 ps excitation pulse at 530 nm was used and cattle rhodopsin was solubilized both in LDAO and octylglucoside as detergents, was published by Kobayashi [124]. This study concluded that hypsorhodopsin is the first photoproduct in the case of cattle rhodopsin as well. However, other workers maintain that bovine hypsorhodopsin may be formed concurrently, in a parellel pathway with bathorhodopsin [123]. Due to the contradictory interpretations by various groups, the role of hypsorhodopsin as a primary intermediate in the bleaching sequence remains uncertain.

(c) Resonance Raman studies of the primary event

In the resonance Raman experiment, selectivity and sensitivity are achieved by the use of tunable lasers which allow the selection of frequencies in resonance with the electronic absorption of the retinal moiety. This permitted the observation of greatly enhanced scattering of the vibrational spectrum of this chromophore above the background of vibrations from the opsin matrix [130]. This technique has been found useful in the studies of a large number of biologically important molecules [131,132,229].

The first conformation-specific information on bathorhodopsin was obtained by low temperature resonance Raman measurements [133]. The spectrum of cattle bathorhodopsin indicated a species in which the Schiff base is protonated, since the C=N vibration frequency in bathorhodopsin was found to be at 1655 cm^{-1}, a value close to the one found for rhodopsin and isorhodopsin, and it shifted to ca. 1630 cm^{-1} upon deuteration. Further, the spectrum indicated that the retinal in bathorhodopsin assumes a highly distorted, *trans* configuration. This conclusion was drawn from the analysis of the lines in the fingerprint region at 1100–1400 cm^{-1} and at 856, 877 and 920 cm^{-1}, which is different from the lines seen in retinal isomers with either 11-*cis*, 9-*cis* or all-*trans* geometry. In these early experiments the problems of photolability of the samples was overcome by freezing to liquid nitrogen temperature, where a photostationary mixture of rhodopsin, isorhodopsin and bathorhodopsin could be maintained and its composition analyzed in the same manner as in low temperature UV/VIS absorption spectroscopy. Raman bands could be assigned to a particular species by simultaneous application of a second pump laser beam, which varied the composition of the mixture while spectra were obtained from the first probe beam. Another technique to eliminate the photolability problem was the rapid flow technique in which the sample travels rapidly through the laser beam, to lower the probability of probing bleached molecules [134,135]. These techniques also permitted performing measurements at room temperature. Resonance Raman spectral data for retinals, Schiff bases and pigments was reviewed through 1977 [79].

More recent work [136,137] confirmed the fully protonated state of the Schiff base linkage and reported additional features in the spectra of both squid and bovine rhodopsin. This led to the proposal of a distorted 11-*cis* or 9-*cis* structure in bathorhodopsin and the conclusion that the primary event does not involve a *cis* to all-*trans* isomerization. The batho intermediate, however, could subsequently relax to an all-*trans* configuration by rotation around the C_9–C_{10} and C_{11}–C_{12} bonds.

Other workers [138,139] are still concerned with the state of protonation of the Schiff base linkage as established by resonance Raman experiments. These authors found that picrates of conjugated Schiff bases have C=N stretching vibration frequencies similar to those of the respective Schiff bases. Therefore, they proposed that bathorhodopsin is the truly protonated Schiff base, not rhodopsin. In rhodopsin the Schiff base would not be fully protonated but would rather be engaged in a hydrogen bond with a proton donor (with the proton remaining closer to the donor). The protonated form of rhodopsin observed in the resonance Raman experiments would then be the result of the absorption of a photon, i.e., increase in the basicity of the excited Schiff base would trigger the transfer of proton onto the nitrogen atom of the Schiff base.

The resonance Raman spectrum of bathorhodopsin has recently been analyzed in detail by using a two laser beam 'pump-probe' technique [140–142]. On the basis of new experimental data as well as theoretical calculations [142] of the C=N stretching and C=N—H bend frequencies and the large shift seen upon deuteration (25 cm^{-1}), it was concluded that the proton in both rhodopsin and bathorhodopsin must be covalently bound to the Schiff base nitrogen. Further, the analysis of the data, while it does not indicate an all-*trans* chromophore in bathorhodopsin, does suggest that some form of geometrical isomerization has occurred in going from rhodopsin to bathorhodopsin. Thus, a structure in which the retinal assumes a twisted transoid configuration seems to be generally accepted. There is currently no published resonance Raman data which would permit to speculate on the geometry of the chromophore in hypsorhodopsin.

(d) Visual pigment analogs and the involvement of the cis–trans *isomerization in the primary event*

(i) Deuterated retinals

Resonance Raman spectra of pigment analogs and batho intermediates containing 5- and 9-deuteriomethyl and 5- and 9-desmethylretinal were measured [143]. These studies indicated that the configuration of the bathorhodopsin chromophore is *trans* and that "the intense low wavenumber lines of bathorhodopsin would originate from the latent hydrogen out-of-plane (HOOP) modes of the all-*trans* protonated Schiff base between 850 and 950 cm^{-1} which are resonantly enhanced by twists about chain single bonds."

In a continuation of the above elegant work [144], the lines near 854, 875 and 922 cm^{-1}, exclusively belonging to bathorhodopsin, were unambiguously assigned by the resonance Raman studies carried out on 10-deuterio- and 11,12-dideuterio-rhodopsins (1 and 2). Thus the 875 cm^{-1} line in bathorhodopsin was assigned to

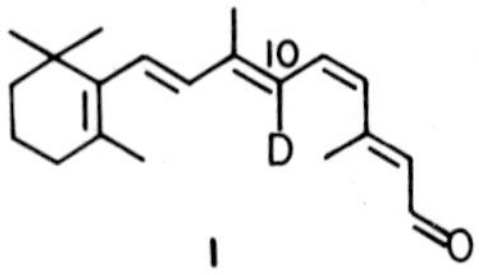

1 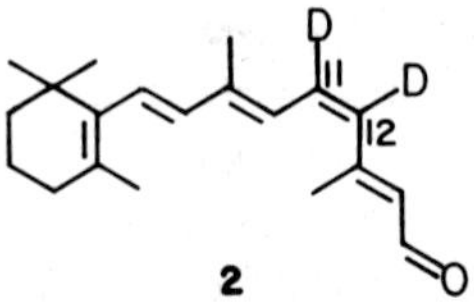 2

$C_{10}H$ HOOP wag and the 922 cm^{-1} line to an HC_{11}–$C_{12}H$ HOOP mode. The line at 854 cm^{-1} was attributed to the $C_{14}H$ HOOP. As the intensities of the lines depend on the resonant excited state geometry, the enhancement of the 922 cm^{-1} line in bathorhodopsin indicates that the twist around the 11,12 bond is different in the excited state from that of the ground state, i.e., a situation expected to occur in *cis–trans* isomerization. Thus the resonance Raman data showed that the chromophore configuration in bathorhodopsin is a perturbed all-*trans*.

(ii) Visual pigment analogs versus proton translocation in the primary event

The main arguments in favor of proton translocation instead of *cis–trans* isomerization in the primary event in vision were based, in addition to the extremely rapid formation of the batho product, on the fact that its rate of formation at low temperatures showed non-Arrhenius behavior, a characteristic of quantum mechanical tunneling, such as translocation of hydrogen [118]. A further support for this proposal was the large isotope effect on the rate of formation of bathorhodopsin in deuterium-exchanged samples.

If proton translocation were involved, it would be important to know where the proton is removed. The 9- and 13-methyl groups of retinal cannot be involved, as 9- and 13-desmethylrhodopsins did form batho intermediates [145]. Synthetic retinals 3 [145], 4 [146,147] and 5 [148] missing the crucial hydrogens assumed to intervene in the proton translocation hypothesis, were prepared and were found to form rhodopsin analogs. Rhodopsin analogs from 3, 4 and 5 all formed red-shifted, batho intermediates at low temperatures [145–148]. Therefore translocation of proton from the ionone ring can also not be implicated in the primary event.

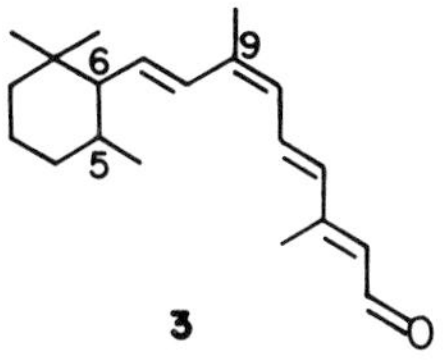

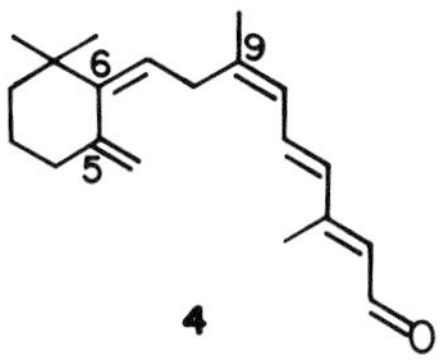

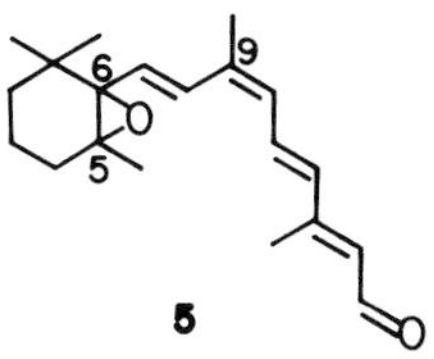

(iii) Non-bleachable rhodopsins retaining the full natural chromophore

Retinals 6, 7, 8 and 9, all containing an 11-*cis*-locked chromophore and mimicking a nonplanar 12-*s-trans*-retinal, were synthesized and were bound to bovine opsin in order to investigate the photochemical properties of the rhodopsin analogs formed [149]. As was expected, since the 11-ene is *cis*-locked, the absorption spectra or the orange color of these pigments did not change when the pigments were exposed to light of wavelength corresponding to their absorption maxima.

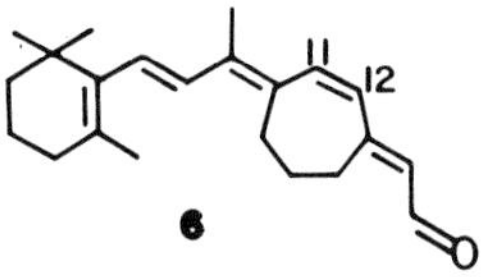

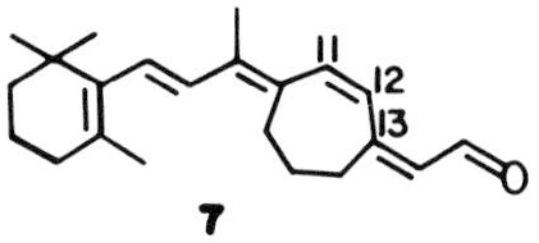

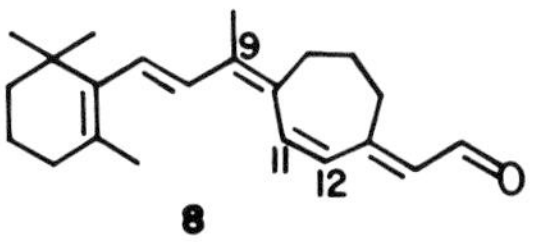

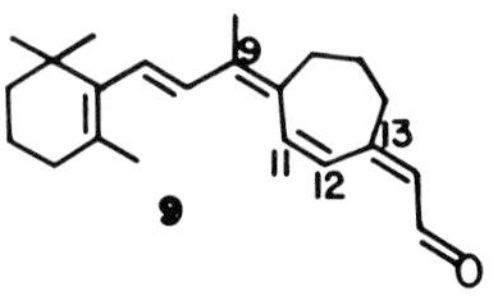

The rhodopsin analog formed from the 11-*cis* isomer (6) was also investigated by flash photolysis technique at low temperatures [150]. No change was observed in the absorption spectrum of the pigment at 77 K when it was irradiated at 460 nm and 540 nm, thus indicating that no formation of bathorhodopsin occurred (Fig. 11). The pigment was also flash-photolysed at 20.8°C using a 0.5 μs 500 nm flash, and the absorption changes were monitored at 380, 480 and 580 nm for the time range of 2 μs to 10 s. In contrast to natural rhodopsin, which in the above photolysis conditions showed the formation of meta-I and its decay to meta-II, no light-induced

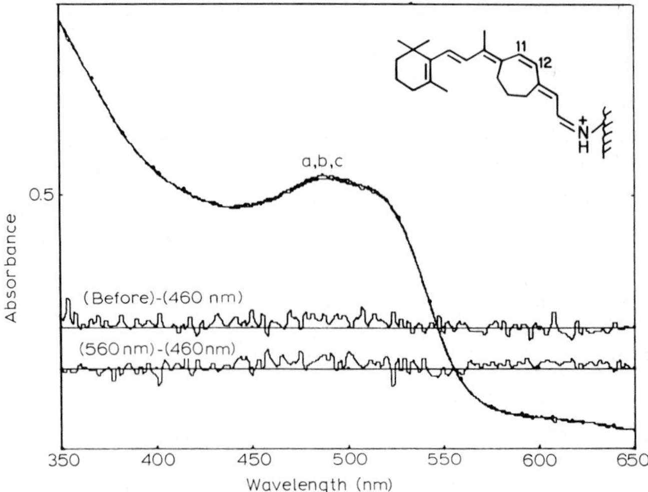

Fig. 11. Absorption spectra of rhodopsin analog 6 [150]. (a) At 77 K before irradiations. (b) After irradiation for 10 min with 460 nm light. (c) After a second irradiation of 10 min with 520 nm light. All of these spectra are superimposable. The two traces in the lower part of the figure are, respectively, the difference spectrum (expanded 10-fold) between curves (a) and (b) (top), and between curves (b) and (c) (bottom).

absorption changes were detected in case of rhodopsin analog from 6. The fact that the artificial pigment does not undergo any light-induced changes establishes that an 11-*cis* to *trans* photoisomerization of the retinal moiety is the essential step in initiating the chain of events in the photolysis of visual pigments.

4. *Conformation of the chromophore*

In all geometric isomers of retinal, steric hindrance precludes the existence of a completely planar π electron system. Interactions which give rise to steric hindrance are relieved by rotation(s) around single bond(s) along the polyene chain in such a way as to preserve the maximum amount of conjugation. This gives rise to conformational isomers, denoted as *s-cis* and *s-trans*. The dihedral angle in *s-cis* conformers can range from 0 to 90° and, in *s-trans* from 90 to 180°.

In all the geometrical isomers of retinal there is a twist around the ring-chain linkage, the 6,7 single bond. The twisted *s-cis* conformation around this bond is the preferred structural feature in retinals. The polyene chain can be close to planar as in the all-*trans* isomer and in some of the less hindered *cis* isomers, such as 9-*cis*-retinal. However, in 11-*cis*-retinal, the natural chromophore, there is, in addition to the ring-chain twist, a twist around the 12,13 single bond. This rotation relieves the steric crowding due to the interaction of 13-methyl group with the hydrogen atom at position 10. Rotation around the 10,11 bond, which could also relieve this crowding, is thought to be less important since this rotation would result in a lesser degree of conjugation than rotation around the 12,13 bond.

Retinals assume a different conformation when in crystalline state as compared to solution. Thus, X-ray studies [151–153] of 11-*cis* retinal crystals have shown that there is a ca. 40° twist around the 6,7 bond and that the polyene chain assumes a 12-*s-cis* conformation with a ca. 39° twist as shown in 10. Theoretical calculations [154], however, suggested a different situation in solution, namely the coexistence of both 12-*s-cis* (10) and 12-*s-trans* (11) conformers, which could be in equilibrium. This idea was further supported by the investigation of the absorption spectrum of retinal isomers, and the corresponding Schiff bases at both room and low temperature [155,156], and NMR studies of 11-*cis* retinal in solution [157,158]. Resonance Raman experiments proposed that the splitting of the $C–CH_3$ stretching vibration (998 and 1018 cm^{-1} in solution but only one line at 1017 cm^{-1} in the crystalline state) was due to the presence of both 12-*s-cis* and 12-*s-trans* conformers in solution [134,135,159], i.e., in the 12-*s-cis* conformer both the 9- and 13-methyl group would have stretching motions at 1018 cm^{-1}, while in the 12-*s-trans* conformer the 9-methyl would scatter at 1018 and the 13-methyl at 998 cm^{-1}. Lewis et al. [160] proposed a different interpretation of the resonance Raman data, i.e., that this splitting was not due to the coexistence of 12-*s-cis* and 12-*s-trans* conformers. Based on the study of 9-, 13-, and 9,13-butyl-substituted retinals [161], the line at 996 cm^{-1} was assigned to $C_{13}–CH_3$ and the line at 1017 cm^{-1} to the $C_9–CH_3$ stretching vibrations.

In another study, aimed at clarifying the role of the 12,13 twist in 11-*cis*-retinal, the absorption properties of 14-methyl- and 13-desmethyl 14-methylretinal analogs were investigated. This proposed a twisted 12-*s-cis* conformation ($C_{12}–C_{13}$ angle set

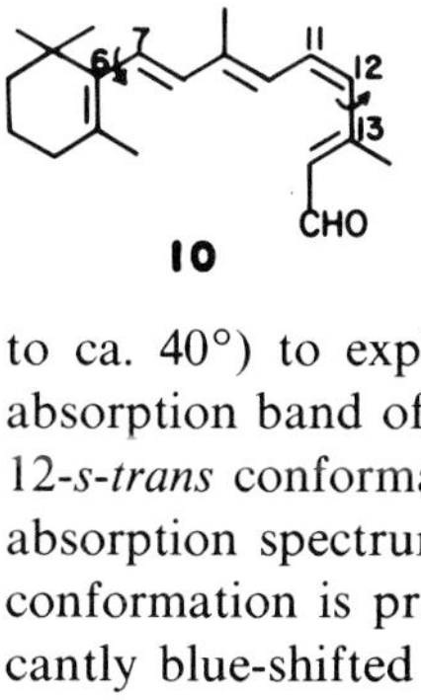

10 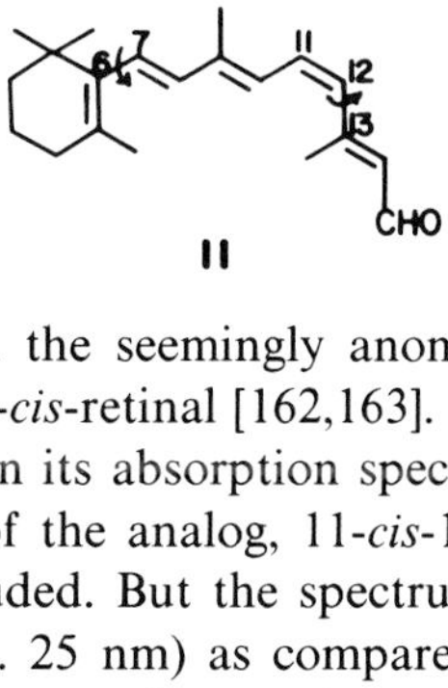 11 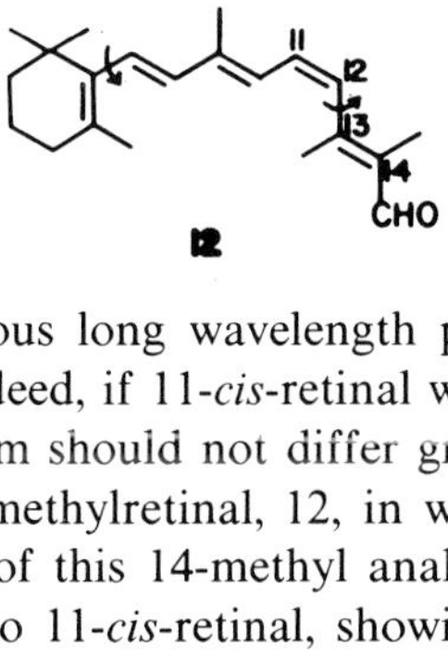 12

to ca. 40°) to explain the seemingly anomalous long wavelength position of the absorption band of 11-*cis*-retinal [162,163]. Indeed, if 11-*cis*-retinal would assume a 12-*s-trans* conformation its absorption spectrum should not differ greatly from the absorption spectrum of the analog, 11-*cis*-14-methylretinal, 12, in which a 12-*s-cis* conformation is precluded. But the spectrum of this 14-methyl analog was significantly blue-shifted (ca. 25 nm) as compared to 11-*cis*-retinal, showing that 11-*cis*-retinal in solution is in another conformation; namely, it is the 12-*s-cis* form 10 which predominates at room temperature.

Studies of retinal isomers and retinal analogs in solution were useful in that they permitted to speculate on the conformation of the opsin-bound 11-*cis*-retinal, an important problem in the stereochemistry of vision. Although there was a priori no reason to believe that the conformation of 11-*cis*-retinal when bound to opsin will be identical either to its solution conformation or to its conformation in the crystalline

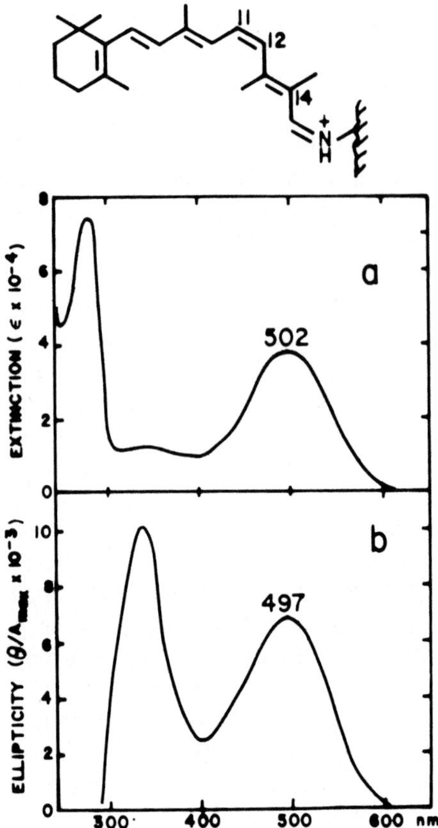

Fig. 12. Absorption (a) and circular dichroism (b) spectra of 14-methylrhodopsin in 2% Ammonyx LO, at room temperature [162].

state, resonance Raman experiments pointed out the close similarities between the solution spectrum of 11-*cis*-retinal and rhodopsin [134,135,164]. Studies of visual pigment analogs also helped to gain information on the conformation of retinal in the binding site. Thus, it was found that the pigment analog formed from 11-*cis*-14-methylretinal [162,163] had very similar absorption and CD spectrum (Fig. 12) as well as photosensitivity to that of rhodopsin. Since in this analog the 12,13 bond has to be in a 12-*s-trans* conformation, the data implies that the binding site is capable of accepting such a conformer, and in rhodopsin itself, 11-*cis*-retinal may be in a 12-*s-trans* form.

Recently, a retinal analog with a blocked 11-*cis* geometry, 13, has been synthesized and formed a visual pigment analog with bovine opsin [149] and also with octopus opsin. (K. Nakanishi, unpublished results). In this retinal analog, the seven-membered ring forces a nonplanar 9,11,13-triene conformation, and again this retinal has to assume a distorted 12-*s-trans* conformation in the binding site.

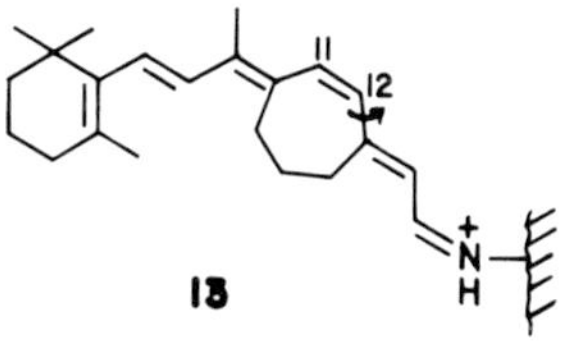

13

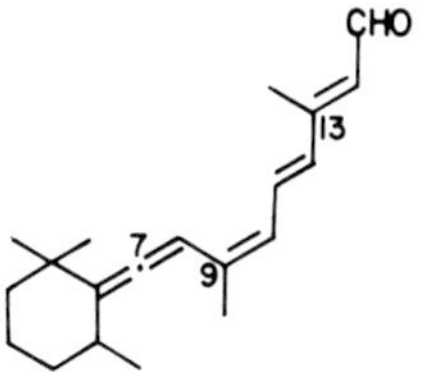

14

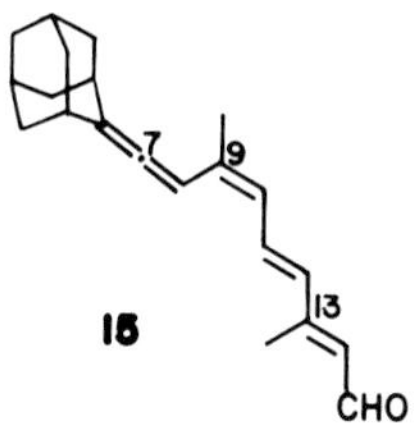

15

Modified retinals also helped to gain information on the requirements of the binding site as far as the 6,7-bond is concerned. Thus, allenic retinals 14 [165] and 15 [87] were synthesized and were combined with bovine opsin to give visual pigment analogs. In these retinals the ring-chain angle is fixed by the allenic bond, i.e., the rings are fixed at right angles with the polyene chain. The fact that these modified retinals are accommodated by the binding site indicated that the ring-chain angle in retinal is not a critical factor in the steric requirements for pigment formation.

5. Visual pigment analogs

The study of visual pigment analogs serves to clarify the nature of chromophore–opsin interactions in visual pigments. Visual pigment analogs are modified retinals bound to opsins, which can be investigated by spectroscopic and biochemical methods.

The structure of modified retinals should be tailored in such a way as to lead to some specific information about the binding site, when they form visual pigment analogs. Information from these studies can lead to the clarification of steric and electronic interactions, which give rise to the very specific properties of visual pigments. This type of study can shed light on the nature of molecular events which take place upon the bleaching of the pigments and lead to transduction, and can reveal the nature of amino acid residues present in, or close to the binding site of retinal.

*(a) Visual pigment analogs from retinal isomers other than 11-*cis*-retinal*

The original concept that the binding site of opsins is highly restricted allowing the binding of only the 11-*cis* isomer [166] has now become obsolete. In addition to the

11-*cis* isomer, eight other isomers were synthesized and were bound to cattle opsin to form visual pigment analogs. These are the following: 9-*cis* [166–168], 9,11-di*cis* [169], 9,13-di*cis* [170], 7-*cis* [171], 7,9-di*cis* [172], 7,13-di*cis* [171], 7,9,13-tri*cis* [171] and 7,11-di*cis* [172]. The binding of 11,13-di*cis* isomer was problematic [166–168]; however, we can now expect it to form a pigment analog since it was shown that a retinal analog in which the 11,12 bond is fixed in a seven-membered ring formed pigments from isomers corresponding to both the 11,13- and the 9,11-di*cis* isomers [149]. The 13-*cis* and all-*trans* isomers are the only ones known with certainty not to combine with opsins.

On the basis of these binding studies a new concept was proposed in which the existence of longitudinal restrictions to binding were held responsible for the observed geometric specificity of opsins [173]. Further refinement of this idea [174], by taking into account more realistic interatomic distances (between the center of the ionone ring and carbon 15), and molecular shapes (in which rotations around crucial single bonds were included in order to obtain low energy conformations) led to an estimated distance of 9.6–10.9 Å between the center of the ring and carbon 15, for retinals which form visual pigment analogs. Retinals in which this distance is longer, such as the 13-*cis* and all-*trans* isomers, do not form pigments. However, some problems remain in predicting binding properties in this manner. According to the above established critical distance, C_{17} aldehyde is expected to bind (distance is 10.0 Å) and the 7,9-di*cis* isomer (distance is 11.1 Å, close to that of 13-*cis* isomer) should not form a pigment analog. The opposite is observed experimentally. Therefore, it is important to appreciate that the interatomic distances estimated from crystal structure data, and the rotation angles from solution studies, may not be identical to the values in the binding site, thus making predictions difficult.

(b) Isotopically labeled retinal derivatives

Elucidation of the structure of a molecule by vibrational spectroscopy is facilitated by observation of frequency shifts and intensity changes between the spectra of the original compound and its derivatives. A study of the resonance Raman spectra of pigments in which there was deuterium substitution in the retinal made crucial contributions in assigning retinal absorptions in the resonance Raman spectra of rhodopsin and its bleaching intermediates. Pigment analogs were prepared from [5-C^2H_3]retinal, [10-^2H]retinal and [11,12-$^2H_2$]retinals for this purpose [143,144]. In addition, resonance Raman study of deuterated pigments and comparison with the corresponding non-deuterated pigments helped to assign a perturbed all-*trans* structure to the batho intermediate formed in the primary event.

The 11,12-dihydro[15-^3H]retinal in which the crucial 11-ene is saturated, could be

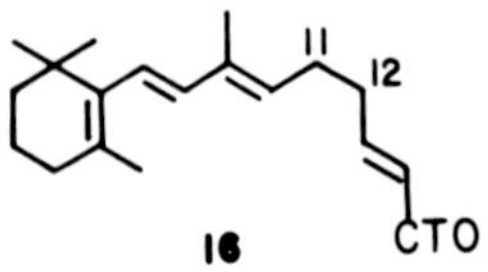

11 12

16 CTO

incorporated in vivo into Vitamin A deficient rats [175]. With the help of this labeled retinal it was shown that this dihydro analog occupies the same binding site as natural 11-*cis*-retinal, and the pigment analog formed cannot restore visual sensitivity to Vitamin A deficient animals since it blocks rhodopsin regeneration. This was attributed to the structure of this retinal analog, 16, in which the absence of *cis–trans* isomerization around the 11,12 bond inhibited its bleaching and therefore its detachment from opsin.

(c) Alkylated and dealkylated retinals

Methyl groups at positions 5, 9 and 13 on retinal are not a necessity for pigment formation. This was shown by synthesis and formation of visual pigment analogs from 5-desmethyl, 17 [145], 9-desmethyl, 18 [70,176–180], 13-desmethyl, 19 [176–180] and 9,13-desmethylretinal, 20 [70,176,177].

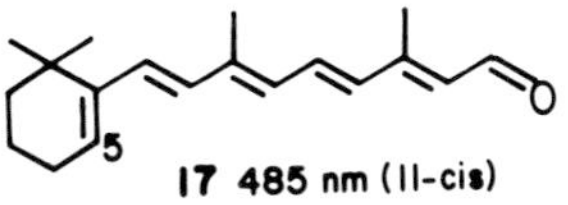

17 485 nm (11-cis)

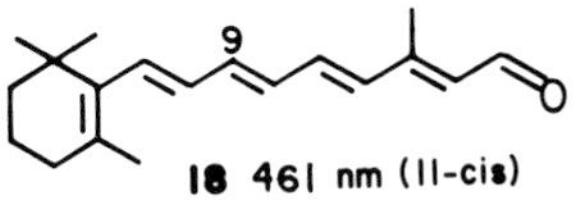

18 461 nm (11-cis)

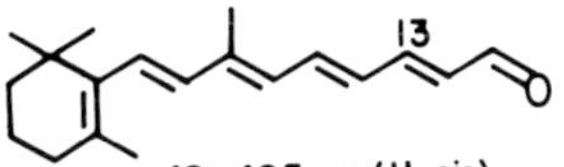

19 495 nm (11-cis)

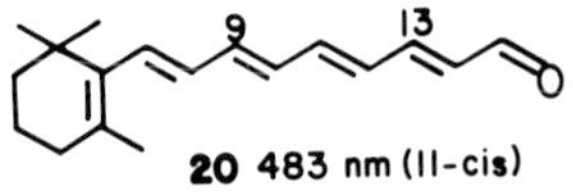

20 483 nm (11-cis)

Changing the position of the methyl group on the side chain, such as in 13-desmethyl-14-methylretinal, 21 [162], or the increase of the bulk of the side chain by additional methyl substitution such as in 10-methyl, 22 (K. Nakanishi, unpublished results), 14-methyl, 23 [162] and 10,14-dimethylretinal, 24 (Nakanishi, unpubl.;

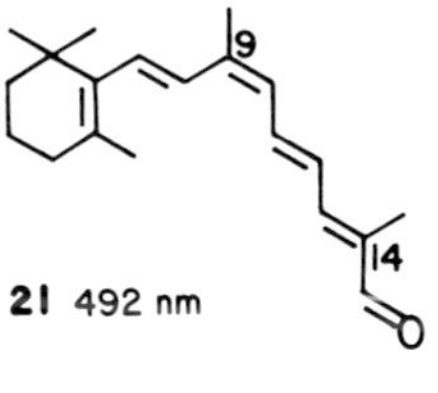

21 492 nm

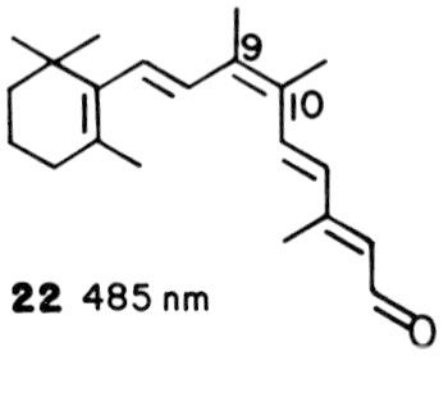

22 485 nm

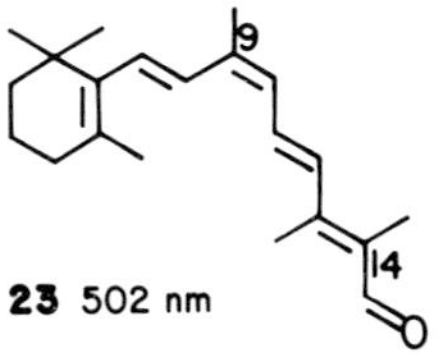

23 502 nm

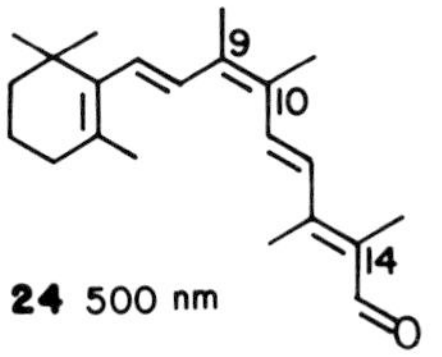

24 500 nm

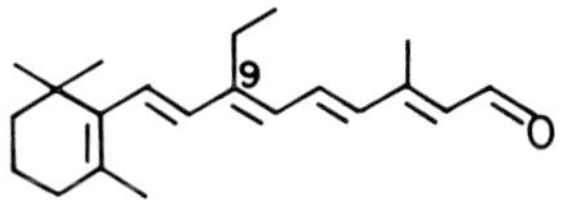

25 497 nm (11-cis)

[181]), or in the retinal analog where the bulk of the alkyl group is increased as in 9-ethylretinal, 25 [182], did not inhibit formation of visual pigment analogs, nor did it have a drastic effect on the absorption maxima or CD spectra of the pigments as compared to rhodopsin or isorhodopsin. However, differences were observed; namely, a slower rate of formation of pigments, lability towards hydroxylamine and lesser stability in detergents.

(d) Halogenated retinals

Brominated and fluorinated retinals 26 [183], 27 [183], 28 [184] and 29 [184] also reacted with cattle opsin but 13-desmethyl-14-bromoretinal and the 7-*cis* isomer of

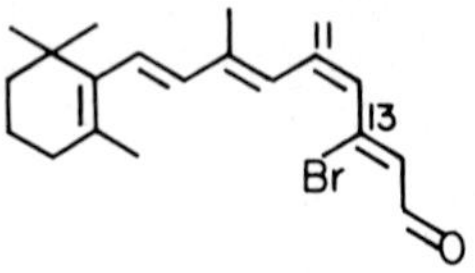

26 520 nm

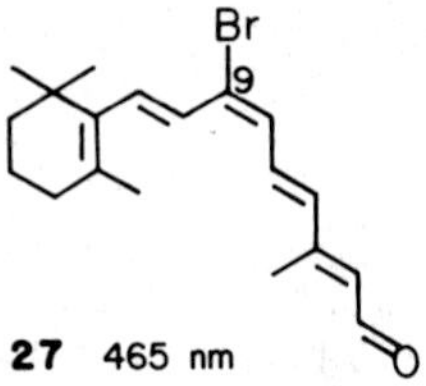

27 465 nm

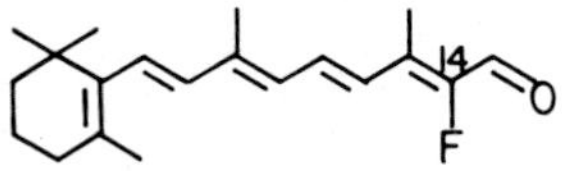

28 511 nm (9-cis)
513 nm (9,13-dicis)
527 nm (11-cis)

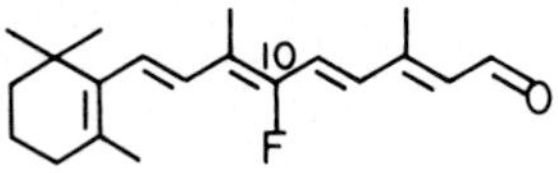

29 484 nm (7-cis)
486 nm (9-cis)
502 nm (11-cis)
464 nm (7,9-dicis)
484 nm (9,13-dicis)

the 14-fluoro derivative did not form visual pigment analogs. The explanation of the red-shifted absorption maxima observed in pigments formed from 26 and 29 as compared to rhodopsin, and the blue-shifted absorption maximum of pigment formed from 27 will be discussed in section 7 in relation to the external point-charge model of visual pigments.

(e) Allenic rhodopsins and the chiroptical requirements of the binding site

Chromophores in which the allenic bond fixes the ring at right angles with the polyene chain were useful not only to study the role of the twist about the 6-*s*-bond in normal retinals (see p. 307), but also to appreciate the steric and chiroptical requirements of opsin towards the chromophore since they contain two chiral centers, one at C-5 and the other at C-6.

Both the 9-*cis* and 9,13-di*cis* isomers of 30 formed visual pigment analogs [165], with λ_{max} at 460 nm and 455 nm (mixed diastereomers), respectively. Moreover, upon incubation of a ca. 1:1 diastereomeric mixture of 30 (the 9-*cis* isomers) followed by extraction of the bound chromophore by the CH_2Cl_2 procedure and

HPLC and CD measurements, it was found that opsin binds equally well with both of the antipodes.

In view of the high specificity usually encountered in binding sites, the leniency of opsin in accepting all four diastereomers and enantiomers was surprising. Therefore, adamantyl allenic retinals, 31, were synthesized [87]. The structure of these com-

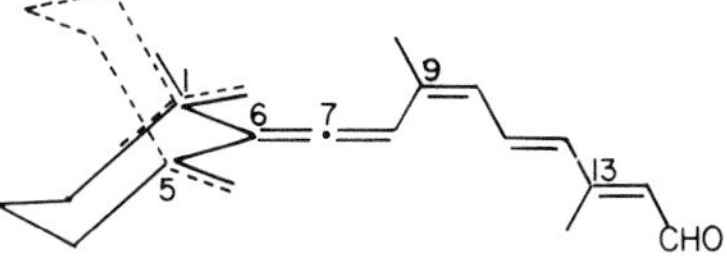

30 5R,6S: ring moiety indicated by solid line
5S,6S: ring moiety indicated by dotted line

31 410 nm (9-cis, 9,13-dicis)

CHO

pounds was designed so as to simulate a situation close to the one encountered if the diastereomers were combined (see 30). Despite their cage structures, both the 9-*cis* and the 9,13-di*cis* isomers formed visual pigments, having λ_{max} at 410 nm, thus confirming that the β-ionone binding site is quite lenient in rhodopsin.

The striking difference in the absorption maxima of the cyclohexyl and adamantyl pigments (460 and 410 nm, respectively) could be explained by the steric interactions prevailing in the retinals from which these pigments were prepared. In the 9-*cis* isomer of 30, the 9-ene is locked into conjugation with the allenic 7-ene owing to steric hindrance between 9-methyl and the annular equatorial methyl groups despite the steric interactions between 8H and 11H. In 31, however, which has no ring methyl groups, the 8,9 single bond is free to rotate, and hence the 8H–11H interaction moves the 9-ene considerably out of conjugation with the allenic 7-ene; hence the blue shift encountered in the adamantyl derivative.

(*f*) *Retinals with modified ring structures*

Pigment analogs from a number of retinals containing oxygenated ionone rings such as 32 [185,186], 33 [148,187], 34 [188] and 35 [187] were also prepared, and rhodopsin analogs were reported from retinals which contain isomerized double bonds involving the ionone ring, such as in 36 and 37 (S.E. Houghton, D.R. Lewin

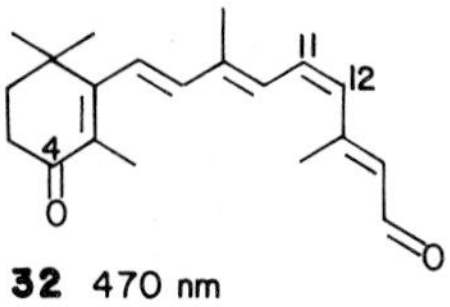

32 470 nm 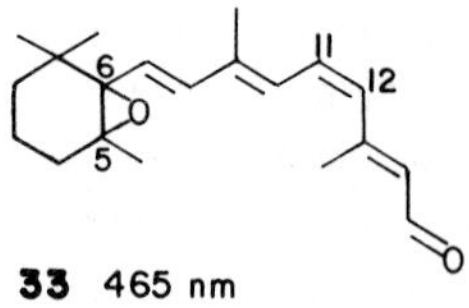 **33** 465 nm 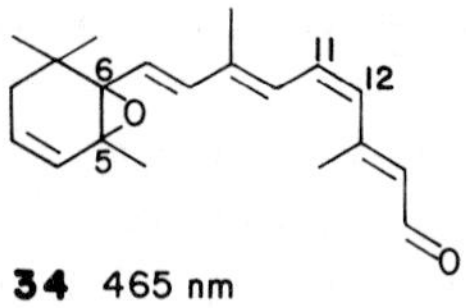 **34** 465 nm

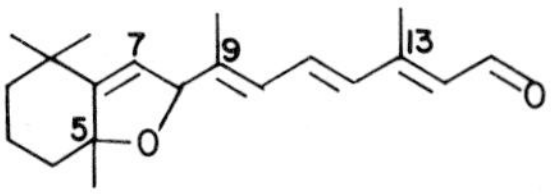

35 411 nm (unknown isomer) 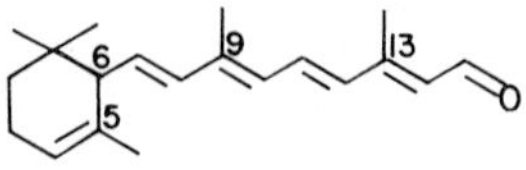 **36** 467 nm (unknown isomer) 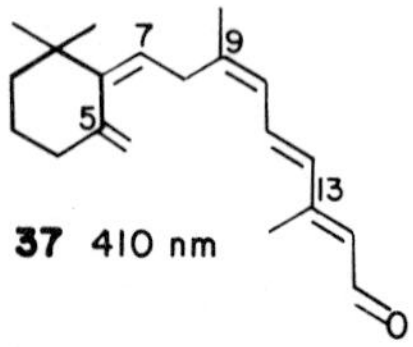 **37** 410 nm

and G.A.J. Pitt, unpublished work) and 37 [146,147]. The pigment obtained from γ-retroretinal, 37, served as a model for studies seeking information on the involvement of proton transfer in the primary event [146,147].

Other retinals, in which the ionone ring was modified by an additional double bond, 38 [189,190], ring contraction, 39 [70], or a retinal with open polyene chain, 40

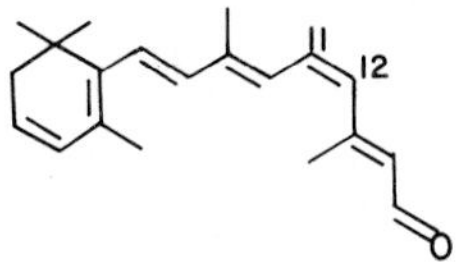

38 517 nm 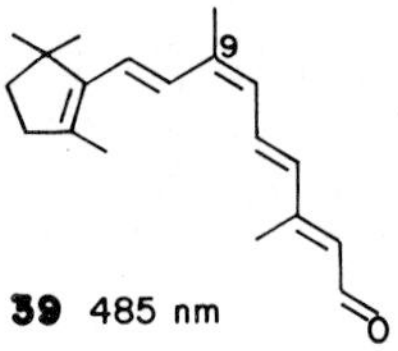 **39** 485 nm 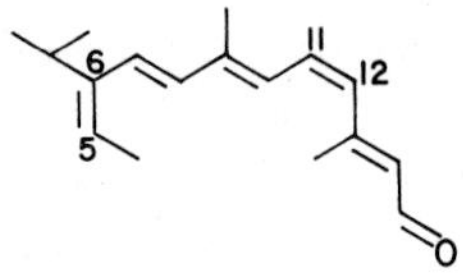 **40** ~490 nm

(A. Kropf, unpublished work), all combined with cattle opsin to give visual pigment analogs.

Recently, even highly modified retinals, in which the trimethylcyclohexenyl ring was replaced with an aromatic ring, did form visual pigment analogs with cattle opsin [183,191]. The pigment analogs from retinals 41, 42, 43 and 44, however, were

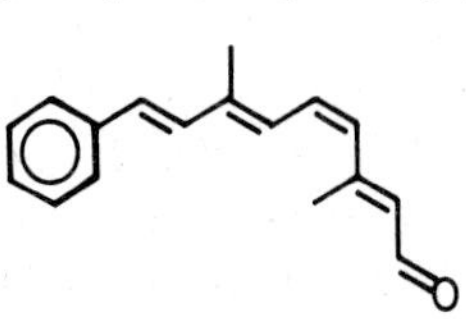

41 494 nm (11-cis)

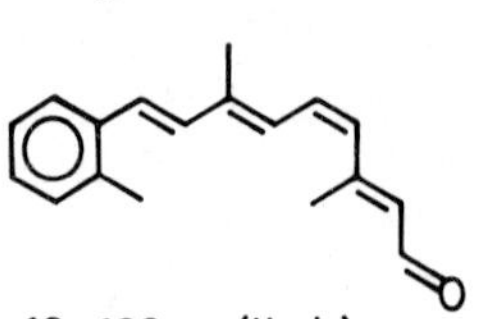

42 496 nm (11-cis)
482 nm (9-cis)

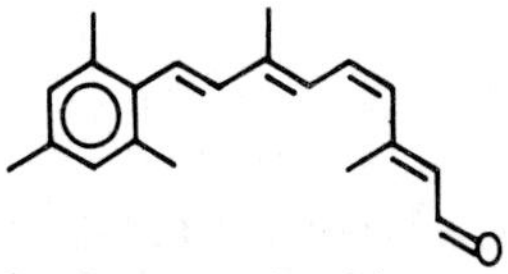

43 480 nm (11-cis)
470 nm (9-cis)

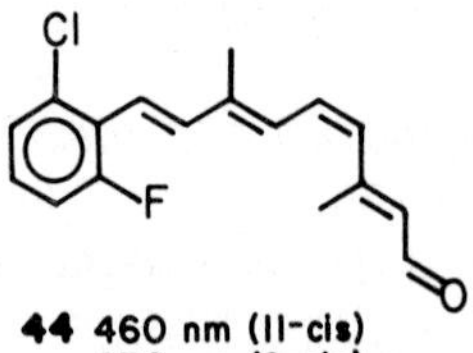

44 460 nm (11-cis)
470 nm (9-cis)

produced in low yields (ca. 10%) and were prone to rapid bleaching. This could be attributed to unfavorable stereo-electronic interactions which reduce the rate of pigment formation; the chromophores then would have to compete with protein degradation during the lengthy incubation times.

The variety of structural modifications carried out on the ionone ring and the observation that pigment analogs could be made from all these derivatives strongly support the idea that the binding site of opsin is indeed very lenient as far as stereo-electronic requirements of the ring binding site are concerned.

(g) Modified retinals for photoaffinity labeling of rhodopsin

Photoaffinity labeling has been used successfully to label covalently active sites in enzymes [192–195] in order to gain information about the molecular details of the active sites, and their mechanisms of action. Such an approach is particularly advantageous with systems like rhodopsin because: (1) the binding of retinal in the active site is via a covalent bond (the protonated Schiff base linkage) which keeps it anchored while activation of photolabile group can be performed; (2) it can be ascertained whether the photoaffinity-labeled retinal is bound to the binding site by carrying out competitive inhibition experiments or by using the CH_2Cl_2 procedure to extract and identify the bound chromophore.

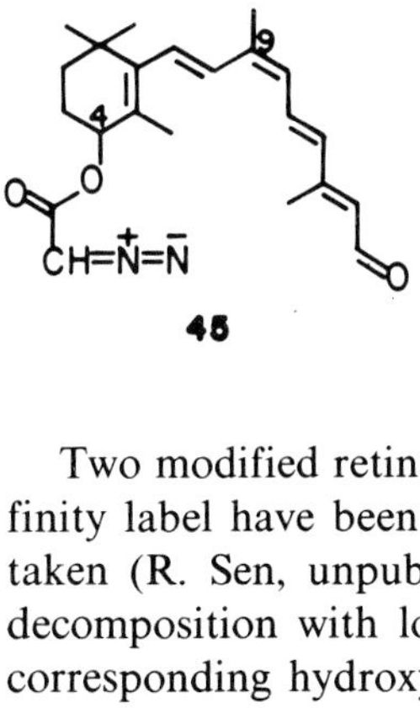

45

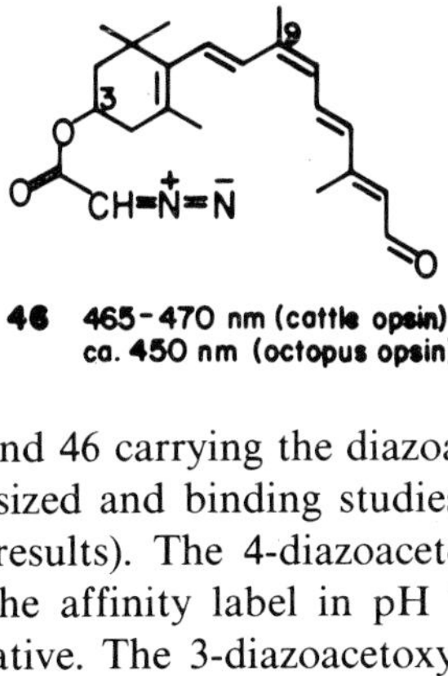

46 465–470 nm (cattle opsin)
ca. 450 nm (octopus opsin)

Two modified retinals 45 and 46 carrying the diazoacetoxy group as the photoaffinity label have been synthesized and binding studies to opsins have been undertaken (R. Sen, unpublished results). The 4-diazoacetoxy derivative 45 underwent decomposition with loss of the affinity label in pH 7.0 buffer, thus yielding the corresponding hydroxy derivative. The 3-diazoacetoxy derivative 46, however, was sufficiently stable and the 9-*cis* isomer was bound to cattle opsin to give a visual pigment analog with λ_{max} at 465–470 nm; it was also bound to octopus opsin and formed a pigment analog absorbing at ca. 450 nm. Selective irradiation of the bovine pigment at 254 nm resulted in the activation of the label (formation of a carbene) and attachment of the retinyl chromophore, in addition to the Schiff base linkage, to site(s) on opsin around the ionone ring. This was deduced from the observation of different bleaching rates upon irradiation of the pigment by $\lambda > 500$ nm light; i.e., in case of the pigment preirradiated at 254 nm (where the activity label absorbs maximally) a much faster bleaching occurred. In addition, a much lesser amount of the bound chromophore was extracted by the methylene chloride procedure, from the pigment preirradiated at 254 nm. During the activation of the affinity label the

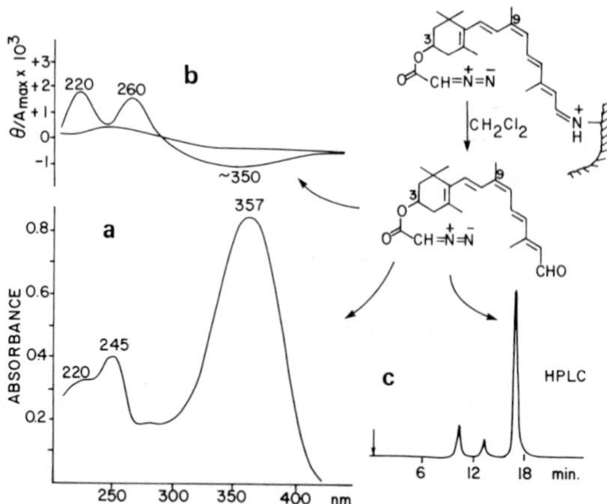

Fig. 13. Absorption (a) and circular dichroism (b) spectra of 3-diazoacetoxy-9-*cis*-retinal in hexane, at room temperature (R. Sen, unpublished results). From the incubation of an enantiomeric mixture of retinals, the opsin preferentially bound one of the enantiomers (at position 3) as shown by the circular dichroism spectrum (in hexane), (b), of the chromophore extracted by the methylene chloride procedure. In (b) and (c) are shown the UV spectrum and the HPLC trace of the extracted chromophore.

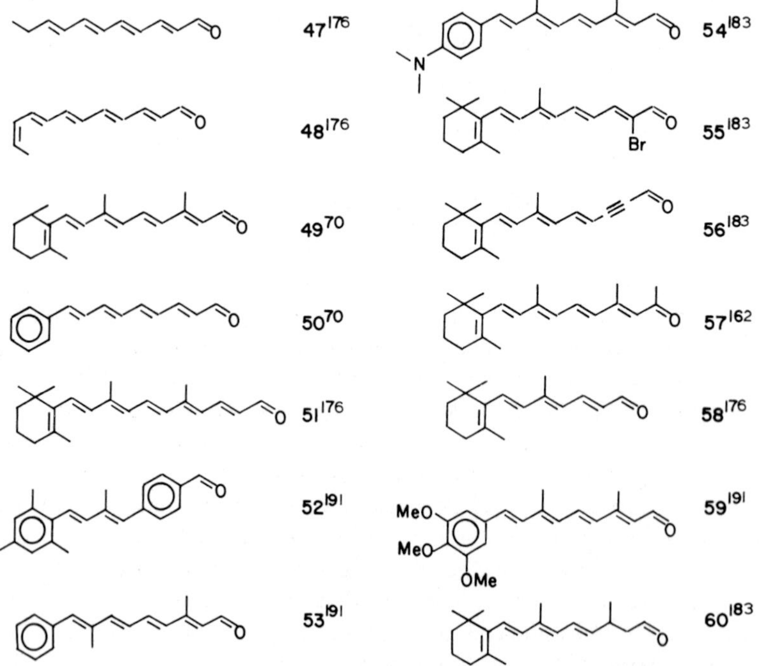

Fig. 14. Retinal analogs of various structures which do not form visual pigment analogs. For the sake of convenience, the figure shows the retinals in the all-*trans* form; however, it is the 11- and/or the 9-*cis* isomers which were tested for binding.

integrity of the binding site did not alter significantly, as was shown by the insignificant (10–20%) reduction of the absorption maximum of the pigment at 470 nm.

By the use of this type of photoaffinity-labeled retinals at various positions on the ring and side chain, and the use of radioactively labeled compounds, and by carrying out hydrolysis of the retinal from the protein into amino acid(s) or short peptides, it will be possible to determine which amino acid(s) or peptide sequences are close to the chromophore in the binding site.

An interesting feature of the binding site became evident the first time by binding studies carried out on 3-diazoacetoxyretinal 46. When a mixture of enantiomers at position 3 were combined with cattle opsin, one of them was preferentially bound. This was shown by retrieval of an optically active chromophore (Fig. 13) from the binding site by the use of the CH_2Cl_2 extraction procedure and circular dichroism measurements. Therefore, while the ring binding site was found to be quite lenient in many respects [87,165], it can also show chiroptical discrimination.

(h) Modified retinals not forming visual pigment analogs

Fig. 14 shows further modifications of the structure of retinal, compounds 47–60, which however, yielded negative results, i.e., these structures, due to unfavorable steric and/or electronic interactions with the binding site did not form rhodopsin analogs.

6. Models proposed to account for molecular changes in the primary event

(a) Proton translocation models directly involving the Schiff base nitrogen

In all these models *cis–trans* isomerization is assumed to follow the light-initiated intramolecular proton translocation.

Proton translocation to the Schiff base nitrogen was proposed to occur by concerted double proton transfer (as shown in Fig. 15) leading to a 'retro retinal' structure in the batho intermediate [127, 196]. However, this model can be eliminated as it is inconsistent with the formation of batho intermediates from pigment analogs based on 5-desmethylretinal [145] and γ-retroretinal [146,147]. It also disagrees with the resonance Raman results.

Models involving single proton transfer towards the Schiff base nitrogen during the formation of bathorhodopsin were also proposed [80,118,138,139]. These models assumed the existence of a hydrogen-bonded Schiff base, rather than the fully protonated form suggested by resonance Raman experiments [133,136,137,140]. The latter concluded that the state of protonation in rhodopsin, bathorhodopsin and also in isorhodopsin is identical (no change observed in the C=N stretching frequency) and that the Schiff base proton is covalently bound (large, ca. 25 cm^{-1} frequency shift seen in deuterated samples).

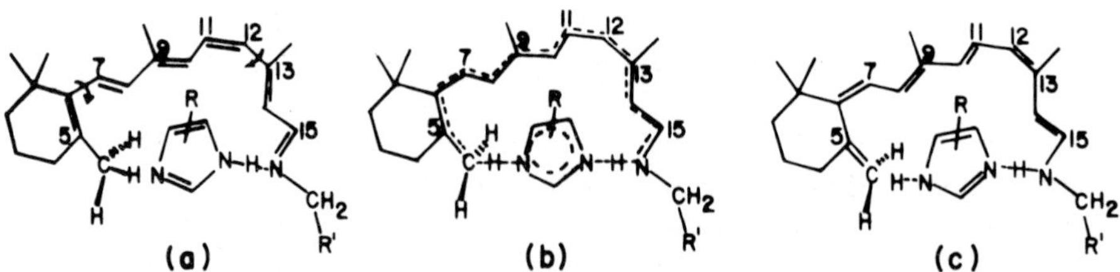

Fig. 15. Structure of the prosthetic group in: (a) rhodopsin, where the retinal Schiff base is in an 11-*cis*,6-*s*-*cis*-12-*s*-*cis* conformation; (b) delocalized excited singlet state; (c) bathorhodopsin, where the chromophore is a hexaene amine–imidazole complex. From van der Meer et al. [127].

A mechanism involving deprotonation of the Schiff base nitrogen [197] as the primary event is now considered unlikely because of resonance Raman experiments and also because deprotonation would lead to blue shift rather than the observed red shift in bathorhodopsin.

Recently, it has also been suggested that proton transfer to the Schiff base nitrogen, either induced by or coupled to the photoisomerization of retinal, could occur upon formation of bathorhodopsin [198]. This suggestion was based on a study investigating the photochemical properties of air-dried films of bovine rhodopsin. The blue-shifted absorption maximum (390 nm) of rhodopsin observed in dehydrated films was explained by the involvement of water in the protonation of the Schiff base linkage, i.e., water in form of hydronium ion would be located near the Schiff base nitrogen and would be stabilized by an adjacent negatively charged side group of opsin. Thus, the blue shift was fully reversible upon hydration of the films. Further, irradiation of dehydrated films at room temperature and at 77 K led to a large red shift (479 nm). This was interpreted as an intramolecular proton transfer from some ionizable amino acid side chain, and it was also hypothesized that proton transfer plays a role in the formation of bathorhodopsin in the hydrated state as well.

(b) Proton translocation models involving charge stabilization

In a model proposed by Lewis [228] the effect of the excited state of retinal on the conformational state of the protein is considered to be the first step of the excitation mechanism. Charge redistribution in the retinal by excitation with light would have the consequence of vibrationally exciting and perturbing the ground state conformation of the protein, i.e., excited retinal would induce transient charge density assisted bond rearrangements (e.g., proton translocation). Subsequently, retinal would assume such an isomeric and conformational state so as to stabilize maximally the new protein structure established. In this model, 11-*cis* to *trans* isomerization would not be involved in the primary process, but would serve to provide irreversibility for efficient quantum detection. It was also proposed that either the 9-*cis*-retinal (in isorhodopsin) or the 11-*cis*-retinal (in rhodopsin) could yield the same, common batho form (structure not defined) but this form would not be an all-*trans* form.

Another mechanism involving charge stabilization [129,199] proposed that interaction between the protein charges and the chromophore can drastically modify the ground state isomerization barrier. Specifically, charge stabilized intermediates would be formed when the light-induced shift of the chromophore positive charge to the ring is stabilized by proton transfer between two properly placed acidic groups on the protein and by partial alternation of the bond lengths of the chromophore. While this approach, based on theoretical calculations, does not exclude isomerization, it envisages the possibility that in the batho intermediate the retinal is still in an 11-*cis* form, although this geometry would differ from that in rhodopsin with respect to the torsional angles $\phi_{5,6}$, $\phi_{6,7}$, $\phi_{11,12}$ and $\phi_{12,13}$. Thus, a very small barrier (ca. 6 kcal/mol) would then exist between rhodopsin and bathorhodopsin. This last prediction of the model is its main difficulty, as it would imply that fast thermal bleaching of rhodopsin can occur, which in vivo would lead to saturation of the photoreceptor response by thermal noise. In addition, both charge stabilization models discussed here are in conflict with resonance Raman data which clearly indicates that bathorhodopsin contains an isomerized chromophore.

(c) Electron transfer model

The extremely rapid rate of formation of bathorhodopsin as compared to isomerization rates observed in model protonated Schiff bases (a factor of 10^3) suggested the idea that electron transfer between an amino acid residue (e.g., tyrosine or tryptophan) in the protein and the chromophore may ‘catalyse’ isomerization. Thus, a photoinduced electron transfer leading to a radical anion chromophore, instead of complete *cis–trans* isomerization, was considered as a plausible alternate mechanism for the primary event [200]. This mechanism, however, is difficult to reconcile with the known photoreversibility but thermal irreversibility of the bleaching process. Thermal irreversibility of the light-induced electron transfer would require geometrical separation of donor and acceptor moieties which would then not allow photoreversibility [201].

(d) Models involving cis–trans *isomerization in the primary event*

The majority of models proposed for the mechanism of the primary event suggest that *cis–trans* isomerization of the chromophore occurs during formation of bathorhodopsin. The strongest evidence favoring the isomerization models comes from the original work of Hubbard and Kropf [94] and Yoshizawa and Wald [96], who demonstrated that at low temperature it is possible to establish a photoequilibrium between rhodopsin and isorhodopsin via bathorhodopsin as a common intermediate. It is now well established that this common intermediate has a transoid chromophore. The isomerization models differ in the manner in which this isomerization can be accomplished so as to be consistent with the following experimental data: (1) bathorhodopsin is still a protonated Schiff base and it is in a perturbed all-*trans* configuration; (2) the same batho product can be formed from rhodopsin, isorhodopsin and also from 7-*cis*-rhodopsin; (3) extremely fast formation of the batho product and its stability at liquid nitrogen temperature; (4) photoreversibility but thermal irreversibility of the primary event; (5) red-shifted absorption maxima of all the batho products compared to the corresponding rhodopsins; (6) visual pigment analogs in which isomerization is precluded by the structural features of the model retinals do not bleach or form a batho intermediate.

A model of 'sudden polarization' leading to partial isomerization was based on theoretical calculations of the charge distribution in the excited state of nonatrienylidene-methyliminium ion, 61 [202]. It proposed that irradiation of rhodopsin would create a strongly polarized twisted singlet excited state. In this excited state, positive charge would move from the Schiff base nitrogen to an average position on C-9, and this charge separation would be maximally effective when the excited singlet molecule would twist around the 11,12 bond from 60° to the fully orthogonal conformation (Fig. 16). The sudden movement of charge, over a length of seven bonds, would trigger a short-living large electrical signal and this was suggested to be the crucial primary event leading to vision. Fig. 16 shows the model in the 11-*cis* 12-*s*-*cis* conformation. However, the authors estimated that a similar charge polarization could also operate in case of the 12-*s*-*trans* conformer. While this model does not propose a specific mechanism for the formation of bathorhodopsin, it suggests that the duration of the electrical signal generated by the singlet state could be consistent with the 6 ps required for the formation of bathorhodopsin.

The so-called torsion model by Kakitani and Kakitani [203] was developed to explain the high quantum yield of photoisomerization in rhodopsins as compared to isomerization yields observed in model protonated Schiff bases. The main feature of this model is the twist around double bonds in retinal when it is bound to opsin, which would make this substrate strained to be near the transition state for isomerization (Fig. 17). The most important twist as estimated by this model would be around the 11,12 double bond, although other double bonds would simultaneously be twisted to some extent ([204] and refs. therein). The protein was considered to be responsible for inducing the selective twist. This model predicted a

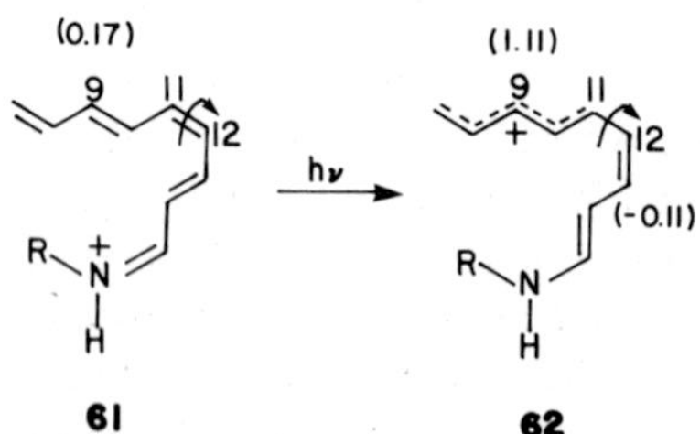

Fig. 16. Excitation and partial isomerization of the protonated Schiff base (calculated net charges are shown in parentheses). In 61, retinal is in an 11-*cis*,12-*s*-*cis* form; in 62, it is in the 11-*orthogonal*,12-*s*-*cis* form. From Salem and Bruckman [202].

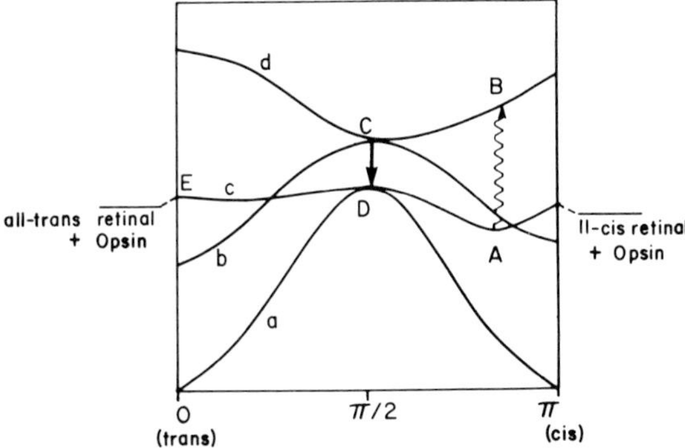

Fig. 17. Adiabatic potential of rhodopsin as a function of torsional angle around the 11,12 bond. Curves a and b represent adiabatic potentials of Schiff base in the ground state and in the excited state. Curves c and d represent adiabatic potentials of rhodopsin in the ground and excited states respectively. Rhodopsin (A) by absorption of photon goes to excited state (B), isomerization occurs (C), it then goes nonradiatively to D and to E and finally it dissociates totally into retinal and opsin. Adapted from Kakitani and Kakitani [203].

barrierless energy surface in the excited state for the isomerization of the 11,12 bond. The major objection to this torsion model which uses twists around double bonds is that single bonds with much smaller twisting force constants are present in the side chain.

A bicycle-pedal model to describe the isomerization process was proposed by Warshel [128]. In this model, a two-bond isomerization involving the concerted rotation about the 11,12 and 15,16 bonds would take place to yield bathorhodopsin in which the chromophore would assume a strained all-*trans* configuration (Fig. 18). Instead of viewing the isomerization process as classical rotation involving bulky groups, the fast formation of the batho product was explained as being due to the small amounts of inertia of the bicycle-pedal motion. This model was mainly criticized because it predicted the energy for bathorhodopsin to be ca. 5 kcal above rhodopsin [201]. It is now established, by the elegant low temperature photocalorimetric measurements of Cooper [205] that the ground-state energy of bathorhodopsin is about 35 kcal higher than rhodopsin.

On the basis of a semiclassical trajectory approach used to study the dynamics of *cis–trans* isomerization it was also proposed that the first step in the vision process is likely to be a *cis–trans* photoisomerization leading to ‘through-space’ charge separation at the Schiff base linkage [208]. Birge et al. [209] investigated the molecular dynamics of *cis–trans* isomerization using CNDO–CISD molecular orbital theory and semiempirical molecular dynamics procedures. These theoretical calculations led to the proposal that a one-bond photochemical *cis–trans* isomerization can occur with high quantum efficiency in ca. 2 ps.

An isomerization model in which arguments were presented to show that rhodopsin and bathorhodopsin are interconvertible via a common, barrierless, thermally re-

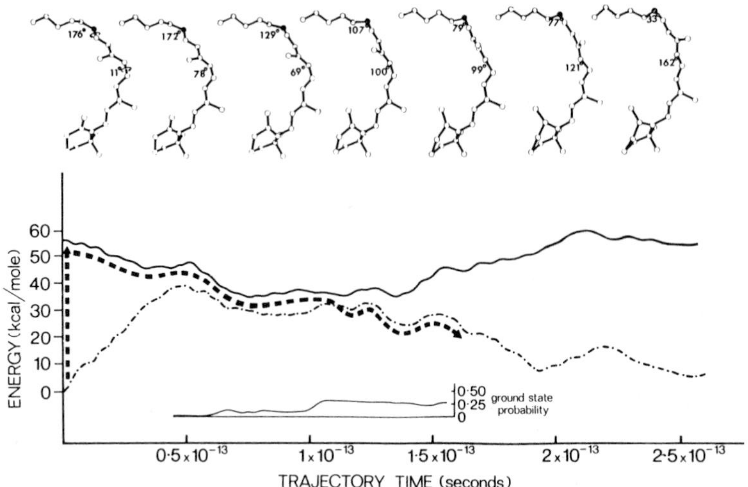

Fig. 18. Computer simulation of the molecular dynamics of the first step of the vision process. The potential energy of the ground and first excited singlet state are described by the broken and solid lines, respectively, as a function of trajectory time while the conformational changes along the trajectory are presented above with the corresponding values of $\phi_{11,12}$ and $\phi_{15,16}$. The simulation shows the motion of the chromophore as a function of the trajectory time. The chromophore starts at the 11-*cis*,12-*s*-*trans* conformation, moves to $\phi_{11,12} = 90°$, crosses to the ground state and forms a strained all-*trans* conformation. The twisting of the 11,12 bond is accommodated by a significant concerted twist of the 9,10 and 15,16 bonds, which results in a small overall displacement of the chromophore atoms, thus minimizing collisions with the protein cavity. The simulation proves that *cis–trans* photo-isomerization reaction can occur in less than 1 picosecond in a restricted active site. Adapted from Warshel [128].

laxed excited state is shown in Fig. 19 [201]. This model, based on the wavelength and temperature independence of the photochemistry of visual pigments [206], and critical evaluations of the quantum yields involved in these processes, suggests that a single potential minimum is populated (both from excited rhodopsin and bathorhodopsin) from which the ground-state 11-*cis* and all-*trans* configurations are formed in a 1:2 ratio (see curve I in Fig. 19). The energy values in Fig. 19 now should be somewhat modified (such as the > 13 kcal/mol value above rhodopsin) in view of the available data by photocalorimetry [205].

A likely result of the photoisomerization would be a charge separation involving the protonated Schiff base and its counterion (Fig. 20) [207]. The cleavage of this salt bridge with the resulting separation of charges in the interior of the protein would be a mechanism by which the energy of a photon is converted into chemical free energy of bathorhodopsin. Isomerization resulting in charge separation would explain the red shift and the observed energy storage in the primary product. This model assumes that proton-dependent processes detected by picosecond measurements would be ground-state relaxation processes following the primary event, bathorhodopsin′ to bathorhodopsin step in Fig. 20. This model is different from other simple *cis–trans* isomerization models in that it views the geometrical change in the

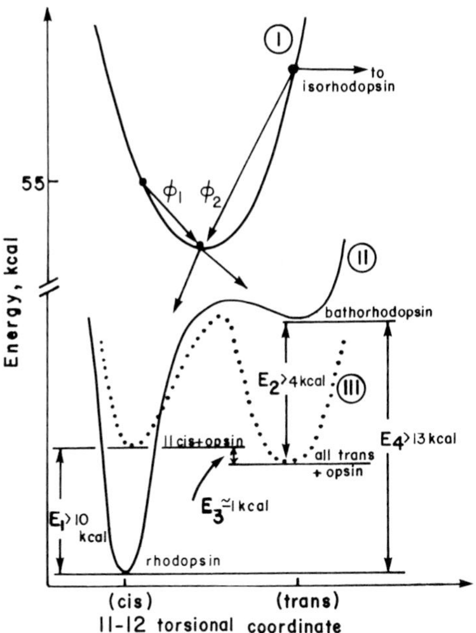

Fig. 19. Potential energy curves and energy relationships in rhodopsin. Curve I: Excited state of rhodopsin and bathorhodopsin. Curve II: Ground state of rhodopsin and bathorhodopsin. Curve III: Ground state of isolated chromophore. Symbols ϕ_1 and ϕ_2 are quantum yields for reaching the single potential minimum along the 11,12 torsional coordinate. From Rosenfeld et al. [201].

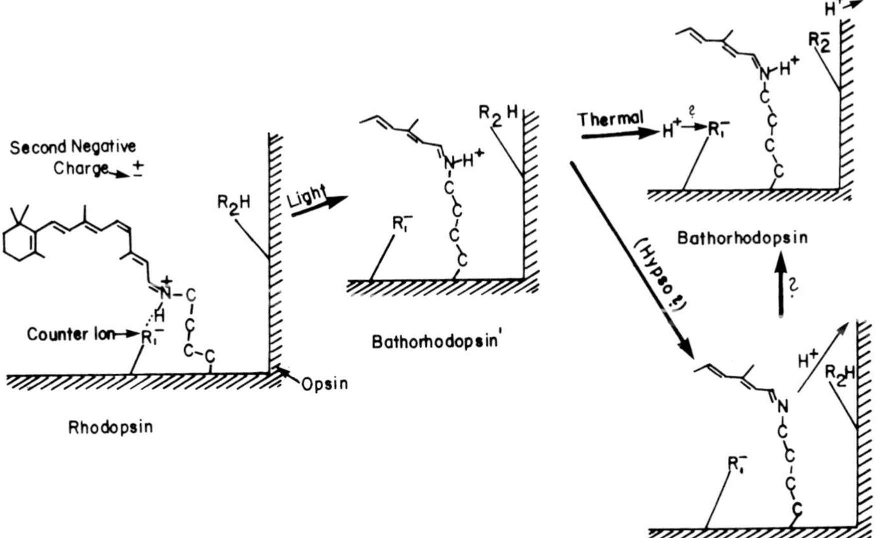

Fig. 20. Model for the primary event in vision. Isomerization of the 11,12 bond leads to charge separation at the Schiff base site. This process, as shown, can possibly be followed by proton transfer, the latter resulting from the charge separation. In rhodopsin, the 'second negative charge' responsible for wavelength regulation is shown close to the 11,12 bond of the polyene chain. This model assumes that hypsorhodopsin is the unprotonated form of the Schiff base, and that it is formed possibly by proton transfer from the Schiff base nitrogen in some pigments. From Honig et al. [207].

chromophore as a means to accomplish charge separation. In the case of rhodopsin, charge separation would be obtained by an 11-*cis* to *trans* isomerization, but the generality of the model lies in the proposal that any isomerization of a flexible chromophore, regardless of which bond rotates and in which direction, could be used for charge separation and energy storage.

(e) Summary

On examination of the data it now appears that the primary event in vision is a *cis–trans* isomerization of the retinal moiety; proton(s) could be translocated on the protein, concomitant with or subsequent to the isomerization process, but not before. The isomerization of the 11-*cis*-retinal would lead to a batho product having a distorted transoid chromophore and a ground-state energy of ca. 35 kcal above rhodopsin.

In spite of the creative and numerous models proposed by the theoreticians to account for the role of the protein in affecting the molecular dynamics of the isomerization process, there is yet no general consensus as to which mechanism is the correct one.

7. Models to account for the color and wavelength regulation in visual pigments

A central problem in vision chemistry is to account for the color and variability of the absorption maxima of different pigments. It is of crucial importance to establish precisely what kind of interactions between retinal and opsins give rise to the so-called bathochromic shift, and how the proteins regulate the absorption maxima of pigments so as to produce pigments most suitable to the perceptual needs of a species. This understanding is also highly relevant to wavelength regulation necessary for color vision.

While 11-*cis*-retinal absorbs at 380 nm and a model protonated Schiff base of retinal with *n*-butylamine in methanol [210,211] absorbs at 440 nm, the absorption maxima of visual pigments based on 11-*cis*-retinal span a wide range of values, from 430 to ca. 600 nm. Therefore, the problem consists of determining the nature of the

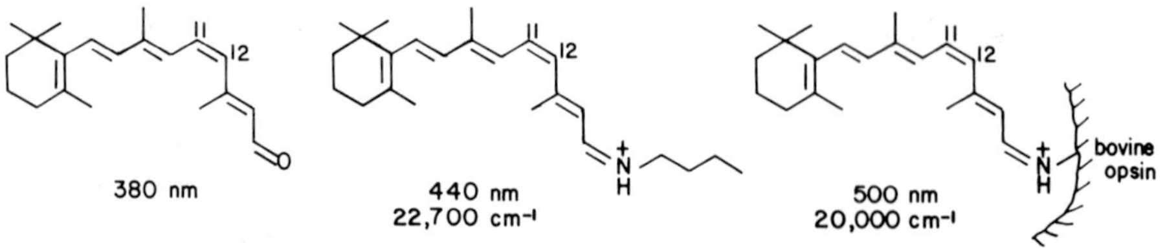

380 nm

440 nm
22,700 cm^{-1}

500 nm
20,000 cm^{-1}

bovine opsin

OPSIN SHIFT: λ_{max} (in cm^{-1}) SBH$^+$ − λ_{max} (in cm^{-1}) Pigment = 2700 cm^{-1}

interactions which cause the 'opsin shifts', i.e., the effect of proteins on the absorption spectra of protonated Schiff bases in the binding sites. 'Opsin shift' is defined [227] as the difference between the absorption maximum of the protonated Schiff base and the absorption maximum of pigment, expressed in reciprocal centimeters.

A large number of both theoretical and experimental models were put forward to account for the opsin shifts encountered in visual pigments:

(a) The retinylic cation

A retinylic cation (λ_{max} = 598 nm) was proposed by Blatz et al. [212] to explain the long wavelength absorption maxima of visual pigments. This model was, however, rejected by the same group later [213] on the basis of further experimental data collected on the absorption spectral properties of polyenyl cations.

(b) Anionic groups close to the ionone ring and a twist of the chromophore

Based on theoretical calculations, Abrahamson and Wiesenfeld [214] proposed anionic groups close to the ionone ring and a twist of the chromophore around the 11,12 double bond to account for the entire spectral range of visual pigments. It was argued, however, that as the torsion angle around the 11,12 bond is negligible, it cannot have any significant contribution to the observed shifts ([215] and refs. therein).

(c) Inductive or field-effect perturbation of the positive charge of the nitrogen in the iminium bond by substituents attached to it

This was considered to affect the absorption spectra of the bound chromophore [216]. This model, based on the absorption spectra of protonated Schiff bases formed from retinal and substituted anilines is, however, unlikely to have any realistic counterpart in the protein.

(d) Microenvironmental polarizability models

Irving and Leermakers [217–219] explained the bathochromic shift in rhodopsin by involving the polarizability of the local microenvironment of the chromophore. Since they observed red shifts in the spectra of protonated Schiff base of retinal in solvents of high polarizability (independent of their polarity), they postulated that suitable polarizable groups, e.g., aromatic amino acid residues, could be in a specific orientation with respect to retinal in the binding site giving rise to shifts even larger than observed in the model protonated Schiff base systems. This proposal to regulate the absorption spectra with induced dipoles was opposed on theoretical grounds [215] which, in accordance with the experimental data (the λ_{max} of SBH^+ in benzene is shifted only to 464 nm), predicted only a weak contribution, even when it was assumed that the groups were favorably oriented.

(e) Distance of the counterion from the protonated Schiff base nitrogen

Blatz et al. [211,220] proposed that the absorption maxima of pigments could be regulated by varying the distance between the protonated Schiff base nitrogen and its counterion. As the isolated protonated Schiff base of retinal in vacuo was estimated to absorb at around 600 nm [221,222], it was also possible to consider the visual pigment spectra to be blue-shifted as compared to SBH^+ in vacuo. Hence it was possible to construct models in which increasing solvent polarity causes blue shift (to be distinguished from increasing polarizability, which causes red shifts). Thus, removal of the counterion by 10 Å would lead to a species absorbing at ca. 580 nm. However, this anion-induced wavelength regulation, which necessitates the removal of counterion to distances greater than 3.5 Å, is unlikely.

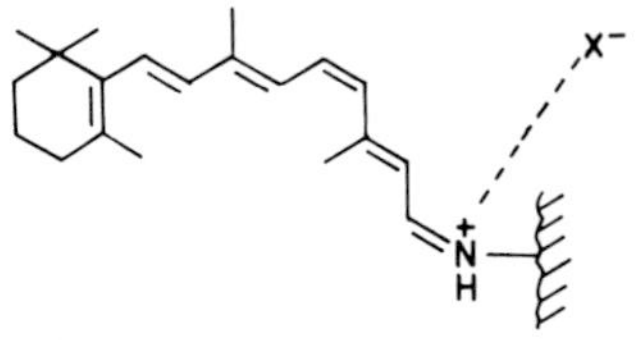

(f) The charge-transfer model

Akhtar et al. [74] proposed that, in rhodopsin, an acceptor group on the protein forms a charge-transfer complex with the unprotonated Schiff base of retinal; furthermore, upon 11-*cis* to *trans* isomerization, separation of donor and acceptor moieties would occur and the Schiff base linkage would be exposed to hydrolysis. This model can now be discarded as unrealistic; the resonance Raman experiments have shown that it is not an unprotonated Schiff base, but a protonated base which is bound to opsin. Further, this and related models were examined by Komatsu and Suzuki [223] using theoretical calculations, who found that charge-transfer type models cannot satisfactorily explain the red shifts seen in visual pigments.

(g) Point-charge perturbation models

Hubbard and Kropf ([224] and refs. therein) were the first to advocate a point-charge perturbation theory, in which it was postulated that some reactive group on opsin would bring about the bathochromic shift by its interaction with the protonated Schiff base of retinal. Such interaction would be able to "increase the mobility of electrons within the molecular orbitals of the chromophore, and to shift the λ_{max} to longer wavelengths". Theoretical calculations by Pullman et al. ([215] and refs. therein) have indicated that the differential stabilization of the ground and excited states, necessary for the creation of the shift, could be accomplished by a negative point-charge on the protein (e.g., a carboxylate) which could alter the charge distribution on the chromophore, essentially by electrostatic interaction. There was however, uncertainty in the exact location of the negative charge with respect to the polyene chain of retinal partly due to the difficulty of estimating an appropriate dielectric constant and partly due to the inherent difficulty of calculations in which a number of parameters are used to describe a single result, the experimental shift.

Other authors [214,222,225], agreeing that a negative charge on the protein can control the observed spectral shift, proposed that this charge is located close to the ionone ring rather than the polyene chain of the chromophore.

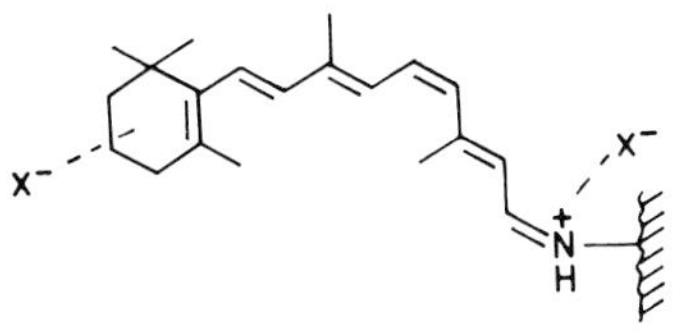

Although the models proposed above were plausible there was no experimental data available to support them, or test their validity. The external point-charge model [82] is the first experimental model to use analog pigments to account for the absorption properties of a specific pigment, bovine rhodopsin. In addition, this model proposes how wavelength regulation could be accomplished in visual pigments.

A series of hydroretinals were synthesized and were bound to bovine opsin to form visual pigment analogs, the hydrorhodopsins [92]. Model protonated Schiff bases were also prepared from each of these retinals, and the magnitudes of the opsin shifts were determined. Table 1 summarizes the data obtained and shows the opsin shift for bovine rhodopsin for comparative purposes. Binding studies were also carried out to ascertain that the hydroretinals occupy the same binding site as 11-*cis*-retinal in rhodopsin. This is required to show the relevance of the data to the natural system.

Analysis of the data in Table 1 shows that as the conjugation of the aldehyde-containing moiety becomes shorter, the absorption maxima of the pigments are blue-shifted. This indicates that the enal moiety is responsible for the observed absorption maxima. Through-space interaction between the chromophore moieties separated by single bonds does not occur as shown by the similarity in the absorption maxima between 11,12-dihydro- and 9,10,11,12-tetrahydrorhodopsins. Most importantly, the largest opsin shift, 5300 cm^{-1}, is observed in the pigment containing the chromophore with the shortest enal moiety, 67. The opsin shifts for the other analogs on the other hand, are close to the value found for rhodopsin. These data strongly suggest the presence of significant electrostatic interactions in the vicinity of the chromophore, from C-11 to nitrogen, i.e., the negative charge must be close to this portion of the polyene chain to yield the largest opsin shift.

In order to place more exactly the charge relative to the polyene chain and to account quantitatively for the large bathochromic shift seen in 11,12-dihydrorhodopsin, semi-empirical π-electron calculations were carried out. Because very little flexibility was available in choosing the position of the external charge

TABLE 1

Absorption maxima of hydroretinals, protonated Schiff bases, hydrorhodopsins and opsin shifts (from Nakanishi et al. [92])

	63	64	65	66	67	68
aldehyde[a]	375	364	342	278	236	234
SBH^{+}[b]	440	425	392	322	270	275
bovine rhodopsin[c]	500	460	420	345	315	310
$\Delta\nu$ (cm^{-1})[d]	2700	1800	1700	2100	5300	4100

Data is also shown for bovine rhodopsin, 63, for comparison. The hydroretinals 66–68 presumably assume 9-*cis* or 11-*cis* like conformations when bound to opsin. Retinals 67 and 68 form non-bleachable pigments, i.e., no change in their λ_{max} occurs upon exposure to room light; irradiation by UV light leads to decomposition products instead of separation of the chromophore from opsin.

[a] In MeOH. In case of split chromophores the absorption maxima of the enal moieties are given.

[b] Protonated Schiff base with *n*-butylamine in MeOH.

[c] In 0.5% digitonin/67 mM phosphate buffer, pH 7.0.

[d] Opsin shift, i.e., difference in λ_{max} between rhodopsin and SBH^{+}, in cm^{-1}.

(due to the shortness of the conjugated moiety), these calculations indicated that the only way to account for the absorption maximum of rhodopsin and also that of these pigment analogs was to place a counterion near the Schiff base nitrogen and, in addition, a negative charge close to carbons 12 and 14, as in Fig. 21. The premise that the regulation of the absorption maximum by through-space interaction of charges was realistic was also substantiated by the shifts found in simple model organic compounds mimicking the situation in 11,12-dihydrorhodopsin [226].

The external point-charge model was further tested by studies on aromatic and bromo analogs of rhodopsin [183]. In the case of the aromatic analog 71, where the stereo-electronic properties of the ring are drastically altered, the similarity of the opsin shift to that found in isorhodopsin, 69, indicates again that the wavelength determining interactions are not close to the ionone ring in rhodopsin. The opsin shift of the 9-*cis*-9-bromo analog, 73, (2190 cm^{-1}) is very close to the opsin shift in 9-*cis*- or isorhodopsin, 69, (2110 cm^{-1}), indicating the tolerance of position 9 to

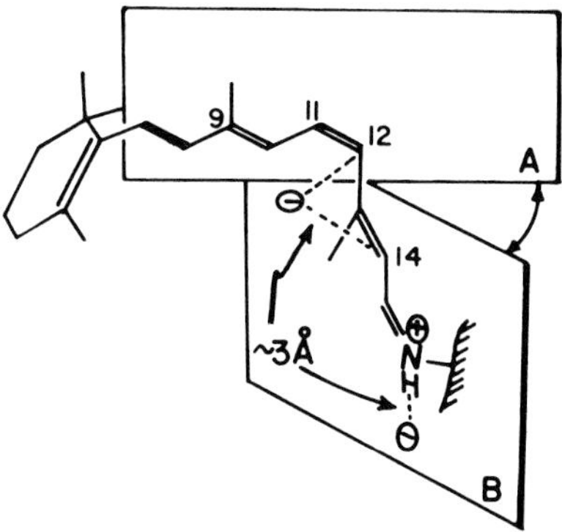

Fig. 21. The external point-charge model showing the electrostatic interactions in the binding site of bovine rhodopsin. In addition to the counterion near the protonated Schiff base linkage (3 Å), a second negative charge is located ca. 3 Å above C-12. This charge could be a member of a charge pair in a salt bridge or possibly the negative end of a neutral dipolar group. From Honig et al. [82].

stereo-electronic effects. This tolerance was not seen with 13-bromo substitution. The shift of 2180 cm^{-1} obtained for 13-bromorhodopsin, 75, is quite different from

69 R = Opsin 485 nm
70 R = n-Butyl 440 nm
Δ = 2110 cm^{-1}

71 R = Opsin 485 nm
72 R = n-Butyl 443 nm
Δ = 1960 cm^{-1}

73 R = Opsin 465 nm
74 R = n-Butyl 422 nm
Δ = 2190 cm^{-1}

75 R = Opsin 520 nm
76 R = n-Butyl 467 nm
Δ = 2180 cm^{-1}

the opsin shift for rhodopsin (2700 cm^{-1}), again indicating the importance of the interaction of the external point-charge with the aldehyde end of the polyene chain.

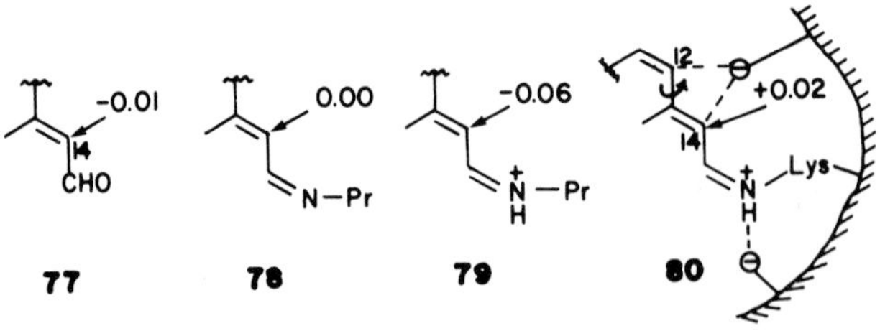

Fig. 22. Calculated π-electron charge densities at C-14 for 11-*cis*-retinal 77, its Schiff base 78, the protonated Schiff base 79, and rhodopsin 80, as indicated by the external point-charge model. From Honig et al. [82].

Further support for positioning the external point-charge close to the end of the polyene chain (as shown in Fig. 21) comes from the reinterpretation of the ^{13}C-NMR chemical shifts reported by Shriver et al. [81]. These authors found the chemical shift for carbon 14 in 11-*cis*-retinal and its Schiff base with propylamine to be at 130 ppm, while the shift in the corresponding protonated Schiff base was at 120.1 ppm. In the rhodopsin prepared from a [14-^{13}C]-labeled retinal, the chemical shift for carbon 14 was at 130.8 ppm. The authors concluded that, in rhodopsin, the Schiff base must be unprotonated, a result contradictory to the resonance Raman experiments. However, the ^{13}C-NMR data can also agree with the existence of a protonated Schiff base linkage in rhodopsin by taking into account the external point-charge model [82]. Calculated π-electron charge densities at carbon 14 for 11-*cis*-retinal 77, its Schiff base 78, the protonated Schiff base 79, and rhodopsin model 80 are shown in Fig. 22. According to this calculation, the point-charge close to carbon 14 could reduce the charge density at this point (by Coulombic repulsion) and thus deshield carbon 14; this leads to the observed value of 130.8 ppm.

More experimental data supporting the external point-charge model comes from the resonance Raman work of Mathies et al. [144], who found that interaction with a negatively charged residue (near $C_{11}{=}C_{12}$) could lead to the low value (922 cm^{-1}) of the $C_{11}{=}C_{12}$ hydrogen out of plane bending mode in rhodopsin, by reduction of the bond order.

An external point-charge model was also developed to explain the color of the purple membrane [227], a pigment formed from the all-*trans* isomer of retinal with bacterial opsin. Photochemistry of this pigment creates a proton gradient, which is used for ATP synthesis. In this pigment however, the point-charge responsible for the color was found to be close to the ionone ring.

External point-charge models could explain the regulation of the absorption maxima in visual pigments, including the visual cone pigments, responsible for color vision. Indeed, the wavelength shifts in rod and cone pigments could be produced by positioning one or more point-charges in other orientations with respect to the chromophore.

References

1 Bownds, M.D. (1980) Photochem. Photobiol. 32, 487–490.
2 Hubbel, W.L. and Bownds, M.D. (1979) Annu. Rev. Neurosci. 2, 17–34.
3 Wald, G. (1933) Nature (London) 132, 316–317.
4 Wald, G. and Brown, P.K. (1950) Proc. Natl. Acad. Sci. U.S.A. 36, 84–92.
5 Hubbard, R. (1954) J. Gen. Physiol. 37, 381–399.
6 Daemen, F.J.M. (1973) Biochim. Biophys. Acta 300, 255–258.
7 Daemen, F.J.M., de Grip, W.J. and Jansen, P.A.A. (1972) Biochim. Biophys. Acta 271, 419–428.
8 Fukuda, M., Papermaster, D.S. and Hargrave, P.A. (1979) J. Biol. Chem. 254, 8201–8207.
9 Liang, C.-J., Yamashita, K., Muellenberg, C.G., Shichi, H. and Kobata, A. (1979) J. Biol. Chem. 254, 6414–6418.
10 Shields, J.E., Dinovo, E.C., Henriksen, R.A., Kimbel, R.L. and Millar, P.G. (1967) Biochim. Biophys. Acta 147, 238–251.
11 Heller, J. (1968) Biochemistry 7, 2906–2913.
12 Shichi, H., Lewis, M.S., Irreverre, F. and Stone, A.L. (1969) J. Biol. Chem. 244, 529–536.
13 Hargrave, P.A. and Fong, S.-L. (1977) J. Supramol. Struct. 6, 559–570.
14 Hargrave, P.A., Fong, S.-L., McDowell, J.H., Mas, M.T., Curtis, D.R., Wang, J.K., Juszczak, E. and Smith, D.P. (1980) Neurochem. Int. 1, 231–244.
15 Bownds, M.D. (1967) Nature (London) 216, 1178–1181.
16 Tsunasawa, S., Narita, K. and Shichi, H. (1979) Am. Soc. Photobiol. Abstr., 71.
17 Wald, G. and Brown, P.K. (1953) J. Gen. Physiol. 37, 189–200.
18 Bridges, C.D.B. (1967) Vision Res. 7, 349–369.
19 Dartnall, H.J.A. (1953) Brit. Med. Bull. 9, 24–30.
20 Knowles, A. and Dartnall, H.J.A. (1977) in H. Dawson (Ed.), The Eye, Vol. 2B, Academic Press, New York, p. 77.
21 Munz, F.W. and Schwanzara, S.A. (1967) Vision Res. 7, 111–120.
22 Knowles, A. and Dartnall, H.J.A. (1977) in H. Dawson (Ed.), The Eye, Vol. 2B, Academic Press, New York, pp. 86–98.
23 Raubach, R.A., Nemes, P.P. and Dratz, E.A. (1974) Exp. Eye Res. 18, 1–12.
24 Wu, C.-W. and Stryer, L. (1972) Proc. Natl. Acad. Sci. U.S.A. 69, 1104–1108.
25 Dewey, M.M., Davis, P.K., Blasie, J.K. and Barr, L. (1969) J. Mol. Biol. 39, 395–405.
26 Jan, L.Y. and Revel, J.-P. (1974) J. Cell Biol. 62, 257–273.
27 Saari, J.C. (1974) J. Cell Biol. 63, 480–491.
28 Trayhurn, P., Mandel, P. and Virmaux, N. (1974) Exp. Eye Res. 19, 259–265.
29 Van Breugel, P.J.G.M., Daemen, F.J.M. and Bonting, S.L. (1975) Exp. Eye Res. 21, 315–324.
30 Chen, Y.S. and Hubbel, W.L. (1973) Exp. Eye Res. 17, 517–532.
31 Corless, J.M., Cobbs III, W.H., Costello, M.J. and Robertson, J.D. (1976) Exp. Eye Res. 23, 295–324.
32 Blasie, J.K., Dewey, M.M., Blaurock, A.E. and Worthington, C.R. (1965) J. Mol. Biol. 14, 143–152.
33 Corless, J.M. (1972) Nature (London) 237, 229–231.
34 Chabre, M. (1975) Biochim. Biophys. Acta 382, 332–335.
35 Yeager, M.J. (1975) Brookhaven Symp. Biol. 27, III, 3–35.
36 Saibil, H., Chabre, M. and Worcester, D. (1976) Nature (London) 262, 266–270.
37 Adams, A.J., Somers, R.L. and Shichi, H. (1979) Photochem. Photobiol. 29, 687–692.
38 Azuma, M. and Kito, Y. (1967) Annu. Rep. Biol. Works Fac. Sci. Osaka Univ. 15, 59–69.
39 Rothschild, K.J., Sanches, R., Hsiao, T.L. and Clark, N.A. (1980) Biophys. J. 31, 53–64.
40 Shichi, H. (1971) Photochem. Photobiol. 13, 499–502.
41 Shichi, H. and Shelton, E. (1974) J. Supramol. Struct. 2, 7–16.
42 Cassim, J.Y. and Lin, T. (1975) J. Supramol. Struct. 3, 510–519.
43 Rafferty, C.N., Cassim, J.Y. and McConnel, D.G. (1977) Biophys. Struct. Mech. 2, 277–320.
44 Waddel, W.H., Yudd, A.P. and Nakanishi, K. (1976) J. Am. Chem. Soc. 98, 238–239.

45 Dartnall, H. (1972) in M.G.F. Fuortes (Ed.), Handbook of Sensory Physiology, Vol. VII/1, Springer-Verlag, Berlin, p. 122.
46 Bridges, C.D.B. (1962) Vision Res. 2, 215–232.
47 Abrahamson, E.W. and Ostroy, S.E. (1967) Prog. Biophys. Mol. Biol. 17, 179–215.
48 Applebury, M.L., Zuckerman, D.M., Lamola, A.A. and Joviin, T.M. (1974) Biochemistry 13, 3448–3458.
49 Knowles, A. and Dartnall, H.J.A. (1977) H. Dawson (Ed.), The Eye, Vol. 2B, Academic Press, New York, pp. 298, 305.
50 Wulff, V.J., Adams, R.G., Linschitz, H. and Kennedy, D. (1958) Ann. N.Y. Acad. Sci. 74, 281–290.
51 Abrahamson, E.W., Marquisee, J.A., Gavuzzi, P. and Roubie, J. (1960) Z. Electrochem. 64, 177–180.
52 Pratt, D.C., Livingston, R. and Grellman, K.-H. (1964) Photochem. Photobiol. 3, 121–127.
53 Hubbard, R., Bownds, M.D. and Yoshizawa, T. (1965) Cold Harbour Symp. Quant. Biol. 30, 301–315.
54 Bownds, M.D. and Wald, G. (1965) Nature (London) 205, 254–257.
55 Yoshizawa, T. and Wald, G. (1967) Nature (London) 214, 566–571.
56 Stewart, J.G., Baker, B.N. and Williams, T.P. (1975) Nature (London) 258, 89–90.
57 Stewart, J.G., Baker, B.N. and Williams, T.P. (1977) Biophys. Struct. Mech. 3, 19–29.
58 Sasaki, N., Tokunaga, F. and Yoshizawa, T. (1980) Febs. Lett. 114, 1–3.
59 Sasaki, N., Tokunaga, F. and Yoshizawa, T. (1980) Photochem. Photobiol. 32, 433–441.
60 Shichi, H., Kawamura, S., Muellenberg, C.G. and Yoshizawa, T. (1977) Biochemistry 16, 5376–5380.
61 Yoshizawa, T. and Horiuchi, S. (1973) in H. Langer (Ed.), Biochemistry and Physiology of Visual Pigments, Springer-Verlag, Berlin, pp. 69–81.
62 Shichida, Y., Tokunaga, F. and Yoshizawa, T. (1978) Biochim. Biophys. Acta 504, 413–430.
63 Chabre, M. and Breton, J. (1979) Vision Res. 19, 1005–1018.
64 Sperling, W. and Rafferty, C.N. (1969) Nature (London) 224, 591–594.
65 Burke, M.J., Pratt, D.C., Faulkner, R.R. and Moscowitz, A. (1973) Exp. Eye Res. 17, 557–572.
66 Ebrey, T.G. and Yoshizawa, T. (1973) Exp. Eye Res. 17, 545–556.
67 Honig, B., Kahn, P. and Ebrey, T.G. (1973) Biochemistry 12, 1637–1643.
68 Waggoner, A.S. and Stryer, L. (1971) Biochemistry 10, 3250–3254.
69 Johnston, E.M. and Zand, R. (1972) Biochim. Biophys. Res. Commun. 47, 712–719.
70 Kropf, A., Whittenberger, P., Goff, S. and Waggoner, A.S. (1973) Exp. Eye Res. 17, 591–606.
71 Azuma, M., Azuma, K. and Kito, Y. (1973) Biochim. Biophys. Acta 295, 520–527.
72 Ebrey, T.G. and Honig, B. (1972) Proc. Natl. Acad. Sci. U.S.A. 69, 1897–1899.
73 Akhtar, M., Blosse, P.T. and Dewhurst, P.B. (1967) Chem. Commun. 631–632.
74 Akhtar, M., Blosse, P.T. and Dewhurst, P.B. (1968) Biochem. J. 110, 693–702.
75 Heller, J. (1968) Biochemistry 7, 2914–2920.
76 Zorn, M. (1971) Biochim. Biophys. Acta 245, 216–220.
77 Fager, R., Sejnowski, P. and Abrahamson, E.W. (1972) Biochim. Biophys. Res. Commun. 47, 1244–1247.
78 Wang, J.K., McDowell, J.H. and Hargrave, P.A. (1980) Biochemistry 19, 5111–5117.
79 Callender, R. and Honig, B. (1977) Annu. Rev. Biophys. Bioeng. 6, 33–55.
80 Favrot, J., Leclerq, J.M., Roberge, R., Sandorfy, C. and Vocelle, D. (1979) Photochem. Photobiol. 29, 99–108.
81 Shriver, J., Mateescu, G., Fager, R., Torchia, D. and Abrahamson, E.W. (1977) Nature (London) 270, 271–274.
82 Honig, B., Dinur, U., Nakanishi, K., Balogh-Nair, V., Gawinowicz, M.A., Arnaboldi, M. and Motto, M.G. (1979) J. Am. Chem. Soc. 101, 7084–7086.
83 Stubbs, G.W. and Litman, B.J. (1978) Biochemistry 17, 220–225.
84 Matsumoto, H., Horiuchi, K. and Yoshizawa, T. (1978) Biochim. Biophys. Acta 501, 257–268.
85 Henselman, R.A. and Cusanowich, M.A. (1974) Biochemistry 13, 5199–5203.
86 Stubbs, G.W., Smith, H.G. and Litman, B.J. (1976) Biochim. Biophys. Acta 426, 46–56.
87 Blatchly, R.A., Carriker, J.D., Balogh-Nair, V. and Nakanishi, K. (1980) J. Am. Chem. Soc. 102, 2495–2497.

88 Shichi, H. (1971) J. Biol. Chem. 246, 6178–6182.
89 Zorn, M. (1975) Exp. Eye Res. 19, 215–221.
90 Pilkiewicz, F.G., Pettei, M.J., Yudd, A.P. and Nakanishi, K. (1977) Exp. Eye Res. 24, 421–423.
91 Rothmans, J.P., Bonting, S.L. and Daemen, F.J.M. (1972) Vision Res. 12, 337–341.
92 Arnaboldi, M., Motto, M.G., Tsujimoto, K., Balogh-Nair, V. and Nakanishi, K. (1979) J. Am. Chem. Soc. 101, 7082–7084.
93 Yoshizawa, T. and Kito, Y. (1958) Nature (London) 182, 1604–1605.
94 Hubbard, R. and Kropf, A. (1958) Proc. Natl. Acad. Sci. U.S.A. 44, 130–139.
95 Grellman, K.H., Livingston, R. and Pratt, D. (1962) Nature (London) 193, 1258–1260.
96 Yoshizawa, T. and Wald, G. (1963) Nature (London) 197, 1279–1286.
97 Wald, G. (1968) Science 162, 230–239.
98 Abrahamson, E.W. and Wiesenfeld, J.R. (1972) in H.J.A. Dartnall (Ed.), Handbook of Sensory Physiology, Vol. VII/1, Springer-Verlag, Berlin, pp. 69–121.
99 Yoshizawa, T. (1972) Ibid., pp. 146–149.
100 Abrahamson, E.W. (1973) in H. Langer (Ed.), Biochemistry and Physiology of Visual Pigments, Springer-Verlag, Berlin, pp. 47–56.
101 Yoshizawa, T. and Wald, G. (1964) Nature (London) 201, 340–345.
102 Yoshizawa, T. and Horiuchi, S. (1969) Exp. Eye Res. 8, 243–244.
103 Tsukamoto, Y., Horiuchi, S. and Yoshizawa, T. (1975) Vision Res. 15, 819–823.
104 Kawamura, S., Tokunaga, F. and Yoshizawa, T. (1977) Vision Res. 17, 991–999.
105 Shichida, Y., Tokunaga, F. and Yoshizawa, T. (1979) Photochem. Photobiol. 29, 343–351.
106 Tsuda, M., Tokunaga, F., Ebrey, T.G., To Yue, K., Marque, J. and Eisenstein, L. (1980) Nature (London) 287, 461–462.
107 Tokunaga, F., Sasaki, N. and Yoshizawa, T. (1980) Photochem. Photobiol. 32, 447–453.
108 Sarai, A., Kakitani, T., Shichida, Y., Tokunaga, F. and Yoshizawa, T. (1980) Photochem. Photobiol. 32, 199–206.
109 Hoffman, W., Siebert, F., Hoffman, K.P. and Kreutz, W. (1978) Biochim. Biophys. Acta 503, 450–461.
110 Uhl, R., Hoffman, K.P. and Kreutz, W. (1978) Biochemistry 17, 5347–5352.
111 Kawamura, S., Miyatani, S., Matsumoto, H., Yoshizawa, T. and Liu, R.S.H. (1980) Biochemistry 19, 1549–1553.
112 Kakitani, K. and Kakitani, H. (1980) Photochem. Photobiol. 32, 707–709.
113 Cone, R.A. (1972) Nat. New Biol. 236, 39–43.
114 Rosenfeld, T., Alchalal, A. and Ottolenghi, M. (1972) Nature (London) 240, 482–483.
115 Bensasson, R., Land, E.J. and Truscott, T.G. (1975) Nature (London) 258, 768–770.
116 Goldschmith, Ch.R., Ottolenghi, M. and Rosenfeld, T. (1976) Nature (London) 263, 169–171.
117 Busch, G.E., Applebury, M.L., Lamola, A.A. and Rentzepis, P.M. (1972) Proc. Natl. Acad. Sci. U.S.A. 69, 2802–2806.
118 Peters, K., Applebury, M.L. and Rentzepis, P.M. (1977) Proc. Natl. Acad. Sci. U.S.A. 74, 3119–3123.
119 Sundstrom, V., Rentzepis, P.M., Peters, K. and Applebury, M.L. (1977) Nature (London) 267, 645–646.
120 Shichida, Y., Yoshizawa, T., Kobayashi, T., Ohtani, H. and Nagakura, S. (1977) Febs. Lett. 80, 214–216.
121 Shichida, Y., Kobayashi, T., Ohtani, H., Yoshizawa, T. and Nagakura, S. (1978) Photochem. Photobiol. 27, 335–341.
122 Green, G.H., Monger, T.G., Alfano, R.R., Aton, B. and Callender, R.H. (1977) Nature (London) 269, 179–180.
123 Monger, T.G., Alfano, R.R. and Callender, R.H. (1979) Biophys. J. 7, 105–115.
124 Kobayashi, T. (1979) Febs. Lett. 106, 313–316.
125 Applebury, M.L., Peters, K. and Rentzepis, P.M. (1978) Biophys. J. 23, 375–382.
126 Applebury, M.L. (1980) Photochem. Photobiol. 32, 425–431.
127 Van der Meer, K., Mulder, J.J.C. and Lughtenburg, J. (1976) Photochem. Photobiol. 24, 363–367.

128 Warshel, A. (1976) Nature (London) 260, 679–683.
129 Warshel, A. (1978) Proc. Natl. Acad. Sci. U.S.A. 75, 2558–2562.
130 Rimai, L., Kilponen, R.G. and Gill, D. (1970) Biochem. Biophys. Res. Commun. 41, 492–497.
131 Spiro, T.G. (1974) Acc. Chem. Res. 7, 339–344.
132 Warshel, A. (1977) Ann. Rev. Biophys. Bioeng. 6, 273–300.
133 Oseroff, A.R. and Callender, R.H. (1974) Biochemistry 13, 4243–4248.
134 Callender, R.H., Doukas, A., Crouch, R. and Nakanishi, K. (1976) Biochemistry 15, 1621–1629.
135 Mathies, R., Oseroff, A.R. and Stryer, L. (1976) Proc. Natl. Acad. Sci. U.S.A. 73, 1–5.
136 Sulkes, M., Lewis, A. and Markus, M.A. (1978) Biochemistry 17, 4712–4722.
137 Lewis, A. (1978) Biophys. J. 24, 249–254.
138 Favrot, J., Vocelle, D. and Sandorfy, C. (1979) Photochem. Photobiol. 30, 417–421.
139 Harosi, F.I., Favrot, J., Leclerq, J.M., Vocelle, D. and Sandorfy, C. (1978) Rev. Can. Biol. 37, 257–271.
140 Eyring, G. and Mathies, R. (1979) Proc. Natl. Acad. Sci. U.S.A. 76, 33–37.
141 Narva, D. and Callender, R.H. (1980) Photochem. Photobiol. 32, 273–276.
142 Aton, B., Doukas, G., Narva, D., Callender, R.H., Dinur, U. and Honig, B. (1980) Biophys. J. 29, 79–94.
143 Eyring, G., Bostick, C., Mathies, R., Fransen, R., Palings, I. and Lughtenburg, J. (1980) Biochemistry 19, 2410–2418.
144 Eyring, G., Bostick, C., Mathies, R., Broek, A. and Lughtenburg, J. (1980) J. Am. Chem. Soc. 102, 5390–5392.
145 Kropf, A. (1976) Nature (London) 264, 92–94.
146 Ito, M.,. Hirata, K., Kodama, A., Tsukida, K., Matsumoto, H., Horiuchi, K. and Yoshizawa, T. (1978) Chem. Pharm. Bull. 26, 925–929.
147 Kawamura, S., Yoshizawa, T., Horiuchi, K., Ito, M., Kodama, A. and Tsukida, K. (1979) Biochim. Biophys. Acta 518, 147–152.
148 Ito, M., Kodama, A., Murata, M., Kobayashi, M., Tsukida, K., Shichida, Y. and Yoshizawa, T. (1979) J. Nutr. Sci. Vitaminol. 25, 343–345.
149 Akita, H., Tanis, S.P., Adams, M., Balogh-Nair, V. and Nakanishi, K. (1980) J. Am. Chem. Soc. 102, 6370–6372.
150 Mao, B., Tsuda, M., Ebrey, T.G., Akita, H., Balogh-Nair, V. and Nakanishi, K. (1981) Biophys. J. 35, 543–546.
151 Gilardy, R., Karle, I.L., Karle, J. and Sperling, W. (1971) Nature (London) 232, 187–188.
152 Gilardy, R., Karle, I.L. and Karle, J. (1972) Acta Crystallogr. Sect. 2B. 28, 2605–2612.
153 Hamanaka, T., Mitsui, T., Ashida, T. and Kakudo, M. (1972) Acta Crystallogr. Sect. B 28, 214–222.
154 Schaeffer, A.H., Waddel, W.H. and Becker, R.S. (1974) J. Am. Chem. Soc. 96, 2063–2068.
155 Honig, B. and Karplus, M. (1971) Nature (London) 229, 558–560.
156 Warshel, A. and Karplus, M. (1974) J. Am. Chem. Soc. 96, 5677–5689.
157 Rowan, R., Warshel, A., Sykes, B.D. and Karplus, M. (1974) Biochemistry 13, 970–980.
158 Becker, R.S., Berger, S., Dalling, D.K., Grant, D.M. and Pugmire, R.J. (1974) J. Am. Chem. Soc. 96, 7008–7014.
159 Gill, D., Heyde, M.E. and Rimai, L. (1971) J. Am. Chem. Soc. 93, 6288–6289.
160 Lewis, A. and Fager, R. (1973) J. Raman Spectrosc. 1, 465–470.
161 Cookingham, R.E. and Lewis, A. (1978) J. Mol. Biol. 119, 569–577.
162 Ebrey, T.G., Govindjee, R., Honig, B., Pollock, E., Chan, W., Crouch, R., Yudd, A.P. and Nakanishi, K. (1975) Biochemistry 14, 3933–3941.
163 Chan, W., Nakanishi, K., Ebrey, T.G. and Honig, B. (1974) J. Am. Chem. Soc. 96, 3642–3644.
164 Mathies, R., Friedman, T.B. and Stryer, L. (1977) J. Mol. Biol. 109, 367–372.
165 Nakanishi, K., Yudd, A.P., Crouch, R., Olson, G.L., Cheung, H.-C., Govindjee, R., Ebrey, T.G. and Patel, D.J. (1976) J. Am. Chem. Soc. 98, 236–238.
166 Hubbard, R. and Wald, G. (1952) J. Gen. Physiol. 36, 269–315.
167 Wald, G., Brown, P.K., Hubbard, R. and Orosnik, W. (1955) Proc. Natl. Acad. Sci. U.S.A. 41, 438–451.

168 Orosnik, W., Brown, P.K., Hubbard, R. and Wald, G. (1956) Proc. Natl. Acad. Sci. U.S.A. 42, 578–580.
169 Kini, A., Matsumoto, H. and Liu, R.S.H. Submitted for publication.
170 Crouch, R., Purvin, V., Nakanishi, K. and Ebrey, T. (1975) Proc. Natl. Acad. Sci. U.S.A. 72, 1538–1542.
171 De Grip, W.J., Liu, R.S.H., Ramamurthy, V. and Asato, A.E. (1976) Nature (London) 262, 416–418.
172 Kini, A., Matsumoto, H. and Liu, R.S.H. (1979) J. Am. Chem. Soc. 101, 5078–5079.
173 Matsumoto, H. and Yoshizawa, T. (1978) Vision Res. 18, 607–609.
174 Matsumoto, H., Liu, R.S.H., Simmons, C.J. and Seff, K. (1980) J. Am. Chem. Soc. 102, 4259–4262.
175 Crouch, R., Katz, S., Nakanishi, K., Gawinowicz, M.A. and Balogh-Nair, V. (1981) Photochem. Photobiol. 33, 91–95.
176 Van den Tempel, P.J. and Huisman, H.O. (1966) Tetrahedron 22, 293–300.
177 Blatz, P.E., Lin, M., Balasubramaniyan, P., Balasubramaniyan, V. and Dewhurst, P.B. (1969) J. Am. Chem. Soc. 91, 5930–5931.
178 Gartner, W., Hopf, H., Hull, W.E., Oesterhelt, D., Scheutzov, D. and Towner, P. (1980) Tetrahedron Lett., 347–350.
179 Nelson, R., de Riel, J.K. and Kropf, A. (1970) Proc. Natl. Acad. Sci. U.S.A. 66, 531–538.
180 Waddel, W.H., Uemura, M. and West, J.L. (1978) Tetrahedron Lett., 3223–3226.
181 Tanis, S.P., Brown, R.H. and Nakanishi, K. (1978) Tetrahedron Lett., 869–872.
182 Kropf, A. (1975) in Abstracts of Annual Meeting of Biophysical Society of Japan, p. 281.
183 Motto, M.G., Sheves, M., Tsujimoto, K., Balogh-Nair, V. and Nakanishi, K. (1980) J. Am. Chem. Soc. 102, 7947–7949.
184 Asato, A., Matsumoto, H., Denny, M. and Liu, R.S.H. (1978) J. Am. Chem. Soc. 100, 5957–5960.
185 Curtis, M.J., Pitt, G.A.J. and Howell, C. (1965) in E.J. Bowen (Ed.), Recent Progress in Photobiology, Blackwell, Oxford, p. 119.
186 Sokolova, N.A., Mitsner, B.I. and Zakis, V.I. (1979) Bioorg. Khim. 5, 1053–1058.
187 Lewin, D.R. and Thomson, N.J. (1967) Biochem. J. 103, 36P.
188 Azuma, M., Azuma, K. and Kito, Y. (1973) Biochim. Biophys. Acta 295, 520–527.
189 Wald, G. (1953) Fed. Proc. 12, 606–611.
190 Liu, R.S.H., Asato, A.E. and Denny, M. (1977) J. Am. Chem. Soc. 99, 8095–8097.
191 Matsumoto, H., Asato, A.E., Denny, M., Baretz, B., Yen, Y.-P., Tong, D. and Liu, R.S.H. (1980) Biochemistry 19, 4589–4594.
192 Knowles, J.R. (1972) Acc. Chem. Res. 5, 155–160.
193 Bailey, H. and Knowles, J.R. (1977) Methods Enzymol. 46, 69–114.
194 Chowdhry, W. and Westheimer, F.H. (1979) Annu. Rev. Biochem. 48, 293–325.
195 Tometsko, A.M. and Richards, F.M. (Eds.) (1980) Ann. N.Y. Acad. Sci. 1–385, 434–474, 491–500.
196 Fransen, M.R., Luyten, W.C.M.M., Van Thuijl, J. and Lughtenburg, J. (1976) Nature (London) 260, 726–727.
197 Thomson, A. (1975) Nature (London) 254, 178–179.
198 Rafferty, C.N. and Shichi, H. (1981) Photochem. Photobiol. 33, 229–234.
199 Warshel, A. and Deakyne, C. (1978) Chem. Phys. Lett. 55, 459–465.
200 Huppert, D., Rentzepis, P.M. and Kliger, D.S. (1977) Photochem. Photobiol. 25, 193–197.
201 Rosenfeld, T., Honig, B. and Ottolenghi, M. (1977) Pure Appl. Chem. 49, 341–351.
202 Salem, L. and Bruckman, P. (1975) Nature (London) 258, 526–528.
203 Kakitani, T. and Kakitani, H. (1975) J. Phys. Soc. Japan 38, 1455–1463.
204 Kakitani, T. (1979) Biophys. Struct. Mech. 5, 293–312.
205 Cooper, A. (1979) Nature (London) 282, 531–533.
206 Hurley, J.B., Ebrey, T.G., Honig, B. and Ottolenghi, M. (1977) Nature (London) 270, 540–542.
207 Honig, B., Ebrey, T.G., Callender, R.H., Dinur, U. and Ottolenghi, M. (1979) Proc. Natl. Acad. Sci. U.S.A. 76, 2503–2507.
208 Weiss, R.M. and Warshel, A. (1979) J. Am. Chem. Soc. 101, 6131–6133.
209 Birge, R.R. and Hubbard, L.M. (1980) J. Am. Chem. Soc. 102, 2195–2205.

210 Erickson, J.O. and Blatz, P.E. (1968) Vision Res. 8, 1367–1374.
211 Blatz, P.E., Mohler, J.H. and Navangul, H.V. (1972) Biochemistry 11, 848–855.
212 Blatz, P.E. and Pippert, D.L. (1968) J. Am. Chem. Soc. 90, 1296–1300.
213 Blatz, P.E., Pippert, D.L. and Balasubramaniyan, V. (1968) Photochem. Photobiol. 8, 309–315.
214 Wiesenfeld, J.R. and Abrahamson, E.W. (1968) Photochem. Photobiol. 8, 487–493.
215 Mantione, M.-J. and Pullman, B. (1971) Int. J. Quant. Chem. 5, 349–360.
216 Rosenberg, B. and Krigas, T.M. (1967) Photochem. Photobiol. 6, 769–773.
217 Irving, C.S. and Leermakers, P.A. (1968) Photochem. Photobiol. 7, 665–670.
218 Irving, C.S., Byers, G.W. and Leermakers, P.A. (1969) J. Am. Chem. Soc. 91, 2141–2143.
219 Irving, C.S., Byers, G.W. and Leermakers, P.A. (1970) Biochemistry 9, 858–864.
220 Blatz, P.E. and Mohler, J.H. (1970) Chem. Commun. 614–615.
221 Suzuki, H., Komatsu, T. and Kitayima, H. (1974) J. Phys. Soc. Japan 37, 177–185.
222 Honig, B., Greenberg, A.D., Dinur, U. and Ebrey, T.G. (1976) Biochemistry 15, 4593–4599.
223 Komatsu, T. and Suzuki, H. (1976) J. Phys. Soc. Japan 40, 1725–1732.
224 Hubbard, R. (1969) Nature (London) 221, 432–435.
225 Waleh, A. and Ingraham, L.L. (1973) Arch. Biochem. Biophys. 156, 261–266.
226 Sheves, M., Nakanishi, K. and Honig, B. (1979) J. Am. Chem. Soc. 101, 7086–7088.
227 Nakanishi, K., Balogh-Nair, V., Arnaboldi, A., Tsujimoto, K. and Honig, B. (1980) J. Am. Chem. Soc. 102, 7945–7947.
228 Lewis, A. (1978) Proc. Natl. Acad. Sci. U.S.A. 75, 549–553.
229 Lewis, A. and Spoonhover, J. (1974) in S. Yip and S. Chen (Eds.), Neutron, X-Ray and Laser Spectroscopy in Biophysics and Chemistry, Academic Press, New York, pp. 347–376.

Subject index

Absolute configuration
of axially chiral molecules 17
of chiral cages 19–20
of hexacoordinate centres 16
of molecular helices and propellors 19
of molecules with a chiral plane 18
of pentracoordinate centres 15
of tetracoordinate centres 12
Absorption spectra of visual
pigments 285–287, 322–328
charge-transfer model 324
counterion and 324
microenvironmental polarizability models 323
nomogram 286–287
point-charge perturbation models 324–328
retinylic cation 323
Achiral molecules 6
carbenium ions 11
carbon radicals 11
diastereomeric 20
meso-forms 23
Achiral point groups 5
Acotinase 87–89, 91–95
Adjacent attack in substitution at phosphorus 203
ADP
analysis for bridging ^{18}O 218–219
α-[^{18}O], assignment of chirality 239–240
α-[^{18}O]phosphorothioate 235
β-[^{18}O]phosphorothioate 210, 216–217, 235
ALA dehydratase 271, 275, 277
ALA synthase 176–177, 271, 275–277
Alanine
proton exchange by transaminases 166–167
racemisation 170–171
Alanine dehydrogenase 134
Alanine racemase 61, 170–171
L-Alanine transaminase 166–167, 268
Alcohol dehydrogenase
active site topography 63, 139, 141, 143–145, 147, 151–153
DMSO complex 142, 146–147
isozymes 113
NADH configuration and 83, 116–117
stereodifferentiation by 74, 117
structure 113, 137–140
Aldehyde reductase 142, 148
Aldose reductase 142
Alkaline phosphatase, stereochemistry of transphosphorylation 232
Allenes 17, 310–311
Amino acid interconversion, PLP-catalysed 178–180
Aminocyclopropanecarboxylic acid (ACC) 192–193
δ-Aminolaevulinic acid (ALA) 271, 273, 275
5-Aminolaevulinate synthetase *see* ALA synthase
AMP [^{18}O]phosphorothioate 211, 216–217, 232
Anisometric molecules 8–9
Anomeric effect 42
Anticlinal conformation 30
Antiperiplanar conformation 30
Apical position 15, 202
Apo-aspartate transaminase 166
Aspartase (aspartate ammonia-lyase) 76, 83
Aspartate-β-decarboxylase 166, 187–188
Aspartate transaminase 167–170
Associative mechanisms in phosphate substitution 202–203
Asymmetric atoms
carbon 6, 12
nitrogen 11
Asymmetric hydrogenation 71, 105–106
Asymmetric organic reactions 104
Asymmetry
reflection 50
rotational 50
ATP
analysis for bridging ^{18}O 217–218
chiral phosphate 206
configuration assignment of ATPαS 214–216
Cr^{3+} and Co^{3+} complexes 227–229, 241–243
cyclisation to cAMP 239
enzymatic bond cleavages 205–206
Mg^{2+} complexes 204–205, 227, 241–243
γ-[^{18}O]phosphorothioate 209–210, 235
Atropisomers 18
Attachment 57
one-point 57, 62–63
three-point 54–55, 57–59, 61–63
two-point 57

Axial–equatorial equilibrium in cyclohexanes 38–39
Axial position 37–38, 41–42
Axis of chirality 17–18
Axis of prochirality 26
Axis of rotation 3, 6, 50
Axis of rotation–reflection 4
Aziridine 11

Barrier of rotation
 about C–O and C–N bonds 35
 about sp^3 C–heteroatom bonds 34
 in dialkyl disulphides, eclipsed and *trans* barriers 36
 of biphenyls 18
 of *n*-butane 31
 of ethylene 21
 of toluene 33
Bathorhodopsin 296–302, 315–322
Benzene 4
Biphenyls 17–18
Bisectional bond 37
Bleaching 288–292
Boat conformer 38
Boat form of cyclohexane 38
Bond rotation 29
Brändén crevice 133, 155

Carbonic anhydrase, environment of zinc in 153
Carboxypeptidase A 58–59, 153
CD spectra
 of metal–nucleotide complexes 229, 240
 of rhodopsin 287–288, 290–292
Centre of chirality 44
 hexacoordinate 15–16
 multiple 20, 22–24, 44
 pentacoordinate 14–15
 tetracoordinate 10, 12–14
 tricoordinate 10–12
Centre of prochirality 26–27
Centre of symmetry 3–4
Chair conformer 38–39, 42
Chair form of cyclohexane 37–38
Chiral cages 19
Chiral catalysts 104–106
Chirality 6–7, 49–50
Chiral methyl groups 181–182
 analysis 101–103
 in *S*-adenosylmethionine 278–279
 synthesis 98–101
Chiral molecules 6–7, 14, 49–50
Chiral phosphate 202, 221–226
 alkaline phosphatase and 232
 configuration assignments 224–226, 232, 239–240
 general stereochemistry in enzyme reactions 243–246
 in ATP 206
 in cyclic phosphodiesters 226
 in 2-phosphoglycerate 222
 in 1,2-propanediol-1-phosphate 222, 224–226
 kinases and 234–237
 synthesis of 222–223
Chiral point groups 5
Chlorochemistry 7
Chloroperoxidase 67–68
Chymotrypsin 151–152
Cinnamyl alcohol dehydrogenase 148
CIP nomenclature *see R* and *S* nomenclature
Cis-band of visual pigments 287
Cis–trans-isomerism
 about a double bond 20–21
 in cyclic molecules 23–24
Cis–trans nomenclature 24
Citric acid configuration 87–97
Clostridium spp. 173, 249–251, 265, 268, 272
Coenzyme B_{12} (AdoCbC) 249–279
 biosynthesis 271–279
 dioldehydratase and 251–261
 ethanol ammonia lyase and 268–270
 glutamate mutase and 250–251
 β-lysine mutase and 265–267
 methylmalonyl-CoA mutase and 261–265
 reactions of 249–271
 structure 250
Cone cells 283–285
Configuration 1–2, 9–10, 50
 determination of 77–97
 effect on conformational equilibrium in hexopyranoses 42
 of chiral methyl groups 101–103
 of chiral phosphate 224–226
 of citric acid 87–97
 of malic acid 80–83
 of metal–nucleotide complexes 227–228
 of NADH and NADPH 83–87
 of tetracoordinate centres 12–13
 prochiral 50
Configurational isomerism 18, 23
Configurational isomers 9–10, 24
Conformation 1–2, 9, 29, 36
 eclipsed 30
 gauche 30
 influence of environment 34
 influence of substituents 31
 of biomolecules 32

of cyclic systems 36–42
of dialkyl disulphides 36
of sp^3-sp^2 systems 33
of sp^3-sp^3 systems 33
staggered 30, 37
trans 30
Conformational equilibrium constant 29
Conformational isomerism 18, 29, 36, 42
Conformation isomers (conformers) 9–10
acyclic 29–30
cyclic 36–42
Conformationally chiral center 11
Conformers *see* Conformational isomers
Constitution 7,10, 24
Constitutional isomerism 7–9, 57
Constitutional isomers 7, 65
Constitutionally heterotopic fragments 10
Constitutionally unsymmetrical molecules 22
Corrin biosynthesis 271–279
Coupled oscilator mechanism 291
Curvularia falcata 76–77
3′,5′-Cyclic AMP (cAMP)
chiral [^{18}O]phosphate 214, 223, 233–234
chiral phosphorothioate 212–214, 232–234
stereochemistry of cyclisation 239
Cyclic GMP 283, 285
2′,3′-Cyclic UMPS 207, 212, 230–232, 239
Cyclodiastereoisomerism 44
Cycloenantiomerism 44
Cyclohexane 37, 40–41
disubstituted 39
monosubstituted 38–39
Cyclostereoisomerism 44
Cystathionine-γ-synthase 189–192

Dark adaptation 290, 292
Decarboxylase 172–174
glutamate, abortive transamination 173
histidine, non-PLP containing 173
meso-diaminopimelate, inversion of configuration by 173–174
retention of configuration in 173
stereochemical mechanism of 173–174
tyrosine, conformational changes on binding substrate 173
Dehydrogenase 113–155
evolution of 154–155
reaction mechanisms 118
structural features of 148–155
Detergents
denaturation of rhodopsin 287–288
effect on bleaching intermediates 290, 297, 300
effect on rhodopsin reconstitution 294–295
Dialkyl amino acid transaminase 166
3, 5-Diaminohexanoic acid (β-lysine) 265–267
Diastereodifferentiation 70
see also Differentiation
Diastereoisomerism 10, 20–24
π-Diastereoisomerism 20
Diastereoisomers 7–10, 28, 53, 55–56, 66
in polysubstituted cyclic systems 24
of hexacoordinate centres 16
of pentacoordinate centres 10, 15
π-diastereoisomers 20–21
Diastereotopic faces 27
Diastereotopic groups 10, 27–28, 70
Diastereotopos 70
Diaxial 1/3 interactions in cyclohexane 38
Dichloroethylene 5, 7, 9
Diequatorial–diaxial equilibrium in 1,2-disubstituted cyclohexanes 39–40
Differentiation 70–75
diastereoface 73
diastereoisomer 74
diastereotopos 73–75
enantioface 71, 74–75
enantiomer 72
enantiotopos 72
Dihydrofolate reductase 121, 123–126, 149
Dihydrofolic acid 121, 123, 126
Dihydrosphingosine synthetase 177–178
Dimethylsulphoxide (DMSO) as dehydrogenase inhibitor 142
glucose-6-phosphate dehydrogenase and 86
Dioldehydratase 251–261
ethylene glycol and 257–261
glycerol and 255–257
1,2-propanediol and 251–255, 261
Dissociative mechanism in phosphate substitution 202–203
Dissymmetric molecules 6–7, 11, 17
D and L nomenclature 12
Double-displacement mechanisms in phosphate substitution 204, 243–245

Eclipsed conformers 30–31
Enantiodifferentiation 70, 105
see also Differentiation
Enantiomerism 10, 42
Enantiomers 7–10, 22, 28, 52
of biphenyls 18
of chiral cages 19–20
of hexacoordinate centres 16
of molecular propellers 19
of pentacoordinate centres 15

of tetracoordinate centres 12
of tricoordinate centres 11
Enanthiomorphic substituents 23
Enantiomorphs 53
Enantiotopic faces 27
Enantiotopic groups 10, 26–28, 69
Enantiotopos 70
Enantiozymes 66
Envelope form of cyclopentane 37
Epimers 22
Epinephrine 54
Equatorial–axial conversion 41, 202–203
Equatorial position 15, 37–38, 41–42, 202
Erythro-isomer 22
Ethanolamine ammonia lyase 268–270
Ethylmalonyl-CoA, conversion to methylsuccinic acid 262–265
E and *Z* conformers 34–35
E and *Z* isomers 20–21

Facial (*fac*) form 16
Flavin coenzymes 114–115, 118–121
Flavodoxins 115, 149
Framework group 6
Fumarase 95, 101–103

GABA-transaminase 170
Galactose-1-phosphate uridylyltransferase 238–239, 243–245
Gauche conformer 31–32
interaction of axial substitutents 38
Gauche-effect 40
Geometrical isomerism 20
Geometry 1–3, 42
Glucose-6-phosphate dehyrogenase 86
Glutamate decarboxylase abortive transamination 173
Glutamate dehydrogenase 85, 134–135
Glutamate mutase 250–251
Glutathionine reductase 119–121
(*R*)-(−)-Glutinic acid 17
Glyceraldehyde-3-phosphate dehydrogenase 150–152
active site 129–130, 150–151
coenzyme binding 133
structure 128, 132
Glycerol, enzymic conversion to 3-hydroxypropionaldehyde 255–257
Glycerol 3-phosphate 130
assignment of phosphate chirality 215
Glycerol-3-phosphate dehydrogenase 130, 134–135
Glycine, as labelled precursor 275–277
Glycolic acid
absolute configuration 78–79
in chiral methyl synthesis 98
Glyoxylate reductase 98
Gyrochiral molecules 20

Half-chair form of cyclopentane 37
Helicines 19
Helicity 19–20
Heterotopic groups 10
High-performance liquid chromatography
of ADPαS isomers 209
of visual pigments 311, 314
Histidine decarboxylase 173
Homochiral reactions 117
Homofacial reactions 117
Homomers 8–9
Homotopic faces 26
Homotopic groups 10, 25–26, 28
Homotopism 25
Hudson Lactone Rule 89
Hydride transfer
in nicotinamide coenzymes 116, 118
in non-enzymic reactions 117
Hydroretinals 325–326
p-Hydroxybenzoate hydroxylase 120–123
4-Hydroxy-2-ketoglutarate aldolase 59
Hydroxymethylbilane 272, 274
β-Hydroxysteroid dehydrogenase 83–84
Hypsorhodopsin 299–300

In-line mechanism in substitution at phosphorus 202, 204, 243
Inversion
as symmetry operation 3–4
at first-row atoms 11, 21
conformer interconversion by 29, 35, 41
inversion barrier 35
ring 38, 40–41
Isobacteriochlorins 272
Isocitrate dehydrogenase 88–89, 91–95
reaction pathway 91
Isoclinal positin 37–38
Isomerization of double bonds 21
in visual pigments 296
Isomers 7–9
separation of 29
torsional 18, 33
Isometric molecules 8–9
Isometry 8–9
Isorhodopsin 297–299
Isotope effect 103, 256–258, 265
Isotopes of hydrogen

chiral methyl groups 98–109
prochiral centres 78–83
Isotopes of oxygen
in dioldehydratase studies 252, 256, 259–260
isotope scrambling in [^{18}O]phosphate 201
isotope shifts in ^{31}P-NMR 218–219, 226
see also Chiral phosphate

2-Ketoglutarate 87, 89
Kinases
specificity for ATP-phosphorothioate isomers 214, 217–218
stereochemistry of phosphoryl group transfer 234–237, 243–245
Klebsiella pneumoniae 251, 255
Kuhn sum rule 291

Lactate dehydrogenase 78–79, 98
active site 130, 151–152
coenzyme binding 133
DMSO complex 142
structure 127–129
Lactic acid, chiral methyl 106
Lignin 142, 148
β-Lysine mutase 265–267
Lysine transaminase 170

Malate dehydrogenase 128, 131
Malate synthese 101–103
Malic acid
configuration and synthesis 80–83
from chiral acetate 101–103
in studying aconitase 93–95
Mandelate racemase 61
Mass spectrometry 86–87, 225
GC–MS 127
metastable ion MS 226
Meridional (*mer*) form 16
Meso form 20, 23
Metal–nucleotide complexes 204–205, 227–229
as enzyme substrates 241–243
configurational assignment 229
configurational designation 227–228
in assigning phosphate configuration 239–240
spectral properties 228–229
synthesis and separation 228
Metaphosphate intermediates in phosphate substitution 202–203
Methoxyisoxazolidine-3,3-dicarboxylic acid bismethylamide 11
Methylmalonyl-CoA mutase 261–265
Methylsuccinic acid ^{1}H-NMR 263

Mirror plane 5
Morphic relationships of molecular fragments 10

NADH/NAD$^+$
as dehydrogenase coenzyme 116–118, 127–148
configuration and stereoselectivity 83–87
selectivity for NAD vs. NADP 154–155
NADH : FMN oxidoreductase 114
NADPH/NADP$^+$
carbonyl reductases and 142
configuration and stereoselectivity 83–87
dehydrogenases and 116, 118, 126–127, 134, 148
dihydrofolate reductase and 121, 123, 125
glutathione reductase and 119
hydroxylases and 120–123
in citric acid cycle 91–94
Neutron diffraction 78
Newman projection 30, 32, 37
Nondissymmetric molecules 6
Nucleotidyltransferases 237–240

Opsin 285
composition 285
conformation 290, 292
delipation of 295
retinal binding 292–296
stability 293
Opsin shift 322–323
Optical activity 7, 14, 23, 50–51, 53
Optical isomers 7
Optical rotation 7
Ororate reductase 114
Oxaloacetate 87–89, 93–96, 128

P and *M* nomenclature 19
Papain 151–152
Paraphane derivatives 18
1-Phenylethanol 7, 98
6-Phosphogluconate dehydrogenase 126–127, 150
Phosphohydrolase reactions, stereospecificity of 203, 230–234
Phosphorothioates 206–221
as enzyme substrates 219–221, 233–234
assignment of configuration 214–219
chiral synthesis of 206–214
[^{16}O,^{17}O,^{18}O]-, enantiomer assignment 218–221
glycerate 211–213
isomer separations 208–209
^{18}O-labelled 206, 209–212

nucleoside 5′- 206
nucleoside 2′,3′-cyclic 207, 212, 230–232
nucleoside 3′,5′-cyclic 212–214
β-thionucleotides 209
Phosphotransferases 234–237, 243
Photoreceptors 283–284
Plane of chirality 18
Plane of prochirality 26
Plane of symmetry 3–4
^{31}P-NMR 201
effect of D-isotopes 218–219
of Co(III) complexes 228
of cyclic phosphate esters 226
of labelled cAMP 233
of nucleotide phosphothioates 214–215
Point group 5–6
Polyaffinity relationship 54
Porphobilinogen (PBG) 271–273, 275–277
Porphyropsins 285–287
Primary event 288, 296–304, 315–322
cis–trans isomerisation models 317–322
electron transfer model 317
proton translocation models 302–303, 315–317
resonance Raman studies 300–302
torsion model 318
visual pigment analogues and 302–304
Prochiral centre 26–27
determination of configuration 78–97
of citric acid 87–97
of glycolic acid 78–79
of keto acids 104
of malic acid 80–83
of NADH/NADPH 83–87
of trihydroxyglutaric acid 23
Prochirality 26–28, 50, 54, 61–63
sp^2 prochirality 70
Proline racemase 61
1,2-Propanediol
enzymic conversion to propionaldehyde 251–255, 261
labelled syntheses 252
Propellers, molecular 19
Propionibacterium shermanii 262, 272
Pro-R, Pro-S 26, 51
Prostereoisomerism 10, 25, 27–29
Proton magnetic resonance
chiral methyl groups and 106–109
of methylsuccinic acid 263
of NAD/NADH 86
Proton translocation in visual pigments 300, 302–303, 315–317
Pseudoasymmetric atom 23
Pseudoaxial position 38
Pseudoequatorial position 38
Pseudorotation of 5-coordinate phosphorus 202, 231, 243
Pseudorotational circuit of cyclopentane 37
Pyridoxal phosphate (PLP) enzymes 161–195
ALA synthase 176–177, 271, 275–277
cleavage and formation of β-hydroxy amino acids 175–176
coenzyme conformation 163–165, 173, 185
α-condensation/decarboxylation 176–178
decarboxylase reactions 172–174
electrophilic displacement at C-β 186–188
α,β-elimination reactions 178–179, 181–182
evolution 194
general reaction stereochemistry 194–195
racemase reactions 170–172
reactions at C-γ 188–193
reaction types 161–163
β-replacement reactions 178–180
transaminase reactions 165–170
Pyridoxamine phosphate (PMP) 163, 165
as cofactor in bacterial cell wall synthesis 193
Pyridoxamine-pyruvate transaminase 166–167
Pyruvate transaminase 170

Quantum topology 3
Quinic acid 89–91, 95

R and *S* nomenclature 12, 15, 18
Racemases 60–61, 170–172
Racemisation 60
by non-racemase PLP enzymes 170–171
Reductases 114, 119–121, 123–126, 142, 148–149
Re-faces, *Si*-faces 27
Reflection 3–5, 49
Reflection symmetry 5–6
Regiochemistry 7
Regiospecificity 65
Resolution of racemic phenylethanol 98–99
Resonance Raman technique 292, 300–302, 305
Retina 283–284
Retinal 285
analogues 302–315
conformation 291, 304–307
extraction 295
isomerisation 288
opsin binding site 292, 295–296
solubilisation 293
Rhodopsin 285, 316
absorption spectra 285–286, 322–328
composition 285
photoaffinity labelling 313–315

photolysis 288–292, 296–304, 315–322
regeneration 292–296
secondary structure 287
spatial arrangement in membrane 287, 297
Rod cells 283–285
Rod out segment (ROS) 283–284
Rossman fold 119–120, 133, 155
Rotation
about double bonds 21
about heteroatom–heteroatom single bonds 36
about sp^2 C–heteroatom single bonds 35
about sp^2–sp^3 and sp^2–sp^2 C–C bonds 33–34
about sp^3–sp^3 C–C bonds 29–30
as symmetry element 3–5
Rotational assymmetry 62
Rotation–reflection 3, 5

Saccharopine dehydrogenase 136
S-Adenosylmethionine, chiral methyl 278–279
S-cis, S-trans conformers 34, 304
Selectivity
diastereo 76
enantio 76
product 76–77
regio 65
stereo 66–68
substrate 75–77
Sequence rule 12–13, 17, 30
Serine transhydroxymethylase 175–177
Shikimic acid 89–90
Single-displacement mechanisms in phosphate substitution 204, 243–245
Sirohydrochlorin (Factor II) 272, 274
Snake venom phosphodiesterase
selectivity for phosphorothioate isomers 214–215
stereochemistry of hydrolysis 233
Sorbitol dehydrogenase 113, 136
Specificity
enzyme 75–76
product 75
regio 65
stereo 66–68
Spiranes 17
Steady-state kinetic analysis in phosphate substitution 204
Stereochemistry , 44
of chelate complexes 16
Stereodifferentiation 68, 70
Stereoheterotopic groups 10, 27
Stereoisomers 7–10, 19, 22–23
biochemical complexity and 42–44
energy differences 29
of pentacoordinate centres 14–15
of tetracoordinate centres 12
Stereoselectivity 66–68
Stereospecificity 66–68
Structural isomers 7–8
Succinate as labelled precursor 277–278
Succinate dehydrogenase 115
'Swinging door' mechanism in alanine racemase 171–172
Symmetry 3–6, 9–10
Symmetry axis 4, 50
Symmetry element 3–5, 50
Symmetry operation 3–5
Synclinal conformation 30
Synperiplanar conformation 30
Synthetases, ATP-dependent 240–241

Tetrahydrofolate
from dihydrofolate reductase 121, 126
in transhydroxymethylase reactions 175–176
Thermolysin, environment of zinc in 153
Three-point attachment *see* Attachment
Threo-isomer 22
Threonine synthetase 191–192
Topic relationships of molecular fragments 10, 24–25
Torsional isomers 18, 33
Transaminases 165–170
Trans-conformer 30–32
Tryptophanase 185–186
Tryptophan synthase 182–185
β_2 subunit, transaminase activity of 166
4,9-Twistadiene 19
Twistane 20
Twist conformer 38
Twist form of cyclohexane 38
Two-plane theory 62–63
Tyrosine decarboxylase 173
Tyrosine phenol-lyase 186

UDP-glucose pyrophosphorylase 237–238, 245–246
Uroporphyrinogen I (urogen I) 272, 274
Uroporphyrinogen III (urogen III) 271–273, 275

Vicinal interchange in B_{12}-mediated reactions 250
intramolecular nature of 261
Visual pigments 285–288
absorption spectra 285–287, 322–328
Visual pigment analogues 285, 307–315
alkylated and dealkylated 306, 309–310
allenic 310–311

cis–trans isomerisation and, 302
double bond isomers 307–308
halogenated 310, 327
hydroretinals 325–326
isotope-labelled 302, 308–309
non-bleachable 303–304
photoaffinity labelling and 313–315
proton translocation and 302
retinal conformation and 305–307
ring-modified 311–314
Vitamin B_{12} *see* Coenzyme B_{12}

X-ray crystallography 78